Thermal Safety of Lithium Ion Batteries

Guest Editor
Mingyi Chen
School of the Environment
and Safety Engineering
Jiangsu University
Zhenjiang
China

Editorial Office
MDPI AG
Grosspeteranlage 5
4052 Basel, Switzerland

This is a reprint of the Special Issue, published open access by the journal *Batteries* (ISSN 2313-0105), freely accessible at: https://www.mdpi.com/journal/batteries/special_issues/45M4Y7205D.

For citation purposes, cite each article independently as indicated on the article page online and as indicated below:

Lastname, A.A.; Lastname, B.B. Article Title. *Journal Name* **Year**, *Volume Number*, Page Range.

ISBN 978-3-7258-3755-7 (Hbk)
ISBN 978-3-7258-3756-4 (PDF)
https://doi.org/10.3390/books978-3-7258-3756-4

Contents

batteries

MDPI

Editorial

Thermal Safety of Lithium-Ion Batteries: Current Status and Future Trends

Mingyi Chen

School of the Environment and Safety Engineering, Jiangsu University, Zhenjiang 212013, China; chenmy@ujs.edu.cn

Research on the thermal safety of lithium-ion batteries (LIBs) is crucial for supporting their large-scale application [1]. With the rapid development of high-energy-density battery systems, the issue of insufficient intrinsic thermal stability of materials has become increasingly prominent. This safety hazard is particularly severe in scenarios involving electric vehicles and energy storage stations, where the thermal runaway (TR) of individual cells within a battery pack may trigger cascading failures through thermal runaway propagation (TRP), leading to significant property damage or even personal injury [2,3].

In this context, an effective battery thermal management system (BTMS) can dissipate heat in time, keeping the LIB within an optimal temperature range, ensuring good temperature consistency, reducing the risk of TR, and improving the safety and stability of battery applications. Existing thermal management methods mainly include air cooling, liquid cooling, phase change material cooling, heat pipe cooling, and their combinations [4,5]. Efficient battery thermal management is an effective means of ensuring the safety of electrochemical energy storage systems, enabling the battery to operate within an acceptable temperature range, with a suitable temperature difference, which plays a key role in preventing TR [6]. To efficiently manage battery heat, a deep understanding of the processes of heat generation, transfer, and dissipation is crucial. Developing thermal models for batteries through theoretical derivation and experimentation is an effective approach, and thermal models also serve as the foundation for the TRP models of battery modules [7].

Excessive temperature can cause TR in batteries, which is the leading cause of battery fires and explosions. Once TR reaches a certain level, the gases and flammable materials produced inside the battery are expelled simultaneously, forming a strong smoke flow, which eventually leads to a fire. The triggering conditions and mechanisms of TR have been comprehensively explored to improve the safety of LIB systems. Researchers have also conducted experimental and numerical simulation studies on the TRP in LIB modules. The propagation mechanism of TR in large-capacity LIB modules has been analyzed, the TRP model has been established, and the temperature distribution and evolution during the TRP of battery modules through experiments and computational fluid dynamics simulations have been demonstrated. On the other hand, after TR and fire occurrence, effective safety measures are also required to block the heat transfer path between batteries or to dissipate heat from runaway batteries in a timely manner, thereby delaying or blocking the TRP. TRP suppression based on existing battery module thermal management has been proposed, achieving good results in phase change material thermal management systems [8,9]. Monitoring and early warning refer to the technology of predicting potential TR disasters in LIB systems by monitoring voltage, temperature, and other signal states during the operation of the LIB system, even in the absence of open flames or smoke. For example, an intelligent early warning system based on gas composition monitoring (such as CO/H_2) can identify the early signs of TR and shorten the accident response time [10].

Received: 26 February 2025
Accepted: 14 March 2025
Published: 15 March 2025

Citation: Chen, M. Thermal Safety of Lithium-Ion Batteries: Current Status and Future Trends. *Batteries* **2025**, *11*, 112. https://doi.org/10.3390/batteries11030112

Currently, research on LIB fire extinguishing technology mainly focuses on fire suppression, thermal management, and optimization of fire extinguishing agents. Researchers are improving water-based, foam, and solid fire extinguishing agents to enhance the extinguishing effect and reduce secondary damage from battery fires [11]. Additionally, various new fire suppression materials have been developed for the special properties of LIBs, such as materials with high thermal conductivity or phase change properties, to enhance thermal management capabilities and prevent overheating or TR. Some studies have also explored fire suppression systems based on automation technologies, using sensors and intelligent control for early fire detection and immediate extinguishing. Despite this progress, the complexity of lithium-ion battery fires still presents a challenge for existing technologies, and research continues to explore more efficient and safer fire suppression solutions [12].

Artificial intelligence (AI) also plays a crucial role in enhancing battery thermal safety [13]. AI is applied in various ways, including modeling and forecasting, where machine learning and deep learning algorithms predict battery behavior and potential risks, such as TR, by analyzing factors like temperature, pressure, and voltage. AI also supports failure mode and effects analysis by identifying potential failure modes through data analysis, helping detect patterns or anomalies in battery behavior. Furthermore, AI-driven, data-centered digital solutions allow for real-time monitoring and adaptive safety strategies by processing large datasets from battery systems. AI is also involved in the optimization of battery cell and pack design, improving heat dissipation and enhancing thermal management to prevent TR. Overall, AI contributes to more accurate predictions, better risk management, and the development of safer, more efficient battery systems [14].

This Special Issue of the journal *Batteries*, "Thermal Safety of Lithium-ion Batteries", brings together 15 research papers on the thermal safety of LIBs, covering a wide range of fields from basic research to applied technology. These studies provide valuable references for improving battery safety and performance. In terms of the TR mechanism and characteristics, Matthew Claassen et al. characterized the particle size distributions and chemical compositions of fine particles (PM2.5) and acidic gases released during TR. The study provided key insights into the emission profiles and potential health hazards, contributing to a better understanding of the risks associated with battery TR [15,16]. Kuntz et al. investigated the impact of LIB aging on safety behavior, comparing the safety performance of fresh and aged cells under various thermal and standardized safety tests [17]. Zhang et al. investigated the TR behavior of a large-capacity ternary lithium battery, providing valuable insights into the temperature profiles, gas emissions, and ejection dynamics and contributing to improved safety design and the suppression of TRP [18]. Wang et al. highlighted the influence of chamber pressure on battery temperature, internal pressure, venting dynamics, and gas composition, contributing insights on the safety assessment and early warning systems for energy storage batteries in high-altitude areas [19]. In terms of TR monitoring and warning, Hebenbrock et al. introduced a novel monitoring method for prismatic lithium-ion cells using a fiber Bragg grating sensor applied to a rupture disk to detect pressure increases [20]. Kisseler et al. developed a method for real-time monitoring of internal and surface temperatures in prismatic lithium-ion cells during TR induced by overcharging [21]. In the field of fire extinguishing technology, Miao et al. demonstrated that water mist effectively extinguished flames and provided significant cooling, offering valuable insights for improving TR suppression in LIBs [22]. Bordes et al. found that run-off waters could pose significant environmental hazards, emphasizing the need for careful assessment of their impact in large-scale battery incidents [23].

In the domain of battery thermal management systems (BTMSs), Liu et al. reviewed recent research on liquid cooling BTMSs, comparing different coolants, liquid channel designs, and system structures for both indirect and direct cooling methods. It also examined the integration of liquid cooling with other techniques, such as air cooling and phase change materials, and highlighted the safety benefits of liquid cooling in managing thermal runaway [24]. Rahmani et al. reviewed the latest developments in BTMSs from 2023 and 2024, highlighting new cooling methods and hybrid designs aimed at improving thermal management efficiency [25]. Yu et al. introduced a novel flame-retardant composite phase change material (CPCM) to enhance thermal management and safety in LIB systems, which effectively controlled battery temperatures and reduced thermal gradients [26]. Murphy et al. provided insights into the optimization of cooling systems, demonstrating that the optimal design achieved superior temperature regulation and uniformity compared to other designs [27]. Graichen et al. introduced a novel thermal management and safety system using polymer-based mini-channel cold plates to delay TR and prevent TRP [28]. Mei et al. proposed a CPCM, specifically with sodium acetate trihydrate, which effectively slowed temperature rise and reduced combustion duration, offering valuable insights for safer thermal management system design [29].

In summary, this Special Issue showcases the latest research results in the field of thermal safety of LIBs, covering a wide range of topics from basic research to applied technology, and provides valuable references for improving battery safety and performance. Looking ahead, with the continuous improvement of battery energy density and the diversification of application scenarios, the thermal safety research of LIBs is expected to encounter greater challenges. Future research should further explore the TR mechanism, develop new thermal management materials and technologies, and improve the monitoring and early warning system to ensure the safety and reliability of batteries under high energy density and complex working conditions. Additionally, AI's integration with advanced sensor networks and real-time data analytics will further enable the development of autonomous, self-learning systems capable of continuously optimizing battery performance and safety under dynamic operating conditions.

Acknowledgments: I am sincerely grateful to everyone who contributed to this special issue.

Conflicts of Interest: The author declares no conflicts of interest.

References

1. Lyu, P.; Liu, X.; Qu, J.; Zhao, J.; Huo, Y.; Qu, Z.; Rao, Z. Recent advances of thermal safety of lithium ion battery for energy storage. *Energy Storage Mater.* **2020**, *31*, 195–220. [CrossRef]
2. Peng, R.; Kong, D.; Ping, P.; Gao, W.; Wang, G.; Gong, S.; Yang, C.; Gao, X.; He, X. Experimental investigation of the influence of venting gases on thermal runaway propagation in lithium-ion batteries with enclosed packaging. *eTransportation* **2025**, *23*, 100388. [CrossRef]
3. Li, H.; Duan, Q.; Zhao, C.; Huang, Z.; Wang, Q. Experimental investigation on the thermal runaway and its propagation in the large format battery module with $Li(Ni_{1}/Co_{1/3}/Mn_{1}/_{3})O_2$ as cathode. *J. Hazard. Mater.* **2019**, *375*, 241–254. [CrossRef] [PubMed]
4. Chavan, S.; Venkateswarlu, B.; Salman, M.; Liu, J.; Pawar, P.; Joo, S.W.; Choi, G.S.; Kim, S.C. Thermal management strategies for lithium-ion batteries in electric vehicles: Fundamentals, recent advances, thermal models, and cooling techniques. *Int. J. Heat Mass Transf.* **2024**, *232*, 125918. [CrossRef]
5. Zhao, L.; Li, W.; Wang, G.; Cheng, W.; Chen, M. A novel thermal management system for lithium-ion battery modules combining direct liquid-cooling with forced air-cooling. *Appl. Therm. Eng.* **2023**, *232*, 120992. [CrossRef]
6. Nasiri, M.; Hadim, H. Advances in battery thermal management: Current landscape and future directions. *Renew. Sustain. Energy Rev.* **2024**, *200*, 114611. [CrossRef]
7. Fu, H.; Wang, J.; Li, L.; Gong, J.; Wang, X. Numerical study of mini-channel liquid cooling for suppressing thermal runaway propagation in a lithium-ion battery pack. *Appl. Therm. Eng.* **2023**, *234*, 121349. [CrossRef]
8. Chen, M.; Yu, Y.; Ouyang, D.; Weng, J.; Zhao, L.; Wang, J.; Chen, Y. Research progress of enhancing battery safety with phase change materials. *Renew. Sustain. Energy Rev.* **2024**, *189*, 113921. [CrossRef]
9. Chen, M.; Zhu, M.; Zhao, L.; Chen, Y. Study on thermal runaway propagation inhibition of battery module by flame-retardant phase change material combined with aerogel felt. *Appl. Energy* **2024**, *367*, 123394. [CrossRef]

10. Kong, D.; Lv, H.; Ping, P.; Wang, G. A review of early warning methods of thermal runaway of lithium ion batteries. *J. Energy Storage* **2023**, *64*, 107073. [CrossRef]
11. Zhang, L.; Jin, K.; Sun, J.; Wang, Q. A review of fire-extinguishing agents and fire suppression strategies for lithium-ion batteries fire. *Fire Technol.* **2024**, *60*, 817–858. [CrossRef]
12. Ghiji, M.; Novozhilov, V.; Moinuddin, K.; Joseph, P.; Burch, I.; Suendermann, B.; Gamble, G. A review of lithium-ion battery fire suppression. *Energies* **2020**, *13*, 5117. [CrossRef]
13. Zhao, J.; Lv, Z.; Li, D.; Feng, X.; Wang, Z.; Wu, Y.; Shi, D.; Fowler, M.; Burke, A.F. Battery engineering safety technologies (BEST): M5 framework of mechanisms, modes, metrics, modeling, and mitigation. *eTransportation* **2024**, *22*, 100364. [CrossRef]
14. Wang, S.; Zhou, R.; Ren, Y.; Jiao, M.; Liu, H.; Lian, C. Advanced data-driven techniques in AI for predicting lithium-ion battery remaining useful life: A comprehensive review. *Green Chem. Eng.* **2024**, *6*, 139–153. [CrossRef]
15. Claassen, M.; Bingham, B.; Chow, J.C.; Watson, J.G.; Wang, Y.; Wang, X. Characterization of Lithium-Ion Battery Fire Emissions—Part 1: Chemical Composition of Fine Particles (PM2.5). *Batteries* **2024**, *10*, 301. [CrossRef]
16. Claassen, M.; Bingham, B.; Chow, J.C.; Watson, J.G.; Chu, P.; Wang, Y.; Wang, X. Characterization of Lithium-ion Battery Fire Emissions—Part 2: Particle Size Distributions and Emission Factors. *Batteries* **2024**, *10*, 366. [CrossRef]
17. Kuntz, P.; Lonardoni, L.; Genies, S.; Raccurt, O.; Azaïs, P. Evolution of safety behavior of high-power and high-energy commercial Li-Ion cells after electric vehicle aging. *Batteries* **2023**, *9*, 427. [CrossRef]
18. Zhang, H. Study on Thermal Runaway Behavior and Jet Characteristics of a 156 Ah Prismatic Ternary Lithium Battery. *Batteries* **2024**, *10*, 282. [CrossRef]
19. Wang, Y.; Wang, Y.; Zhao, J.; Li, H.; Xu, C.; Li, Y.; Wang, H.; Lu, L.; Dai, F.; Yu, R. Experimental Research on Thermal-Venting Characteristics of the Failure 280 Ah LiFePO4 Battery: Atmospheric Pressure Impacts and Safety Assessment. *Batteries* **2024**, *10*, 270. [CrossRef]
20. Hebenbrock, A.; Orazov, N.; Benger, R.; Schade, W.; Hauer, I.; Turek, T. Innovative early detection of high-temperature abuse of prismatic cells and post-abuse degradation analysis using pressure and external fiber bragg grating sensors. *Batteries* **2024**, *10*, 92. [CrossRef]
21. Kisseler, N.; Hoheisel, F.; Offermanns, C.; Frieges, M.; Heimes, H.; Kampker, A. Monitoring of Thermal Runaway in Commercial Prismatic High-Energy Lithium-Ion Battery Cells via Internal Temperature Sensing. *Batteries* **2024**, *10*, 41. [CrossRef]
22. Miao, B.; Lv, J.; Wang, Q.; Zhu, G.; Guo, C.; An, G.; Ou, J. The suppression effect of water mist released at different stages on lithium-ion battery flame temperature, heat release, and heat radiation. *Batteries* **2024**, *10*, 232. [CrossRef]
23. Bordes, A.; Papin, A.; Marlair, G.; Claude, T.; El-Masri, A.; Durussel, T.; Bertrand, J.-P.; Truchot, B.; Lecocq, A. Assessment of Run-Off Waters Resulting from Lithium-Ion Battery Fire-Fighting Operations. *Batteries* **2024**, *10*, 118. [CrossRef]
24. Liu, J.; Chen, H.; Huang, S.; Jiao, Y.; Chen, M. Recent progress and prospects in liquid cooling thermal management system for lithium-ion batteries. *Batteries* **2023**, *9*, 400. [CrossRef]
25. Rahmani, A.; Dibaj, M.; Akrami, M. Recent advancements in battery thermal management systems for enhanced performance of Li-ion batteries: A comprehensive review. *Batteries* **2024**, *10*, 265. [CrossRef]
26. Yu, Y.; Zhang, J.; Zhu, M.; Zhao, L.; Chen, Y.; Chen, M. Experimental investigation on the thermal management for lithium-ion batteries based on the novel flame retardant composite phase change materials. *Batteries* **2023**, *9*, 378. [CrossRef]
27. Murphy, M.; Akrami, M. Advanced Thermal Management of Cylindrical Lithium-Ion Battery Packs in Electric Vehicles: A Comparative CFD Study of Vertical, Horizontal, and Optimised Liquid Cooling Designs. *Batteries* **2024**, *10*, 264. [CrossRef]
28. Graichen, H.-C.; Boye, G.; Sauerhering, J.; Köhler, F.; Beyrau, F. The impact of a combined battery thermal management and safety system utilizing polymer mini-channel cold plates on the thermal runaway and its propagation. *Batteries* **2023**, *10*, 1. [CrossRef]
29. Mei, J.; Shi, G.; Liu, H.; Wang, Z. Organic and Inorganic Hybrid Composite Phase Change Material for Inhibiting the Thermal Runaway of Lithium-Ion Batteries. *Batteries* **2023**, *9*, 513. [CrossRef]

batteries

MDPI

Article

Characterization of Lithium-Ion Battery Fire Emissions—Part 2: Particle Size Distributions and Emission Factors

Matthew Claassen [1,2], Bjoern Bingham [1,3], Judith C. Chow [1], John G. Watson [1], Pengbo Chu [4], Yan Wang [2] and Xiaoliang Wang [1,*]

1 Division of Atmospheric Sciences, Desert Research Institute, Reno, NV 89512, USA; matt.claassen@dri.edu (M.C.); bjoern.bingham@dri.edu (B.B.); judith.chow@dri.edu (J.C.C.); john.watson@dri.edu (J.G.W.)
2 Department of Mechanical Engineering, University of Nevada, Reno, NV 89557, USA; yanwang@unr.edu
3 Atmospheric Sciences, Department of Physics, University of Nevada, Reno, NV 89557, USA
4 Department of Mining and Metallurgical Engineering, University of Nevada, Reno, NV 89557, USA; pengboc@unr.edu
* Correspondence: xiaoliang.wang@dri.edu; Tel.: +1-775-674-7177

Abstract: The lithium-ion battery (LIB) thermal runaway (TR) emits a wide size range of particles with diverse chemical compositions. When inhaled, these particles can cause serious adverse health effects. This study measured the size distributions of particles with diameters less than 10 μm released throughout the TR-driven combustion of cylindrical lithium iron phosphate (LFP) and pouch-style lithium cobalt oxide (LCO) LIB cells. The chemical composition of fine particles ($PM_{2.5}$) and some acidic gases were also characterized from filter samples. The emission factors of particle number and mass as well as chemical components were calculated. Particle number concentrations were dominated by those smaller than 500 nm with geometric number mean diameters below 130 nm. Mass concentrations were also dominated by smaller particles, with PM_1 particles making up 81–95% of the measured PM_{10} mass. A significant amount of organic and elemental carbon, phosphate, and fluoride was released as $PM_{2.5}$ constituents. The emission factor of gaseous hydrogen fluoride was 10–81 mg/Wh, posing the most immediate danger to human health. The tested LFP cells had higher emission factors of particles and HF than the LCO cells.

Keywords: Li-ion battery; fire; smoke; particulate matter; thermal runaway; ultrafine particles; HF; $PM_{2.5}$; emission factor; particle size distribution

Citation: Claassen, M.; Bingham, B.; Chow, J.C.; Watson, J.G.; Chu, P.; Wang, Y.; Wang, X. Characterization of Lithium-Ion Battery Fire Emissions—Part 2: Particle Size Distributions and Emission Factors. *Batteries* **2024**, *10*, 366. https://doi.org/10.3390/batteries10100366

Academic Editor: Wojciech Mrozik

Received: 24 August 2024
Revised: 17 September 2024
Accepted: 15 October 2024
Published: 16 October 2024

1. Introduction

Lithium-ion batteries (LIB) can generate significant gaseous and particulate emissions when they experience thermal failure, through venting, thermal runaway (TR), fire, and explosion [1,2]. The detailed characterization of particle size distribution (PSD), chemical composition, emission factor, temporal evolution, and thermal stability is important for LIB safety, including understanding the health effects of inhaled smoke particles, proper personal protection, the mechanisms of fire origination and propagation, fire detection and suppression, post-fire cleanup, and material recycling. For example, recent studies of soot particles from LIB fires show higher toxicity to human cells than wood smoke [3,4]. However, as shown in a recent review [2], few studies have examined particle emissions from LIB fires.

Size distribution is an important parameter that describes the transport behavior, atmospheric residence time, and inhalation risk of particles. LIB TR generates a wide range of particle sizes, varying from several nanometers to several millimeters. Most previous studies have focused on larger particles that settled in the combustion chamber after experiments. Zhang et al. [5] collected particles after burning a nickel manganese cobalt (NMC) LIB cell and separated them by sieving into four size fractions. The mass percentages were

44% for 0–0.85 mm, 36% for 0.85–1.70 mm, 9% for 1.70–2.00 mm, and 11% for 2.00–8.00 mm. Laser diffraction particle sizing showed that the volume distribution had a median diameter of 198 μm for the 0–0.85 mm size fraction. Several other studies focusing on larger particles also showed volume distribution median diameters >100 μm [6–8]. Due to their high temperature and large thermal mass, the larger sparking particles may contribute to igniting flammable gases and TR propagation [9]. As they dominate the emitted particle mass, the chemical composition of these larger particles provides useful information about particle origin and TR reactions [5,8,10]. When released into the environment, they may cause air, water, and soil contamination [11–13]. However, these particles represent a low inhalation risk to humans as their large size causes them to settle quickly by gravity and makes them unlikely to be inhaled while suspended in air [14].

Particles with aerodynamic diameters less than 10 μm (PM_{10}), 2.5 μm ($PM_{2.5}$), and even smaller (e.g., ultrafine particles with diameters <100 nm [$PM_{0.1}$]) can remain suspended in air for longer than larger particles. When inhaled, they will deposit at different locations in the human respiratory track depending on the particle size, causing respiratory, cardiovascular, and other diseases [15]. Very few studies have measured PM_{10} size distributions from LIB fires. Premnath et al. [16] used an Engine Exhaust Particle Sizer (EEPS) [17,18] to measure particle number concentrations in the size range of 5.6–560 nm in real time from lithium iron phosphate (LFP) and NMC cells undergoing TR. The geometric number mean diameters (GNMD) of the PSDs [19] ranged from 54 to 69 nm with either monomodal or bimodal distributions. The emission rate varied from 1.6×10^{15} to 2.1×10^{17} particles/hour, 5–6 orders of magnitude higher than modern heavy-duty diesel engines. While the tests evaluated the effects of the TR triggering mechanism (nail penetration vs. overcharging) and LIB chemistry (LFP vs. NMC), the number of tests were small, and the influence of the cell's state of charge (SOC) was not examined. Goupil et al. [20] measured PSDs from an NMC cell fire using a scanning mobility particle sizer (SMPS; 17.5–532.8 nm) and an optical particle spectrometer (250 nm–32 μm). The number distributions were bimodal, with one peak below 50 nm and one centered around 125 nm, while the mass distribution had a peak around 10 μm. The slow SMPS scan time (76 s) may cause inaccuracy in the PSD, and the conversion from optical to aerodynamic diameter is affected by the particle's optical properties, morphology, and density [21,22]. Several other studies have estimated PSDs from image analysis [2,11]. However, this method is prone to multiple artifacts, such as particle loss during collection and sample preparation, a compromise between field of view and size resolution, and potential overlaps between particles [23].

Emission factors (EFs) are commonly used in air quality management to estimate pollutant emissions based on emission activities [24]. Because the total emission amount depends on LIB SOCs and the number of LIBs burned, the proper activity metric for estimating LIB fire emissions is the total energy capacity of the LIB cells burned [25]. Therefore, the particle number EFs are expressed in particles/Wh (watt–hours), and the size-segregated PM mass and $PM_{2.5}$ constituent EFs are expressed in mg/Wh. EFs in these units are easier to use for estimating the total emissions than the emission rates in particles/hour reported by Premnath et al. [16]. To our knowledge, no studies have reported EFs of LIB fire particle chemical constituents.

In companion with the particle chemical characteristics presented by Claassen et al. [26], the objectives of this paper are to: (1) characterize the size distribution of particles in the diameter range of 6 nm–10 μm; (2) quantify the EFs of particle number, particle mass, and $PM_{2.5}$ constituents emitted from LIB combustion; and (3) evaluate the dependence of PSDs and EFs on cell type and SOC.

2. Materials and Methods

The experiments and data analysis have been described in the companion paper [26] and only a brief description is provided here. Two commonly used LIB types were tested: a cylindrical, 18650-style, LFP cell and a pouch-style lithium cobalt oxide (LCO) cell [27,28]. Each cell type was tested at five SOC levels with three to six tests per SOC. The cells used for these tests were purchased shortly before the experimental campaign and were received at proper storage voltages. It was found that the LFP cells only vented at 0% and 30% SOCs, flamed at 50% and 75% SOCs, and had either venting or flaming at 100% SOCs. In contrast, LCO tests flamed at all SOCs [26]. While these two cell types will be referred to by their cell chemistry (LFP vs. LCO), they also differ in form-factor and casing, both of which can affect emissions. However, this study did not attempt to isolate the effects of these factors. The LIB combustion tests were conducted inside an ~8 m^3 burn chamber [29]. The exhaust flow rate (Q) was set to ~3300 L/min, as determined by thermal anemometer measurements in the exhaust duct [30]. Before each test, the chosen LIB cell was charged to the desired SOC by a programable charger. The cell was then placed in a ceramic crucible and heated to failure by an electric hot plate. A type K thermocouple was used to measure external cell and flame temperatures.

A sample of LIB combustion emissions was extracted from the chamber exhaust vent and directed to a suite of particle analyzers and the filter sampling system. Real-time particle mass concentrations were measured by two DustTrak DRX (TSI Inc., Shoreview, MN, USA) aerosol monitors in five size fractions (i.e., PM_1, $PM_{2.5}$, PM_4, PM_{10}, and PM_{15}) [21,22]. The DustTraks were placed before and after sample line dilution to determine the dilution factors.

Finer resolution PSDs in the range of 6 nm–10 μm were measured every second by an electrical low-pressure impactor (ELPI+; Dekati Ltd., Kangasala, Finland) [31–34] which was placed on a diluted sample line to prevent impactor overloading. In an ELPI+, particles are first introduced into a unipolar diffusion charger to achieve a stable charge distribution, and then are collected on 13 stages of electrically conducting cascade impactors and a final filter, depending on their aerodynamic diameters. The electrical charges carried by the deposited particles are measured as current by sensitive electrometers. An inversion algorithm is used to convert the 14-channel electric current data to particle number distributions with up to 500 bins. In this study, the particle number distributions were converted to mass distributions by assuming spherical particle shape and a density of 1 g/cm^3. The high-resolution data were integrated over the size fractions of <0.1 μm ($PM_{0.1}$), 0.1–1 μm ($PM_{0.1-1}$), 1–2.5 μm ($PM_{1-2.5}$), and 2.5–10 μm ($PM_{2.5-10}$). There are multiple potential artifacts (e.g., deposition of smaller particles at upper stages, size-dependent particle density, and image current) that could lead to errors in the mass concentration calculation [35]. Therefore, the ELPI $PM_{2.5}$ mass was normalized by gravimetric $PM_{2.5}$ mass to account for these artifacts.

Two filter channels were used to collect $PM_{2.5}$ and acidic gases for analysis of mass by gravimetry, ions and acidic gases by ion chromatography (IC), organic and elemental carbon (OC and EC) by thermal/optical analysis, and elements by x-ray fluorescence (XRF) and inductively coupled plasma mass spectrometry (ICP-MS) [24,36]. The backup potassium hydroxide (KOH)-impregnated cellulose-fiber filters behind the Teflon-membrane front filters were analyzed for acidic gases, including hydrogen fluoride (HF), hydrochloric acid (HCl), nitric acid (HNO_3), and sulfuric acid (H_2SO_4) as their corresponding ions (i.e., F^-, Cl^-, NO_3^-, and SO_4^{2-}) by IC [37,38].

Emission factors (EF_i in particles/Wh for particle number and mg/Wh for mass) were calculated for each species i by Equation (1):

$$EF_i = C_i \times Q \times \Delta t / E \quad (1)$$

where C_i (in particles/m^3 for number and mg/m^3 for mass) is the mean stack concentration for species i, Q (in m^3/s) is the mean stack exhaust flow rate, Δt (in s) is the sampling duration, and E (in Wh) is the nominal LIB cell energy capacity.

3. Results

3.1. Particle Size Distributions

Figure 1 shows the heat maps of PSD evolution for the representative tests for each cell type and SOC. The y-axis represents particle aerodynamic diameter (D_p), and the coloring represents particle number concentration normalized by the size bin width (dlog (D_p)). While most particles had diameters below 500 nm, significant variations in PSD evolution were seen. Low SOC (0 and 30%) LFP tests in this study, which had no flaming combustion, emitted high concentrations of particles over a wide size range, but concentrations never exceeded 10^7 particles/cm^3. These emissions occurred for a long period of time, with maximum concentrations above 10^5 particles/cm^3 for over 30 min for both low SOCs.

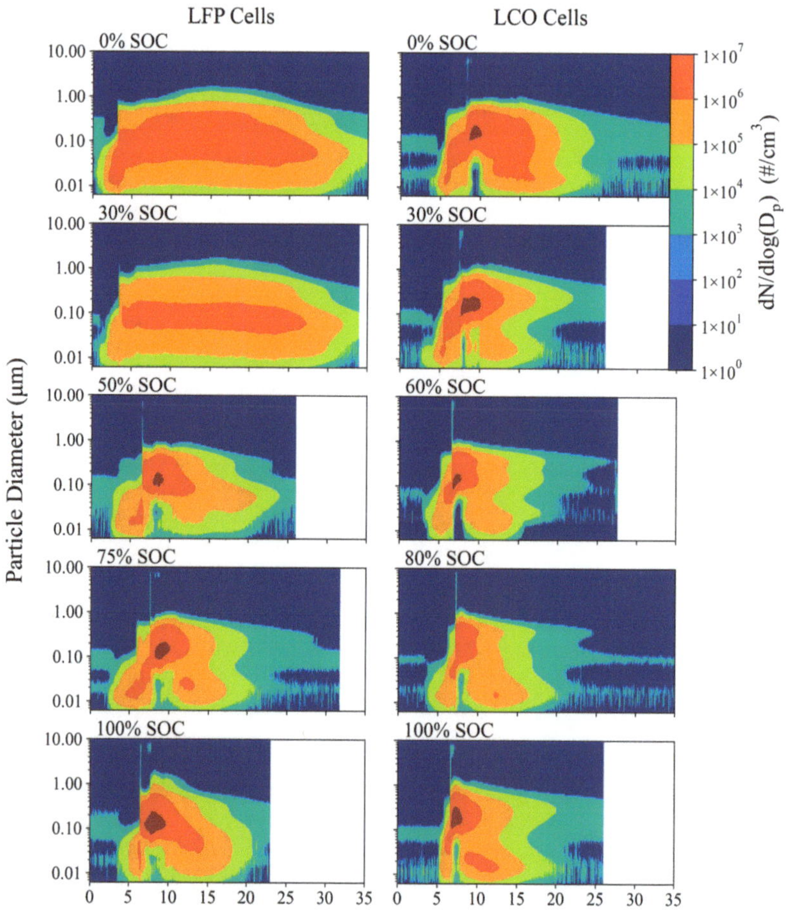

Figure 1. Particle number distribution heatmaps for representative LFP and LCO tests at each SOC.

Particle emissions for other LFP SOCs and LCO tests reached peak concentrations above 10^7 particles/cm^3, but concentrations above 10^5 particles/cm^3 only lasted for 10–18 min. A PSD discontinuity is seen in Figure 1 for these tests, occurring between five and ten minutes of elapsed time and representing abrupt concentration increases and particle size changes. This event coincided with visual and temperature-based TR indicators, suggesting it was caused by the onset of TR. A short 5–10 s long emission of coarse particles (up to 10 µm) can be seen to occur simultaneously, likely corresponding to the ejection of spark particles. Maximum particle concentrations occurred just after TR began. This is where most of the particle mass was emitted as particle sizes peaked here as well, with a notable decrease in particles below 50 nm in diameter. After this, peak concentrations and particle sizes decreased gradually, often forming a bimodal pattern of nanoparticles (<100 nm) and larger particles between 100 nm and 1 μm.

Figures 2 and 3 show individual PSD "snapshots" for a 0% SOC LFP test and a 60% SOC LCO test, each at four time intervals. The 0% SOC LFP test only vented, without flaming combustion, while the 60% LCO test flamed vigorously. A detailed discussion of the observed combustion behavior for all tests can be found in [26]. The intervals include background concentrations (t_1), when the test cell is outgassing (t_2), after TR (t_3), and after combustion is complete and the test cell is smoldering (t_4). These time intervals were selected to represent similar combustion stages despite the differences in combustion duration between tests. The GNMDs and geometric number standard deviations (GNSD) were calculated to characterize the PSD at each time interval [19].

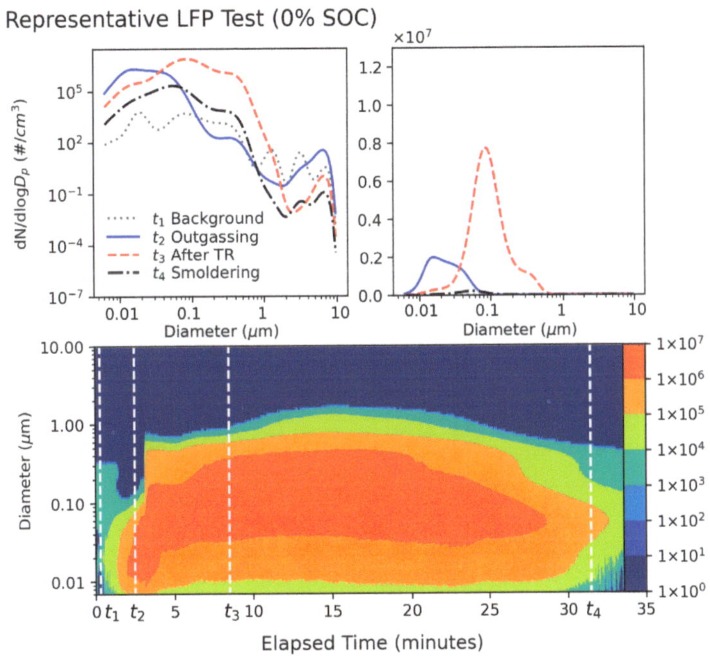

Figure 2. Particle number distribution snapshots (**left**: log scale and **right**: linear scale) and heatmap (**bottom**) for a representative 0% SOC LFP test. GNMDs for t_{1-4} are 59 nm, 22 nm, 91 nm, and 49 nm, respectively.

The 0% SOC LFP "venting" test and the 60% SOC LCO "flaming" test had similar peak outgassing (t_2) concentrations at ~2×10^6 particles/cm^3 but the particle diameters were smaller for the venting test, with a GNMD of 22 nm and a GNSD of 1.7. The flaming outgassing concentrations peaked at 59 nm with a GNSD of 1.9. While different cell types are compared here, the intent is to contrast the PSD differences due to the varied combustion behavior of the tests, not to infer differences due to the cell type itself. Concentrations after TR (t_3) peaked at four and six times the outgassing concentration for the venting and flaming tests, respectively, with the flaming test PSD being generally larger-sized and narrower. GNMDs at this time were 91 nm and 195 nm for the venting and flaming tests, with GNSDs of 1.9 and 1.6, respectively.

Figure 4a,b show the evolution of PM_1, $PM_{2.5}$, and PM_{10} for the tests shown in Figures 2 and 3, respectively. PM_1 accounted for most of the emission mass. Large spikes of PM_{10} were observed at the onset of TR, corresponding to the emission of coarse particles ($PM_{2.5-10}$), as was discussed for Figure 1. Interestingly, small spikes of coarse particles also occurred just before TR and before $PM_{2.5}$ emissions began. This behavior may be useful for early TR detection by LIB pack monitoring systems. The venting test (Figure 4a) shows $PM_{2.5}$ mass concentrations rising above PM_1 starting when emissions peaked around 15 min of elapsed time and lasting until combustion ended. This behavior occurred for all venting-only tests (LFP tests with 0% and 30% SOCs as well as half of the 100% SOC tests [26]). For tests where flaming occurred, PM_1 dominated PM mass emissions outside of TR onset.

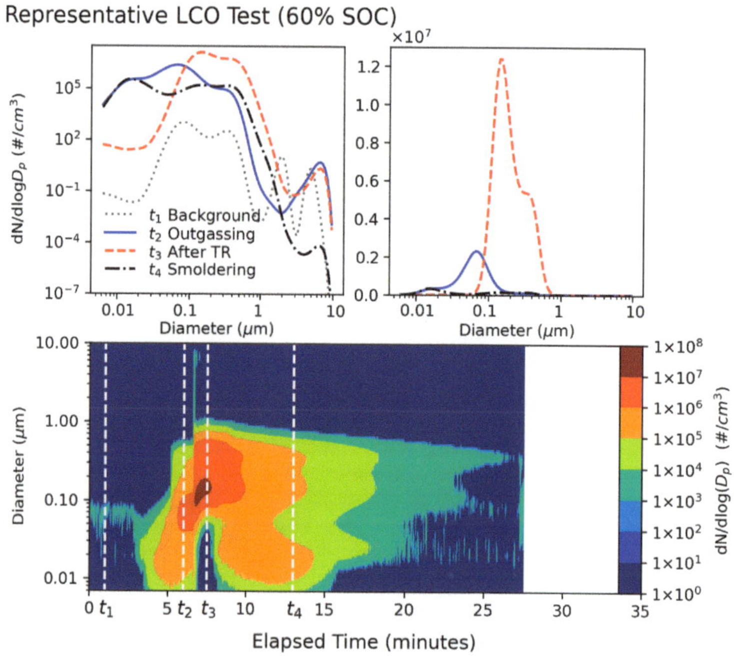

Figure 3. Particle number distribution snapshots (**left**: log scale and **right**: linear scale) and heatmap (**bottom**) for a representative 60% SOC LCO test. GNMDs for t_{1-4} are 117 nm, 59 nm, 195 nm, and 52 nm, respectively.

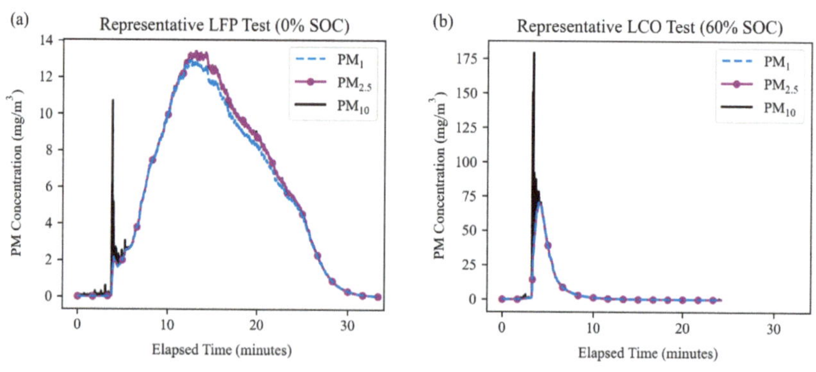

Figure 4. Mass concentrations of PM_1, $PM_{2.5}$, and PM_{10} for the representative tests: (**a**) LFP at 0% SOC and (**b**) LCO at 60% SOC (same as those in Figures 2 and 3). The coarse particles ($PM_{2.5-10}$) are released predominantly during TR.

Figure 5 shows the average number-based (a and b) and mass-based (c and d) PSDs for each SOC and cell type. The number distributions show a dominant mode centered around 70–140 nm, with additional modes centered around 20 nm and 300 nm. These distributions are similar to those reported by Goupil et al. [20]. For LFP tests, particles were smaller and were released in lower concentrations at low SOCs, increasing in size and concentration at mid-SOCs. Tests with 100% SOC fell in the middle due to having disparate combustion behavior. This is consistent with the combustion behavior of LFP tests: they only vented at 0% and 30% SOCs, intensely flamed at 50% and 75% SOCs, and had more variable combustion at 100% SOCs [26]. Particle sizes also increased with SOC for LCO tests, but there was no consistent dependence of concentrations on SOC, likely because flaming combustion was observed at all SOCs. The mass distributions show that LFP tests with 0% and 30% SOCs only have one mode centered around 460 nm, with elevated concentrations between 1 and 2.5 μm. All other tests had bimodal distributions, with one fine-particle mode centered around 400 nm and a coarse-particle mode centered around 8 μm. This indicates that the coarse particles were generated from flaming rather than cell venting. The coarse particle mode mass increased with SOC for both cell types, being insignificant for low SOC LFP tests. The fine particle mode mass also increased with SOC for LFP tests as the PSD shifted to larger particle sizes. A coarse mode centered around 10 μm was also observed by Goupil et al. [20].

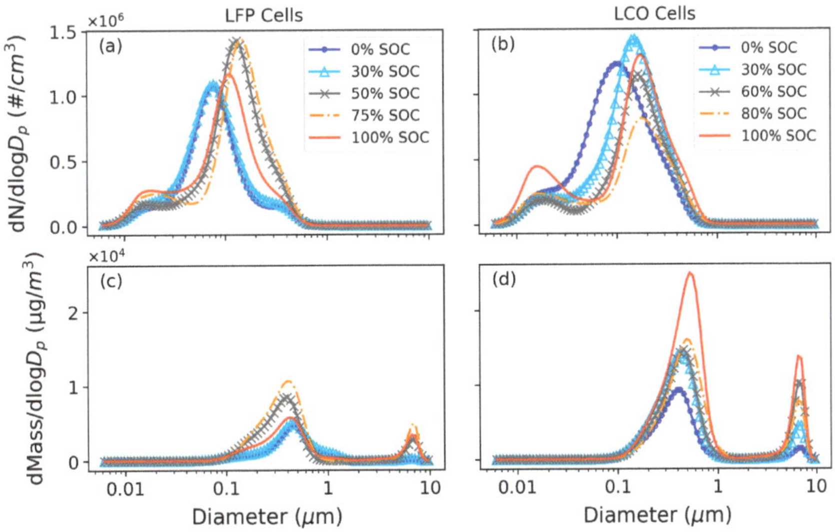

Figure 5. Particle number (top panels) and mass (bottom panels) distributions for LFP (**a**,**c**) and LCO (**b**,**d**) tests. An outlier was removed from some SOC groups to better show the prevailing trends.

Table 1 shows the GNMD and geometric mass mean diameters (GMMD) [19] for each cell type and SOC. The GNMD varied from 74 to 114 nm for LFP tests and 90 to 130 nm for LCO tests, showing that particles from LCO tests tended to be larger. GMMDs were larger than GNMDs due to mass being proportional to the cube of particle diameter. Interestingly, GNMDs peaked at mid-range SOCs while GMMDs increased with SOC, peaking at 100%. This increase at high SOCs, along with the previously discussed emission of large particles at TR, confirms that $PM_{2.5\text{–}10}$ is generated by the energetic combustion and ejection of material that occurs in a highly charged LIB cell during TR. Premnath et al. [16] found that GNMD and GMMD ranges were 54–69 nm and 97–204 nm for LFP and NMC cells, respectively, at 100% SOCs. These diameters are smaller than those in Table 1, especially for GMMD, which is likely caused by the lower upper size limit of the EEPS (560 nm) [17,18] than the ELPI+ (10 μm).

Table 1. Average and range (in parentheses) of geometric number mean diameters (GNMD) and geometric mass mean diameters (GMMD) for each SOC and cell type.

Cell Type	SOC (%)	GNMD (nm)	GMMD (nm)
LFP	0	74 (72–77)	415 (409–424)
	30	74 (72–77)	426 (403–453)
	50	110 (109–111)	425 (415–435)
	75	114 (106–129)	479 (434–530)
	100	84 (71–93)	568 (536–611)
LCO	0	90 (78–102)	387 (371–405)
	30	120 (111–137)	474 (470–477)
	60	130 (128–132)	663 (553–751)
	80	118 (109–128)	673 (651–704)
	100	116 (96–129)	668 (602–778)

3.2. *Emission Factors (EFs)*

EFs for particle number, mass, $PM_{2.5}$ constituents, and acidic gases are listed in Tables S1–S5. The following sections describe EFs for particle numbers, mass, carbon, PO_4^{3-}, selected metals and ions, and acidic gases, as well as the relationship between EF and combustion temperature.

3.2.1. EFs for Particle Number and Mass

Figure 6a shows particle number EFs by cell SOC for each size fraction. The EFs for ultrafine particles ($PM_{0.1}$) ranged between 1.5×10^{13} and 3.6×10^{13} particles/Wh for LFP tests and 3.2×10^{12} and 9.6×10^{12} particles/Wh for LCO tests (Table S1). Total particle numbers decreased with SOC, and LFP tests released several times more particles than LCO tests. $PM_{0.1}$ and $PM_{0.1\text{–}1}$ together accounted for >99.9% of PM_{10} numbers, at 42–84% and 16–58% of PM_{10}, respectively. Particles larger than 1 µm had negligible contributions to particle number. $PM_{0.1}$ tended to make up a higher proportion of PM_{10} when particle emissions were high. On the other hand, Figure 6b shows $PM_{0.1}$ made up a much smaller proportion of the PM_{10} mass due to having very small particle volumes, at 1–8%, while $PM_{0.1\text{–}1}$ dominated the PM_{10} mass at 77–89% due to their high concentrations. PM_1 particles made up 81–95% of the measured PM_{10} mass, as also shown in Figure 4. Larger particles made up significant portions of PM_{10} mass despite their low number concentrations, with $PM_{1\text{–}2.5}$ and $PM_{2.5\text{–}10}$ accounting for up to 4% and 18% of PM_{10} mass, respectively. $PM_{1\text{–}2.5}$ EFs were high for low SOC LFP tests and venting-only 100% SOC LFP tests while all other tests had higher $PM_{2.5\text{–}10}$ EFs instead.

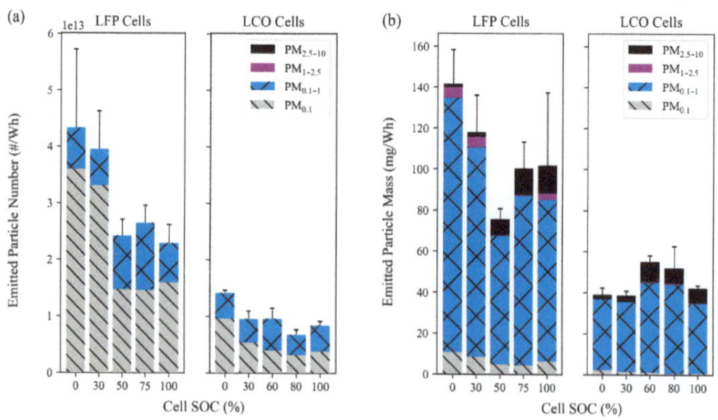

Figure 6. Emission factors for (**a**) particle number and (**b**) particle mass by size fraction for LFP and LCO tests. Error bars represent the total PM_{10} standard error (including all smaller particle sizes) and are symmetric.

3.2.2. EFs for $PM_{2.5}$, OC, EC, PO_4^{3-}, and Toxic Metals

EFs for $PM_{2.5}$, OC, EC, and PO_4^{3-} are shown in Figure 7a. EFs for $PM_{2.5}$ mass were between 35 and 140 mg/Wh. This equates to the emission of 1.4–5.6 g of $PM_{2.5}$ from the combustion of a standard laptop-sized LIB (40 Wh), or 2.1–8.4 kg of $PM_{2.5}$ from the combustion of an electric vehicle battery pack (60 kWh) if all cells are burned in a similar way as those in this study.

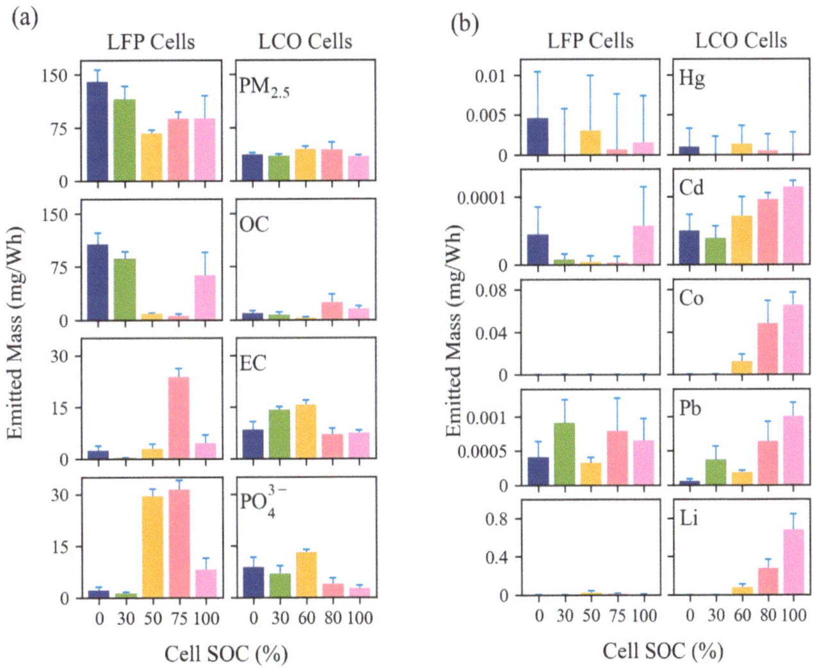

Figure 7. Emission factors for (**a**) $PM_{2.5}$ mass, OC, EC, and PO_4^{3-} and (**b**) selected metals. The error bars represent the larger of the propagated analytical uncertainty or the standard error within each SOC and are symmetric.

LFP tests had $PM_{2.5}$ EFs of 67–140 mg/Wh, with the least mass emitted at 50% SOC and increased variability at 100% SOC. OC EFs were 6–9 mg/Wh for 50–75% SOC and increased to 63–106 mg/Wh otherwise. OC variabilities at 100% SOC were similarly high to $PM_{2.5}$. Low OC emissions at mid-range SOCs corresponded to high EC emissions for 75% SOC only, where EC EFs spiked from <5 mg/Wh for all other SOCs to 24 mg/Wh. This resulted in total carbon EFs dropping from 109 mg/Wh (0% SOC) to 12 mg/Wh (50% SOC), before increasing back to 68 mg/Wh (100% SOC). PO_4^{3-} emissions were also significant for mid-range SOCs, at approximately 30 mg/Wh, but only 1–8 mg/Wh otherwise.

LCO tests had more consistent $PM_{2.5}$ mass EFs of 35–45 mg/Wh, lower than LFP tests by 1.5–3.8 times. EFs for OC, EC, and PO_4^{3-} had much less dependence on SOC for LCO tests than LFP tests. OC EFs were 3–24 mg/Wh and peaked at 75% SOC, where LFP OC was lowest. EC EFs were 7–16 mg/Wh, higher than those of LFP tests except for at 75%/80% SOC. PO_4^{3-} EFs were 3–13 mg/Wh with a similar SOC dependance to EC.

Figure 7b shows EFs for four hazardous air pollutant metals [39] and Li. Variability within SOCs was high but several features are notable. LCO tests had Cd, Co, Pb, and Li emissions that increase with SOC, reaching $(10 \pm 1) \times 10^{-5}$ mg/Wh for Cd, 0.07 ± 0.01 mg/Wh for Co, $(10 \pm 2) \times 10^{-4}$ mg/Wh for Pb, and 0.7 ± 0.2 mg/Wh for Li, all at 100% SOC. All of these elements were near detection limits at low SOCs except for Cd. Only Pb was consistently present for LFP tests at between $(30 \pm 8) \times 10^{-5}$ mg/Wh and (9 ± 3) × 10^{-4} mg/Wh, with no dependence on SOC. Hg and Cd were detected sporadically. EFs for major compositions and elements are summarized in Tables S2 and S5, respectively.

3.2.3. EFs for Anions and Acidic Gases

EFs for acidic gases HF, HCl, HNO_3, and H_2SO_4 and their corresponding particulate anions are shown in Figure 8. These acidic gases are toxic and corrosive and can cause adverse effects on human health and material integrity if not neutralized after emission.

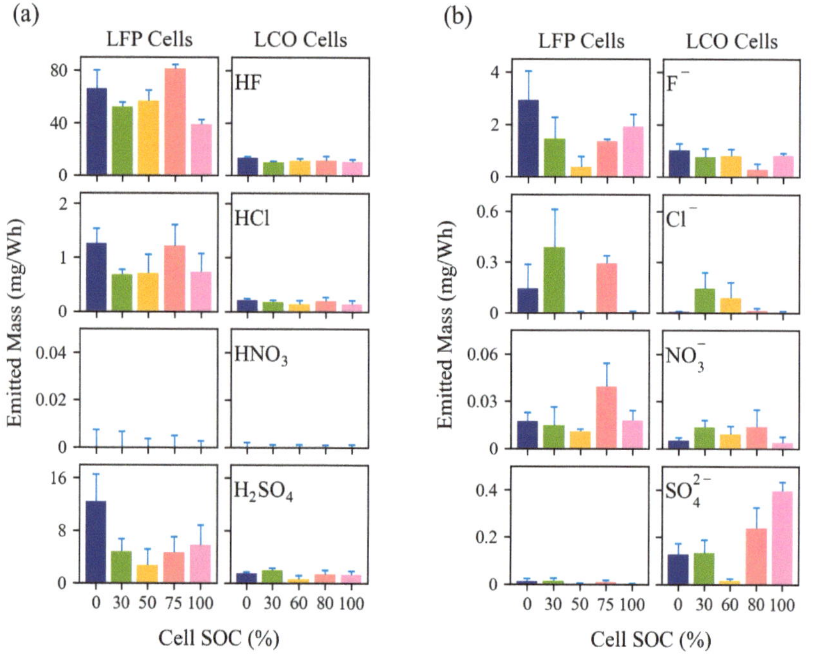

Figure 8. Emission factors for (**a**) selected acidic gases and (**b**) corresponding particulate anions. The error bars represent the larger of the propagated analytical uncertainty or the standard error within each SOC and are symmetric.

Gaseous EFs were higher than that of particulate emissions (by cell type) for all compounds shown except for NO_3^-, where HNO_3 emission was not detected. Gaseous HF and particulate F^- had the highest EFs for all LIB types and SOCs. Gaseous HF EFs ranged from 39 to 81 mg/Wh for LFP tests and 10 to 14 mg/Wh for LCO tests. Neither cell type showed a relationship between SOC and HF emissions. Particulate F^- EFs ranged from 0.4 to 3 mg/Wh for LFP tests and 0.3 to 1 mg/Wh for LCO tests. Particulate F^- had a distinct minimum at 50% SOC for LFP tests, corresponding to the minimum $PM_{2.5}$ EF (Figure 7a). Gaseous H_2SO_4 was the second most emitted acidic compound, with much higher gaseous emissions than particulate. Gaseous EFs were 3–12 mg/Wh for LFP tests and 0.6–2 mg/Wh for LCO tests. Unlike gaseous H_2SO_4 and all other acidic emissions shown, particulate SO_4^{2-} emissions were much higher for LCO tests than LFP tests, where only trace concentrations were detected. LCO tests had EFs of 0.1–0.4 mg/Wh for most SOCs, but only trace amounts for 60% SOC. Both gaseous and particulate emissions were low for 50/60% SOC, but this was more pronounced for LFP tests. Particle SO_4^{2-} emissions for LCO tests increased with SOC with the exception of 60% SOC. Gaseous HCl was consistently detected, averaging 0.9 mg/Wh for LFP tests and 0.2 mg/Wh for LCO tests. Particulate Cl^- was present sporadically for both cell types with EFs of up to 0.4 mg/Wh.

Only particulate NO_3^- was detected at between 0.005 and 0.04 mg/Wh, with higher emissions from LFP tests.

Table 2 compares the gaseous EFs of HF and HCl with the values reported in the literature. LFP HF emissions in this study (39–81 mg/Wh) compared well with the literature values which range at 12–350 mg/Wh. Emissions from LCO tests (10–14 mg/Wh) were lower than the 30–40 mg/Wh found in the literature, but this may be due to the limited number of studies [25,40]. Other cell types were reported to release a comparable 1–200 mg/Wh of HF. HCl emissions were reported by Diaz et al. [40] at 8–125 mg/Wh for LFP and LCO cells, significantly higher than those measured in this study (0.1–1.3 mg/Wh). No published EF values for HNO_3 or H_2SO_4 were found. Acidic gas and particulate ion EFs for all cell types and SOCs are listed in Tables S3 and S4, respectively.

Table 2. Comparison of gaseous emissions of HF and HCl to the literature sources, grouped by cell chemistry. EFs for HNO_3 and H_2SO_4 were not found in the literature.

Source	Cell Type	HF (mg/Wh)	HCl (mg/Wh)
This study	LFP	39–81	0.7–1.3
[40]	LFP	350	125
[25]	LFP	12–24	
[41]	LFP	40–125	
This study	LCO	10–14	0.14–0.21
[40]	LCO	30	8
[25]	LCO	30–40	
[42]	Unknown	20–200	
[43]	LMO	40–70	
[44]	NMC/LFP	23–36	
[45]	NMC	1–10	
[46]	NMC	4.2	

3.2.4. Relationship between Emission Factors and Combustion Temperatures

Figure 9 shows the relationship between the EF and the maximum measured combustion temperature for LFP tests. Cell SOC is also labeled by symbol coloring to show groupings. With the exception of two outlier tests (Tests C1 and C8 in Figure 9a,b), EFs for $PM_{2.5}$ and OC were found to decrease with increasing combustion temperature. This is expected as higher temperatures indicate more complete combustion that will release more gaseous CO_2 instead of carbonaceous particulate. In contrast, PO_4^{3-} increased with combustion temperatures, indicating more efficient conversion of P-containing electrolyte and cathode materials to particle emissions. Low EC emissions were found at lower combustion temperatures as expected; however, EC was not necessarily higher at high temperatures. Figure 9c shows that the 50% SOC tests had temperatures >600 °C but the EC emissions were comparable to those of low temperature tests. Particulate F^- and gaseous HF (Figure 9e,h) had the opposite trend, with F^- decreasing (for SOCs > 30%) and HF increasing with increasing temperature. This is possibly due to more HF remaining in the gas phase at higher temperatures rather than condensing to form particles. Li^+ emissions were hardly detected below 600 °C and increased rapidly thereafter, indicating a potential exponential relationship. More tests with temperatures reaching >700 °C are needed to explore this. Figure 9g shows that the part of $PM_{2.5}$ mass with unidentified composition increased with temperature, a correlation that could help determine what additional components (e.g., unmeasured phosphorous compounds) [26] need to be analyzed for.

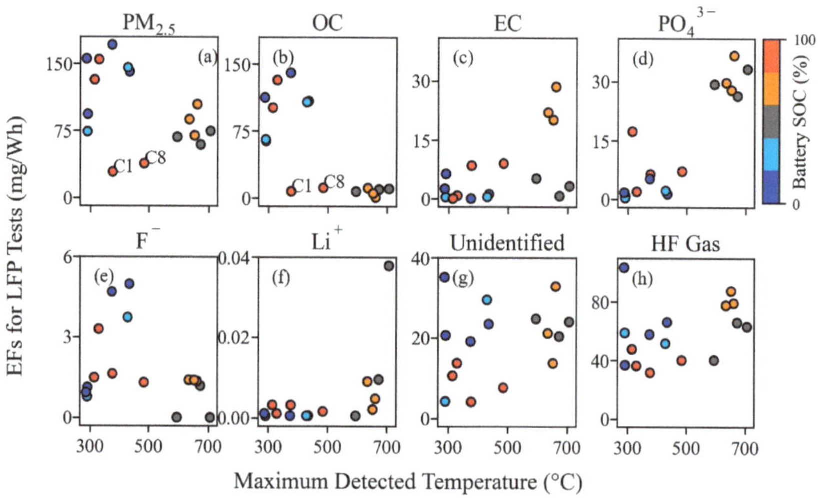

Figure 9. Relationship between EFs and maximum detected combustion temperature for LFP tests. LCO tests showed little correlation, possibly due to poor temperature measurement, and are not shown.

3.3. Cell Mass Losses

Each cell was weighed before and after combustion to determine the cell mass that was lost. LFP cells lost 2.2–2.4 g/Wh during combustion for all except 50% SOC, which lost noticeably less at 1.8–2 g/Wh. LCO cells lost roughly half the mass per Wh of cell capacity at 0.95–1.2 g/Wh. As shown in Table S6, this equates to a loss of ~20% of the cell mass for both cell types. However, only approximately 0.6–1.2% of the total cell mass, or 3–6% of the lost cell mass, was emitted as $PM_{2.5}$ that reached the filter assemblies. The proportion of lost cell mass that was emitted as $PM_{2.5}$ was highest for low SOC LFP tests, where 5–6% was converted to $PM_{2.5}$. All other SOC groups averaged 3–4%. By subtraction, this means that 94–97% of the lost mass, or roughly 16–22% of total cell mass, was emitted as either settleable PM or as gas emissions. As shown in Table S6, cell mass losses from previous studies range from 15% to 60%, with settleable particle mass accounting for 5–50% of the original cell mass, depending on the cell chemistry, casing, and SOCs [5,6,10,47]. Even though $PM_{2.5}$ only accounted for a small fraction of the total PM mass emitted from LIB fires, it accounts for most of the particle numbers that can be inhaled by humans and presents great health risks.

4. Discussion and Conclusions

This study conducted a detailed characterization of the size distribution of particles with diameter less than 10 μm emitted from LIB fires and determined EFs for size-segregated particle number, particle mass, and $PM_{2.5}$ chemical constituents. The analysis covered two cell types at a range of SOCs. While emissions may vary significantly depending on the specific cell types or LIB pack sizes, the cells tested represent two common cell chemistries and form-factors, in terms of both cell construction and energy capacity, that are used widely in modern LIB applications. As such, the findings offer valuable insights into the characteristics of emission from LIB combustion. The key conclusions are as follows:

(1) LIB fires emit high concentrations of fine and ultrafine particles. The particle number distributions showed a dominant mode centered around 100 nm, with additional modes centered around 20 nm and 300 nm. The particle mass distributions showed that the venting-only LFP tests in this study with 0% and 30% SOCs had one mode centered around 460 nm, while all other tests had bimodal distributions, with one fine particle mode centered around 400 nm and a coarse particle mode centered

around 8 μm. $PM_{0.1}$ and $PM_{0.1–1}$ together accounted for >99.9% of PM_{10} numbers, while $PM_{0.1–1}$ dominated the PM_{10} mass at 77–89%. Super-micron particles have non-negligible mass contributions, with $PM_{1–2.5}$ and $PM_{2.5–10}$ accounting for up to 4% and 18% of PM_{10} mass, respectively. $PM_{1–2.5}$ was emitted during cell venting, after TR and when flaming did not occur, while $PM_{2.5–10}$ was emitted just before and during TR onset for all tests.

(2) Venting-only combustion (LFP tests at 0% SOC, 30% SOC, and some at 100% SOC) showed smooth evolution of PSDs, while the onset of TR caused a discontinuity in particle size distribution by increasing both mode diameter and concentration and releasing coarse mode particles (1–10 μm). Particles from LCO tests in this study (GNMD 90–130 nm; GMMD 387–673 nm) were generally larger than those from LFP tests (GNMD 74–114 nm; GMMD 415–568 nm). The GMMD increased with SOC.

(3) LFP tests had higher EFs for particle number and mass than LCO tests. The EFs for ultrafine particles ($PM_{0.1}$) ranged from 1.5×10^{13} to 3.6×10^{13} particles/Wh for LFP and 3.2×10^{12} to 9.6×10^{12} particles/Wh for LCO tests. LFP tests had $PM_{2.5}$ EFs between 67 and 140 mg/Wh with more variability at different SOCs, whereas LCO tests had lower and more consistent $PM_{2.5}$ mass EFs between 35 and 45 mg/Wh. Even though $PM_{2.5}$ only accounted for 0.25–1.5% of the total cell mass and a small fraction of total PM mass emitted from LIB fires, it accounts for the most particle numbers that can be inhaled by humans and presents great health risks.

(4) LIB fires emit acidic gases, such as HF, HCl, and H_2SO_4. Gaseous HF ranged from 39 to 81 mg/Wh for LFP tests and 10 to 14 mg/Wh for LCO tests. These toxic and corrosive gases may represent great hazards to people and properties.

(5) Emissions are highly dependent on cell type, SOC, and combustion temperatures. The emitted $PM_{2.5}$ mass can depend on cell SOC by a factor of two, while the emitted OC, EC, and PO_4^{3-} can differ by a factor of 10 or more. PO_4^{3-} increased with combustion temperature, particularly when it reached >500 °C. Toxic metal emissions increased with SOC, but only for LCO tests, with no trend for LFP tests. Acidic gas emissions depended primarily on cell type, indicating that cell design is crucial to lowering emissions of HF and other corrosive compounds. The emission dependence on LIB cell properties should be considered when evaluating the overall hazard that each LIB pack presents. For example, while LFP cells are known to be more thermally stable, results from this study show that they may release more HF and can generate higher particulate concentrations than LCO cells.

The emissions documented here have implications for health and safety as well as environmental contamination. Considering that modern LIB packs can contain hundreds to millions of Wh of energy capacity, the effects of significant emissions of volatile organics, ultrafine and fine particles, hazardous metals, and acidic compounds must be taken into consideration when responding to and cleaning up after LIB fires.

LIB fires are becoming more common as LIB use expands. The combustion process can occur quickly and without warning, limiting the effectiveness of fire suppression. Therefore, a detailed understanding of the resulting emissions is imperative to allow for effective fire response and cleanup. This paper has detailed the expected fine particulate emissions in a way that allows extrapolation to any size LIB pack if the burned LIB energy capacity is known. This resource will allow first responders, LIB manufacturers, and responsible authorities to plan for and respond to LIB fires.

Supplementary Materials: The following supporting information can be downloaded at: https://www.mdpi.com/article/10.3390/batteries10100366/s1, Table S1. EFs for four size fractions in particles/Wh and mg/Wh for particle number and particle mass, respectively; Table S2. EFs in mg/Wh for $PM_{2.5}$ chemical constituents; Table S3. EFs in mg/Wh for acidic gases for all cell types and SOCs; Table S4. EFs in mg/Wh for particulate anions and cations for all cell types and SOCs; Table S5. EFs in mg/Wh for elements for all cell types and SOCs; Table S6. A comparison of cell mass loss rates to the literature values [5,6,10,47].

Author Contributions: Conceptualization, X.W.; Methodology, M.C. and X.W.; Software, M.C.; Validation, M.C. and X.W.; Formal Analysis, M.C.; Investigation, M.C., B.B., and X.W.; Resources, X.W.; Data Curation, M.C.; Writing—Original Draft Preparation, M.C. and X.W.; Writing—Review and Editing, X.W., J.C.C., J.G.W., P.C., and Y.W.; Visualization, M.C.; Supervision, X.W. and Y.W.; Project Administration, X.W.; Funding Acquisition, X.W. All authors have read and agreed to the published version of the manuscript.

Funding: This research was funded by the U.S. National Aeronautics and Space Administration's Established Program to Stimulate Competitive Research, CAN Grant No. 80NSSC19M0152, and Nevada Space Grant No. 80NSSC20M0043 22–24.

Data Availability Statement: The raw data supporting the conclusions of this article will be made available by the authors on request. The calculated emission factors for all measured species are included in the Supplementary Materials.

Acknowledgments: The authors thank Hans Moosmüller for the use of the burn chamber where experiments were performed and DRI personnel for support and filter analysis.

Conflicts of Interest: The authors declare no conflicts of interest.

Abbreviations

Δt	sampling duration
C	concentration
Cd	cadmium
Cl^-	chloride
Co	cobalt
D_p	particle diameter
E	battery cell energy capacity
EC	elemental carbon
EEPS	engine exhaust particle sizer
EF	emission factor
ELPI	electrical low-pressure impactor
F^-	fluoride
GMMD	geometric mass mean diameter: mean diameter of a particle mass distribution in logarithmic scale
GNMD	geometric number mean diameter: mean diameter of a particle number distribution in logarithmic scale
GNSD	geometric number standard deviation: standard deviation of a particle number distribution in logarithmic scale
HCl	hydrochloric acid
Hg	mercury
HF	hydrofluoric acid
HNO_3	nitric acid
H_2SO_4	sulfuric acid
IC	ion chromatography
ICP-MS	inductively coupled plasma mass spectrometry
KOH	potassium hydroxide
LCO	lithium cobalt oxide
Li	lithium
Li^+	lithium ion
LIB	lithium-ion battery
LFP	lithium iron phosphate
NMC	nickel manganese cobalt
NO_3^-	nitrate
OC	organic carbon
P	phosphorus
Pb	lead
PM	particulate matter
PM_x	particles with aerodynamic diameters $\leq$x μm

PO_4^{3-}	phosphate
PSD	particle size distribution
Q	flow rate
SMPS	scanning mobility particle sizer
SO_4^{2-}	sulfate
SOC	state of charge
TR	thermal runaway
Wh	watt hours
XRF	X-ray fluorescence

References

1. Bugryniec, P.J.; Resendiz, E.G.; Nwophoke, S.M.; Khanna, S.; James, C.; Brown, S.F. Review of gas emissions from lithium-ion battery thermal runaway failure—Considering toxic and flammable compounds. *J. Energy Storage* **2024**, *87*, 111288. [CrossRef]
2. Li, W.; Xue, Y.; Feng, X.; Rao, S.; Zhang, T.; Gao, Z.; Guo, Y.; Zhou, H.; Zhao, H.; Song, Z.; et al. Characteristics of particle emissions from lithium-ion batteries during thermal runaway: A review. *J. Energy Storage* **2024**, *78*, 109980. [CrossRef]
3. Xu, Y.; Wang, Y.; Chen, X.; Pang, K.; Deng, B.; Han, Z.; Shao, J.; Qian, K.; Chen, D. Thermal runaway and soot production of lithium-ion batteries: Implications for safety and environmental concerns. *Appl. Therm. Eng.* **2024**, *248*, 123193. [CrossRef]
4. Xu, Y.; Wang, Y.; Chen, D. Soot formation and its hazards in battery thermal runaway. *J. Aerosol Sci.* **2024**, *181*, 106420. [CrossRef]
5. Zhang, Y.; Wang, H.; Li, W.; Li, C.; Ouyang, M. Size distribution and elemental composition of vent particles from abused prismatic Ni-rich automotive lithium-ion batteries. *J. Energy Storage* **2019**, *26*, 100991. [CrossRef]
6. Zhang, Y.; Wang, H.; Li, W.; Li, C. Quantitative identification of emissions from abused prismatic Ni-rich lithium-ion batteries. *eTransportation* **2019**, *2*, 100031. [CrossRef]
7. Wang, H.; Zhang, Y.; Li, W.; Li, C.; Ouyang, M. Particles released by abused prismatic Ni-rich automotive lithium-ion batteries. *WSEAS Trans. Syst. Control* **2020**, *15*, 30–38. [CrossRef]
8. Wang, H.; Wang, Q.; Jin, C.; Xu, C.; Zhao, Y.; Li, Y.; Zhong, C.; Feng, X. Detailed characterization of particle emissions due to thermal failure of batteries with different cathodes. *J. Hazard. Mater.* **2023**, *458*, 131646. [CrossRef]
9. Wang, G.; Kong, D.; Ping, P.; Wen, J.; He, X.; Zhao, H.; He, X.; Peng, R.; Zhang, Y.; Dai, X. Revealing particle venting of lithium-ion batteries during thermal runaway: A multi-scale model toward multiphase process. *eTransportation* **2023**, *16*, 100237. [CrossRef]
10. Essl, C.; Golubkov, A.W.; Gasser, E.; Nachtnebel, M.; Zankel, A.; Ewert, E.; Fuchs, A. Comprehensive Hazard Analysis of Failing Automotive Lithium-Ion Batteries in Overtemperature Experiments. *Batteries* **2020**, *6*, 30. [CrossRef]
11. Yang, Y.; Fang, D.; Maleki, A.; Kohzadi, S.; Liu, Y.; Chen, Y.; Liu, R.; Gao, G.; Zhi, J. Characterization of Thermal-Runaway Particles from Lithium Nickel Manganese Cobalt Oxide Batteries and Their Biotoxicity Analysis. *ACS Appl. Energy Mater.* **2021**, *4*, 10713–10720. [CrossRef]
12. Held, M.; Tuchschmid, M.; Zennegg, M.; Figi, R.; Schreiner, C.; Mellert, L.D.; Welte, U.; Kompatscher, M.; Hermann, M.; Nachef, L. Thermal runaway and fire of electric vehicle lithium-ion battery and contamination of infrastructure facility. *Renew. Sustain. Energy Rev.* **2022**, *165*, 112474. [CrossRef]
13. Bordes, A.; Papin, A.; Marlair, G.; Claude, T.; El-Masri, A.; Durussel, T.; Bertrand, J.-P.; Truchot, B.; Lecocq, A. Assessment of Run-Off Waters Resulting from Lithium-Ion Battery Fire-Fighting Operations. *Batteries* **2024**, *10*, 118. [CrossRef]
14. ICRP. Human Respiratory Tract Model for Radiological Protection. ICRP Publication 66. Ann. ICRP 24 (1-3). 1994. Available online: https://www.icrp.org/publication.asp?id=icrp%20publication%2066 (accessed on 2 July 2024).
15. Thangavel, P.; Park, D.; Lee, Y.C. Recent Insights into Particulate Matter (PM(2.5))-Mediated Toxicity in Humans: An Overview. *Int. J. Env. Res. Public Health* **2022**, *19*, 7511. [CrossRef] [PubMed]
16. Premnath, V.; Wang, Y.; Wright, N.; Khalek, I.; Uribe, S. Detailed characterization of particle emissions from battery fires. *Aerosol Sci. Technol.* **2022**, *56*, 337–354. [CrossRef]
17. Wang, X.L.; Grose, M.A.; Avenido, A.; Stolzenburg, M.R.; Caldow, R.; Osmondson, B.L.; Chow, J.C.; Watson, J.G. Improvement of Engine Exhaust Particle Sizer (EEPS) size distribution measurement—I. Algorithm and applications to compact-shape particles. *J. Aerosol Sci.* **2016**, *92*, 95–108. [CrossRef]
18. Wang, X.L.; Grose, M.A.; Caldow, R.; Osmondson, B.L.; Swanson, J.J.; Chow, J.C.; Watson, J.G.; Kittelson, D.B.; Li, Y.; Xue, J.; et al. Improvement of Engine Exhaust Particle Sizer (EEPS) size distribution measurement—II. Engine exhaust particles. *J. Aerosol Sci.* **2016**, *92*, 83–94. [CrossRef]
19. Hinds, W.C. *Aerosol Technology, Properties, Behavior, and Measurement of Airborne Particles*, 2nd ed.; Wiley: Los Angeles, CA, USA, 1999.
20. Goupil, V.; Gaya, C.; Boisard, A.; Robert, E. Effect of the heating rate on the degassing and combustion of cylindrical Li-Ion cells. *Fire Saf. J.* **2022**, *133*, 103648. [CrossRef]
21. Wang, X.L.; Chancellor, G.; Evenstad, J.; Farnsworth, J.E.; Hase, A.; Olson, G.M.; Sreenath, A.; Agarwal, J.K. A Novel Optical Instrument for Estimating Size Segregated Aerosol Mass Concentration in Real Time. *Aerosol Sci. Technol.* **2009**, *43*, 939–950. [CrossRef]

22. Wang, X.L.; Zhou, H.; Arnott, W.P.; Meyer, M.E.; Taylor, S.; Firouzkouhi, H.; Moosmüller, H.; Chow, J.C.; Watson, J.G. Evaluation of gas and particle sensors for detecting spacecraft-relevant fire emissions. *Fire Saf. J.* **2020**, *113*, 102977. [CrossRef]
23. Lee, J.; He, S.; Song, G.; Hogan, C.J. Size distribution monitoring for chemical mechanical polishing slurries: An intercomparison of electron microscopy, dynamic light scattering, and differential mobility analysis. *Powder Technol.* **2022**, *396*, 395–405. [CrossRef]
24. Wang, X.L.; Firouzkouhi, H.; Chow, J.C.; Watson, J.G.; Carter, W.; De Vos, A.S. Characterization of gas and particle emissions from open burning of household solid waste from South Africa. *Atmos. Chem. Phys.* **2023**, *23*, 8921–8937. [CrossRef]
25. Larsson, F.; Andersson, P.; Blomqvist, P.; Mellander, B.-E. Toxic fluoride gas emissions from lithium-ion battery fires. *Sci. Rep.* **2017**, *7*, 10018. [CrossRef] [PubMed]
26. Claassen, M.; Bingham, B.; Chow, J.C.; Watson, J.G.; Wang, Y.; Wang, X. Characterization of Lithium-Ion Battery Fire Emissions—Part 1: Chemical Composition of Fine Particles (PM2. 5). *Batteries* **2024**, *10*, 301. [CrossRef]
27. Lithiumwerks. APR18650M1B Nanophosphate® Technology Data Sheet. Austin, Texas. 2023. Available online: https://lithiumwerks.com/products/lithium-ion-18650-cells/ (accessed on 2 July 2024).
28. AA Portable Power Corp. Model 544792 Polymer Lithium-Ion Battery Data Sheet. Available online: https://www.batteryspace.com/prod-specs/3175.pdf (accessed on 2 July 2024).
29. Tian, J.; Chow, J.C.; Cao, J.; Han, Y.; Ni, H.; Chen, L.-W.A.; Wang, X.; Huang, R.; Moosmu, H.; Watson, J.G. A biomass combustion chamber: Design, evaluation, and a case study of wheat straw combustion emission tests. *Aerosol Air Qual. Res.* **2015**, *15*, 2104–2114. [CrossRef]
30. U.S. EPA. Method 1—Sample and Velocity Traverses for Stationary Sources. Code of Federal Regulations, Title 40, Part 60, Appendix A-1. Available online: https://www.ecfr.gov/current/title-40/chapter-I/subchapter-C/part-60/appendix-Appendix%20A-1%20to%20Part%2060 (accessed on 6 August 2024).
31. Järvinen, A.; Aitomaa, M.; Rostedt, A.; Keskinen, J.; Yli-Ojanperä, J. Calibration of the new electrical low pressure impactor (ELPI+). *J. Aerosol Sci.* **2014**, *69*, 150–159. [CrossRef]
32. Saari, S.; Arffman, A.; Harra, J.; Rönkkö, T.; Keskinen, J. Performance evaluation of the HR-ELPI+ inversion. *Aerosol Sci. Technol.* **2018**, *52*, 1037–1047. [CrossRef]
33. Marjamäki, M.; Keskinen, J.; Chen, D.-R.; Pui, D.Y.H. Performance Evaluation of the Electrical Low-Pressure Impactor (ELPI). *J. Aerosol Sci.* **2000**, *31*, 249–261. [CrossRef]
34. Keskinen, J.; Pietarinen, K.; Lehtimäki, M. Electrical low pressure impactor. *J. Aerosol Sci.* **1992**, *23*, 353–360. [CrossRef]
35. Maricq, M.M.; Podsiadlik, D.H.; Chase, R.E. Size Distributions of Motor Vehicle Exhaust PM: A Comparison Between ELPI and SMPS Measurements. *Aerosol Sci. Technol.* **2000**, *33*, 239–260. [CrossRef]
36. Wang, X.; Firouzkouhi, H.; Chow, J.C.; Watson, J.G.; Ho, S.S.H.; Carter, W.; De Vos, A.S. Chemically speciated air pollutant emissions from open burning of household solid waste from South Africa. *Atmos. Chem. Phys.* **2023**, *23*, 15375–15393. [CrossRef]
37. Eldering, A.; Solomon, P.A.; Salmon, L.G.; Fall, T.; Cass, G.R. Hydrochloric acid: A regional perspective on concentrations and formation in the atmosphere of Southern California. *Atmos. Environ. Part A Gen. Top.* **1991**, *25*, 2091–2102. [CrossRef]
38. Sturges, W.T.; Harrison, R.M. The use of nylon filters to collect HCl: Efficiencies, interferences and ambient concentrations. *Atmos. Environ. (1967)* **1989**, *23*, 1987–1996. [CrossRef]
39. U.S. EPA. *Initial List of Hazardous Air Pollutants with Modifications*; Air Toxics Assessment Group, U.S. Environmental Protection Agency: Research Triangle Park, NC, USA, 2020. Available online: https://www.epa.gov/haps/initial-list-hazardous-air-pollutants-modifications (accessed on 6 August 2024).
40. Diaz, F.; Wang, Y.; Weyhe, R.; Friedrich, B. Gas generation measurement and evaluation during mechanical processing and thermal treatment of spent Li-ion batteries. *Waste Manag.* **2019**, *84*, 102–111. [CrossRef]
41. Andersson, P.; Blomqvist, P.; Lorén, A.; Larsson, F. *Investigation of Fire Emissions from Li-Ion Batteries*; SP Technical Research Institute of Sweden: Borås, Sweden, 2013.
42. Larsson, F. *Lithium-Ion Battery Safety-Assessment by Abuse Testing, Fluoride Gas Emissions and Fire Propagation*; Chalmers Tekniska Hogskola: Göteborg, Sweden, 2017.
43. Ribière, P.; Grugeon, S.; Morcrette, M.; Boyanov, S.; Laruelle, S.; Marlair, G. Investigation on the fire-induced hazards of Li-ion battery cells by fire calorimetry. *Energy Environ. Sci.* **2012**, *5*, 5271–5280. [CrossRef]
44. Sturk, D.; Rosell, L.; Blomqvist, P.; Ahlberg Tidblad, A. Analysis of li-ion battery gases vented in an inert atmosphere thermal test chamber. *Batteries* **2019**, *5*, 61. [CrossRef]
45. Bordes, A.; Marlair, G.; Zantman, A.; Herreyre, S.; Papin, A.; Desprez, P.; Lecocq, A. New insight on the risk profile pertaining to lithium-ion batteries under thermal runaway as affected by system modularity and subsequent oxidation regime. *J. Energy Storage* **2022**, *52*, 104790. [CrossRef]
46. Willstrand, O.; Pushp, M.; Andersson, P.; Brandell, D. Impact of different Li-ion cell test conditions on thermal runaway characteristics and gas release measurements. *J. Energy Storage* **2023**, *68*, 107785. [CrossRef]
47. Koch, S.; Fill, A.; Birke, K.P. Comprehensive gas analysis on large scale automotive lithium-ion cells in thermal runaway. *J. Power Sources* **2018**, *398*, 106–112. [CrossRef]

batteries

MDPI

Article

Characterization of Lithium-Ion Battery Fire Emissions—Part 1: Chemical Composition of Fine Particles ($PM_{2.5}$)

Matthew Claassen [1,2], Bjoern Bingham [1,3], Judith C. Chow [1], John G. Watson [1], Yan Wang [2] and Xiaoliang Wang [1,*]

1 Division of Atmospheric Sciences, Desert Research Institute, Reno, NV 89512, USA; matt.claassen@dri.edu (M.C.); bjoern.bingham@dri.edu (B.B.); judy.chow@dri.edu (J.C.C.); john.watson@dri.edu (J.G.W.)
2 Department of Mechanical Engineering, University of Nevada, Reno, NV 89557, USA; yanwang@unr.edu
3 Atmospheric Sciences, Department of Physics, University of Nevada, Reno, NV 89557, USA
* Correspondence: xiaoliang.wang@dri.edu; Tel.: +1-775-674-7177

Abstract: Lithium-ion batteries (LIB) pose a safety risk due to their high specific energy density and toxic ingredients. Fire caused by LIB thermal runaway (TR) can be catastrophic within enclosed spaces where emission ventilation or occupant evacuation is challenging or impossible. The fine smoke particles ($PM_{2.5}$) produced during a fire can deposit in deep parts of the lung and trigger various adverse health effects. This study characterizes the chemical composition of $PM_{2.5}$ released from TR-driven combustion of cylindrical lithium iron phosphate (LFP) and pouch-style lithium cobalt oxide (LCO) LIB cells. Emissions from cell venting and flaming combustion were measured in real time and captured by filter assemblies for subsequent analyses of organic and elemental carbon (OC and EC), elements, and water-soluble ions. The most abundant $PM_{2.5}$ constituents were OC, EC, phosphate (PO_4^{3-}), and fluoride (F^-), contributing 7–91%, 0.2–40%, 1–44%, and 0.7–3% to the $PM_{2.5}$ mass, respectively. While OC was more abundant during cell venting, EC and PO_4^{3-} were more abundant when flaming combustion occurred. These freshly emitted particles were acidic. Overall, particles from LFP tests had higher OM but lower EC compared to LCO tests, consistent with the higher thermal stability of LFP cells.

Keywords: Li-ion battery; fire; smoke; toxicity; particle emission; chemical composition; $PM_{2.5}$

Citation: Claassen, M.; Bingham, B.; Chow, J.C.; Watson, J.G.; Wang, Y.; Wang, X. Characterization of Lithium-Ion Battery Fire Emissions—Part 1: Chemical Composition of Fine Particles ($PM_{2.5}$). *Batteries* **2024**, *10*, 301. https://doi.org/10.3390/batteries10090301

Academic Editor: Wojciech Mrozik

Received: 27 June 2024
Revised: 6 August 2024
Accepted: 20 August 2024
Published: 27 August 2024

1. Introduction

Lithium-ion batteries (LIB) are a ubiquitous component in modern consumer products and commercial systems. Fire safety is a critical concern for LIBs due to their high energy densities [1,2]. Failing LIBs possess all three components needed for a fire (i.e., fuel, oxygen, and heat) in close contact, thereby allowing combustion to occur even when deprived of ambient air [1]. LIB electrolytes, separator and binder materials, and anodes are all combustible under suitable conditions. Oxygen can be released from cathode decomposition or is available from ambient air once the cell casing is compromised. Heat can be generated internally from a short-circuit or from exothermic chemical reactions, both of which occur during initial cell breakdown. Heat can also be applied externally by fire, neighboring cell failure, or other overheating components. As a result, when LIBs overheat, are used improperly, or are defective, they can enter an exothermic chain reaction known as thermal runaway (TR), which may further lead to violent combustion and even explosion.

Numerous fire incidents have been reported for LIBs of various sizes and in different applications. Between the years 2010 and July 2023, 393 electrical vehicle (EV) fire incidents were reported globally, most of them occurring after 2020 due to increasing EV prevalence [3]. Several large fires have occurred at stationary storage facilities. The 250 MW Gateway Energy Storage facility fire in San Diego, California, burned for over a week in 2024. The fire risk and consequences for light EVs (e.g., electric scooters and bikes) are much higher than those for passenger EVs owing to lower quality LIB cells and battery management systems, device wear and tear, and variable domestic charging methods. Over 500 LIB fires, resulting in 36 fatalities, were caused by light EVs between January and July 2023 alone [3]. LIBs also pose risks to transportation. The Federal Aviation Administration (FAA) has recorded over 250 incidents of smoke, fire, extreme heat, or explosion involving LIBs aboard aircraft in the last 10 years. Three incidences suspected of being caused by LIBs carried as cargo resulted in the loss of the aircraft and crew [4]. While a major LIB fire has not yet occurred during a manned space mission, NASA test centers have experienced catastrophic LIB failures, even in advanced, custom-designed applications that had seen prolonged use without incident [5]. Fire risks persist even at the end of a LIB's useful life [6]. The U.S. Environmental Protection Agency (U.S. EPA) found that 245 fires across 64 waste facilities in the U.S. were definitely or probably caused by lithium (Li) metal, or LIBs, between the years 2013 and 2020 [7].

In addition to intense heat and flame, LIB fires can generate large quantities of combustible, toxic, and corrosive gaseous and particulate emissions. Past studies on LIB fire emissions focus on combustible and/or toxic gaseous compounds [8–16]. Fewer studies have examined smoke particle emissions from LIB fires [17–23]. Several studies measured the elemental composition of large particles (up to several millimeters in diameter) deposited on nearby surfaces after a LIB fire and found that these particles are enriched with toxic and corrosive chemicals, including Li, transition metals, fluoride compounds, and carbonates [17,20,21,24]. However, many of these elemental measurements used scanning electron microscopy-energy dispersive X-ray analysis (SEM-EDX) under vacuum conditions, causing volatile particulate compounds to evaporate. Therefore, the volatile compounds are not reported. While these large particles (>100 μm) may cause soil and water contamination, they settle out quickly and are not likely to penetrate deep into the human respiratory track.

Very few studies have examined the smaller particles (<10 μm) that pose significant inhalation hazards. Barone et al. [18] pointed out that little information was available on the size, composition, and morphology of respirable particles. They used image and elemental analyses of particles from the combustion of three types of LIBs: nickel manganese cobalt oxide (NMC), lithium iron phosphate (LFP), and lithium titanate oxide (LTO) and found abundant elements originating from LIB anode, cathode, and separator materials. NMC and LTO particles had higher abundances of transition metals (e.g., nickel [Ni], manganese [Mn], cobalt [Co], and titanium [Ti]) than LFP particles, causing greater health and environmental contamination concerns. However, that study did not report particle concentrations, and the composition focused mostly on elements that can be analyzed by EDX. Premnath et al. [19] measured total and solid $PM_{2.5}$ (particles with aerodynamic diameters ≤ 2.5 μm) concentrations, black carbon levels, and organic and elemental carbon (OC and EC) for NMC and LFP cells at 100% state of charge (SOC). OC and EC were 17–47% and 7–33% of $PM_{2.5}$, respectively; however, elements and ions were not measured. Padilla et al. [22] showed high smoke concentrations and obscurations from laptop computer fires, highlighting the potential catastrophic hazards of LIB fires in spacecraft. Xu et al. [25,26] characterized the morphology, elemental composition, surface chemistry, and cytotoxicity of soot collected from combusting LFP and ternary cells. Particles were collected inside or close to the flame without size classification. To the best of our knowledge, no studies have reported the detailed chemical composition of $PM_{2.5}$ from LIB fires.

The objectives of this paper are to: (1) determine the chemical composition of $PM_{2.5}$ emitted from LIB combustion events; and (2) evaluate the dependence of $PM_{2.5}$ compositions on cell chemistry and SOC. The $PM_{2.5}$ size fraction of particles from LIB fires is of particular concern in terms of human health. While larger particles quickly settle on surrounding surfaces due to gravity, $PM_{2.5}$ can stay airborne much longer; when inhaled, these particles can penetrate deep into the lung [27]. Toxic constituents of $PM_{2.5}$ can either accumulate in the lung or enter the bloodstream, leading to respiratory, cardiovascular, and other diseases [28]. Depending on exposure level and individual susceptibility, $PM_{2.5}$ can cause both acute and chronic health effects. A recent study by Xu et al. found that soot particles from LIB fires have significant cytotoxicity to human cells, causing greater damage than smoke particles from wood combustion [26]. In addition, the chemical characteristics of emitted $PM_{2.5}$ are useful for fire detection, personal protection, fire suppression, and postfire cleanup. A companion paper [29] presents particle size distributions and emission factors, and a future paper will describe toxic, combustible, and corrosive gas emissions.

2. Materials and Methods

2.1. Tested LIBs

LIBs are often categorized by their cathode composition. Two commonly used cathode types (i.e., "cell chemistries") were tested: a cylindrical, 18650-style, LFP cell manufactured by LithiumWerks (Enschede, The Netherlands), and a pouch-style lithium cobalt oxide (LCO) cell manufactured by AA Portable Power Corp. (Richmond, CA, USA) (as shown in Figure S1 of Supplementary Materials). These cells were commercially available and were safe to handle, charge, and discharge with minimal specialty equipment. The cells were purchased shortly before testing and were received at proper storage voltages. Specifications for each LIB type are shown in Table 1.

Table 1. Specifications for the two cell types tested [30,31].

Cell Type	Cell Format	Cathode Chemistry	Dimensions (mm)	Nominal Voltage (V)	Nominal Capacity (Ah)	Average Cell Mass (g)
LFP	Cylinder	$LiFePO_4$	18 × 18 × 65	3.3	1.1	42
LCO	Pouch	$LiCoO_2$	5.4 × 47 × 95	3.7	2.5	47

Cell chemistry can affect the combustion behavior of a compromised LIB cell. As illustrated in Figure S2, LFP cells have increased thermal stability and reduced heat generation during cell breakdown [32–35], but they are disadvantaged by having a lower energy density. LCO cells are the least thermally stable of common cathode chemistries, but they are commonly manufactured due to their high energy density, discharge voltage, cycling performance, and legacy as the first mass-produced LIB type. LCO cells are mostly used for personal electronics (e.g., phones, laptops, cameras, etc.) and are common in space applications due to their high energy density [36], including use in the primary international space station power storage system [37]. LCO cells were selected for this study due to their wide use in personal electronics and space applications and the possibility of unique emission profiles due to their thermal instability. LFP cells were selected due to their increasing adoption and contrasting thermal stability to LCO cells. NMC and other Ni-based chemistries have been represented in the current literature on TR emissions and are commonly used in power tools, EVs, and other micromobility devices, such as electric bikes and scooters. However, LFP cell prevalence is also increasing in these markets due to lower costs and increased thermal stability.

2.2. Experimental Setup and Procedures

LIB combustion tests were conducted inside a burn chamber at the Desert Research Institute [38]. The chamber is ~8 m^3 and is sealed except for an air intake and a flow-controlled exhaust port (Figure 1). At the beginning of each testing day, all instruments were turned on and allowed to warm up before use. Instrument consumables were checked, along with all sample line ducting connections. The clean air dilution flow rate, discussed below, was adjusted and measured. All particle analyzers had zero-calibrations performed daily. Before each test, the chosen LIB cell was charged to the desired SOC by a programable charger. The cell was then placed in a ceramic crucible heated by an electric hot plate. A type K thermocouple was used to measure the external cell and flame temperatures. For LFP tests, the thermocouple was placed next to the cell overpressure vents on the cylinder endcap, where vented gases exit the failing cell. For LCO tests, the thermocouple was placed above the flat side of the prone cell (see Figure S1). Filter flow rates were measured for each filter channel, and each instrument was configured for data collection. After completing the test setup, the burn chamber door was closed, filter pumps were turned on, and the hotplate power supply was turned on to begin heating the ceramic crucible containing the LIB cell to 450 °C. Different heating rates did not affect emissions, as thermal runaway will occur independently of heating once it has been triggered [39]. The hotplate was left on until the cell had entered TR, combusted, and ceased emitting particles. The time when each event occurred (power on pumps and the heater, smoke visible, flames visible, and flames ending) was recorded. This approach simulates a TR propagation event where a LIB cell is subject to continual heating due to the failure of adjacent cells in a LIB pack. Such a failure mode is responsible for most emissions from large LIB packs, as the failure of any single cell or group of cells spreads through the entire pack.

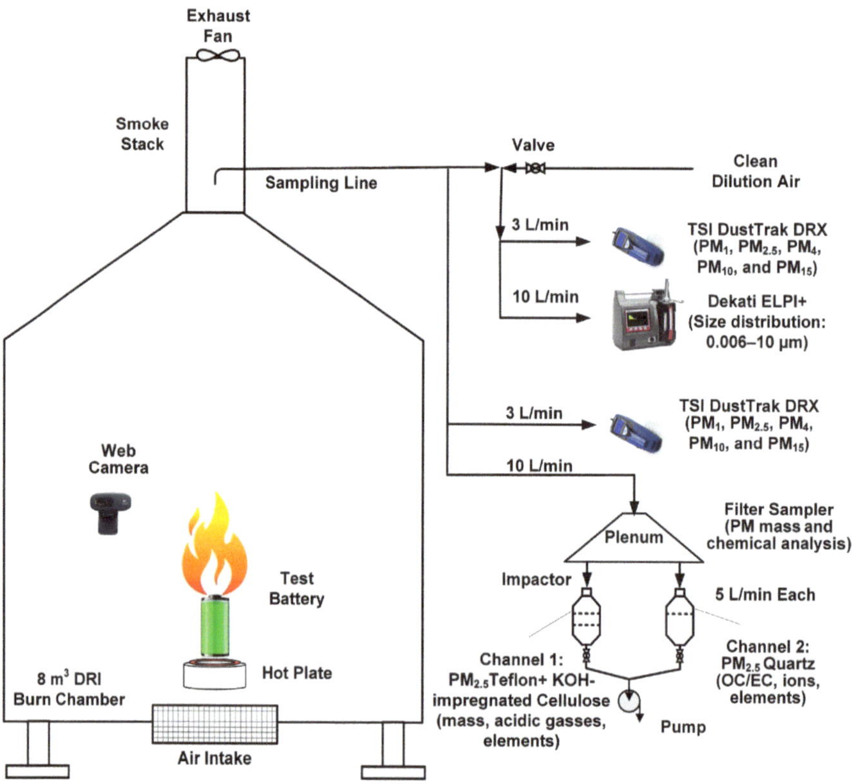

Figure 1. Burn chamber, instrument, and filter assembly layout.

The LIB fire emission sampling setup is similar to that used in previous studies [40–42]. As shown in Figure 1, a sample of LIB combustion emissions was extracted from the exhaust vent and directed to a suite of particle analyzers, as well as the filter sampling system. Particle mass concentrations were measured by two DustTrak DRX (Model 8534, TSI Inc., Shoreview, MN, USA) aerosol monitors in five size fractions (i.e., PM_1, $PM_{2.5}$, PM_4, PM_{10}, and PM_{15}) [43]. Measurements were made before and after dilution to determine the dilution factors for other instruments. Finer-resolution particle size distributions were measured by an electrical low-pressure impactor (Dekati ELPI+, Kangasala, Finland; 6 nm–10 μm) on the diluted sample line, to prevent impactor overloading [44,45]. Two filter channels with different sampling media were used to collect $PM_{2.5}$ at a flow rate of 5 L/min each and submitted for detailed chemical analyses [46]. Each test started with a 1–2 min background measurement and ended when particle concentrations returned to background levels after combustion ended. A webcam (Model C270, Logitech, San Jose, CA, USA) was used to record combustion behavior. Gas concentrations were also measured, which will be reported in a future paper.

Each type of LIB was tested at five SOC levels: 0%, 30%, 50%, 75%, and 100% for LFP and 0%, 30%, 60%, 80%, and 100% for LCO. Each SOC condition was tested between three and six times, with some tests discarded due to instrument malfunction.

2.3. Chemical Analysis

The three filters from each test were analyzed by various instruments to determine the composition of the deposited $PM_{2.5}$. The cellulose back filters in Channel 1 were for collecting acidic gases, and the data will be reported in a future publication. The Channel 1 Teflon filters were analyzed for deposited mass and elements, while the Channel 2 Quartz filters were analyzed for elements and carbon compounds. The Teflon filters were first equilibrated in a clean room at a controlled temperature (21.5 ± 1.5 °C) and relative humidity (35 ± 5%) before gravimetric analysis [47]. Filters were weighed before and after sampling using an XP6 microbalance (Mettler Toledo Inc., Columbus, OH, USA) with a sensitivity of ±1 μg. This allowed for the calculation of the $PM_{2.5}$ mass collected on the filter as well as the average mass concentration of the smoke emissions using the filter sampling volume. After being weighed, the Teflon filters were analyzed for 51 elements (sodium to uranium) by an X-ray fluorescence (XRF) analyzer (Model Epsilon 5, PANalytical, Eindhoven, The Netherlands) [48].

Half of the Channel 2 quartz filters were extracted in distilled and deionized water, and analyzed by ion chromatography (IC; Dionex ICS-6000; Thermo Fisher Scientific, San Jose, CA, USA) [49] for 11 water-soluble ions, including: lithium (Li^+), ammonium (NH_4^+), sodium (Na^+), magnesium (Mg^{2+}), potassium (K^+), calcium (Ca^{2+}), fluoride (F^-), chloride (Cl^-), nitrate (NO_3^-), sulfate (SO_4^{2-}), and phosphate (PO_4^{3-}). A 0.5 cm^2 punch was taken from the remaining quartz filter halves and analyzed for OC, EC, and eight thermal fractions (OC1–OC4, pyrolyzed carbon [OP], and EC1–EC3) following the IMPROVE_A thermal/optical protocol using the DRI Model 2015 Multiwavelength Carbon Analyzer (Magee Scientific, Berkeley, CA, USA) [50–52]. The remaining quartz filter material was digested in a solution of 2% nitric acid (HNO_3) and 5% hydrochloric acid (HCl) and analyzed by inductively coupled plasma mass spectrometry (ICP-MS; NexIon 2000B; Perkin-Elmer, Waltham, MA, USA) for 44 elements, including ones that could not be quantified by XRF (i.e., Li, Be, Na, Mg, Pr, Nd, Pt, Bi, and Th). ICP-MS complements XRF as it has lower detection limits for rare-earth elements [48].

2.4. Data Analysis and Quality Assurance

All raw $PM_{2.5}$ chemical composition data were first converted to filter mass loadings in μg/filter and stack mass concentrations in $\mu g/m^3$. Concentration uncertainties were calculated from the propagation of analytical and sample flow volume uncertainties [53].

Mass closure was conducted to determine the proportion of the collected filter deposits accounted for by chemical analyses, which is an indicator of the data quality [54,55]. The sum of the measured chemical species was first compared to the gravimetric mass. As shown in Figure 2a,b, these two masses were highly correlated with coefficients of determination (R^2) of 0.99. The linear regression slopes were near but lower than 1, being 0.84 and 0.86 for the LFP and LCO tests, respectively, indicating that most chemical species have been measured.

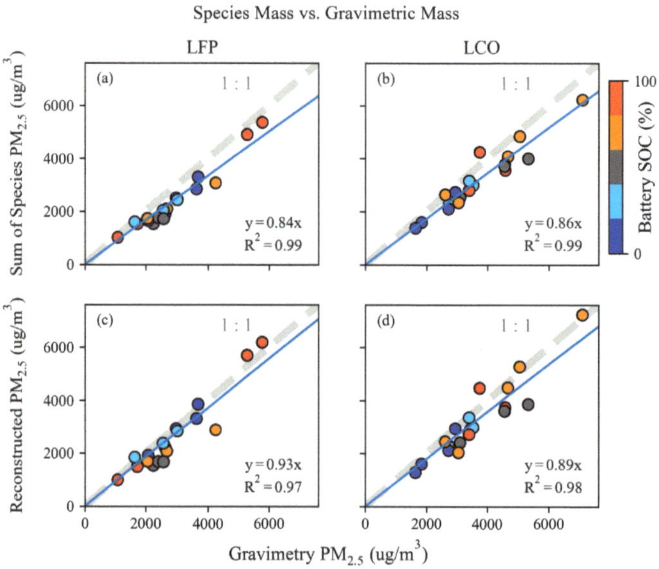

Figure 2. Comparison of chemical analysis sum of species (**a**,**b**) and reconstructed mass (**c**,**d**) with gravimetric mass for LFP (**a**,**c**) and LCO (**b**,**d**) tests.

Next, the "reconstructed" mass was compared to the measured gravimetric mass. The measured $PM_{2.5}$ species were grouped into six major compositions: (a) organic matter (OM = OC × multiplier), (b) EC, (c) elemental phosphorus (P) that is not in PO_4^{3-} (termed P*, calculated as P from ICP-MS minus $PO_4^{3-}/3.07$ from IC), (d) PO_4^{3-}, (e) F^-, and (f) other ions and elements without double counting elements present in multiple species (using the same method as for P*). The multiplier for calculating OM from OC was taken as 1.2 for fresh combustion emissions [56]. The sum of these six composition groups is defined as the reconstructed mass. The difference between the gravimetric and reconstructed masses is reported as the "unidentified" mass. Mass reconstruction attempts to account for common elements not being measured in chemical analysis, such as oxygen (e.g., in mineral oxides) and hydrogen. The slopes for reconstructed vs. gravimetric mass were 0.93 for LFP tests (Figure 2c) and 0.89 for LCO tests (Figure 2d). These relationships have similarly high R^2 values as the sum of species. The regression slopes being slightly below 1.0 indicates that the reconstruction assumptions mentioned above were not completely accurate for all samples. Figure S3 shows that the unidentified mass was correlated with OM (inversely) and PO_4^{3-} (directly), indicating that these species may have been the main sources of uncertainty. The reconstructed-to-gravimetric mass ratio for all tests was, however, within the acceptable range of 0.60 to 1.32 set by the U.S. EPA [57]. Mass fractions (chemical profiles) were calculated by dividing the mass of each species by the total $PM_{2.5}$ gravimetric mass.

The balance between anions and cations was examined via ion microequivalent mole concentrations (μeq/m³), which were calculated by dividing the ion mass concentration (in μg/m³) by the ratio of atomic or molecular weight of the chemical species and the species' charge, as shown in Equations (1) and (2) [54,56]:

$$\mu eq/m^3 for\ anions = \frac{[F^-]}{19} + \frac{[Cl^-]}{35.5} + \frac{[NO_3{}^-]}{62} + \frac{[SO_4{}^{2-}]}{96/2} + \frac{[PO_4{}^{3-}]}{95/3} \quad (1)$$

$$\mu eq/m^3 for\ cations = \frac{[Li^+]}{6.9} + \frac{[NH_4{}^+]}{18} + \frac{[Na^+]}{23} + \frac{[Mg^{2+}]}{24.3/2} + \frac{[K^+]}{39.1} + \frac{[Ca^{2+}]}{40.1/2} \quad (2)$$

3. Results

3.1. LIB Combustion Behavior

LIB combustion behavior ranged from only venting (with no visible flames throughout the test) to mostly flaming. Most cells vented for a period immediately after the start of TR and then autoignited for the remainder of combustion. Visual examples of the two combustion phases for an LCO test are shown in Figure 3.

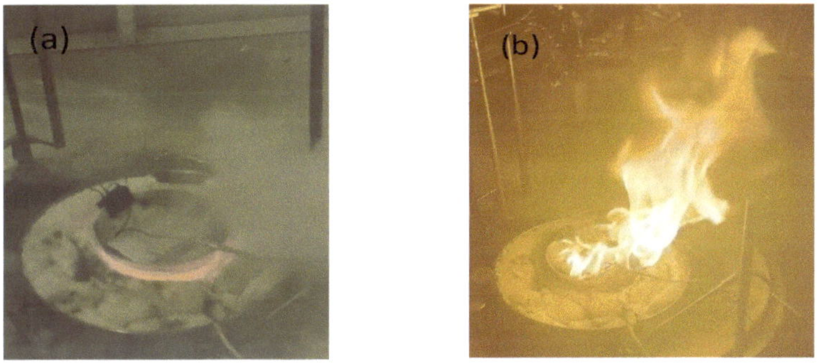

Figure 3. An example of combustion phases from an LCO test: (**a**) the LIB vented vigorously; and (**b**) the LIB burst into large flames.

Both webcam footage (Figure 3) and $PM_{2.5}$ concentration time series measured by the ELPI+ were used to characterize the combustion behavior of each cell and estimate the proportion of $PM_{2.5}$ emissions from venting and flaming phases, as shown in Figure 4. A threshold of 10% of the maximum $PM_{2.5}$ concentration was used to signal the start of TR. The webcam recording was then used to determine if and when flaming started. Flaming combustion often began vigorously (as seen in Figure 3b) but quickly diminished in intensity, ending as a small, candle-sized flame that persisted well into the overall decay of $PM_{2.5}$ concentrations. Figure 4 shows the venting and flaming phase peaks. Emissions from each phase were integrated, and the proportions of each phase to total emissions were calculated. This metric is a qualitative measure of emission proportion to characterize cell combustion characteristics and explain variations in emission profiles.

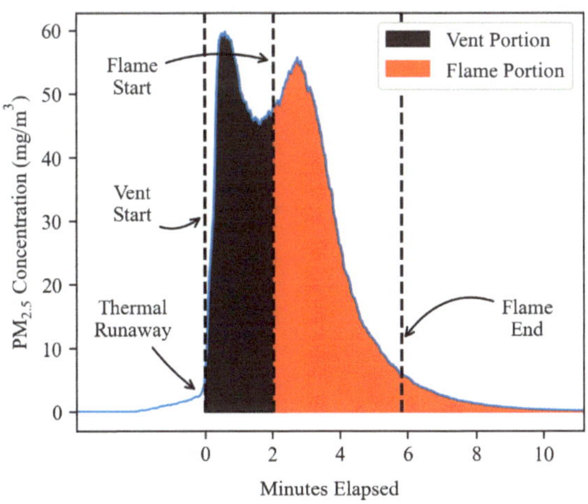

Figure 4. An example of emission allocation for the same LCO test in Figure 3, showing 44% and 56% of $PM_{2.5}$ emissions were from venting and flaming phases, respectively. Particle concentrations are measured by ELPI+ and adjusted for sample dilution.

Figure 5a shows that LFP tests did not flame at 0% and 30% SOCs, while LCO tests flamed at all SOCs (Figure 5b). This is consistent with LFP cells having higher thermal stability compared to LCO cells. LCO tests had an increasing proportion of emissions from flaming as SOC increased, while LFP tests had the highest proportion of flaming emission at middle (50–75%) SOCs, corresponding to higher detected flame temperatures (Figure 5c). All LCO tests except one had >50% of $PM_{2.5}$ emissions come from flaming combustion.

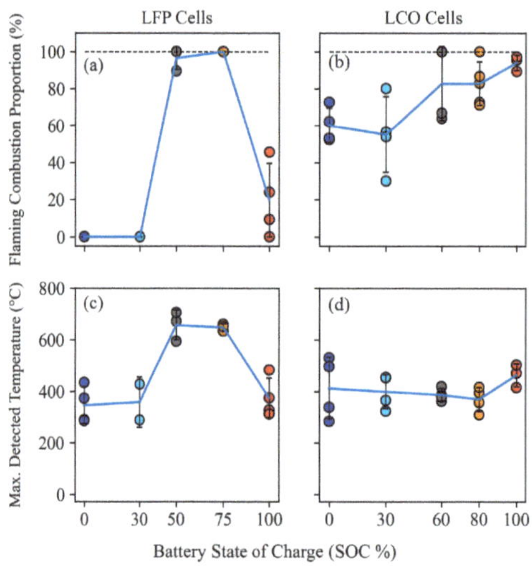

Figure 5. Proportions of $PM_{2.5}$ emitted by flaming combustion (**a**,**b**) and maximum detected flame temperatures (**c**,**d**) for different SOCs and cell types. Symbol color represents cell SOC for each test.

3.2. Chemical Characteristics

3.2.1. Major Compositions

The six major composition groups as a percentage of gravimetric $PM_{2.5}$ mass are shown in Figure 6 for each cell type and SOC. OM, EC, and PO_4^{3-} were the three most abundant compositions. Overall, particles from LFP tests had higher OM and lower EC compared to LCO tests. P* was mostly present in LCO tests.

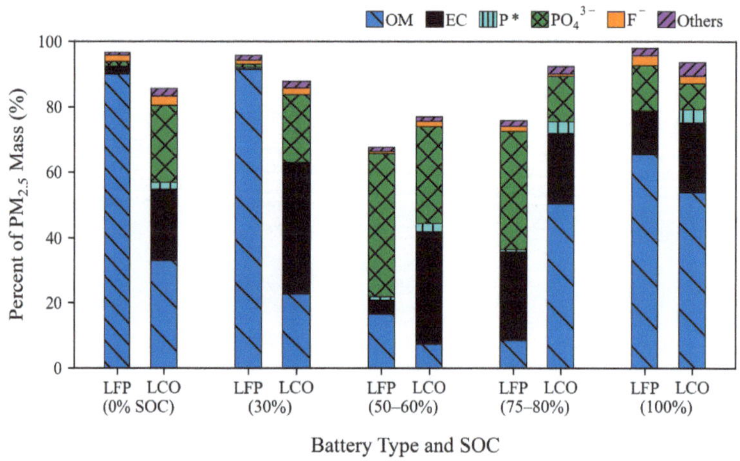

Figure 6. Average $PM_{2.5}$ compositions as a percent of $PM_{2.5}$ mass for LFP and LCO tests. The number of tests varied from three to five for each SOC group. P* is calculated as the difference between the total P and the P in PO_4^{3-}.

The $PM_{2.5}$ chemical compositions of individual tests are presented in Figure 7, grouped by SOCs and cell type. Figure 7a shows that $PM_{2.5}$ composition variation for LFP tests was low within each SOC, except among 100% SOC tests. $PM_{2.5}$ was characterized by high OM for 0% and 30% SOC levels. EC, most dominantly produced from flaming combustion, was low for these low SOC tests but increased for SOCs above 50%. PO_4^{3-} was similarly low at 0% and 30% SOC, but other ions, mostly F^- and Cl^-, were comparable among all SOCs.

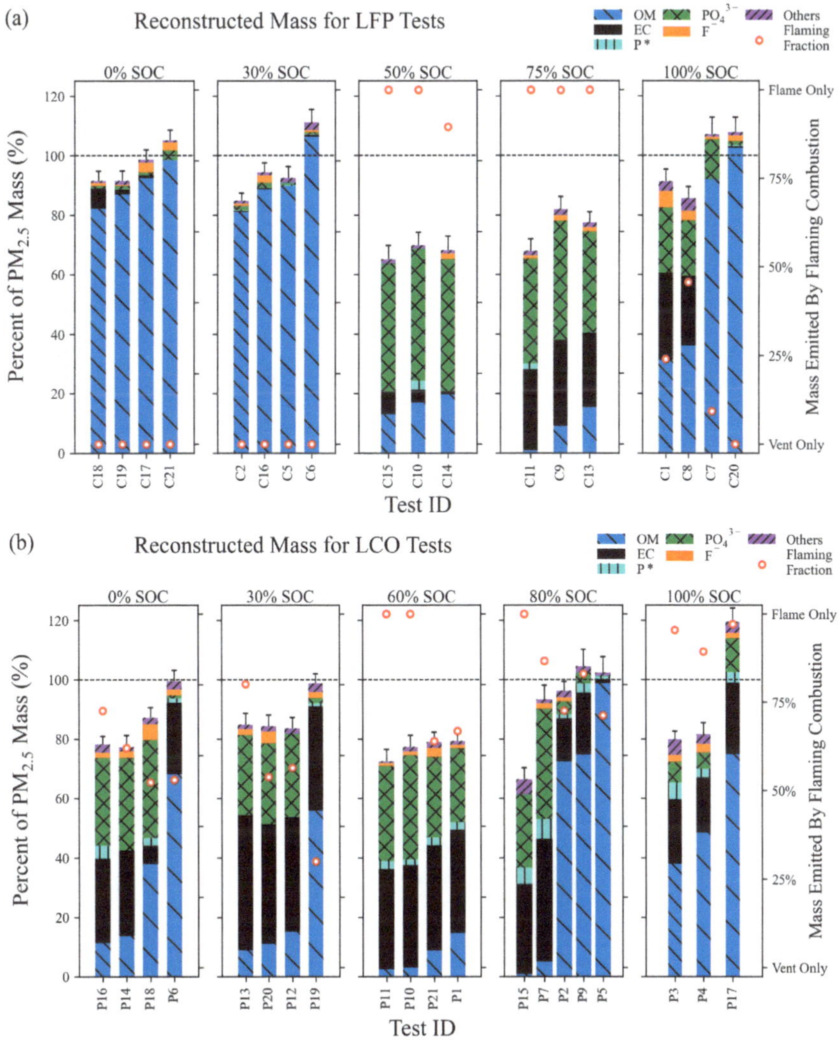

Figure 7. $PM_{2.5}$ compositions as a percent of $PM_{2.5}$ mass for each test of: (**a**) LFP and (**b**) LCO tests. The percentages of $PM_{2.5}$ mass emitted in flaming phase are indicated by the red circles (right y-axis). Error bars indicate the combined uncertainty of all mass fractions in each test.

LFP tests with 50% and 75% SOCs were dominated by flaming combustion. The reconstructed mass percentages were the lowest of any test group, at ~60–80%. Compared to the 0% and 30% SOC cases, the OM fractions were much lower (1–20%), while the PO_4^{3-} fraction, practically absent in the 0% and 30% SOC cases, dominated with 34–44% of $PM_{2.5}$ mass. The average EC fraction increased from 0–7% for 0–50% SOC to 25–29% for 75% SOC.

The shift in OM, EC, and PO_4^{3-} mass fractions was likely caused by a more energetic TR process due to increased cell energy as well as the ignition of the emissions stream.

LFP tests with 100% SOC exhibited two different combustion behaviors: Tests C1 and C8 had 24–46% emissions from flaming, while emissions from Tests C7 and C20 were dominated by venting (0–9% of emissions from flaming). The 100% SOC $PM_{2.5}$ compositions had more variability but adhered to the trend that a higher proportion of flaming combustion resulted in higher EC and PO_4^{3-} mass fractions and lower OM mass fractions. The two flaming tests (Tests C1 and C8) showed similar EC abundance to the 75% SOC tests but with higher OM, F^-, and others, and lower PO_4^{3-}. The two venting tests (Tests C7 and C20), however, resembled the vented emissions from low SOCs with no EC and little PO_4^{3-} present, although C7 (13%) had over four times the PO_4^{3-} fraction of any low SOC test. This indicates that ignition of the emissions stream, rather than cell SOC, is mostly responsible for the general composition of emitted $PM_{2.5}$ particles.

$PM_{2.5}$ mass composition for LCO tests (Figure 7b) differed from that of LFP tests, (Figure 7a), with less dependence of emissions composition on cell SOC. Two groups existed within 0–80% SOCs: one with high OM (56–99%), low PO_4^{3-} (0.2–4%), and variable EC (1–35%), resembling low SOC LFP tests; and the other with lower OM (1–15%), high PO_4^{3-} (25–37%), and consistently high EC (29–45%), resembling mid-range SOC LFP tests. As with LFP tests, the mass closure was lower for this low-OM group. Higher P* emissions occurred for most SOCs. An exception to this grouping is the 0% SOC LCO Test P18, where high PO_4^{3-} suggests high EC should be present but high OM was seen instead, a composition similar to 50% SOC LFP tests. As with LFP tests, these groupings were related to combustion behavior, with the high EC group having more flaming combustion than the high OM group. However, compared to LFP tests, this grouping was due to smaller differences in combustion behavior, as significant flaming occurred for all LCO tests. For example, significant flaming during 80% SOC LCO Test P5 resulted in a similar emissions profile to low SOC LFP tests with no flaming. This is likely due to differences in cell casing between LCO and LFP cells. The LCO cell's thin wrapper likely results in multiple venting locations, only some of which are ignited. Since all emissions are attributed to flaming once any flames are seen, this could result in a misallocation of the emission source. The LFP cell's steel case and relief vents, on the other hand, ensure venting is concentrated into a single stream, which was either ignited or not.

LCO tests with 100% SOC are somewhere between these two groups, emitting mostly OM (38–75%) but with moderate EC (19–24%) and PO_4^{3-} (5–11%) fractions. This was in contrast to tests with high OM at SOCs below 100% (i.e., Tests P2, P5, P6, P9, and P19, with P18 being the exception), where PO_4^{3-} mass fractions dropped to $\leq$2%. Ignoring SOC dependence, the major combustion products from the LFP and LCO tests were broadly similar, either consisting of high OM and low EC, PO_4^{3-}, and unidentified fractions (with relatively less flaming combustion); or low OM and high EC, PO_4^{3-}, and unidentified fractions (from relatively more flaming combustion). All other ions and elements only contributed to 1.4–13% (averaging 5%) of the gravimetric mass.

3.2.2. Carbon Fractions

Emitted carbon compounds were split into seven thermal fractions depending on their volatility and thermal stability, four for OC, and three for EC [50]. As shown in Figure S4, the dominant OC fractions were OC1, OC2, and OC3, accounting for 25% ± 8%, 50% ± 11%, and 23% ± 12% of total OC on average for both cell types. OC4 was relatively low for all tests, accounting for 1.4% ± 1.2% of OC on average. EC2 was the dominant EC fraction, accounting for 76% of total EC on average. However, while EC1 and EC3 account for only 17% and 7% of total EC on average, they could individually account for over 50% of EC for certain tests.

The SOC dependence of OC1–4 and EC1–3 is shown in Figure 8. For LFP tests (Figure 8a–d), each EC fraction was emitted by a specific combustion behavior. EC1 was emitted during venting-dominated combustion during low SOC tests and half of 100% SOC tests. EC2 was emitted across all SOCs but was most abundant for flaming 75% and 100% SOC tests. EC3 was emitted for 50% SOC tests but much less so for 75% SOC tests, despite both being heavily flaming. LCO tests (Figure 8e–h) showed a similar pattern for EC3 as LFP tests but differed for EC1 and EC2. EC2 and EC3 were more abundant in LCO tests than in LFP tests. EC2 had the largest increase in both mass percentage and its presence for a range of SOCs, with only three tests (i.e., P2, P5, and P18) having EC2 abundances <18%. In contrast, only five LFP tests (i.e., C1, C8, C9, C11, and C13) had EC2 abundances >18%, with the rest having abundances <5%. OC1–4 varied similarly with SOC, despite large variations in total OC. This dependence of OC1–4 (Figure 8a,e) is similar to EC1 for both cell types, an understandable result given that EC1 is measured at the closest analysis temperature to the OC fractions.

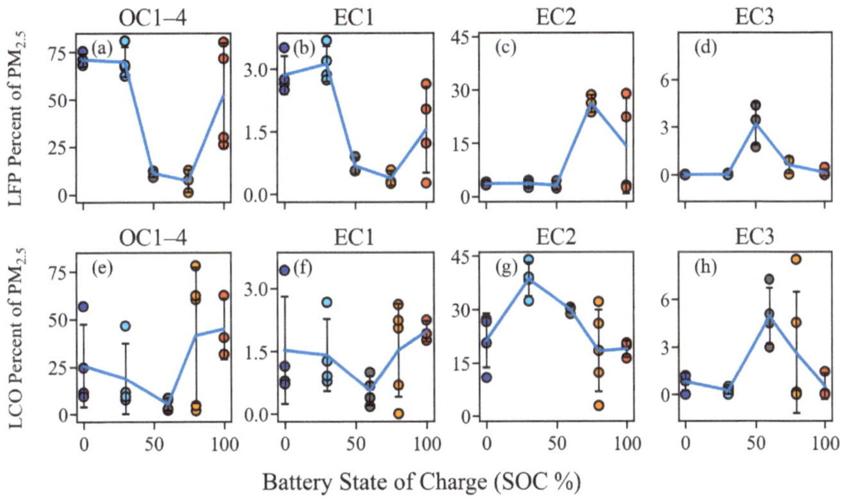

Figure 8. Dependence of total OC and EC fractions on SOC for LFP (**a–d**) and LCO (**e–h**) tests. Symbol color represents cell SOC for each test. Four OC fractions that evolved at different temperatures in a pure helium (He) carrier gas include: OC1 (140 °C), OC2 (280 °C), OC3 (480 °C), and OC4 (580 °C); three EC fractions that evolved in 98% He/2% O_2 include: EC1 (580 °C), EC2 (740 °C), and EC3 (840 °C) [50]. These fractions are indications of carbon volatility and pyrolysis and have been used to distinguish contributions to $PM_{2.5}$ from different sources.

Carbonaceous aerosols can be generated from the combustion of graphite anodes, electrolytes, polymer separators, wrappers, and other carbon-containing components in LIBs. As electrolytes are volatile, unburnt electrolytes can contribute to OC fractions. EC2 and EC3 are refractory carbon with graphite-like structures, which may originate from the decomposition and recondensation of graphite material, as well as from high-temperature combustion [58]. Electrolyte and cathode materials may react to form carbonate, which could also contribute to various carbon fractions, depending on their decomposition temperatures. Previous studies show that polycyclic aromatic hydrocarbons (PAHs), polychlorinated biphenyls, alcohols, aldehydes, alkenes, and esters are present in particles ejected or deposited from LIB fires [20,24]. Future studies should examine the detailed composition of carbonaceous $PM_{2.5}$ components from LIB fire emissions to assess their health effects and infer particle sources.

3.2.3. Elemental Abundances

Figures 9 and 10 show the mass fractions of 24 elements and 10 ions for the LFP and LCO tests, respectively. Elements with an average abundance of ≥0.001% of $PM_{2.5}$ mass for all tests were plotted. Unlike for P* and the "other" components in Figures 6 and 7, the elemental masses within ionic compounds were not subtracted from the total measured elemental mass here. Therefore, measured elements (primarily Li, Na, Mg, P, and Ca) may have been in ionic, elemental, or oxide forms.

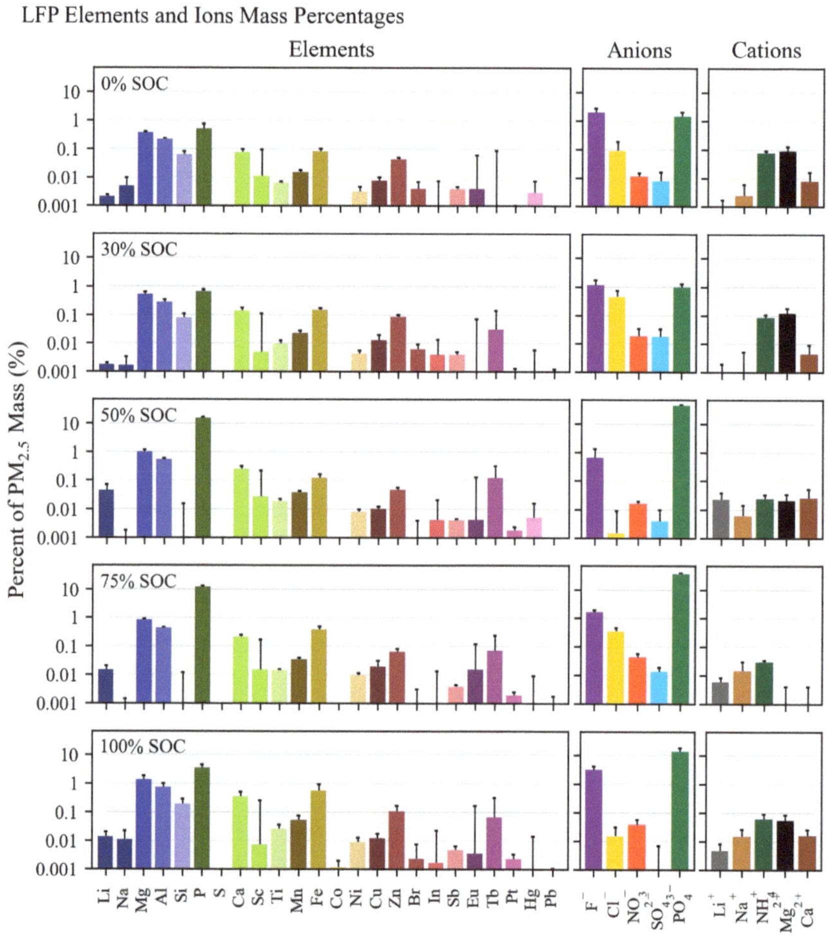

Figure 9. Average abundance (in log-scale) of elements, anions, and cations for each LFP SOC. Error bars indicate the larger of the propagated analytical uncertainty or the standard error within each SOC.

Elemental emissions for both LIB types were dominated by P, consisting of 6.4% and 8.5% of $PM_{2.5}$ mass on average for LFP and LCO tests, respectively. PO_4^{3-} accounted for virtually all detected elemental P for LFP tests, and 77% of elemental P for LCO tests (Figure S5c,g). Other elements with significant abundance (0.1–1%) for both cell types included Mg, Al, Ca, and Fe. Li and Zn were also within this abundance range, but for LCO tests only, with Co falling just below it.

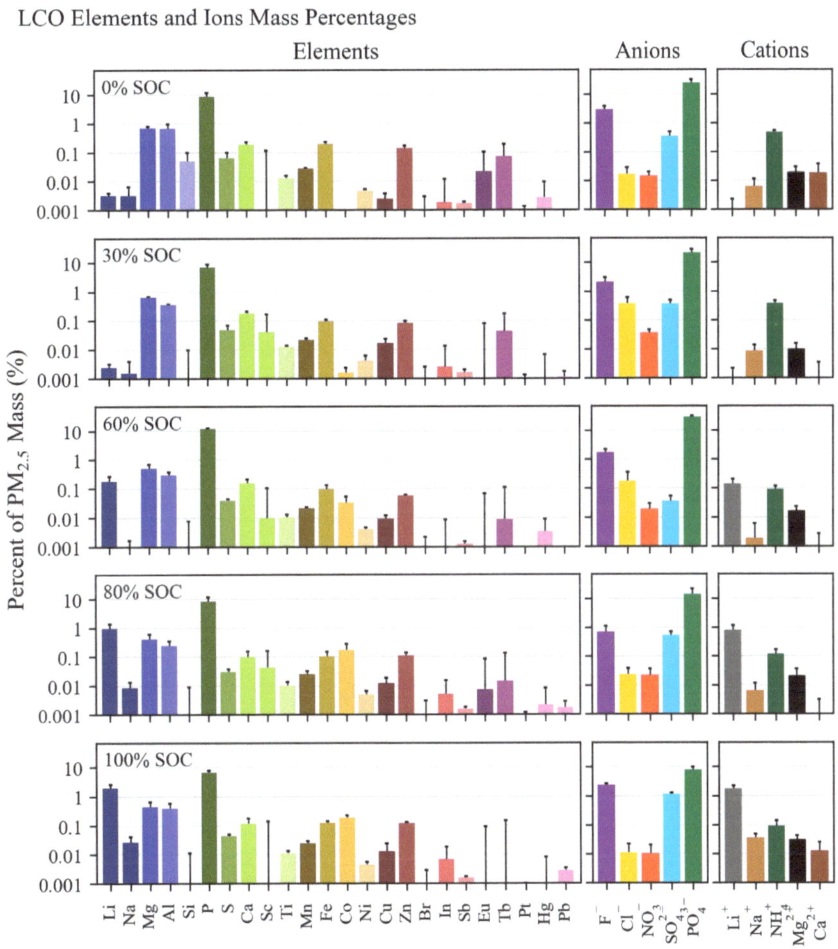

Figure 10. Average abundance (in log-scale) of elements, anions, and cations for each LCO SOC. Error bars indicate the larger of the propagated analytical uncertainty or the standard error within each SOC.

Mg was the second most abundant element after P for LFP tests (Figure 9), at 0.4–1.3%. Light metals are the next most abundant set of elements, with up to 0.7% Al, 0.6% Fe, 0.3% Ca, and 0.1% Zn. Si had maximum abundances of 0.2% but was absent from mid-SOC tests. Rare earth Tb was the only other element reaching 0.1%, while Li, Na, Sc, Ti, Mn, Ni, Cu, and Eu had maximum abundances between 0.01% and 0.1%. S, Co, Br, In, Sb, Pt, Hg, and Pb all had maximum abundances of ≤0.01%.

LCO tests (Figure 10) had higher Li abundances compared to LFP tests, reaching 2% at higher SOCs. Mg was the third most abundant element, at 0.4–0.7%. Light metal abundances were similar to LFP tests, with up to 0.7% Al, 0.2% Fe, 0.2% Ca, and 0.1% Zn. Due to its presence in the LCO cathode, Co was more abundant than for LFP tests, at up to 0.2%. The elements with maximum abundances between 0.01% and 0.1% were also similar (Na, Sc, Ti, Mn, Cu, and Eu) but included Si, S, and Tb for LCO tests. The maximum Ni abundance was half that of LFP tests (0.005%), while Br, In, Sb, Pt, Hg, and Pb were all ≤0.01%.

Table 2 compares the maximum detected abundances by SOC to literature values for total carbon (TC) and 16 elements. The measured and literature values that are most similar are compared in the right column, giving a "best case" comparability. Measured values for Mn, Co, Ni, and Cu were much lower at 0.1–3% of the literature values. This difference may have been due to several factors. First, the relevant literature mostly involved NMC cells, so high levels of Ni, Mn, and Co are expected. Second, the larger, settleable particles that were analyzed in previous studies were likely to have metallic compositions instead of the majority of carbonaceous particles collected in this study, which were lighter and suspended in air for a longer time. This resulted in a higher average TC abundance of ~57% in this study as compared to 23–30% in most literature sources, leaving less mass to be composed of other compounds. Xu et al. [25,26], in contrast, reported much higher TC, at 90–94%, and lower metals than other sources, more similar to this study. The particles were likely in the submicron range due to a diffusion- and thermophoretic-dominated collection mechanism. The soot-dominated composition from Xu et al. could be caused by the particles being collected close to the flame before volatile compounds (e.g., phosphate) cooled down and condensed in the particle phase, which is evident from much lower P abundances than in this study. This is less representative of the composition of inhaled particulate, however, as the emission stream will most likely have cooled before inhalation.

Table 2. Comparison of maximum measured species abundance to literature values. Literature values are from NMC cell combustion with maximum collected particle sizes ranging from 200 to 15,000 μm [17,21,59,60] and LFP cell combustion with unspecified particle sizes [25,26].

Element	This Study		Literature [17,21,25,26,59,60]	Closest Factor Difference
Measure LIB Type	% of $PM_{2.5}$ Mass (≤2.5 μm) LFP	 LCO	% of Mass (200–15,000 μm) * NMC and LFP	
TC	58 (avg.)	56 (avg.)	23–94	0.6 × lit. values
Li	0.044	1.98	2.9–4.0	1.3 × lit. values
Na	0.011	0.027	0.06–0.1	0.45 × lit. values
Mg	1.33	0.719	0.001–0.006	120 × lit. values
Al	0.720	0.688	3.6–13.5	0.2 × lit. values
Si	0.193	0.050	0.025	2 × lit. values
P	15.1	12.2	0.02–2.2	5.6 × lit. values
S	0	0.066	0.3–0.9	0.22 × lit. values
Ca	0.354	0.191	0.04–0.05	3.8 × lit. values
Ti	0.026	0.013	0.003–0.01	1.3 × lit. values
Mn	0.051	0.027	5.2–13.1	0.01 × lit. values
Fe	0.553	0.201	0.07–0.5	1.1 × lit. values
Co	0.001	0.191	5.9–10.5	0.032 × lit. values
Ni	0.010	0.005	18–51	0.001 × lit. values
Cu	0.020	0.016	2.2–9.5	0.009 × lit. values
Zn	0.108	0.141	0.001–0.008	14 × lit. values
Br	0.006	0.001	0.01–0.04	0.61 × lit. values

* Refs. [25,26] did not specify particle size. Since particles were collected near the flame with the dominant deposition mechanisms of diffusion and thermophoresis, the particles are expected to be in the submicron size range.

Values for Li, Na, Al, S, Ti, Fe, Br, and Sb were more similar at 20–130% of literature values. Finally, some light elements had larger abundances than literature values, with Mg, Si, P, Ca, and Zn having 120-, 2-, 6-, 4-, and 14-times higher abundances, respectively. The differences in elemental composition from different studies can be caused by several factors, including cell chemistry/age/SOC, combustion conditions, sampling methods, particle size ranges, and analytical methods.

Figures 11 and 12 show SOC dependence for abundant elements or those with interesting behavior for LFP and LCO tests, respectively. Both show the same species except that Ni is shown for LFP tests, while Co is shown for LCO tests. As shown in Figure 11, many elements have a positive correlation with SOC for LFP tests. Mg, Al, Ca, Fe, Ni, Zn, as well as Ti, Mn, Pt, and, to a lesser extent, Li, Co, and Cu, all increased with SOC. Variability was high at 100% SOC, but the trend is still clear. SOC dependence of P was quite different, with none found at low SOCs and then the abundance peaking at 50% SOC before dropping below one-third the peak value at 100% SOC. The positive correlation of many elements is interesting, as both the maximum temperature detected and the proportion of flaming combustion (Figure 5) peaked abruptly at mid-range SOCs and then dropped for 100% SOC, matching the behavior of P. The SOC dependence of P matching that of cell temperature indicates that the thermally stable LFP cathode decomposed at the expected ~500 °C [35]. The difference in SOC dependence for the other elements indicates that internal changes other than just hotter or more flaming combustion may be occurring with increasing SOC.

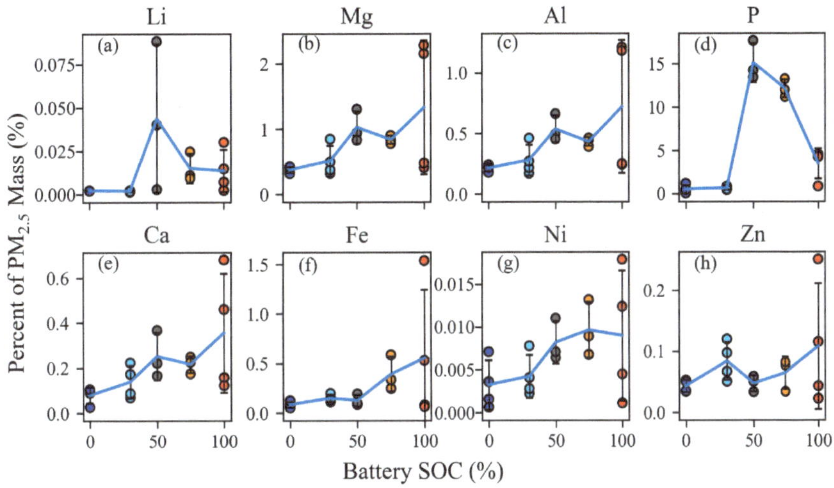

Figure 11. Dependence of element abundance on SOC for LFP tests for Li (**a**), Mg (**b**), Al (**c**), P (**d**), Ca (**e**), Fe (**f**), Ni (**g**), and Zn (**h**). Symbol color represents SOC.

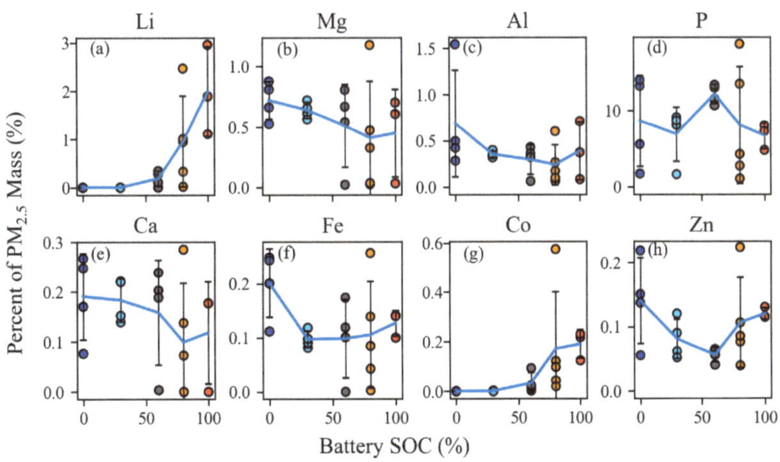

Figure 12. Dependence of element abundance on SOC for LCO tests for Li (**a**), Mg (**b**), Al (**c**), P (**d**), Ca (**e**), Fe (**f**), Co (**g**), and Zn (**h**). Symbol color represents SOC.

For LCO tests, Figure 12 shows that fewer elements (Li and Co) exhibited a positive correlation with SOC. Instead, many elements (Mg, Al, Ca, Fe, and Zn) were most abundant at 0% SOC and least abundant at 60% SOC, increasing again thereafter. P abundance is variable for LCO cell tests, with no clear dependence on SOC due to the variable combustion behavior effects discussed in Section 3.2.1. That Li and Co increased with SOC indicates that the LIB anode and cathode underwent more decomposition at higher SOCs despite constant flame temperatures of ~400 °C (Figure 5), contrasting with the abrupt temperature changes for LFP tests. This may be due to the primary flame jet missing the thermocouple during LCO cell tests, resulting in an undermeasurement of peak flame temperatures. The fraction of $PM_{2.5}$ emitted from flaming combustion, however, increased with SOC (Figure 5b).

3.2.4. Water-Soluble Ions

Water-soluble anions and cations are shown alongside element abundances in Figures 9 and 10. PO_4^{3-} was by far the dominant ion in $PM_{2.5}$ for LFP tests (14–44% abundance) at 50% SOC and above, and LCO tests (8–30% abundance) for all SOCs. For LFP tests at low SOCs, PO_4^{3-} and F^- abundances were similar at 1–2%. F^- was the second most abundant ion at 0.7–3% for both cell types, followed by Cl^- for LFP tests (up to 0.5%) and Li^+ for LCO tests (up to 1.7%). Other notable abundances were SO_4^{2-} (0.5–1.1%) for LCO tests at high SOC and Cl^- and ammonium (NH_4^+) for LCO tests (up to 0.4–0.5%). All other ions had maximum abundances of $\leq$0.1%.

The ratios of ion abundance over that of corresponding elements were variable. Li^+ made up only 40% of total Li for LFP tests but reached 87% for LCO tests (Figure S5, ratio does depend on SOC as seen in Figure S6). Mg was predominantly nonionic, while the P contained in PO_4^{3-} accounted for ~100% and 77% of the total P for LFP and LCO tests, respectively. The S in SO_4^{2-} was consistently higher than total elemental S for LCO tests, especially at high SOCs (Figure S5h). This suggests that some sulfur compounds might have evaporated in the vacuum of the XRF instrument.

Abundance dependences on SOC are shown for some ions in Figure 13a–d for LFP and Figure 13e–h for LCO tests. F^- variability was high, but abundance tended to be low for mid-range SOCs and higher for low, and high, SOCs. SO_4^{2-} increased significantly for >60% SOC in LCO tests but had no SOC dependence for LFP tests. PO_4^{3-} dependence on SOC resembled that of P seen in Figures 11 and 12, except that a noticeable downward correlation was seen for LCO tests. NH_4^+ abundance was highest at low SOCs for both cell types and decreased for mid-range SOCs, especially in LCO tests, likely because NH_4^+ decomposes at high flaming temperatures.

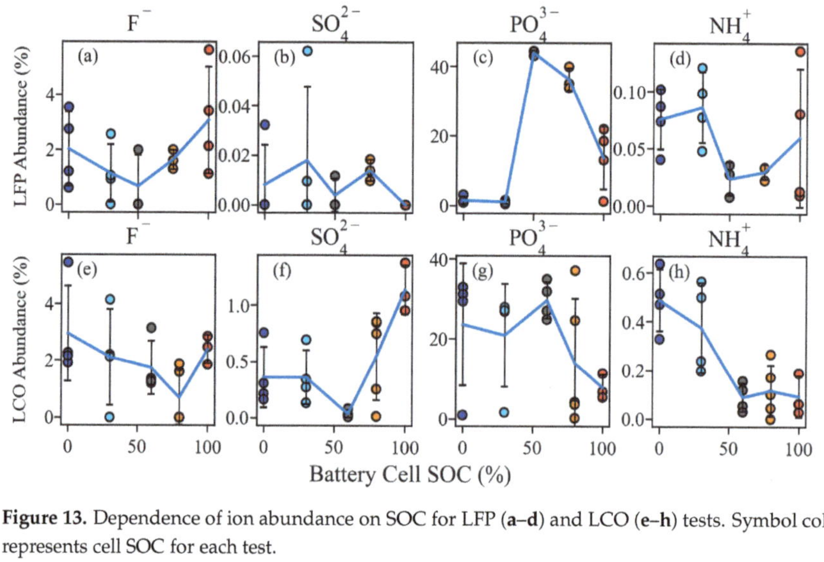

Figure 13. Dependence of ion abundance on SOC for LFP (**a–d**) and LCO (**e–h**) tests. Symbol color represents cell SOC for each test.

Figure 14 shows the anion and cation balances for each test. The ratio of anion to cation abundance indicates degrees of acidity or alkalinity since H^+ or OH^- were not directly measured [54,56]. For LFP tests, the anion/cation ratio ranges depended on combustion behavior, with ~10:1 for venting tests and ~200:1 for tests with significant flaming. It initially appears there is a correlation to SOC in Figure 14a, but the two 100% SOC outliers (labeled Tests C20 and C7, which are nonflaming tests with high OM as shown in Figure 7a) show the underlying trend was actually due to combustion behavior (Figure 14c). In contrast, LCO anion/cation ratios were directly dependent on SOC. With one exception (Test P7), 80–100% SOCs had ratios just above 1:1, while 0–60% SOCs had ratios near 20:1. Heavily flaming LCO tests had lower ratios on average, but could be present in both groups.

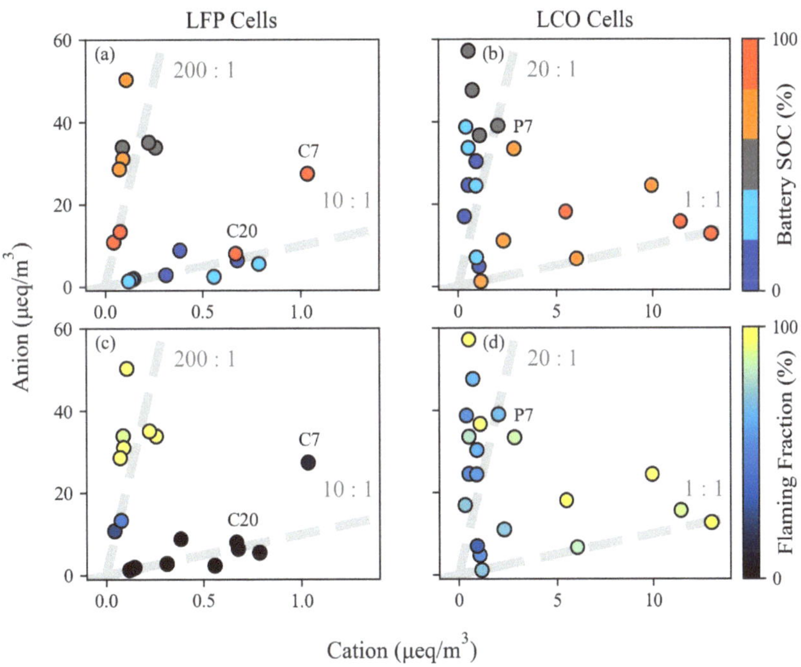

Figure 14. Anion and cation microequivalent (μeq) concentration relationships for LFP (**a**,**c**) and LCO (**b**,**d**) tests. Each test is color coded for SOC (**a**,**b**) or flaming fraction (**c**,**d**).

Anion/cation ratios > 1.0 indicate that the particles were acidic. Held et al. [20] analyzed ion concentrations in sprinkler water used to extinguish an NMC LIB fire and in storage water that the LIB was submerged in after extinguishing. They found that the sprinkling water was moderately alkaline, with a pH value of 8, while the storage water was highly alkaline, with a pH value of 12. The different acidity between the $PM_{2.5}$ measured in this study and the water analysis by Held et al. [20] could be caused by different battery chemistry (Held et al. did not detect PO_4^{3-}) as well as the different compositions of particles and process water.

Figure 15 shows a much higher abundance of particulate anions than cations. PO_4^{3-} was excluded from the anions to allow the relative abundances of other ions to be seen. F^- dominated non-PO_4^{3-} anion abundance, while Li^+ was only present in high SOC LCO tests. Mg^{2+} and NH_4^+ were the dominant cations except for high SOC LCO tests, which have elevated Li^+. A greater SO_4^{2-} abundance for LCO than LFP tests can also be seen.

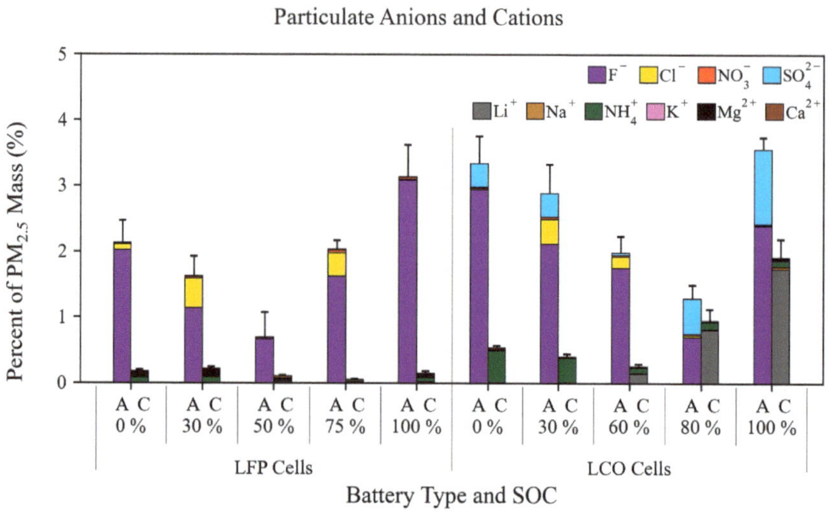

Figure 15. Abundance comparisons of particulate anions and cations ("A" and "C") for each cell type and SOC. The error bars represent the larger of the propagated analytical uncertainty or the standard error within each SOC.

3.2.5. Comparison with Previous Studies

The mass fractions of major emission components are compared to the literature values in Table 3. Average values for LFP tests at 100% SOC were used to most closely match tests in the literature. Only one study [19] reported carbon fractions (also for LFP cells), while others [17,20,21,24,59] reported the mass fraction of EC (all for NMC cells). PO_4^{3-} and F^- were reported by Held et al. [20], while elemental P and F were reported by others [17,21,26,59–61]. The collected particle sizes were much larger than $PM_{2.5}$ in most literature, resulting in much higher metal mass fractions due to enriched compositions from electrodes, current collectors, and casings [23]. The smaller particles are expected to contain more combustion products, including organic compounds, EC, and PO_4^{3-}.

Table 3. Mass fraction comparison to literature values.

Reference	SOC and Cell Type	Max Particle Size (μm)	Total Carbon	EC/OC	Total P	F and F^-	Other Metals
This study	100% LFP	2.5	68	16/52	3.4	3.1 (F^-)	2
[17]	100% NMC	850	30		0.55	0.002	42
[19]	100% LFP	2.5	63	20/43			
[20]	100% NMC	Settleable	10–15			2.5 (F^-)	24
[21]	100% NMC	850	23		0.81	0.60	63
[24]	Var. NMC	Settleable	68				27
[25,26]	100% LFP	Soot	90–94		0.02–0.1	1–2	<2
[59]	100% NMC	Settleable	28		0.31	0.34	57
[60]	100% NMC	200			2.2	4.1	92
[62]	100% NMC	50					53

OC and EC carbon fractions from this study compared well to those from Premnath et al. [19], who also measured $PM_{2.5}$ particles. Held et al. [20] found that the carbon content of soot increased from 1–5% at a distance of 1 m to 10–15% at 50 m from the battery fire, indicating that smaller particles, which remain suspended longer after being emitted, contained more carbon. The nucleation and condensation of carbonaceous materials upon cooling will also generate smaller particles further away from the fire. Xu et al. [25,26] found very high carbon content (90–94%) for soot particles collected near the flame, likely dominated by EC. F and total metal contents are more similar to our findings than other studies. Total P and F^- from this study compared well with literature values, but no published airborne PO_4^{3-} mass fractions were found.

4. Discussion and Conclusions

LIB fires emit a large number of particles over a wide size range. While several past studies have examined the chemical properties of larger particles, few studies have examined the properties of $PM_{2.5}$. These fine particles are of special concern because they can be inhaled into the human respiratory track and cause adverse health effects [26]. The chemical compositions of $PM_{2.5}$ are also important for improving fire detection, designing proper personal protection equipment for people near fires, such as emergency responders, and implementing proper postfire cleanup. This study fills the chemical composition knowledge gap for $PM_{2.5}$ emitted from the combustion of two types of LIBs at various SOCs.

Key characteristics of LIB combustion behaviors and the composition of emitted $PM_{2.5}$ are:

(1) Consistent with the higher thermal stability of LFP than LCO cells, LFP tests did not flame at 0% and 30% SOCs, while LCO tests flamed at all SOCs. LCO tests had more emissions from flaming as SOC increased, while LFP tests had the highest proportion of flaming emissions at 50% and 75% SOCs when higher flame temperatures were detected.

(2) The chemical composition of the $PM_{2.5}$ emitted from LIB combustion was dominated by OM, EC, and PO_4^{3-}. While OM was mostly emitted through cell venting, EC and PO_4^{3-} were generated from flaming combustion. Particles from LFP tests had higher OM but lower EC than those from LCO tests.

(3) OC1–3 fractions that volatilize at a range of temperatures (140–480 °C) were abundant, indicating the presence of a variety of organic compounds. The high abundance of EC2 and the presence of EC3 in some tests are indicators of high combustion temperatures.

(4) Metals were present in small proportions of $PM_{2.5}$ mass, with the most abundant being Li, Mg, Al, Ca, Fe, and Zn. LCO tests had higher Li and Co abundances than LFP tests, and both elements exhibited a positive correlation with SOC. LFP had more elements (e.g., Mg, Al, Ca, Fe, Ni, and Zn) that increased in abundance with SOC. The metal abundances in $PM_{2.5}$ were much lower than those reported for larger, settleable particles.

(5) Ionic compounds other than PO_4^{3-} were detected, primarily F^- with lower abundances of Cl^- and SO_4^{2-}. The balance between water-soluble anion and cation species indicates that the freshly emitted particles were strongly acidic.

(6) While multiple tests within the same nominal conditions (LIB cell type and SOC) showed mostly consistent $PM_{2.5}$ characteristics, some variations were observed. In addition to cell chemistry and SOC, the $PM_{2.5}$ composition was also affected by the combustion behavior, such as the fraction of vented emission streams that were ignited and combustion temperatures.

The detailed $PM_{2.5}$ chemical compositions and their variation with cell chemistry, SOC, and combustion behavior will allow emergency responders, LIB manufacturers, and responsible authorities to plan for and respond to LIB fires. The toxicity of the emitted organics, ionic compounds, and phosphorus and lithium compounds warrants further research. Future research should also address the interaction between fire suppression methods and LIB fires, for example, water mist might temporarily intensify fires and increase hydrogen fluoride emissions [63,64]. Finally, using data from this study to investigate cell chemistry, electrochemical reactions, and combustion processes may lead to improved LIB designs.

Supplementary Materials: The following supporting information can be downloaded at: https://www.mdpi.com/article/10.3390/batteries10090301/s1, Supplementary Captions. Figure S1: The two LIB types tested; Figure S2: A comparison of LFP and LCO cell characteristics and performance metrics; Figure S3: Correlation between unidentified portion of reconstructed mass and selected mass fractions; Figure S4: Thermal carbon fractions as a percent of $PM_{2.5}$ mass; Figure S5: Proportions of total elemental abundance attributed to ionic compounds; and Figure S6: Dependence of ion/element ratios on cell SOC [50,65,66].

Author Contributions: Conceptualization, X.W.; Methodology, M.C. and X.W.; Software, M.C.; Validation, M.C. and X.W.; Formal Analysis, M.C.; Investigation, M.C., B.B. and X.W.; Resources, X.W.; Data Curation, M.C.; Writing—Original Draft Preparation, M.C. and X.W.; Writing—Review and Editing, X.W., J.C.C., J.G.W. and Y.W.; Visualization, M.C.; Supervision, X.W. and Y.W.; Project Administration, X.W.; Funding Acquisition, X.W. All authors have read and agreed to the published version of the manuscript.

Funding: This research was funded by the U.S. National Aeronautics and Space Administration's Established Program to Stimulate Competitive Research, CAN Grant No. 80NSSC19M0152 and Nevada Space Grant No. 80NSSC20M0043 22–24.

Data Availability Statement: The raw data supporting the conclusions of this article will be made available by the authors on request. Calculated emission factors for all measured species are included in the Supplementary Material.

Acknowledgments: The authors thank Hans Moosmüller for the use of the burn chamber where experiments were performed and DRI personnel for support and filter analysis.

Conflicts of Interest: The authors declare no conflicts of interest.

Abbreviations

μeq	microequivalent
Al	aluminum
Br	bromine
Ca	calcium
Ca^{2+}	calcium ion
Cl^-	chloride
Co	cobalt
Cu	copper
DRI	Desert Research Institute
EC	elemental carbon
EC1	elemental carbon evolved at 580 °C
EC2	elemental carbon evolved at 740 °C
EC3	elemental carbon evolved at 840 °C
EDX	energy dispersive X-ray analysis
ELPI	electrical low-pressure impactor
EPA	Environmental Protection Agency
Eu	europium
EV	electric vehicle
F^-	fluoride
FAA	Federal Aviation Administration
Fe	iron
HCl	hydrochloric acid
Hg	mercury
HNO_3	nitric acid
IC	ion chromatography
ICP-MS	inductively coupled plasma mass spectrometry
In	indium
K^+	potassium ion
LCO	lithium cobalt oxide
Li	lithium

Li^+	lithium ion
LIB	lithium-ion battery
LFP	lithium iron phosphate
Mg	magnesium
Mg^{2+}	magnesium ion
Mn	manganese
MW	megawatt
Na	sodium
Na^+	sodium ion
NASA	National Aeronautics and Space Administration
$NH_4{}^+$	ammonium
Ni	nickel
NMC	nickel manganese cobalt oxide
$NO_3{}^-$	nitrate
OC	organic carbon
OC1	organic carbon evolved at 140 °C
OC2	organic carbon evolved at 280 °C
OC3	organic carbon evolved at 480 °C
OC4	organic carbon evolved at 580 °C
OP	pyrolyzed carbon
OM	organic matter
P	phosphorus
PAH	polycyclic aromatic hydrocarbons
Pb	lead
PM	particulate matter
PM_x	particles with aerodynamic diameters $\leq$x μm
$PO_4{}^{3-}$	phosphate
Pt	platinum
S	sulfur
Sb	antimony
Sc	scandium
Si	silicon
$SO_4{}^{2-}$	sulfate
SOC	state of charge
SEM	scanning electron microscopy
TC	total carbon
Tb	terbium
Ti	titanium
TR	thermal runaway
XRF	X-ray fluorescence
Zn	zinc

References

1. Doughty, D.H.; Roth, E.P. A General Discussion of Li Ion Battery Safety. *Electrochem. Soc. Interface* **2012**, *21*, 37–44. [CrossRef]
2. Feng, X.; Ouyang, M.; Liu, X.; Lu, L.; Xia, Y.; He, X. Thermal runaway mechanism of lithium ion battery for electric vehicles: A review. *Energy Storage Mater.* **2018**, *10*, 246–267. [CrossRef]
3. EVFireSafe. Source: evfiresafe.com. Available online: https://www.evfiresafe.com/ (accessed on 31 October 2023).
4. FAA. Events with Smoke, Fire, Extreme Heat or Explosion Involving Lithium Batteries Federal Aviation Administration. Available online: https://www.faa.gov/hazmat/resources/lithium_batteries/media/Battery_incident_chart.pdf (accessed on 8 February 2020).
5. NASA. Lithium-Ion Battery Fire; 3516; National Aeronautics and Space Administration: Public Lessons Learned System. Available online: https://llis.nasa.gov/lesson/3516 (accessed on 1 February 2024).
6. Nigl, T.; Baldauf, M.; Hohenberger, M.; Pomberger, R. Lithium-Ion Batteries as Ignition Sources in Waste Treatment Processes—A Semi-Quantitate Risk Analysis and Assessment of Battery-Caused Waste Fires. *Processes* **2020**, *9*, 49. [CrossRef]
7. U.S. EPA. *An Analysis of Lithium-Ion Battery Fires in Waste Management and Recycling; Office of Resource Conservation and Recovery;* U.S. Environmental Protection Agency: Washington, DC, USA, 2021. Available online: https://www.epa.gov/system/files/documents/2021-08/lithium-ion-battery-report-update-7.01_508.pdf (accessed on 1 February 2024).

8. Yuan, L.; Dubaniewicz, T.; Zlochower, I.; Thomas, R.; Rayyan, N. Experimental study on thermal runaway and vented gases of lithium-ion cells. *Process Saf. Environ. Prot.* **2020**, *144*, 186–192. [CrossRef]
9. Nedjalkov, A.; Meyer, J.; Köhring, M.; Doering, A.; Angelmahr, M.; Dahle, S.; Sander, A.; Fischer, A.; Schade, W. Toxic gas emissions from damaged lithium ion batteries—Analysis and safety enhancement solution. *Batteries* **2016**, *2*, 5. [CrossRef]
10. Larsson, F.; Andersson, P.; Blomqvist, P.; Mellander, B.-E. Toxic fluoride gas emissions from lithium-ion battery fires. *Sci. Rep.* **2017**, *7*, 10018. [CrossRef]
11. Essl, C.; Seifert, L.; Rabe, M.; Fuchs, A. Early Detection of Failing Automotive Batteries Using Gas Sensors. *Batteries* **2021**, *7*, 25. [CrossRef]
12. Koch, S.; Fill, A.; Birke, K.P. Comprehensive gas analysis on large scale automotive lithium-ion cells in thermal runaway. *J. Power Sources* **2018**, *398*, 106–112. [CrossRef]
13. Fernandes, Y.; Bry, A.; De Persis, S. Identification and quantification of gases emitted during abuse tests by overcharge of a commercial Li-ion battery. *J. Power Sources* **2018**, *389*, 106–119. [CrossRef]
14. Willstrand, O.; Pushp, M.; Andersson, P.; Brandell, D. Impact of different Li-ion cell test conditions on thermal runaway characteristics and gas release measurements. *J. Energy Storage* **2023**, *68*, 107785. [CrossRef]
15. Baird, A.R.; Archibald, E.J.; Marr, K.C.; Ezekoye, O.A. Explosion hazards from lithium-ion battery vent gas. *J. Power Sources* **2020**, *446*, 227257. [CrossRef]
16. Sun, J.; Li, J.; Zhou, T.; Yang, K.; Wei, S.; Tang, N.; Dang, N.; Li, H.; Qiu, X.; Chen, L. Toxicity, a serious concern of thermal runaway from commercial Li-ion battery. *Nano Energy* **2016**, *27*, 313–319. [CrossRef]
17. Zhang, Y.; Wang, H.; Li, W.; Li, C.; Ouyang, M. Size distribution and elemental composition of vent particles from abused prismatic Ni-rich automotive lithium-ion batteries. *J. Energy Storage* **2019**, *26*, 100991. [CrossRef]
18. Barone, T.L.; Dubaniewicz, T.H.; Friend, S.A.; Zlochower, I.A.; Bugarski, A.D.; Rayyan, N.S. Lithium-ion battery explosion aerosols: Morphology and elemental composition. *Aerosol Sci. Technol.* **2021**, *55*, 1183–1201. [CrossRef] [PubMed]
19. Premnath, V.; Wang, Y.; Wright, N.; Khalek, I.; Uribe, S. Detailed characterization of particle emissions from battery fires. *Aerosol Sci. Technol.* **2022**, *56*, 337–354. [CrossRef]
20. Held, M.; Tuchschmid, M.; Zennegg, M.; Figi, R.; Schreiner, C.; Mellert, L.D.; Welte, U.; Kompatscher, M.; Hermann, M.; Nachef, L. Thermal runaway and fire of electric vehicle lithium-ion battery and contamination of infrastructure facility. *Renew. Sustain. Energy Rev.* **2022**, *165*, 112474. [CrossRef]
21. Wang, H.; Wang, Q.; Jin, C.; Xu, C.; Zhao, Y.; Li, Y.; Zhong, C.; Feng, X. Detailed characterization of particle emissions due to thermal failure of batteries with different cathodes. *J. Hazard. Mater.* **2023**, *458*, 131646. [CrossRef]
22. Padilla, R.E.; Meyer, M.; Dietrich, D.L.; Ruff, G.A.; Urban, D.L. Hazardous Effercts of Li-Ion Battery Based Fires. In Proceedings of the 50th International Conference on Environmental Systems, Boston, MA, USA, 12–16 July 2020.
23. Li, W.; Xue, Y.; Feng, X.; Rao, S.; Zhang, T.; Gao, Z.; Guo, Y.; Zhou, H.; Zhao, H.; Song, Z.; et al. Characteristics of particle emissions from lithium-ion batteries during thermal runaway: A review. *J. Energy Storage* **2024**, *78*, 109980. [CrossRef]
24. Chen, S.; Wang, Z.; Yan, W. Identification and characteristic analysis of powder ejected from a lithium ion battery during thermal runaway at elevated temperatures. *J. Hazard. Mater.* **2020**, *400*, 123169. [CrossRef]
25. Xu, Y.; Wang, Y.; Chen, X.; Pang, K.; Deng, B.; Han, Z.; Shao, J.; Qian, K.; Chen, D. Thermal runaway and soot production of lithium-ion batteries: Implications for safety and environmental concerns. *Appl. Therm. Eng.* **2024**, *248*, 123193. [CrossRef]
26. Xu, Y.; Wang, Y.; Chen, D. Soot formation and its hazards in battery thermal runaway. *J. Aerosol Sci.* **2024**, *181*, 106420. [CrossRef]
27. ICRP. *Human Respiratory Tract Model for Radiological Protection, ICRP Publication 66*; International Commission on Radiological Protection (ICRP): Ottawa, ON, Canada, 1994; Volume 24, pp. 1–3.
28. Thangavel, P.; Park, D.; Lee, Y.C. Recent Insights into Particulate Matter (PM(2.5))-Mediated Toxicity in Humans: An Overview. *Int. J. Env. Res. Public Health* **2022**, *19*, 7511. [CrossRef]
29. Claassen, M.; Bingham, B.; Chow, J.C.; Watson, J.G.; Wang, Y.; Wang, X.L. Characterization of Lithium-ion Battery Fire Emissions—Part 2: Particle Size Distributions and Emission Factors. *Batteries*, 2024; *Submitted.*
30. Lithiumwerks. APR18650M1B Nanophosphate® Technology Data Sheet. Austin, Texas. 2023. Available online: https://lithiumwerks.com/products/lithium-ion-18650-cells/ (accessed on 1 February 2024).
31. AA Portable Power Corp. *Model 544792 Polymer Lithium-Ion Battery Data Sheet*; AA Portable Power Corp.: Richmond, CA, USA; Available online: https://www.batteryspace.com/prod-specs/3175.pdf (accessed on 1 February 2024).
32. Chombo, P.V.; Laoonual, Y. Prediction of the onset of thermal runaway and its thermal hazards in 18650 lithium-ion battery abused by external heating. *Fire Saf. J.* **2022**, *129*, 103560. [CrossRef]
33. Peng, P.; Jiang, F. Thermal safety of lithium-ion batteries with various cathode materials: A numerical study. *Int. J. Heat Mass Transf.* **2016**, *103*, 1008–1016. [CrossRef]
34. Nitta, N.; Wu, F.; Lee, J.T.; Yushin, G. Li-ion battery materials: Present and future. *Mater. Today* **2015**, *18*, 252–264. [CrossRef]
35. Barkholtz, H.M.; Preger, Y.; Ivanov, S.; Langendorf, J.; Torres-Castro, L.; Lamb, J.; Chalamala, B.; Ferreira, S.R. Multi-scale thermal stability study of commercial lithium-ion batteries as a function of cathode chemistry and state-of-charge. *J. Power Sources* **2019**, *435*, 226777. [CrossRef]
36. Anand, M.D.; Sasidharakurup, R.; Mercy, T.; Jacob, T.M.; Devi, S.A. Lithium-ion cells for space applications: Aspects of durability. *Adv. Space Res.* **2023**, *72*, 2948–2958. [CrossRef]

37. Dalton, P.J.; Schwanbeck, E.; North, T.; Balcer, S. International Space Station Lithium-Ion Battery. In Proceedings of the NASA Aerospace Battery Workshop, Huntsville, AL, USA, 15–17 November 2016.
38. Tian, J.; Chow, J.C.; Cao, J.; Han, Y.; Ni, H.; Chen, L.-W.A.; Wang, X.; Huang, R.; Moosmu, H.; Watson, J.G. A biomass combustion chamber: Design, evaluation, and a case study of wheat straw combustion emission tests. *Aerosol Air Qual. Res.* **2015**, *15*, 2104–2114. [CrossRef]
39. Goupil, V.; Gaya, C.; Boisard, A.; Robert, E. Effect of the heating rate on the degassing and combustion of cylindrical Li-Ion cells. *Fire Saf. J.* **2022**, *133*, 103648. [CrossRef]
40. Wang, X.L.; Zhou, H.; Arnott, W.P.; Meyer, M.E.; Taylor, S.; Firouzkouhi, H.; Moosmüller, H.; Chow, J.C.; Watson, J.G. Characterization of smoke for spacecraft fire safety. *J. Aerosol Sci.* **2019**, *136*, 36–47. [CrossRef]
41. Wang, X.L.; Firouzkouhi, H.; Chow, J.C.; Watson, J.G.; Carter, W.; De Vos, A.S. Characterization of gas and particle emissions from open burning of household solid waste from South Africa. *Atmos. Chem. Phys.* **2023**, *23*, 8921–8937. [CrossRef]
42. Wang, X.L.; Zhou, H.; Arnott, W.P.; Meyer, M.E.; Taylor, S.; Firouzkouhi, H.; Moosmüller, H.; Chow, J.C.; Watson, J.G. Evaluation of gas and particle sensors for detecting spacecraft-relevant fire emissions. *Fire Saf. J.* **2020**, *113*, 102977. [CrossRef]
43. Wang, X.L.; Chancellor, G.; Evenstad, J.; Farnsworth, J.E.; Hase, A.; Olson, G.M.; Sreenath, A.; Agarwal, J.K. A Novel Optical Instrument for Estimating Size Segregated Aerosol Mass Concentration in Real Time. *Aerosol Sci. Technol.* **2009**, *43*, 939–950. [CrossRef]
44. Järvinen, A.; Aitomaa, M.; Rostedt, A.; Keskinen, J.; Yli-Ojanperä, J. Calibration of the new electrical low pressure impactor (ELPI+). *J. Aerosol Sci.* **2014**, *69*, 150–159. [CrossRef]
45. Saari, S.; Arffman, A.; Harra, J.; Rönkkö, T.; Keskinen, J. Performance evaluation of the HR-ELPI+ inversion. *Aerosol Sci. Technol.* **2018**, *52*, 1037–1047. [CrossRef]
46. Wang, X.; Firouzkouhi, H.; Chow, J.C.; Watson, J.G.; Ho, S.S.H.; Carter, W.; De Vos, A.S. Chemically speciated air pollutant emissions from open burning of household solid waste from South Africa. *Atmos. Chem. Phys.* **2023**, *23*, 15375–15393. [CrossRef]
47. Watson, J.G.; Tropp, R.J.; Kohl, S.D.; Wang, X.L.; Chow, J.C. Filter processing and gravimetric analysis for suspended particulate matter samples. *Aerosol Sci. Eng.* **2017**, *1*, 193–205. [CrossRef]
48. Watson, J.G.; Chow, J.C.; Frazier, C.A. X-ray fluorescence analysis of ambient air samples. In *Elemental Analysis of Airborne Particles*; Landsberger, S., Creatchman, M., Eds.; Gordon and Breach Science: Amsterdam, The Netherlands, 1999; Volume 1, pp. 67–96.
49. Chow, J.C.; Watson, J.G. Enhanced Ion Chromatographic Speciation of Water-Soluble PM_2.5 to Improve Aerosol Source Apportionment. *Aerosol Sci. Eng.* **2017**, *1*, 7–24. [CrossRef]
50. Chow, J.C.; Watson, J.G.; Chen, L.-W.A.; Chang, M.C.O.; Robinson, N.F.; Trimble, D.; Kohl, S. The IMPROVE_A temperature protocol for thermal/optical carbon analysis: Maintaining consistency with a long-term database. *J. Air Waste Manag. Assoc.* **2007**, *57*, 1014–1023. [CrossRef]
51. Chen, L.-W.A.; Chow, J.C.; Wang, X.L.; Robles, J.A.; Sumlin, B.; Lowenthal, D.H.; Zimmermann, R.; Watson, J.G. Multi-wavelength optical measurement to enhance thermal/optical analysis for carbonaceous aerosol. *Atmos. Meas. Tech.* **2015**, *8*, 451–461. [CrossRef]
52. Chow, J.C.; Wang, X.L.; Sumlin, B.J.; Gronstal, S.B.; Chen, L.-W.A.; Hurbain, M.J.; Zimmermann, R.; Watson, J.G. Optical Calibration and Equivalence of a Multiwavelength Thermal/Optical Carbon Analyzer. *Aerosol Air Qual. Res.* **2015**, *15*, 1145–1159. [CrossRef]
53. Watson, J.G.; Turpin, B.J.; Chow, J.C. The measurement process: Precision, accuracy, and validity. In *Air Sampling Instruments for Evaluation of Atmospheric Contaminants*, 9th ed.; Cohen, B.S., McCammon, C.S., Jr., Eds.; American Conference of Governmental Industrial Hygienists: Cincinnati, OH, USA, 2001; pp. 201–216.
54. Wang, X.; Chow, J.C.; Kohl, S.D.; Percy, K.E.; Legge, A.H.; Watson, J.G. Characterization of PM2. 5 and PM10 fugitive dust source profiles in the Athabasca Oil Sands Region. *J. Air Waste Manag. Assoc.* **2015**, *65*, 1421–1433. [CrossRef] [PubMed]
55. Chow, J.C.; Lowenthal, D.H.; Chen, L.-W.A.; Wang, X.L.; Watson, J.G. Mass reconstruction methods for $PM_{2.5}$: A review. *Air Qual. Atmos. Health* **2015**, *8*, 243–263. [CrossRef]
56. Wang, X.; Gronstal, S.; Lopez, B.; Jung, H.; Chen, L.-W.A.; Wu, G.; Ho, S.S.H.; Chow, J.C.; Watson, J.G.; Yao, Q. Evidence of non-tailpipe emission contributions to PM2. 5 and PM10 near southern California highways. *Environ. Pollut.* **2023**, *317*, 120691. [CrossRef]
57. U.S. EPA. *Quality Assurance Guidance Document—Quality Assurance Project Plan: $PM_{2.5}$ Chemical Speciation Sampling at Trends, NCore, Supplemental and Tribal Sites*; Ambient Air Monitoring Group, Air Quality Assessment Division, US EPA, Office of Air Quality Planning and Standards: Research Triangle Park, NC, USA, 2012. Available online: https://www3.epa.gov/ttnamti1/files/ambient/pm25/spec/CSN_QAPP_v120_05-2012.pdf (accessed on 26 June 2024).
58. Chow, J.C.; Watson, J.G.; Crow, D.; Lowenthal, D.H.; Merrifield, T. Comparison of IMPROVE and NIOSH Carbon Measurements. *Aerosol Sci. Technol.* **2001**, *34*, 23–34. [CrossRef]
59. Zhang, Y.; Wang, H.; Li, W.; Li, C. Quantitative identification of emissions from abused prismatic Ni-rich lithium-ion batteries. *eTransportation* **2019**, *2*, 100031. [CrossRef]
60. Yang, Y.; Fang, D.; Maleki, A.; Kohzadi, S.; Liu, Y.; Chen, Y.; Liu, R.; Gao, G.; Zhi, J. Characterization of Thermal-Runaway Particles from Lithium Nickel Manganese Cobalt Oxide Batteries and Their Biotoxicity Analysis. *ACS Appl. Energy Mater.* **2021**, *4*, 10713–10720. [CrossRef]

61. Bordes, A.; Marlair, G.; Zantman, A.; Herreyre, S.; Papin, A.; Desprez, P.; Lecocq, A. New insight on the risk profile pertaining to lithium-ion batteries under thermal runaway as affected by system modularity and subsequent oxidation regime. *J. Energy Storage* **2022**, *52*, 104790. [CrossRef]
62. Wang, H.; Zhang, Y.; Li, W.; Li, C.; Ouyang, M. Particles released by abused prismatic Ni-rich automotive lithium-ion batteries. *WSEAS Trans. Syst. Control* **2020**, *15*, 30–38. [CrossRef]
63. Krebs, R.; Owens, J.; Luckarift, H. Formation and detection of hydrogen fluoride gas during fire fighting scenarios. *Fire Saf. J.* **2022**, *127*, 103489. [CrossRef]
64. Yuan, S.; Chang, C.; Yan, S.; Zhou, P.; Qian, X.; Yuan, M.; Liu, K. A review of fire-extinguishing agent on suppressing lithium-ion batteries fire. *J. Energy Chem.* **2021**, *62*, 262–280. [CrossRef]
65. Saldaña, G.; San Martín, J.I.; Zamora, I.; Asensio, F.J.; Oñederra, O. Analysis of the current electric battery models for electric vehicle simulation. *Energies* **2019**, *12*, 2750. [CrossRef]
66. Chow, J.C.; Watson, J.G.; Chen, L.-W.A.; Rice, J.; Frank, N.H. Quantification of PM2.5 organic carbon sampling artifacts in US networks. *Atmos. Chem. Phys.* **2010**, *10*, 5223–5239. [CrossRef]

batteries

MDPI

Article

Study on Thermal Runaway Behavior and Jet Characteristics of a 156 Ah Prismatic Ternary Lithium Batter

Huipeng Zhang

Department of Mechanical and Electronic Engineering, Yuncheng University, Yuncheng 044000, China; zhanghuipeng2024@126.com

Abstract: Ternary lithium batteries have been widely used in transportation and energy storage due to their high energy density and long cycle life. However, safety issues arising from thermal runaway (TR) need urgent resolution. Current research on thermal runaway in large-capacity ternary lithium batteries is limited, making the study of hazard indicators during the thermal runaway ejection process crucial. This study places a commercial 156 Ah prismatic battery (positive electrode material: $Li(Ni_{0.8}Mn_{0.1}Co_{0.1})O_2$, negative electrode material: graphite) in a nitrogen-filled sealed container, triggering thermal runaway through lateral heating. The experimental results show that the battery's maximum surface temperature can reach 851.8–943.7 °C, exceeding the melting point of aluminum. Temperature surge inflection points at the battery's bottom and near the small side of the negative electrode coincide with the inflection point on the heated surface. The highest jet temperatures at three monitoring points 50 mm, 150 mm, and 250 mm above the safety valve are 356.9 °C, 302.7 °C, and 216.5 °C, respectively. Acoustic signals reveal two ejection events. The average gas production of the battery is 0.089 mol/Ah, and the jet undergoes three stages: ultra-fast ejection (2 s), rapid ejection (32 s), and slow ejection (47 s). Post-thermal runaway remnants indicate that grooves from internal jet impacts are mainly located at ±45° positions. This study provides valuable insights for the safety design of batteries and the suppression of thermal runaway propagation.

Keywords: ternary lithium batteries; thermal runaway; jet characteristics

Citation: Zhang, H. Study on Thermal Runaway Behavior and Jet Characteristics of a 156 Ah Prismatic Ternary Lithium Battery. *Batteries* **2024**, *10*, 282. https://doi.org/10.3390/batteries10080282

Academic Editor: Mingyi Chen

Received: 16 June 2024
Revised: 22 July 2024
Accepted: 29 July 2024
Published: 6 August 2024

1. Introduction

In recent years, with the depletion of global fossil energy resources, new energy sources represented by lithium-ion batteries (LIBs) have been widely applied in the fields of transportation and energy storage. Among various types of lithium batteries, ternary lithium batteries, whose cathode materials consist of nickel, cobalt, and manganese, have become an essential part of modern energy storage technology due to their high energy density, long cycle life, and low self-discharge rate [1–4]. As battery technology advances, the application scope of ternary lithium batteries is expected to expand further [5].

However, there are still many safety issues with ternary lithium batteries that need urgent resolution [6–9]. In recent years, incidents of fires and explosions caused by battery thermal runaway have occurred frequently. For instance, in 2023, the New York Fire Department handled 267 fire incidents caused by thermal runaway of lithium-ion batteries, resulting in 150 injuries and 18 deaths [10]. Accident investigations have shown that lithium batteries are highly likely to trigger thermal runaway when subjected to mechanical crush, overcharging, internal short circuits, and high-temperature environments [11–17]. Ternary lithium batteries have high energy density and generate significant heat during thermal runaway, with faster heat propagation rates within the battery modules, leading to more severe fire incidents [18]. Lithium batteries are composed of cathode materials, anode materials, copper current collectors, aluminum current collectors, separators, electrolytes, aluminum casings, and safety valves, with the electrolyte and electrode materials being flammable [16,19–21]. During thermal runaway, reactions between the cathode and anode materials and the electrolyte produce large amounts of flammable gases such as CO, H_2, CH_4, and C_2H_4 [22–24], releasing a significant amount of heat that causes the safety valve to open. This results in combustible gases, electrolyte, and electrode materials coming into contact with air, leading to intense combustion and flame jets [25–27].

Currently, numerous scholars are conducting extensive research on the mechanisms of thermal runaway and fire behaviors of ternary lithium batteries. Wei et al. studied the effects of different abuse methods on thermal runaway behavior using an embedded thermocouple and an accelerating rate calorimeter (EV-ARC) with $Li(Ni_{0.5}Co_{0.2}Mn_{0.3})O_2$ as the cathode material and graphite as the anode material. They triggered thermal runaway using three methods: puncture, overcharging, and side heating, with internal battery temperatures reaching 994.8, 964.3, and 1020 °C, respectively. Observations indicated that overcharging-induced thermal runaway caused the most severe battery damage [28].

Ohneseit et al. conducted a study on the thermal runaway behavior of 21,700 batteries at different states of charge (SOC) using an EV-ARC. The batteries had a cathode material of $Li(Ni_{0.8}Co_{0.1}Mn_{0.1})O_2$ and a capacity of 4950 mAh. The results showed that at 0% SOC, the batteries did not undergo thermal runaway, while at 30% SOC, 50% SOC, and 100% SOC, thermal runaway occurred [29].

Han et al. performed thermal runaway propagation experiments on aged batteries using a self-made box. The subjects were 1880 mAh ternary lithium batteries. The results indicated that, at room temperature, the thermal runaway onset temperature and temperature change rate of batteries with a state of health (SOH) of 90% were lower than those of fresh batteries, but the mass loss was significantly greater [30].

Previous studies mainly focused on methods for triggering thermal runaway in small-capacity batteries [31,32], the impact of SOC [33–36], and SOH [31,37–41] on thermal runaway. There is relatively less research on large-capacity high-nickel NCM batteries, and jet studies specifically targeting NCM batteries are also rare. Therefore, this study investigates the surface temperature and jet characteristics of large-capacity NCM batteries during thermal runaway. The experiments were conducted in a nitrogen atmosphere with 99.75% concentration. The battery's cathode material was $Li(Ni_{0.8}Co_{0.1}Mn_{0.1})O_2$, the anode material was graphite, and the capacity was 156 Ah. The experiments were carried out in a 1000 L container, with thermocouples monitoring the surface temperatures at five positions (front, back, left, right, and bottom) of the battery, jet temperatures at three positions (50, 150, and 250 mm above the safety valve), and ambient temperatures at four positions within the test chamber. Pressure sensors were used to monitor the gas production in real time, and cameras and recording devices were employed to observe the characteristics during the battery ejection process. This study aims to provide references for battery safety design and fire warning systems.

2. Materials and Methods

2.1. Battery Samples

The subject of this experiment is a large-capacity commercial prismatic lithium-ion battery with a rated capacity of 156 Ah. The cathode material is $Li(Ni_{0.8}Co_{0.1}Mn_{0.1})O_2$, and the anode material is graphite. The cathode current collector is aluminum foil, while the anode current collector is copper foil. The energy density is 248.00 Wh/kg, and the discharge energy is 572.00 Wh. The nominal voltage is 3.65 V. The charging cut-off voltage is 4.2 V, and after resting for 24 h, the stabilized voltage is 4.146 V. The discharging cut-off voltage is 2.8 V. The operating voltage range is 2.8 to 4.2 V. The internal resistance of the battery is less than or equal to 0.6 mΩ. The operating temperature range under discharge conditions is −30 to 55 °C, and under charging conditions, it is −20 to 55 °C. The weight of the battery is 2287 ± 25 g. The main performance parameters of the battery provided by the manufacturer are shown in Table 1 below.

Table 1. Basic Parameters of the Battery to be tested.

Item	Specification	Condition
Nominal Capacity	156 Ah	25 °C, 52 A(1/3C) DC to 2.8 V
Energy	572 Wh	25 °C, 52 A(1/3C) DC to 2.8 V
Specific Energy	248.00 Wh/Kg	25 °C, 52 A(1/3C) DC to 2.8 V
Operating voltage	2.8~4.2 V	-30 °C $\leq$ T $\leq$ 55 °C
Cathode	$Li(Ni_{0.8}Co_{0.1}Mn_{0.1})O_2$	N.A.
Anode	Graphite	N.A.
Standard Voltage	3.637 $\pm$ 0.01 V	25 °C, BOL, 40% SOC
Operating temperature (Charge)	-20~55 °C	N.A.
Operating temperature (discharge)	-30~55 °C	N.A.
Cycle life	$\geq$1500	25 °C, 1C/1C, 5–97% SOC, 80% SOH
discharge power	$\geq$2100 W	25 °C, 50% SOC, 10 s
Discharge power density	$\geq$912 W/Kg	25 °C, 50% SOC, 10 s
SOC	100%	N.A.
Cell weight	2287 $\pm$ 25 g	N.A.
Cell dimension	220 $\times$ 102 $\times$ 45 mm	N.A.
Shell Material	Aluminum alloy	N.A.

2.2. *Experimental Equipment*

The primary equipment used in this study is a self-made sealed pressure vessel with a capacity of 1000 L. Its basic components include the chamber, signal acquisition system, gas intake and exhaust system, and thermal runaway trigger system. Figure 1 shows the composition of the experimental equipment.

The sealed chamber is cylindrical in shape. It has a circular sealing door on the side, which is secured to the chamber body with bolts. Both the chamber body and the sealing door are made of Q345R material, with a maximum pressure capacity of 2 MPa.

The signal acquisition system includes pressure sensors inside the chamber, temperature sensors, voltage signal sensors, and a camera system. An absolute pressure sensor is mounted on the equipment door, with a range of 0–200 KPa and an accuracy of $\pm$0.2%FS. Its function is to measure pressure changes in the container before, during, and after the battery thermal runaway. Fifteen K-type armored thermocouples are installed on the sealed chamber door, with a range of -200 °C to 1300 °C. The thermocouples measure the temperature at various positions on the battery, the ambient temperature inside the chamber, and the temperature of the ejected gases. Voltage acquisition lines with alligator clips are also installed on the sealed chamber door to clamp onto the battery terminals and collect voltage signals in real time. The TP700 multi-channel data recorder displays and stores data from the absolute pressure sensor, K-type armored thermocouples, and voltage acquisition lines, and uploads the data to a computer for analysis. A camera with charging capabilities and recording equipment is installed inside the chamber. The camera has six fill lights on its side, a battery capacity of 2500 mAh, and can operate continuously for 18 h. It has 128 Gb of storage space, a resolution of 1920 P, and can capture 30 frames per second. Its function is to record the ejection phenomena during the battery thermal runaway process. The camera can be reused. The specific protection method involves attaching a 3 mm glass sheet to the camera using Teflon tape to protect the camera lens. In addition to the lens area, the rest of the camera is covered with two layers of Teflon tape to isolate heat and protect the camera body.

The gas intake and exhaust system includes a nitrogen source, intake valve, exhaust valve, and vacuum pump. The nitrogen source is provided by a pure nitrogen cylinder with a purity of 99.9%. When the intake valve is manually opened, nitrogen can be charged into the sealed chamber. After manually closing the intake valve and opening the exhaust valve, the vacuum pump can evacuate the mixed gases from the chamber.

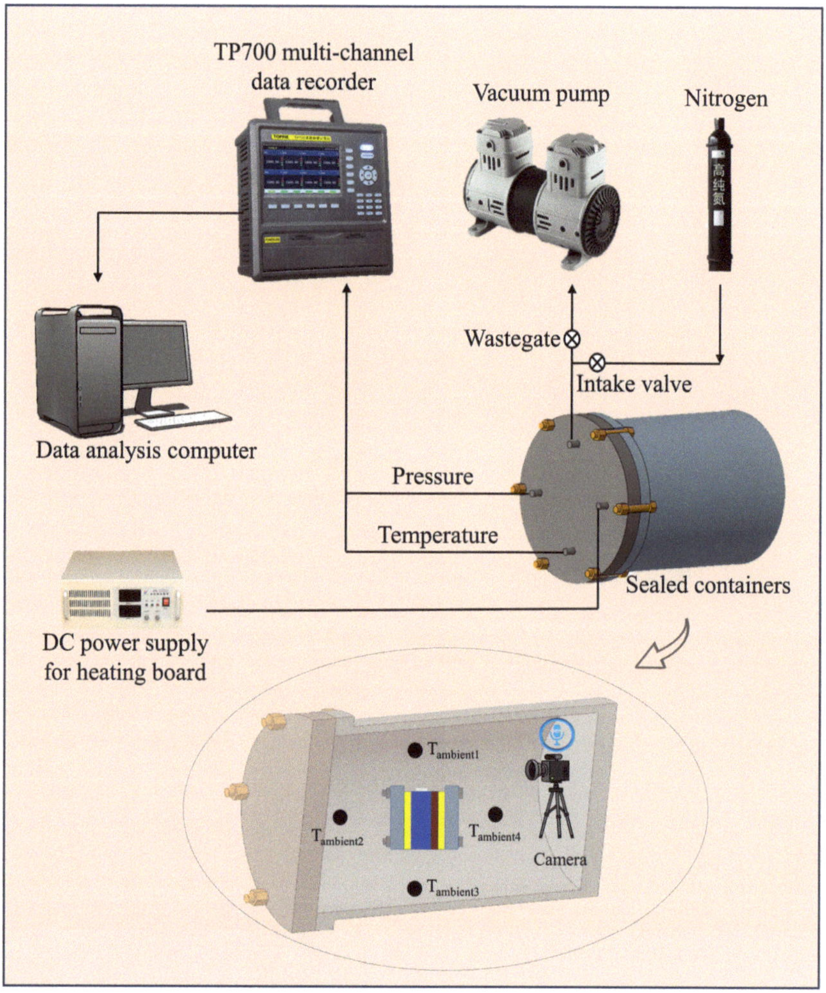

Figure 1. Experimental equipment.

The thermal runaway trigger system includes specialized aluminum alloy clamps, mica insulating boards, electric heating plates, bolts, and nuts. The functions of the specialized aluminum alloy clamps include: (1) According to China's GB 38031-2020 "Electric vehicles traction battery safety requirements", the electric heating plate is closely attached to the battery, allowing the heat generated by the electric heating plate to quickly transfer into the battery, triggering thermal runaway. (2) securely installing thermocouples via bolt connections between the two aluminum alloy clamp planes, preventing them from falling off during thermal runaway. The mica insulating board, with a thickness of 3 mm, insulates heat transfer, ensuring most of the heat generated by the heating plate is conducted into the battery. The resistance of the electric heating plate is between 91.8 Ω and 93.8 Ω. A DC power supply is used to power the heating plate, maintaining a heating power of 400 W. According to the research by Jin and Ouyang et al., when lateral heating triggers thermal runaway in a battery, the greater the power of the heating plate per unit area, the faster the thermal runaway trigger rate [42]. In this study, a lower power 400 W heating plate was selected. The battery can absorb more heat from the heating plate before thermal runaway, resulting in a more intense thermal runaway process. This simulates the maximum hazard level during thermal propagation between adjacent batteries.

2.3. Experimental Procedure

1. Preparation: Charge the 156 Ah NCM811 battery to 100% SOC (State of Charge) at 0.1 C using a battery charge-discharge machine (model NEWARE BTS4000-5V30A) and let it rest for 24 h;
2. Thermocouple Installation: Fix K-type armored thermocouples to the battery surface using polyimide and Teflon tapes at the temperature monitoring points shown in Figure 2b,c. Special attention should be paid to the thermocouple near the heating plate. First, fix it with a layer of polyimide tape, and then attach two layers of Teflon tape to prevent the thermocouple from coming into direct contact with the heating plate. This ensures that the thermocouple measures the battery surface temperature rather than the surface temperature of the heating plate. Clamp the battery as shown in Figure 2a and use a torque wrench to pre-tighten the fastening bolts to 1 N·m.

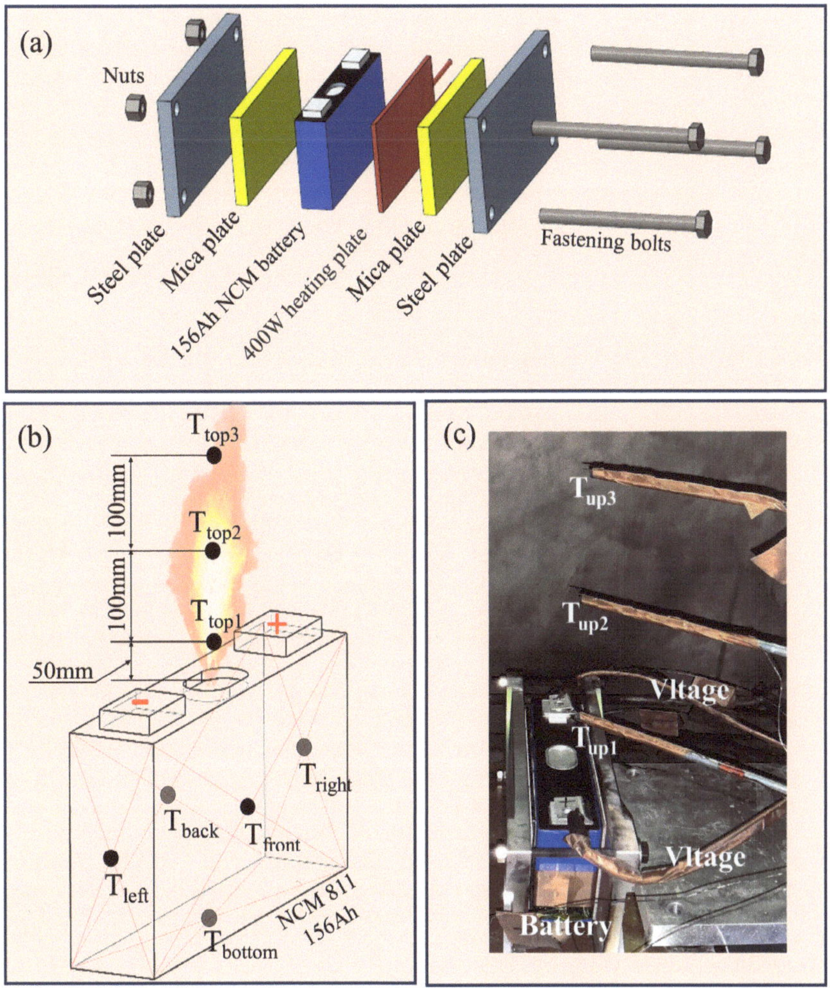

Figure 2. Battery installation. (**a**) Battery clamping structure. (**b**) Thermocouple placement locations. (**c**) Actual thermocouple placement.

1. Place the clamped battery in a custom-made sealed pressure vessel with a capacity of 1000 L. The top of the battery should be approximately 800 mm from the top of the experimental chamber. Adjust the positions of the four environmental monitoring thermocouples on all sides (top, bottom, left, and right), and start the camera. Close the pressure vessel door and lock the bolts;
2. Nitrogen Purging: Ensure both the intake and exhaust valves are initially closed. At the start of the test, open the exhaust valve and turn on the vacuum pump to evacuate the air inside the chamber. When the pressure drops to 5 Kpa, close the exhaust valve and turn off the vacuum pump. Open the intake valve to fill the chamber with nitrogen until the pressure reaches 101 Kpa, then close the intake valve. Repeat this process three times to reduce the oxygen concentration in the chamber to 0.0025%, ensuring the chamber is filled with 99.75% nitrogen;
3. Triggering Thermal Runaway: Turn on the heating plate power supply and maintain the heating plate at a power of 400 W. Monitor the battery voltage data; when the voltage drops to 0 V, immediately turn off the heating plate power supply, indicating that thermal runaway has been triggered;
4. Post-Experiment Procedure: After the battery temperature inside the chamber drops to room temperature, open the chamber door, collect the solid substances ejected during thermal runaway, and clean the chamber;
5. Repetition: Repeat the above process for a total of three experiments.

3. Experimental Results and Analysis

3.1. Battery Surface Temperature and Voltage Changes

During thermal runaway, the battery surface temperature undergoes significant changes [18,43,44]. To measure the temperature at various surfaces during thermal runaway, thermocouples were attached to the center of the front, back, left, right, and bottom surfaces of the battery.

Figure 3a shows the changes in surface temperature and voltage during the first experiment with thermal runaway. The battery was heated with a 400 W heating plate, causing a gradual accumulation of heat. The temperatures at all monitoring points increased steadily. When the temperature on the front surface (T_{front}) reached 257.2 °C, the surface temperature began to rise rapidly, indicating the onset of severe thermal runaway. The internal battery generated a large amount of gas, leading to the opening of the safety valve, and the highest recorded temperature reached 943.7 °C. This suggests that a significant amount of heat was generated during the decomposition of the electrolyte and the short-circuiting process inside the battery during thermal runaway [45–47]. Concurrently, the battery voltage dropped sharply to 0. Compared to the findings of Shen et al. [5,48], the maximum temperature observed in this study was higher, likely due to the larger battery capacity, which resulted in longer chemical reaction times and more heat generation, leading to higher temperatures. Therefore, it can be inferred that the larger the battery capacity, the greater the heat generation during thermal runaway and the higher the temperature.

Figure 3b illustrates the surface temperature of the battery during the thermal runaway ejection phase. At 1107 s, the Tfront rapidly increased, marking the start of the severe thermal runaway phase. At 1136 s, the temperature at the back surface (T_{back}) began to rise quickly, indicating that the thermal runaway front had reached the back of the battery [49]. At 1140 s, the voltage dropped to 0, signifying the complete reaction of the active materials within the battery. The propagation of thermal runaway from the front surface to the back surface took a total of 29 s. Given the battery's thickness of 45 mm, the propagation speed of thermal runaway within the battery can be calculated using Equation (1) as 0.00155 m/s.

$$v_{incell} = \frac{h}{t} \times 0.001 \quad (1)$$

V_{incell} represents the propagation speed of thermal runaway within the battery, measured in m/s; h represents the battery thickness, measured in mm; t represents the duration of thermal runaway, measured in seconds.

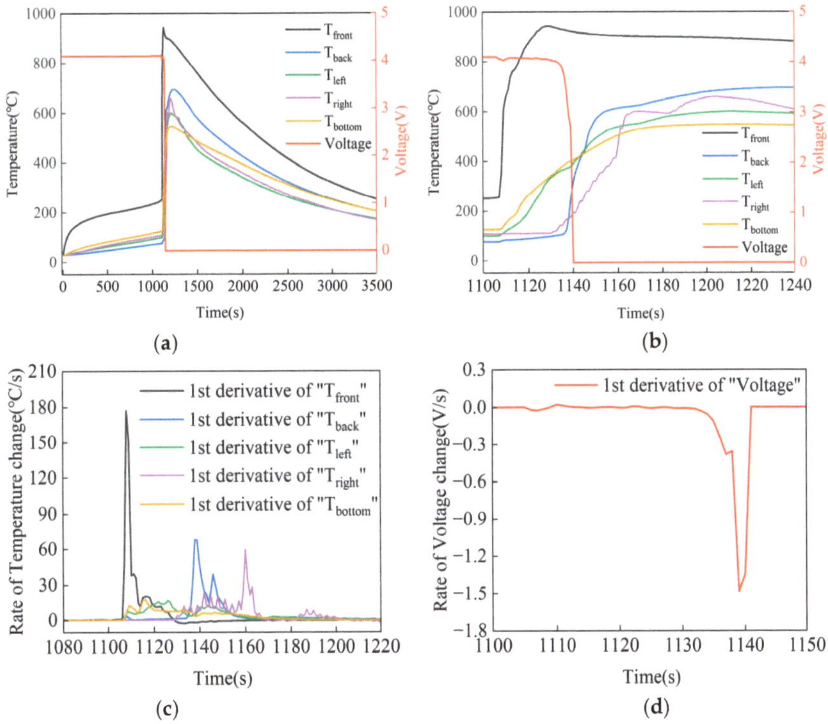

Figure 3. Battery surface temperature during the first experiment. (**a**) Battery surface temperature. (**b**) Battery surface temperature during thermal runaway ejection. (**c**) Rate of change in battery surface temperature. (**d**) Rate of change in voltage.

Figure 3c shows the temperature rise rate at various monitoring points, with the highest rate observed at Derivative (T_{front}), reaching 177.15 °C/s. This is because the continuous heat input from the heating plate causes more intense thermal runaway reactions at high temperatures, resulting in a higher heat release rate. The lowest temperature rise rate is at Derivative (T_{bottom}), only 17.8 °C/s, indicating good temperature uniformity. This is because the electrolyte inside the battery evenly distributes the heat generated by thermal runaway to the bottom surface, and the evaporation of the electrolyte also carries away a significant amount of heat. The order of the temperature rise rates at the monitoring points is: Derivative (T_{front}) > Derivative (T_{back}) > Derivative (T_{right}) > Derivative (T_{left}) > Derivative (T_{bottom}). The Derivative (T_{left}) and Derivative (T_{right}) exhibit two peak groups due to the two wound cores inside the battery, with thermal runaway propagating from one core to the other sequentially. The order of the appearance of the thermal runaway inflection points $T_{x\text{-}TR}$ (where the temperature rise rate exceeds 3 °C/s) is: Time(T_{front}) = Time(T_{bottom}) = Time(T_{left}) < Time(T_{right}) < Time(T_{back}), indicating that the direction of thermal runaway propagation is from the heated side to the non-heated side, and from the anode side to the cathode side.

Figure 3d shows the rate of voltage change during the battery's thermal runaway. At the start of the thermal runaway ejection phase, the voltage fluctuated for 33 s but remained mostly unchanged. When the thermal runaway front reached the non-heated side, the voltage began to drop sharply, decreasing from 4.0 V to 0 V within 6 s. This indicates that the battery voltage remains relatively stable during the ejection phase of thermal runaway and only experiences a sharp decline when the thermal runaway front reaches the back of the battery [50].

- $T_{X\text{-}TR}$: Inflection point temperature at the monitoring point (temperature rise rate greater than 3 °C/s).
- $T_{X\text{-}max}$: Maximum temperature at the monitoring point.

Figure 4 shows the inflection point temperatures and maximum temperatures at each monitoring point across three experiments [51], along with the calculated average values for the three experiments. The average maximum inflection point temperature on the heated side (front), $T_{front\text{-}TR}$, is 252.5 °C, due to the direct heating effect of the heating plate. The inflection point temperature at the bottom of the battery ranks second at 126.7 °C. This may be because the bottom of the battery is directly placed on a mica plate, resulting in no heat conduction with the air. The temperatures on the left and right sides of the battery are lower because they are exposed to air, creating thermal convection and causing some heat loss. The back of the battery has a relatively low temperature due to its distance from the heat source, leading to significant heat loss during heat transfer. The inflection point temperatures at the back, left, and right sides of the battery are nearly identical, at 103.7 °C, 102.7 °C, and 107.6 °C, respectively. The order of the highest temperatures at the six monitoring points from highest to lowest is: $T_{front\text{-}max} > T_{back\text{-}max} > T_{right\text{-}max} > T_{left\text{-}max} > T_{bottom\text{-}max}$. This indicates that the largest surfaces of the battery have the highest temperatures due to the maximum contact area with the wound cores, resulting in efficient heat conduction. Since the copper current collector has better thermal conductivity than the aluminum current collector, the small surface near the cathode has a higher temperature than the small surface near the anode. The bottom surface of the battery has the lowest $T_{bottom\text{-}max}$ temperature due to the heat conduction and heat absorption by the electrolyte's evaporation.

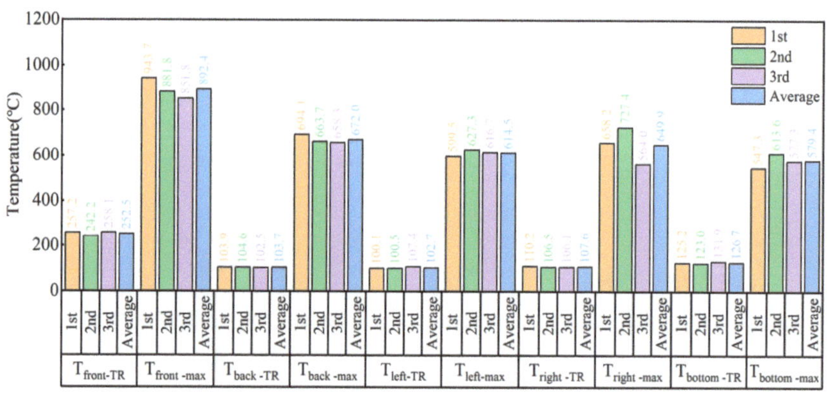

Figure 4. Inflection point temperature and maximum temperature at each monitoring point in three experiments.

3.2. *Battery Thermal Runaway Jet Temperature*

To measure the temperature of the thermal runaway jet of the battery, three temperature monitoring points were set up at 50 mm, 150 mm, and 250 mm above the safety valve. Figure 5 shows the temperature monitoring of the thermal runaway jet of the battery.

Figure 5 shows the temperature monitoring of the battery thermal runaway jet during the first experiment. Figure 5a–c display the temperatures at three monitoring points across three experiments. The temperature is highest at the 50 mm position from the nozzle, ranging from 351.6 °C to 364.2 °C. The temperature is lowest at the 250 mm position from the nozzle, ranging from 175.3 °C to 216.5 °C. The temperature at the middle monitoring point fluctuates between 229.8 °C and 302.7 °C. This indicates that the further away from the outlet, the lower the jet temperature. This is because there are no continuous heat-producing substances within the jet, causing the heat within the jet to dissipate rapidly into the environment, thus gradually lowering the jet temperature [52–55].

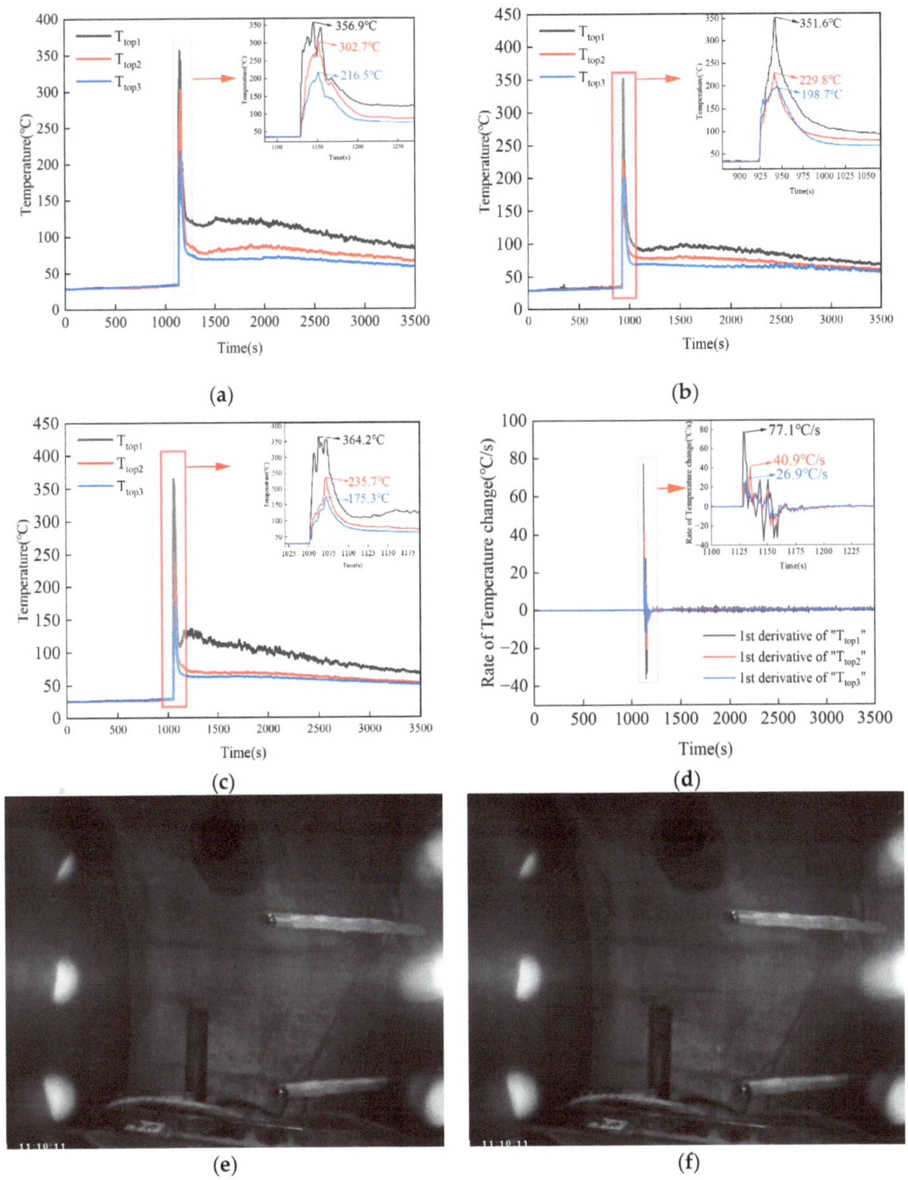

Figure 5. Temperature of Battery Thermal Runaway Jet (**a**) Temperature at monitoring points 50 mm, 150 mm, and 250 mm above the battery nozzle during the first experiment. (**b**) Temperature at monitoring points 50 mm, 150 mm, and 250 mm above the battery nozzle during the second experiment. (**c**) Temperature at monitoring points 50 mm, 150 mm, and 250 mm above the battery nozzle during the third experiment. (**d**) Rate of temperature change at monitoring points 50 mm, 150 mm, and 250 mm above the battery nozzle during the first experiment. (**e**) Image before thermal runaway. (**f**) Image of the jet at the initial stage of thermal runaway.

Figure 5d shows the rate of temperature change at the three monitoring points during the first experiment. The maximum temperature rise rates at the positions 50 mm, 150 mm, and 250 mm from the safety valve were 77.1 °C/s, 40.9 °C/s, and 26.9 °C/s, respectively. The temperature change rate is inversely proportional to the distance from the outlet. This is because, during the jet ejection process, the jet expands in multiple directions. The further the distance from the outlet, the less thermal energy-containing jet reaches the monitoring point, resulting in lower temperatures [56,57].

Measuring the jet temperature during thermal runaway is challenging because the shape of the jet often changes. Golubkov et al. (2018) used a fixed device for directional jetting and placed a thermocouple offset from the centerline, resulting in higher measured temperatures compared to most other studies on inert gas emissions [58]. This study focuses on the thermal runaway jet of batteries in a nitrogen atmosphere. Experiments revealed that during the initial phase of battery thermal runaway ejection, the jet direction is not directly upward due to the internal structure of the battery.

As seen in the images captured by the camera in Figure 5e,f, the initial jet forms an angle of approximately 45 degrees with the battery. Thus, thermocouples placed directly above the battery cannot effectively capture the initial jet temperature. As thermal runaway progresses and the internal structure of the battery is damaged, the jet eventually ejects directly upwards, allowing the thermocouples to measure the temperature during the latter part of the ejection process.

3.3. *Video and Audio Signals during the Battery Thermal Runaway Ejection Process*

To investigate the changes in video and audio signals during the battery ejection process [59], night vision cameras and recording equipment were used to monitor the thermal runaway process of the battery. The camera recorded images in night vision mode, resulting in black and white images. Figure 6 shows the audio signals and thermal runaway images.

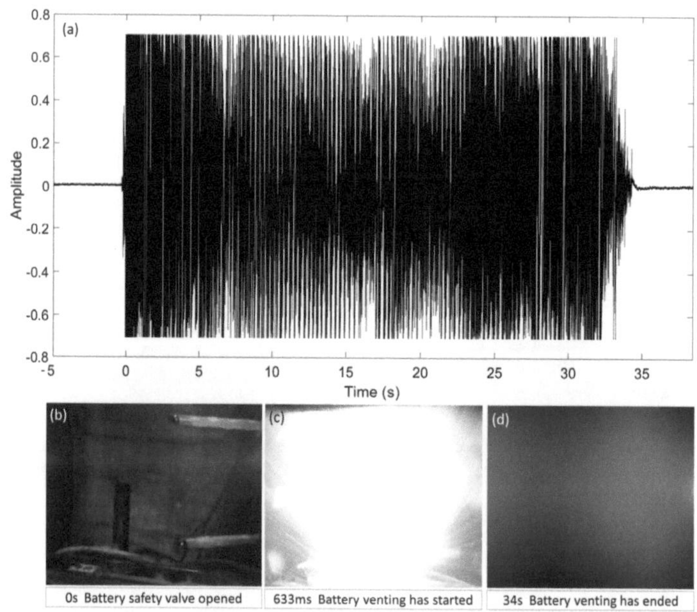

Figure 6. Sound and video signals during thermal runaway. (**a**) Amplitude of sound signal during thermal runaway. (**b**) Image of safety valve opening during thermal runaway. (**c**) Image of flame generation at the start of thermal runaway combustion. (**d**) Image after completion of thermal runaway ejection.

Figure 6a illustrates the process from the opening of the safety valve to the complete ejection of the battery. The moment when the sound starts is defined as time 0. During the time intervals of 0–6 s and 24–32 s, the sound amplitude is larger, with a relatively smaller amplitude in between. This is because, during the thermal runaway process, the heating plate first triggers thermal runaway in the first wound core, and then the thermal runaway propagates to the second wound core, resulting in two high-amplitude periods. The total duration of the sound signal is 34 s. Sound is only produced during high-speed ejection; low-speed ejection generally does not produce sound.

Figure 6b shows the image inside the chamber at the moment the sound starts, indicating the initial opening of the safety valve. Figure 6c shows the image when the thermal runaway ejection products begin to burn. Figure 6d shows the image inside the chamber at the moment the ejection sound disappears, indicating that the ejection has essentially ended and the chamber is filled with solid particle smoke. Therefore, the thermal runaway process involves the accumulation of battery heat to a certain threshold, partial opening of the safety valve, the beginning of gas ejection through the safety valve, generation of airflow noise, and as the reaction progresses, intense reactions and jet combustion occur. After a period of sustained combustion, the airflow noise ceases, the thermal runaway ends, and the flames extinguish. By capturing images of the battery and monitoring the light intensity of the flames, a low-cost early warning method for battery thermal runaway can be achieved [60–62].

3.4. Pressure Changes and Gas Production

To investigate the gas production rate of the NCM811 battery during thermal runaway in an oxygen-free environment, the battery was tested under a nitrogen atmosphere, and the pressure changes inside the chamber were recorded, as shown in Figure 7.

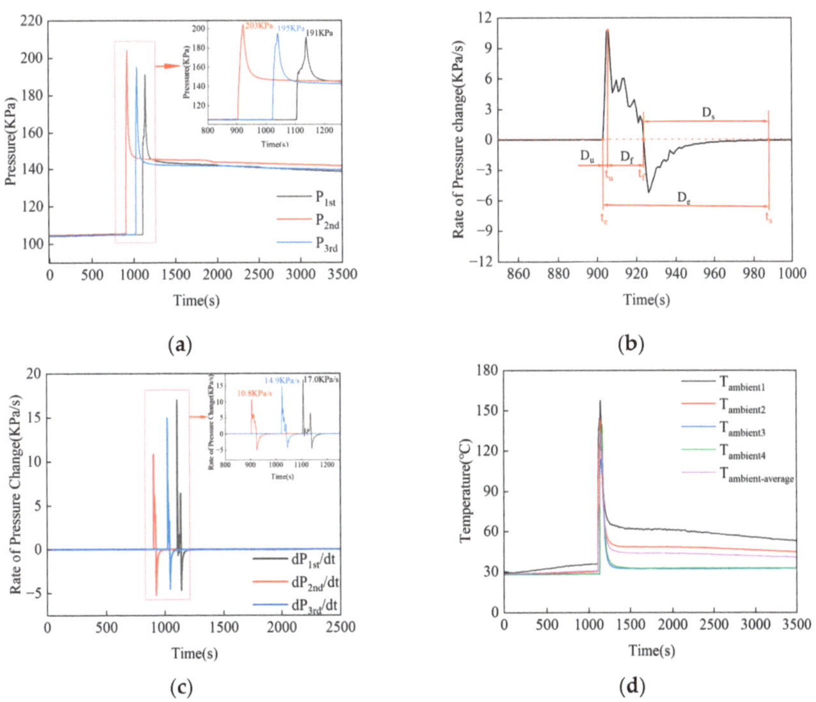

Figure 7. Pressure and temperature inside the experimental chamber. (**a**) Pressure changes during three experiments. (**b**) Segmentation of ejection phases. (**c**) Rate of pressure change inside the chamber. (**d**) Ambient temperature inside the chamber.

Figure 7a shows the pressure changes inside the chamber during the three experiments. During the triggering of thermal runaway, the pressure inside the chamber increased rapidly [35,63–66], reaching a peak of 203 KPa before quickly dropping and stabilizing at 143 KPa. The presence of a single pressure peak in this experiment indicates that the thermal runaway process involved only one ejection event.

Figure 7b illustrates the rate of pressure change inside the chamber during the first experiment. According to the segmentation of the ejection process proposed in the reference [67], the gas ejection stage (D_e) of thermal runaway is divided into three phases:

1. Ultra-High-Speed Ejection Phase (D_u): The time from the start of thermal runaway ejection (t_e) to the moment of maximum ejection rate (t_u);
2. High-Speed Ejection Phase (D_f): The time from the moment of maximum ejection rate (tu) to the moment when the ejection rate drops to zero (t_f);
3. Slow Ejection Phase (D_s): The time from the moment when the ejection rate drops to zero (t_f) to the end of the ejection (t_s).

Figure 7c shows the rate of pressure change during the three experiments. Based on the segmentation of the ejection phases, the durations of the ejection phases for the three experiments are listed in Table 2.

Table 2. Duration of each ejection phase and maximum pressure change rate in the three experiments.

Experiment	Du (s)	Df (s)	Ds (s)	De (s)	$(dP/dt)_{max}$ (Kpa/s)
1st	2	32	47	81	17
2nd	3	18	53	74	10.8
3rd	2	17	53	72	14.8
Average	2.3	22.3	51	75.7	14.2

To calculate the gas production during thermal runaway, the ideal gas law Equation (2) is introduced in this study. By rearranging and transforming the equation, Formula (3) can be obtained.

$$PV = nRT \tag{2}$$

$$n = \frac{PV}{RT_{ambient-average}} - \frac{P_0 V}{RT_0} \tag{3}$$

In the formula:

- P is the pressure inside the chamber, in units of Pa;
- P_0 is the initial pressure inside the chamber, in units of Pa;
- V is the volume of the experimental chamber, in units of m^3;
- n is the amount of gas in the chamber, in units of mol;
- R is the ideal gas constant, with a value of 8.31441 J/(mol·K);
- T_0 is the ambient temperature at the start of the experiment, in units of K;
- $T_{ambient\text{-}average}$ is the average temperature of the four environmental monitoring points inside the experimental chamber, in units of K.

Based on the above formulas and the ambient temperature shown in Figure 7d, the amount of gas produced during the battery thermal runaway can be calculated. The data from the three experiments are shown in Table 3.

Table 3. Gas production and parameters from the three experiments.

Experiment	P (Pa)	$T_{ambient\text{-}average}$ (K)	$T_{ambient\text{-}max}$ (K)	n (mol)	n_{Ah} (mol/Ah)
1st	143,200	321.45	465.7	13.3	0.085
2nd	144,889	318.87	453.2	14.4	0.092
3rd	141,911	314.62	438.6	14.0	0.090
Average	143,333	318.31	452.5	13.9	0.089

Table 2. Characteristics of TR in different stages.

Stage	I	II	III	IV	V
Time (s)	0–800	800–958	958–1220	1220–1493	1493–2000
Basic features	1. Heating period. 2. T_v and T_c increase slowly.	1. The safety valve opens. 2. T_v increases dramatically, and the first crest occurs, respectively.	1. TR is triggered. 2. T_v and T_c increase dramatically, and the second crest occurs, respectively.	1. The second venting occurs. 2. T_v and T_c decrease after the third peak.	1. Cooling stage. 2. The temperature drops.
Temperature range (°C)	T_v	26.6–54 °C	49.6–74.6 °C	49.6–176.2 °C	141.8–245.9 °C
	T_s	112.7–138.4 °C	112.7–140.9 °C	136.3–302.8 °C	242.1–254.5 °C
	T_c	27–37 °C	37–42.9 °C	37.4–63.1 °C	63–90.4 °C

From Table 2, it can be seen that in Stage I, the battery is in the heating stage, and the battery temperature is rising. The main chemical reaction that occurs in this stage is the decomposition of the SEI film [32], as shown in reaction Equations (1) and (2).

$$(CH_2OCO_2Li)_2 \rightarrow Li_2CO_3 + C_2H_4 + CO_2 + 1/2O_2 \quad (1)$$

$$2Li + C_3H_4O_3 \rightarrow Li_2CO_3 + C_2H_4 \quad (2)$$

Stage II: The battery safety valve opens at 831 s and the T_s is 140.9 °C. After the safety valve opens, a large amount of chemical reaction mixture and electrolyte vapor is ejected from the safety valve, carrying a large amount of heat into the chamber. Both of which reach their first peak in this stage. The T_s decreases rapidly due to heat exchange. The main chemical reaction that occurs in this stage is that the lithium metal embedded in the negative electrode reacts with the electrolyte [33], as shown in reaction Equations (3)–(5).

$$2Li + C_3H_4C_3\ (EC) \rightarrow Li_2CO_3{+}C_2H_4 \quad (3)$$

$$2Li + C_5H_{10}O_3\ (DEC) \rightarrow LiCO_3{+}C_2H_4{+}C_2H_6 \quad (4)$$

$$2Li + C_3H_6O_3\ (DMC) \rightarrow LiCO_3{+}C_2H_6 \quad (5)$$

Stage III: The range of T_v is between 49.8 °C and 176.2 °C. The temperature change rate reaches 1 °C/s at 958 s, defining the moment of TR triggering, which corresponds to a T_v of 49.6 °C. T_v reaches its peak change rate at 964 s, which is 1.68 °C/s. At this moment, a large amount of heat is transferred to the chamber, causing a rapid increase in temperature. T_c reaches a second peak at 1084 s, corresponding to a temperature of 71.6 °C. Subsequently, the temperature decreases to 57.8 °C after 1084 s. T_s reaches its peak at 1123 s, which is 302.8 °C. At this stage, the main decomposition process of the battery is film melting, and the direct contact of the positive and negative electrodes causes an internal short circuit, releasing a large amount of heat and emitting gases such as O_2, C_2H_4, HF, and PF_5, as shown in Equations (6)–(8).

$$LiPF_6 \rightarrow LiF + PF_5 \quad (6)$$

$$PF_5 + H_2O \rightarrow POF_3 + 2HF \quad (7)$$

$$C_3H_6O_3\ (DMC) + PF_5 \rightarrow CH_3OCOOPF_4 + CH_3F \quad (8)$$

Stage IV: As the chemical reaction progresses, T_c reaches its third peak at 1275 s, at 90.4 °C, and then rapidly decreases. T_v continues to rise rapidly during this stage, reaching its third peak at 245.9 °C at 1424 s. As the reaction progresses and the internal active material decreases, the chemical reaction rate gradually weakens, and the temperature begins to decrease. T_s fluctuates around 245 °C, with little increase in temperature. The chemical reaction occurring in this stage involves the decomposition of the positive elec-

In the experiment, the distance from the top of the battery to the top of the chamber is approximately 800 mm. During thermal runaway, the jet reflects off the chamber walls and fills the entire container, causing the ambient temperature to rise. In the three experiments, the highest ambient temperature reached 465.7 K (192.5 °C), which is within the operating temperature range of the pressure sensor (−40 °C to 200 °C). Therefore, the 1000 L experimental chamber can accurately measure the pressure changes within the chamber.

Because gas production is closely related to battery capacity, a parameter called unit capacity molar amount, n_{Ah}, is defined to evaluate the gas production per unit capacity. The calculation formula is as follows:

$$n_{Ah} = \frac{n}{Capacity} \tag{4}$$

In the formula:

- n_{Ah} represents the molar amount of gas produced per unit capacity, in mol/Ah;
- n represents the total amount of gas produced by the battery, in mol;
- *Capacity* represents the battery capacity, in Ah.

Based on the above formula, the unit capacity molar amount for the 156 Ah NCM811 battery from the three experiments is 0.089 mol/Ah.

Based on the research by Shen and Yang et al. [5,48], the main gas components produced during the thermal runaway of 811 ternary lithium batteries in a nitrogen atmosphere are CO_2 (30.75%), H_2 (23.96%), CO (22.92%), C_2H_4 (15.71%), and CH_4 (5.3%).

Figure 8 illustrates the relationship between gas generation and temperature during the thermal runaway of the battery. The primary processes involved are as follows:

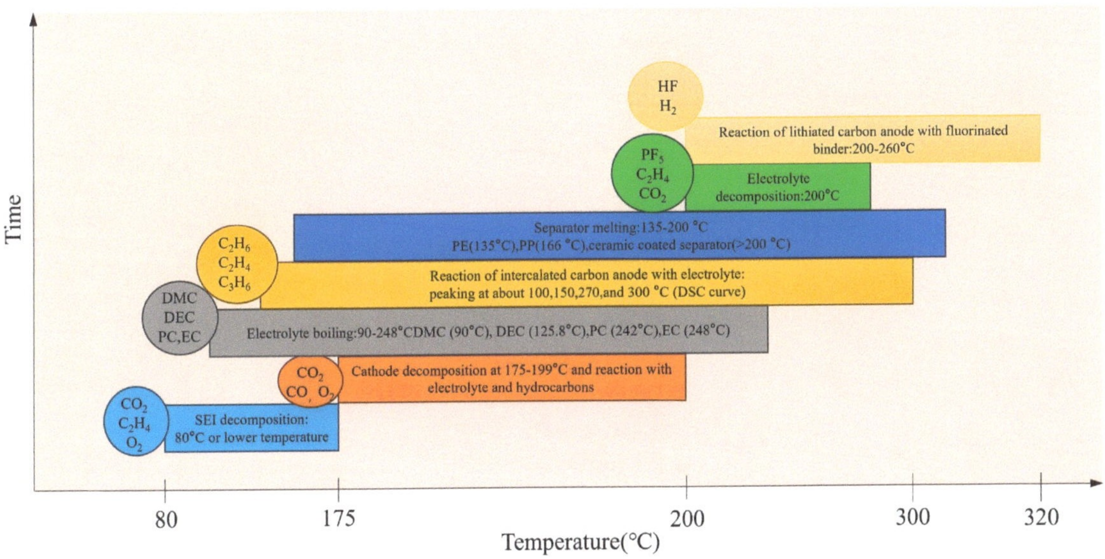

Figure 8. Time Series of Gas Generation [68].

SEI Film Decomposition:

$$(CH_2OCO_2Li)_2 \rightarrow LiCO_3 + C_2H_4 + CO_2 + \frac{1}{2}O_2 \tag{5}$$

As the temperature increases, the active lithium in the electrodes begins to react with the electrolyte:

$$2Li + C_3H_4O_3\ (EC) \rightarrow Li_2CO_3 \tag{6}$$

$$2Li + C_4H_6O_3\ (PC) \rightarrow Li_2CO_3 + C_3H_6 \tag{7}$$

$$2Li + C_3H_6O_3\ (DMC) \rightarrow Li_2CO_3 + C_2H_6 \tag{8}$$

The electrolyte simultaneously undergoes its own decomposition reactions:

$$C_2H_5OCOOC_2H_5 + PF_5 \rightarrow C_2H_5OCOOPF_4 + HF + C_2H_4 \tag{9}$$

$$C_2H_4 + HF \rightarrow C_2H_5F \tag{10}$$

$$C_2H_5OCOOC_2F_4 \rightarrow PF_3O + CO_2 + HF + C_2H_4 \tag{11}$$

$$C_2H_5OCOOC_2H_5 + PF_5 \rightarrow PF_3O + CO_2 + C_2H_5F \tag{12}$$

$$C_2H_5OCOOC_2H_5 + HF \rightarrow PF_4OH + CO_2 + C_2H_5F \tag{13}$$

According to Li et al.'s theory on the ignition triangle of gas generation during battery thermal runaway, the upper flammable limit (UFL) and lower flammable limit (LFL) curves of gases produced by ternary lithium batteries exhibit a peninsula shape [69]. The flammable range remains relatively constant before sharply decreasing, forming a triangular shape. In this study, experiments conducted in a nitrogen atmosphere effectively extend the safety threshold during the thermal runaway process of lithium batteries.

3.5. *Battery Morphological Characteristics*

NCM811 lithium batteries have high energy density, and their thermal runaway generates a significant amount of heat. This heat leads to chemical reactions between the electrode materials and the electrolyte, producing a large amount of flammable gas, causing the battery casing to become damaged. To investigate the morphological characteristics of the battery after thermal runaway [70,71], the remnants of the battery from the three experiments were analyzed. Figure 8 shows the battery's morphology before and after thermal runaway.

Figure 9a shows the external appearance of the battery before thermal runaway. Figure 9b–d display the morphology of the battery after thermal runaway. It can be observed that 2/3 of the large surface area of the battery casing is damaged and melted, exposing the internal copper foils. The casing near the safety valve is completely broken, and jet grooves formed by ejection are present near the safety valve, with jet angles between 45 and −45 degrees.

Experimental data indicate that the inflection point of the pressure increase inside the experimental chamber caused by the valve opening and the rapid temperature rise occur almost simultaneously. At the 1106 s mark, the battery's safety valve opens, and the highest temperature at the Tfront position is 257.2 °C, which is far below the melting point of the battery's aluminum casing at 660.3 °C. Therefore, the battery casing will not be damaged before the safety valve opens. After the safety valve opens, using the aluminum casing's melting point of 660.3 °C as the dividing line:

1. When the T_{front} position temperature is below 660.3 °C, the battery casing remains intact, and all the jet is expelled through the safety valve;
2. When the T_{front} position temperature exceeds 660.3 °C, the battery casing will be damaged, and the jet may be expelled from both the safety valve and the damaged area.

The damage and melting of the battery casing indicate that the intense reactions of the internal electrode active materials and the electrolyte produced a large amount of heat, with temperatures exceeding the melting point of aluminum (660.4 °C), causing the aluminum casing and internal aluminum foils to melt. The exposed internal copper foils show no signs of melting, indicating that the internal thermal runaway temperature did not reach the melting point of copper (1083.4 °C). Therefore, the internal thermal runaway temperature is between 660.4 °C and 1083.4 °C. The jet grooves are mainly distributed at angles of ±45 degrees, indicating that the jet direction is within this range. Placing temperature monitoring points directly above the ejection outlet may result in inaccurate jet temperature measurements. The jet grooves are mainly distributed at angles of ±45 degrees, indicating that the jets are ejected at these angles. This occurs because, during the initial phase of the ejection, the internal structure of the battery restricts the jet to a 45° angle. As the reaction progresses, the temperature inside the battery rapidly increases, causing damage to the internal structure. This results in the airflow starting to damage the central part of the battery as well. Therefore, setting the temperature measurement point directly above the nozzle may result in inaccurate jet temperature measurements.

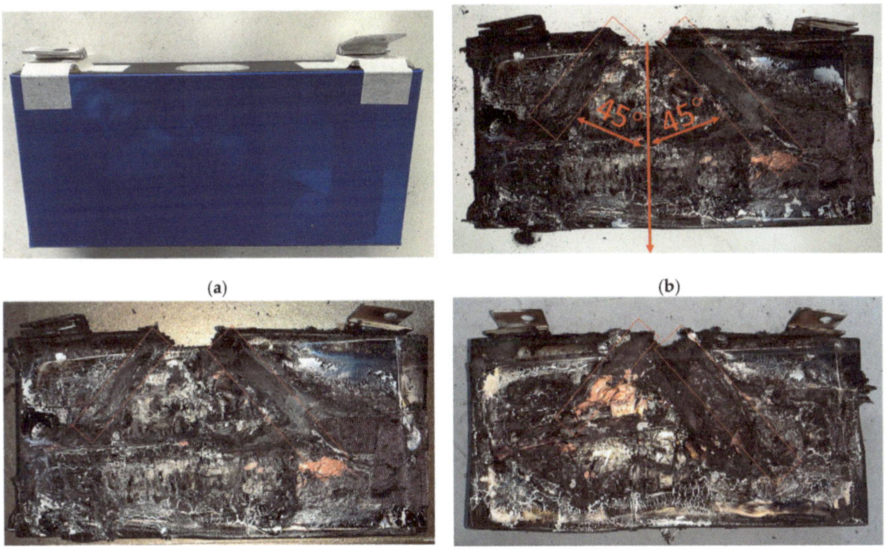

(a) (b) (c) (d)

Figure 9. Battery photographs. (**a**) Battery before thermal runaway. (**b**) Battery remains after thermal runaway in the first experiment. (**c**) Battery remains after thermal runaway in the second experiment. (**d**) Battery remains after thermal runaway in the third experiment.

4. Conclusions

In this study, the aim was to investigate the jetting behavior of ternary lithium batteries at 100% SOC under a nitrogen atmosphere (99.75%) during thermal runaway. This was achieved by laterally heating the battery with a 400 W heating plate, triggering thermal runaway in a ternary lithium battery with a cathode material of $Li(Ni_{0.8}Co_{0.1}Mn_{0.1})O_2$ and a capacity of 156 Ah. Through data analysis, the battery surface temperature, jet temperature, pressure changes, acoustic signal changes, and post-thermal runaway morphological characteristics were obtained. The main conclusions are as follows:

1. During the thermal runaway of the 811 ternary lithium battery, the front surface temperature of the battery can reach 851.8–943.7 °C, and the back surface temperature can reach 658.3–694.1 °C, both exceeding the melting point of the aluminum casing (660.4 °C). This melting can cause inaccuracies in temperature measurements. The temperatures of the bottom surface and the small side surfaces are lower than the aluminum melting point, and the order of the appearance of thermal runaway temperature inflection points is: Time(Tfront) = Time(Tbottom) = Time(Tleft) < Time(Tright) < Time(Tback). From the perspective of thermal runaway early warning, monitoring the temperature at the bottom surface and the small side surface near the battery anode is more accurate;
2. The highest jet temperatures at 50 mm, 150 mm, and 250 mm above the safety valve were 356.9 °C, 302.7 °C, and 216.5 °C, respectively. This indicates that the further from the outlet, the lower the jet temperature. The temperature rise rates at these

three points were 77.1 °C/s, 40.9 °C/s, and 26.9 °C/s, respectively, showing that the temperature rise rate is inversely proportional to the distance from the safety valve;

3. Based on acoustic and image signals, it was found that there are two intensive sound amplitude stages during the ternary battery ejection process, which are closely related to the number of battery wound cores;
4. Under the conditions of this experiment, the average gas production of the battery was 0.089 mol/Ah. Based on the pressure data from the first experiment, one ejection process was observed, which included ultra-high-speed ejection (2 s), high-speed ejection (32 s), and slow ejection (47 s) stages. Combined with the ejection process captured by the acoustic signals (34 s), it indicates that exhaust sounds are produced during the ultra-high-speed and high-speed ejection stages due to high-speed airflow impact, while no exhaust sound is produced during the slow ejection stage. This has significant research value for using acoustic signals to provide early warnings of battery thermal runaway;
5. Based on the post-thermal runaway remnants, it was found that the grooves caused by airflow impact are mainly located at ±45°. Therefore, when measuring the jet temperature during thermal runaway, monitoring points should be arranged within the ±45° range.

Funding: This research received no external funding.

Data Availability Statement: The data are available through appropriate requests.

Conflicts of Interest: The author declares no conflicts of interest.

References

1. Dong, Y.; Meng, J.; Sun, X.; Zhao, P.; Sun, P.; Zheng, B. Experimental Study on Effects of Triggering Modes on Thermal Runaway Characteristics of Lithium-Ion Battery. *World Electr. Veh. J.* **2023**, *14*, 270. [CrossRef]
2. Du, Q.; Fang, Z. Analysis of the Thermal Behavior of a Lithium Cell Undergoing Thermal Runaway. *Fluid Dyn. Mater. Process.* **2021**, *17*, 887–898. [CrossRef]
3. Gao, F.; Yang, K.; Wang, C.; Liu, W.; Zhu, Y. Experimental Study on Thermal Runaway of $LiFePO_4$/C Battery under Heating Condition. *IOP Conf. Ser. Earth Environ. Sci.* **2020**, *546*, 042025. [CrossRef]
4. Ji, C.; Zhang, S.; Wang, B.; Sun, J.; Zhang, Z.; Liu, Y. Study on Thermal Safety of the Overcharged Lithium-Ion Battery. *Fire Technol.* **2023**, *59*, 1089–1114. [CrossRef]
5. Shen, H.; Wang, H.; Li, M.; Li, C.; Zhang, Y.; Li, Y.; Yang, X.; Feng, X.; Ouyang, M. Thermal Runaway Characteristics and Gas Composition Analysis of Lithium-Ion Batteries with Different LFP and NCM Cathode Materials under Inert Atmosphere. *Electronics* **2023**, *12*, 1603. [CrossRef]
6. Lai, Y.; Yang, K.; Liu, H.; Gao, F.; Zhang, M.; Xu, W. A Probe into the Accuracy of Thermal Runaway Simulation Model of Lithium-ion Battery under Adiabatic Condition. *IOP Conf. Ser. Earth Environ. Sci.* **2021**, *680*, 012009. [CrossRef]
7. Li, Q.; Yu, J.; Liu, G.; Ma, X.; Si, W.; Hu, X.; Zhu, G.; Liu, T. Study on the Effectiveness of Water Mist on Suppressing Thermal Runaway in $LiFePO_4$ Batteries. *Crystals* **2023**, *13*, 1346. [CrossRef]
8. Li, X.; Liu, X.; Zhang, Y. Research and Development of Fire Alarm Detection Device for Lithium Ion Battery Based on Strain Measurement. *IOP Conf. Ser. Earth Environ. Sci.* **2020**, *605*, 012005. [CrossRef]
9. Li, Y.; Jiang, L.; Huang, Z.; Jia, Z.; Qin, P.; Wang, Q. Pressure Effect on the Thermal Runaway Behaviors of Lithium-Ion Battery in Confined Space. *Fire Technol.* **2023**, *59*, 1137–1155. [CrossRef]
10. Available online: https://homeland.house.gov/2024/02/15/subcommittee-chairman-desposito-delivers-opening-statement-in-hearing-on-fire-hazards-lithium-ion-battery-risks/ (accessed on 16 June 2024).
11. Miao, H. Research on accurate fire detection & early warning model for lithium-ion battery packs. *J. Phys. Conf. Ser.* **2024**, *2703*, 012068. [CrossRef]
12. Qian, F.; Wang, H.; Li, M.; Cheng, L.; Shen, H.; Wang, J.; Li, Y.; Ouyang, M. Thermal Runaway Vent Gases from High-Capacity Energy Storage $LiFePO_4$ Lithium Iron. *Energies* **2023**, *16*, 3485. [CrossRef]
13. Seyed Saeed, M.; Ziebert, C.; Mousa, M. Thermal Characteristics and Safety Aspects of Lithium-Ion Batteries: An In-Depth Review. *Symmetry* **2023**, *15*, 1925. [CrossRef]
14. Sun, J.; Li, G.; Xie, S.; He, Y. The Influence of Airflow Rate on the Thermal Runaway Propagation Characteristics of Lithium-Ion Batteries in a Low-Pressure Environment. *Fire Technol.* **2022**, *58*, 3553–3576. [CrossRef]
15. Wang, K.; Ouyang, D.; Qian, X.; Yuan, S.; Chang, C.; Zhang, J.; Liu, Y. Early Warning Method and Fire Extinguishing Technology of Lithium-Ion Battery Thermal Runaway: A Review. *Energies* **2023**, *16*, 2960. [CrossRef]

16. Wang, L.; Yang, X.; Wang, Q.; Tian, H. Research on Thermal Out of Control of Lithium Battery in New Energy Vehicles. *IOP Conf. Ser. Earth Environ. Sci.* **2020**, *446*, 022038. [CrossRef]
17. Wang, Z.; Tang, X.; Zhou, Y.; Huang, H.; Dai, H. Experimental and Modeling Analysis of Thermal Runaway for $LiNi_{0.5}Mn_{0.3}Co_{0.2}O_2$/Graphite Pouch Cell Triggered by Surface Heating. *Energies* **2024**, *17*, 826. [CrossRef]
18. Zhang, G.; Li, Z.; Wang, H.; Yuan, D. Study on the Suppression Effect of Cryogenic Cooling on Thermal Runaway of Ternary Lithium-Ion Batteries. *Fire* **2022**, *5*, 182. [CrossRef]
19. Li, J.; Gao, P.; Tong, B.; Cheng, Z.; Cao, M.; Mei, W.; Wang, Q.; Sun, J.; Qin, P. Revealing the mechanism of pack ceiling failure induced by thermal runaway in NCM batteries: A coupled multiphase fluid-structure interaction model for electric vehicles. *eTransportation* **2024**, *20*, 100335. [CrossRef]
20. Liu, Z.; Zhu, Y.; Li, R.; Tao, C.; Chen, Z.; Liu, T.; Li, Y. The experimental investigation of thermal runaway characteristics of lithium battery under different concentrations of heptafluoropropane and air. *J. Energy Storage* **2024**, *84*, 110828. [CrossRef]
21. Xie, H.J.; Sun, J.; Li, J.G.; Zhou, T.; Wei, S.P.; Yi, Z.H. Lithium-Ion Battery Thermal Runaway Electro-Thermal Triggering Method and Toxicity Analysis. *IOP Conf. Ser. Earth Environ. Sci.* **2021**, *701*, 012007. [CrossRef]
22. Bugryniec, P.J.; Resendiz, E.G.; Nwophoke, S.M.; Khanna, S.; James, C.; Brown, S.F. Review of gas emissions from lithium-ion battery thermal runaway failure—Considering toxic and flammable compounds. *J. Energy Storage* **2024**, *87*, 111288. [CrossRef]
23. Wei, G.; Huang, R.; Zhang, G.; Jiang, B.; Zhu, J.; Guo, Y.; Han, G.; Wei, X.; Dai, H. A comprehensive insight into the thermal runaway issues in the view of lithium-ion battery intrinsic safety performance and venting gas explosion hazards. *Appl. Energy* **2023**, *349*, 121651. [CrossRef]
24. Zhang, Q.; Yang, K.; Niu, J.; Liu, T.; Hu, J. Research on the lower explosion limit of thermal runaway gas in lithium batteries under high-temperature and slight overcharge conditions. *J. Energy Storage* **2023**, *79*, 109976. [CrossRef]
25. Chen, M.; He, Y.; De Zhou, C.; Richard, Y.; Wang, J. Experimental Study on the Combustion Characteristics of Primary Lithium Batteries Fire. *Fire Technol.* **2014**, *52*, 365–385. [CrossRef]
26. Jia, L.; Wang, D.; Yin, T.; Li, X.; Li, L.; Dai, Z.; Zheng, L. Experimental Study on Thermal-Induced Runaway in High Nickel Ternary Batteries. *ACS Omega* **2022**, *7*, 14562–14570. [CrossRef]
27. Thomas, F.; Mills, G.; Howe, R.; Zobell, J. Lithium Battery Fires: Implications for Air Medical Transport. *Air Med. J.* **2012**, *31*, 242–248. [CrossRef]
28. Wei, D.; Zhang, M.; Zhu, L.; Hu, C.; Huang, W.; Yao, J.; Yuan, Z.; Xu, C.; Feng, X. Study on Thermal Runaway Behavior of Li-Ion Batteries Using Different Abuse Methods. *Batteries* **2022**, *8*, 201. [CrossRef]
29. Ohneseit, S.; Finster, P.; Floras, C.; Lubenau, N.; Uhlmann, N.; Seifert, H.J.; Ziebert, C. Thermal and Mechanical Safety Assessment of Type 21700 Lithium-Ion Batteries with NMC, NCA and LFP Cathodes—Investigation of Cell Abuse by Means of Accelerating Rate Calorimetry (ARC). *Batteries* **2023**, *9*, 237. [CrossRef]
30. Han, Z.; Zhao, L.; Zhao, J.; Xu, J.; Liu, H.; Chen, M. An Experimental Study on the Thermal Runaway Propagation of Cycling Aged Lithium-Ion Battery Modules. *Fire* **2024**, *7*, 119. [CrossRef]
31. Deng, J.; Chen, B.; Lu, J.; Zhou, T.; Wu, C. Thermal runaway and combustion characteristics, risk and hazard evaluation of lithium-iron phosphate battery under different thermal runaway triggering modes. *Appl. Energy* **2024**, *368*, 123451. [CrossRef]
32. Zhou, Z.; Li, M.; Zhou, X.; Li, L.; Ju, X.; Yang, L. Investigating thermal runaway triggering mechanism of the prismatic lithium iron phosphate battery under thermal abuse. *Renew. Energy* **2023**, *220*, 119674. [CrossRef]
33. Li, C.; Wang, H.; Han, X.; Wang, Y.; Wang, Y.; Zhang, Y.; Feng, X.; Ouyang, M. An Experimental Study on Thermal Runaway Behavior for High-Capacity $Li(Ni_{0.8}Co_{0.1}Mn_{0.1})O_2$ Pouch Cells at Different State of Charges. *J. Electrochem. Energy Convers. Storage* **2021**, *18*, 021012. [CrossRef]
34. Liao, Z.; Zhang, S.; Li, K.; Zhao, M.; Qiu, Z.; Han, D.; Zhang, G.; Habetler, T.G. Hazard analysis of thermally abused lithium-ion batteries at different state of charges. *J. Energy Storage* **2019**, *27*, 101065. [CrossRef]
35. Sun, Q.; Liu, H.; Zhi, M.; Chen, X.; Lv, P.; He, Y. Thermal characteristics of thermal runaway for pouch lithium-ion battery with different state of charges under various ambient pressures. *J. Power Sources* **2022**. [CrossRef]
36. Wu, H.; Chen, S.; Hong, Y.; Xu, C.; Zheng, Y.; Jin, C.; Chen, K.; He, Y.; Feng, X.; Wei, X.; et al. Thermal safety boundary of lithium-ion battery at different state of charge. *J. Energy Chem.* **2023**, *91*, 59–72. [CrossRef]
37. Doose, S.; Hahn, A.; Fischer, S.; Müller, J.; Haselrieder, W.; Kwade, A. Comparison of the consequences of state of charge and state of health on the thermal runaway behavior of lithium ion batteries. *J. Energy Storage* **2023**, *62*, 106837. [CrossRef]
38. Liu, J.; Wang, Z.; Bai, J.; Gao, T.; Mao, N. Heat generation and thermal runaway mechanisms induced by overcharging of aged lithium-ion battery. *Appl. Therm. Eng.* **2022**, *212*, 118565. [CrossRef]
39. Xiao, Y.; Wen, J.; Yao, L.; Zheng, J.; Fang, Z.; Shen, Y. A comprehensive review of the lithium-ion battery state of health prognosis methods combining aging mechanism analysis. *J. Energy Storage* **2023**, *65*, 107347. [CrossRef]
40. Zhang, L.; Liu, L.; Terekhov, A.; Warnberg, D.; Zhao, P. Thermal runaway of Li-ion battery with different aging histories. *Process Saf. Environ. Prot.* **2024**, *185*, 910–917. [CrossRef]
41. Zhang, L.; Liu, L.; Yang, S.; Xie, Z.; Zhang, F.; Zhao, P. Experimental investigation on thermal runaway suspension with battery health retention. *Appl. Therm. Eng.* **2023**, *225*, 120239. [CrossRef]
42. Jin, C.; Sun, Y.; Wang, H.; Zheng, Y.; Wang, S.; Rui, X.; Xu, C.; Feng, X.; Wang, H.; Ouyang, M. Heating power and heating energy effect on the thermal runaway propagation characteristics of lithium-ion battery module: Experiments and modeling. *Appl. Energy* **2022**, *312*, 118760. [CrossRef]

43. Meng, D.; Wang, X.; Chen, M.; Wang, J. Effects of environmental temperature on the thermal runaway of lithium-ion batteries during charging process. *J. Loss Prev. Process Ind.* **2023**, *83*, 105084. [CrossRef]
44. Parhizi, M.; Ahmed, M.B.; Jain, A. Determination of the core temperature of a Li-ion cell during thermal runaway. *J. Power Sources* **2017**, *370*, 27–35. [CrossRef]
45. Liu, C.; Huang, Q.; Zheng, K.; Qin, J.; Zhou, D.; Wang, J. Impact of Lithium Salts on the Combustion Characteristics of Electrolyte under Diverse Pressures. *Energies* **2020**, *13*, 5373. [CrossRef]
46. Teslenko, V.S.; Drozhzhin, A.P.; Medvedev, N.; Manzhalei, V.I. Initiation of combustion of a gas mixture by an electric explosion of an electrolyte. *Combust. Explos. Shock. Waves* **2012**, *48*, 730–733. [CrossRef]
47. Yang, W.; Zhang, Y.; Deng, J.; Chen, J.; Ji, X.; Wu, H.; Zhao, J. Experimental study on combustion characteristics of electrolyte pool fire. *J. Energy Storage* **2024**, *93*, 112214. [CrossRef]
48. Yang, X.; Wang, H.; Li, M.; Li, Y.; Li, C.; Zhang, Y.; Chen, S.; Shen, H.; Qian, F.; Feng, X.; et al. Experimental Study on Thermal Runaway Behavior of Lithium-Ion Battery and Analysis of Combustible Limit of Gas Production. *Batteries* **2022**, *8*, 250. [CrossRef]
49. Zhang, F.; Feng, X.; Xu, C.; Jiang, F.; Ouyang, M. Thermal runaway front in failure propagation of long-shape lithium-ion battery. *Int. J. Heat Mass Transf.* **2021**, *182*, 121928. [CrossRef]
50. Ma, Z.; Huo, Q.; Wang, W.; Zhang, T. Voltage-temperature aware thermal runaway alarming framework for electric vehicles via deep learning with attention mechanism in time-frequency domain. *Energy* **2023**, *278*, 127747. [CrossRef]
51. Zhang, Q.; Liu, T.; Wang, Q. Experimental study on the influence of different heating methods on thermal runaway of lithium-ion battery. *J. Energy Storage* **2021**, *42*, 103063. [CrossRef]
52. Liu, P.; Li, Y.; Mao, B.; Chen, M.; Huang, Z.; Wang, Q. Experimental study on thermal runaway and fire behaviors of large format lithium iron phosphate battery. *Appl. Therm. Eng.* **2021**, *192*, 116949. [CrossRef]
53. Wang, G.; Ping, P.; Zhang, Y.; Zhao, H.; Lv, H.; Gao, X.; Gao, W.; Kong, D. Modeling thermal runaway propagation of lithium-ion batteries under impacts of ceiling jet fire. *Process Saf. Environ. Prot.* **2023**, *175*, 524–540. [CrossRef]
54. Zhao, R.; Lai, Z.; Li, W.; Ye, M.; Yu, S. Development of a coupled model of heat generation and jet flow of lithium-ion batteries during thermal runaway. *J. Energy Storage* **2023**, *63*, 107048. [CrossRef]
55. Zhou, W.; Zhao, H.; Wu, D.; Yang, Y.; Yuan, C.; Li, G. Inhibition effect of inert gas jet on gas and hybrid explosions caused by thermal runaway of lithium-ion battery. *J. Loss Prev. Process Ind.* **2024**, *90*, 105336. [CrossRef]
56. Zhou, Z.; Zhou, X.; Wang, D.; Li, M.; Wang, B.; Yang, L.; Cao, B. Experimental analysis of lengthwise/transversal thermal characteristics and jet flow of large-format prismatic lithium-ion battery. *Appl. Therm. Eng.* **2021**, *195*, 117244. [CrossRef]
57. Zou, K.; Chen, X.; Ding, Z.; Gu, J.; Lu, S. Jet behavior of prismatic lithium-ion batteries during thermal runaway. *Appl. Therm. Eng.* **2020**, *179*, 115745. [CrossRef]
58. Golubkov, A.W.; Planteu, R.; Krohn, P.; Rasch, B.; Brunnsteiner, B.; Thaler, A.; Hacker, V. Thermal runaway of large automotive Li-ion batteries. *RSC Adv.* **2018**, *8*, 40172–40186. [CrossRef]
59. Su, T.; Lyu, N.; Zhao, Z.; Wang, H.; Jin, Y. Safety warning of lithium-ion battery energy storage station via venting acoustic signal detection for grid application. *J. Energy Storage* **2021**, *38*, 102498. [CrossRef]
60. Kong, D.; Lv, H.; Ping, P.; Wang, G. A review of early warning methods of thermal runaway of lithium ion batteries. *J. Energy Storage* **2023**, *64*, 107073. [CrossRef]
61. Nishant Ranjan Sinha, K.; Ranjan, D.; Raza, M.Q.; Kumar, N.; Kaner, S.; Thakur, A.; Raj, R. In-situ acoustic detection of critical heat flux for controlling thermal runaway in boiling systems. *Int. J. Heat Mass Transf.* **2019**, *138*, 135–143. [CrossRef]
62. Wang, Z.; Zhu, L.; Liu, J.; Wang, J.; Yan, W. Gas Sensing Technology for the Detection and Early Warning of Battery Thermal Runaway: A Review. *Energy Fuels* **2022**, *36*, 6038–6057. [CrossRef]
63. Chen, S.; Wang, Z.; Yan, W.; Liu, J. Investigation of impact pressure during thermal runaway of lithium ion battery in a semi-closed space. *Appl. Therm. Eng.* **2020**, *175*, 115429. [CrossRef]
64. Dubaniewicz, T.H.; Zlochower, I.; Barone, T.; Thomas, R.; Yuan, L. Thermal Runaway Pressures of Iron Phosphate Lithium-Ion Cells as a Function of Free Space within Sealed Enclosures. *Min. Metall. Explor.* **2020**, *38*, 539–547. [CrossRef]
65. Thomas, H.D.; Teresa, L.B.; Connor, B.B.; Richard, A.T. Comparison of thermal runaway pressures within sealed enclosures for nickel manganese cobalt and iron phosphate cathode lithium-ion cells. *J. Loss Prev. Process Ind.* **2022**, *76*, 104739. [CrossRef]
66. Xu, L.; Wang, S.; Li, Y.; Li, Y.; Sun, J.; Zhao, F.; Wang, H.; Wang, Y.; Xu, C.; Feng, X. Thermal runaway propagation behavior and gas production characteristics of NCM622 battery modules at different state of charge. *Process Saf. Environ. Prot.* **2024**, *185*, 267–276. [CrossRef]
67. Zhang, Y.; Wang, H.; Li, W.; Li, C.; Ouyang, M. Quantitative analysis of eruption process of abused prismatic Ni-rich automotive batteries based on in-chamber pressure. *J. Energy Storage* **2020**, *31*, 101617. [CrossRef]
68. Wang, H.; Zhang, Y.; Li, W.; Gao, Z.; Zhang, B.; Ouyang, M. Experimental study on the cell-jet temperatures of abused pris-matic Ni-rich automotive batteries under medium and high states of charge. *Appl. Therm. Eng.* **2022**, *202*, 117859. [CrossRef]
69. Li, W.; Wang, H.; Zhang, Y.; Ouyang, M. Flammability characteristics of the battery vent gas: A case of NCA and LFP lithium-ion batteries during external heating abuse. *J. Energy Storage* **2019**, *24*, 100775. [CrossRef]

70. Gong, Z.; Gu, C.; Sun, J.; Wang, H.; Li, Y.; Zhou, X.; Jia, Y.; Han, D. Experimental study on thermal runaway characteristic and residue of $Li(Ni_{0.8}Co_{0.1}Mn_{0.1})O_2$ lithium-ion batteries induced by overcharge. *J. Energy Storage* **2023**, *68*, 107705. [CrossRef]
71. Wei, N.; Li, M. Experimental study of thermal runaway process of 256Ah prismatic nickel-rich battery. *Front. Energy Res.* **2023**, *11*, 1230429. [CrossRef]

batteries

MDPI

Article

Experimental Research on Thermal-Venting Characteristics of the Failure 280 Ah $LiFePO_4$ Battery: Atmospheric Pressure Impacts and Safety Assessment

Yu Wang [1], Yan Wang [1,*], Jingyuan Zhao [2], Hongxu Li [3], Chengshan Xu [3], Yalun Li [3], Hewu Wang [3,*], Languang Lu [3], Feng Dai [4], Ruiguang Yu [1] and Feng Qian [5]

[1] School of Mechanical and Automotive Engineering, Qingdao University of Technology, Qingdao 266520, China; w17660239776@126.com (Y.W.); ruiguang.yu@bjtu.edu.cn (R.Y.)
[2] Institute of Transportation Studies, University of California Davis, Davis, CA 95616, USA; jyzhao@ucdavis.edu
[3] State Key Laboratory of Automotive Safety and Energy, Tsinghua University, Beijing 100084, China; ooaxus@163.com (H.L.); xcs_pcg@mail.tsinghua.edu.cn (C.X.); liyalun@tsinghua.edu.cn (Y.L.); lulg@tsinghua.edu.cn (L.L.)
[4] Sichuan New Energy Vehicle Innovation Center Co., Ltd., Yibin 644005, China; df12366784@163.com
[5] School of Mechanical Engineering, Dalian Jiaotong University, Dalian 116028, China; 13842821050@163.com
* Correspondence: wang_yan@qut.edu.cn (Y.W.); wanghw@tsinghua.edu.cn (H.W.)

Abstract: With the widespread application of lithium-ion batteries (LIBs) energy storage stations in high-altitude areas, the impact of ambient pressure on battery thermal runaway (TR) behavior and venting flow characteristics have aroused wide research attention. This paper conducts a lateral heating experiment on 280 Ah lithium iron phosphate batteries (LFPs) and proposes a method for testing battery internal pressure using an embedded pressure sensor. This paper analyzes the battery characteristic temperature, internal pressure, chamber pressure, and gas components under different chamber pressures. The experiment is carried out in a N_2 atmosphere using a 1000 L insulated chamber. At 40 kPa, the battery experiences two instances of venting, with a corresponding peak in temperature on the battery's side of 136.3 °C and 302.8 °C, and gas generation rates of 0.14 mol/s and 0.09 mol/s, respectively. The research results indicate that changes in chamber pressure significantly affect the center temperature of the battery side (T_s), the center temperature of the chamber (T_c), the opening time of the safety valve (t_{open}), the triggering time of TR (t_{TR}), the time difference (Δt), venting velocity, gas composition, and flammable limits. However, the internal pressure and gas content of the battery are apparently unaffected. Considering the TR characteristics mentioned above, a safety assessment method is proposed to evaluate the TR behavior and gas hazard of the battery. The results indicate that the risk at 40 kPa is much higher than the other three chamber pressures. This study provides theoretical references for the safe use and early warning of energy storage LIBs in high-altitude areas.

Keywords: torage; lithium-ion batteries; thermal runaway; flammable limit; gas content; six-dimensional radar chart; embedded pressure sensor

Citation: Wang, Y.; Wang, Y.; Zhao, J.; Li, H.; Xu, C.; Li, Y.; Wang, H.; Lu, L.; Dai, F.; Yu, R.; et al. Experimental Research on Thermal-Venting Characteristics of the Failure 280 Ah $LiFePO_4$ Battery: Atmospheric Pressure Impacts and Safety Assessment. *Batteries* **2024**, *10*, 270. https://doi.org/10.3390/batteries10080270

Academic Editor: Carlos Ziebert

Received: 1 July 2024
Revised: 20 July 2024
Accepted: 24 July 2024
Published: 29 July 2024

1. Introduction

Lithium iron phosphate batteries (LFPs) have lots of advantages, such as a high voltage, small size, high energy density, no memory effect, and a long lifespan [1]. They are widely used in electrochemical energy storage and other fields [2]. However, security is the bottleneck restricting the large-scale application of high-specific-energy LFPs [3].

The reliability and quality of a power supply cannot be guaranteed effectively because of the shortage of conventional energy and the difficulty of supplying the plateau area. But renewable energy sources such as hydro, solar, and wind power are widely distributed. However, renewable energy is intermittent and unstable, and electrochemical energy storage power stations are the main solution. More and more energy storage stations are landing at high altitudes [4]. In the high-altitude energy storage field, accidents caused by LFPs thermal runaway (TR) are increasing, as reported in previous studies [5]. Therefore, studying the TR characteristics and mechanisms of large-capacity LFPs at low pressure, especially for high-altitude areas, is of great importance [6].

When the internal heat generation rate of lithium-ion batteries (LIBs) is higher than the heat dissipation rate, the internal chemical reaction will be aggravated [7], which causes the battery to burn and explode. In recent years, many researchers have explored the causes and characteristics of battery TR. Feng et al. [8] found mechanical abuse, thermal abuse and electrical abuse were the main causes of TR of LIBs. Tran et al. [9] discussed the reasons for the large-scale internal short circuit caused by TR, focusing on the collapse of the diaphragm, and summarized the existing methods for establishing TR models. Li et al. [10] studied the mechanism of thermal runaway after rapid charging of batteries. The results showed that the reaction between lithium plating and electrolyte was the trigger for thermal runaway. Guo et al. [11] studied the impact of the heat-transfer coefficient on the TR response. They found that an increase in the heat transfer coefficient led to a shorter heat absorption phase in the battery, resulting in a shorter triggering time for TR, and the critical temperature for triggering TR in the battery also increased. The maximum battery temperature remained 430 K when the heat-transfer coefficient was higher than 25 $W/(m^2 \cdot K)$. Xu et al. [12] studied the impact of charge and discharge rates on the TR of LIBs. The results indicated that the higher the charge rate, the higher the peak temperature of TR. When the charging rate was 1C, the peak temperature was 362.15 °C and the time was 283 s. When the charging rate was 3C, the peak temperature was 364.62 °C and the time was 284 s. Paster et al. [13] studied the differences in TR between aged and normal LIBs. The results indicated that the heat released during TR in aged batteries is significantly lower than in normal ones, due to the reduction in lithium in the anode and electrolyte. Li et al. [14] have created a model of an aluminum heat sink to study the impact of different structural arrangements on the TR of LIBs. When there is no thermal management structure, the average temperature of the battery is greater than 500 °C, and when there is a thermal management structure, the temperature of the four adjacent batteries is lower than 120 °C.

TR of LIBs produces flammable gases [15], and it is important to study the gas content and flammability for TR protection. Chen et al. [16] studied the TR gas composition and used the Le Chatelier Formula to calculate the lower flammability limit (LEL). They found that with the increase in the state of charge (SOC), the LEL showed a trend of initially rising and then declining. When the battery SOC was 60%, the LEL was the largest at 21.10%. When the battery SOC was 100%, the LEL was the smallest at 5.08%. Zhang et al. [17] used gas chromatography (GC) analysis of produced gas components, and they found that an increase in the multi-chain gas component proportion in mixed gas was the cause of the increase in the combustible limit range. Baird et al. [18] compared the flammability limits of Lithium Nickel Manganese Cobalt Oxide (NMC), LFP, Lithium Cobalt Oxide (LCO), and Nickel Cobalt Aluminum (NCA), and found that among the four types of batteries, LCO has the lowest LEL at 6.1%, while NCA has the highest LEL at 11.8%.

The venting gases of the LIBs safety valve after opening also enhances the thermal radiation [19] and the impact on the battery TR behavior, and the resulting harm to the environment cannot be ignored [20]. Zhou et al. [21] used a high-speed camera to study the gas venting behavior of LIBs during TR, and LIBs exhibited two distinct venting streams with speeds of 55 m/s and 40 m/s, respectively. Kang et al. [22] studied the thermal characteristics that led to thermal runaway of LPF batteries of different capacities under overcharge conditions. The experimental results showed that the danger caused by TR increases with the increase in capacity, and the maximum temperature of a 140 Ah battery is 476 °C, and the maximum temperature of an 86 Ah battery is 374 °C.

With the extensive application of LIBs in plateau energy storage, research on the TR characteristics and mechanism of batteries under low pressures is a hotspot at present [23]. Liu et al. [24] studied the effect of low-pressure environments on the TR of LIBs. Their research found that the propagation velocity of TR increased with the chamber pressure. When the chamber pressure increased from 0 atm to 1 atm, the propagation velocity increased by 170%. Li et al. [25] studied the effects of low pressure on gas content components. They found that as the pressure decreased, the content of CO increased, while the content of CO_2 decreased. It was caused by incomplete flammability under low pressure. Liu et al. [26] studied the temperature changes of LIBs under different chamber pressures and found that the peak TR temperature is lower at low pressure. Ding et al. [27] studied several key parameters, including the opening time of the safety valve (t_{open}), the triggering time of TR (t_{TR}), and the time difference (Δt). The results showed that as the chamber pressure decreased from 101 kPa to 30 kPa, t_{TR} exhibited an increasing trend. When the chamber pressure decreased from 101 kPa to 30 kPa, Δt decreased to 22 s. Sun et al. [28] used $LiNi_{0.5}Co_{0.2}Mn_{0.3}O_2$ soft-pack batteries with a capacity of 10,000 mAh to perform TR experiments in the air at different low pressures. The researchers found that as the pressure in the test chamber increased, a significant flame appeared and the temperature of the battery increased. Zhao et al. [29] introduced a hierarchical framework that takes advantage of emerging trends in practical, cloud-based artificial intelligence technologies. The core of this approach is comprehensive monitoring, early diagnosis, and risk prediction at the battery, package, and system level to improve battery safety.

There have been many studies on the thermal runaway characteristics of LFPs at low pressure, mainly from several perspectives such as gas production and temperature change [30]. The innovation of this paper is that the battery is modified and the pressure sensor is embedded to detect the battery pressure change. In order to further explore the influence of different chamber pressures on the thermal runaway of the battery in inert gases, the relative most dangerous chamber pressure is determined to provide a reference for the safe use of high-altitude or aviation LFPs. Therefore, this study conducts lateral heating experiments on a 280 Ah LIBs and proposes a method for testing internal pressure using an embedded pressure sensor. The study analyzes the battery's characteristic temperature, internal pressure and chamber pressure, gas content characteristics, gas mixture composition, and flammable limits under chamber pressures of 40 kPa, 60 kPa, 80 kPa, and 101 kPa. This research explains the mechanism of the influence of chamber pressure on battery TR. It combines with characteristic parameters, quantitatively predicts its risk, aiming to provide theoretical reference for the safe use of LFPs in high-altitude energy storage and aviation fields.

2. Thermal Runaway Experimental Equipment and Methods

2.1. Methodology and Procedure

As shown in Figure 1a, the overall experiment has four parts. Firstly, the battery is modified with an embedded pressure sensor. Nitrogen is then used to replace the air in the sealed chamber to reach different pressure. Finally, the thermal runaway experiment is conducted to obtain the experimental data. As shown in Figure 1b, the experimental chamber is cylindrical. The internal volume of the chamber is 1000 L, the highest operating temperature is 1500 K, and the maximum pressure is 5 MPa. The rear gas pump is connected to charge nitrogen. After the replacement, the chamber is left to stand for 2 h to observe whether the air tightness is good. Figure 1c shows the battery comparison before and after the experiment, and the fixture uses 2 N·m clamping force to fix the battery. As shown in Figure 1d, the data in the chamber are collected by a specific device, and the data of the battery are obtained by the data collector model HIOKI LR8431-30. As shown in Figure 1d, the model is a TRACE 1300 gas chromatograph (GC). The gas bag is used to collect the gas generated by the thermal runaway of the battery, and then the gas is sent into the GC to analyze the gas composition. This process is repeated three times. It is worth noting that many researchers use the accelerated rate calorimeter (ARC) for TR experiments, but this paper is not tested with ARC. The reason is that ARC has a pressure reduction process, the lumen pressure is difficult to change, and ARC cannot meet the needs of replacing the inert gas. Therefore, this paper is conducted in the equipment developed by itself. Qian et al. [15] also used the same equipment to study TR gas generation behavior. The data acquisition system includes sensors for temperature and pressure measurements within the chamber.

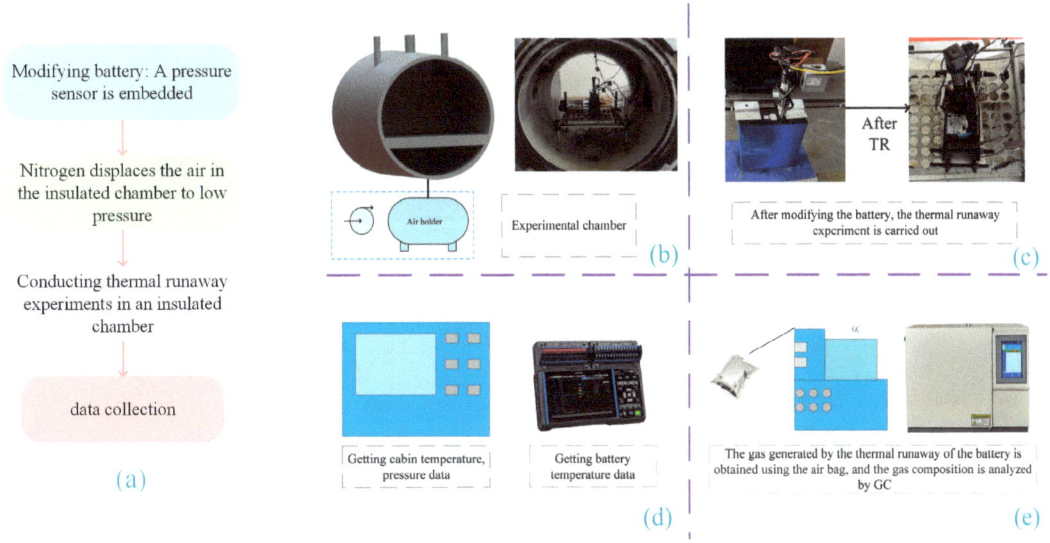

Figure 1. (**a**) Experimental flow chart. (**b**) Experimental chamber. (**c**) Battery changes after thermal runaway experiment. (**d**) Data collection. (**e**) GC analysis.

2.2. Battery with Embedded Sensor

The experimental sample utilizes capacity-type LFPs with a rated capacity of 280 Ah. The battery shell dimensions are 207 × 72 × 174 mm, with an internal volume of 0.25 L. The battery cover is equipped with a safety valve measuring 12 mm in length and 20mm in diameter. The electrolyte of the battery is $LiPF_6$, and the solvent for the electrolyte comprises ethylene carbonate (EC), methyl ethyl carbonate (EMC), and dimethyl carbonate (DMC) in a ratio of 1.9:2.8:1. Their respective evaporation temperatures are measured at 248 °C, 107 °C, and 90.1 °C, with a saturated vapor pressure of 5.54 kPa and an evaporation temperature of 151.1 °C.The battery used in this paper is provided by Xiamen Haichen New Energy Technology Co. Ltd.,Xiamen, China, as detailed in Table 1, which showcases comprehensive battery parameters.

Table 1. Basic parameters of the sample battery.

Parameters	Value
Weight	5.34 ± 0.3 kg
End of charge voltage	3.65 V
End of discharge voltage	2.5 V
Specific heat	1029.49 J/(kg·°C)

Zhang et al. [31] used the embedded sensor to obtain the internal thermal state of the battery. This paper also uses the same method. The embedded pressure sensor schematic is shown in Figure 2. Firstly, the battery is discharged to 0% SOC, and then the hole is drilled by a hand drill in the glove box. The pressure sensor is installed using hot melt adhesive. The modified battery is tested for charging and discharging to determine its capacity state. The capacity of the battery is recorded for three charge–discharge cycles. The difference between the three results and the battery capacity before the reform is within 5%, which proves that the modification method is feasible.

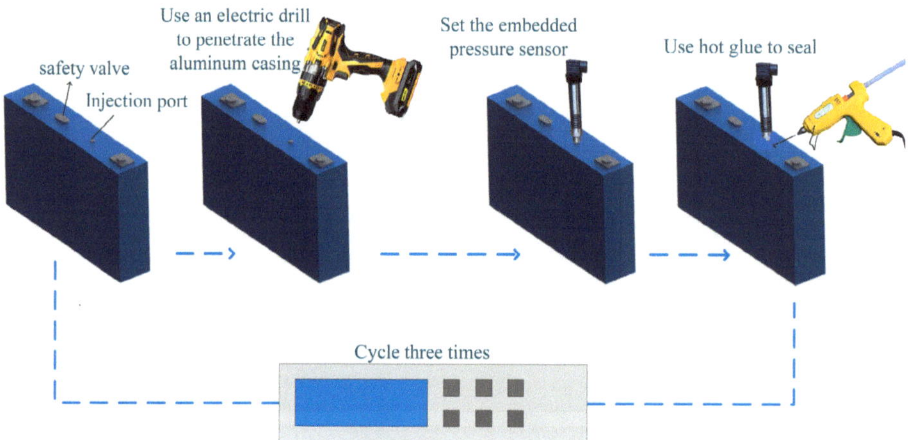

Figure 2. Battery embedded sensor method.

2.3. Thermal Runaway Experiment

In order to measure the temperature and pressure, pressure sensors and thermocouples are installed. To ensure the accuracy of the data, the thermocouples are firmly secured to the battery's surface using insulation tape. The sensor placement is shown in Figure 3. Thermocouple (T3) and pressure sensor (P2) are arranged in the center of the chamber. Thermocouples (T1) and (T2) are positioned in the center of the battery's side and on the safety valve. The embedded sensor (P1) is placed 2–3 cm below the battery's upper housing. The thermocouples are ETA GG-K-30; the pressure sensors are HM90.

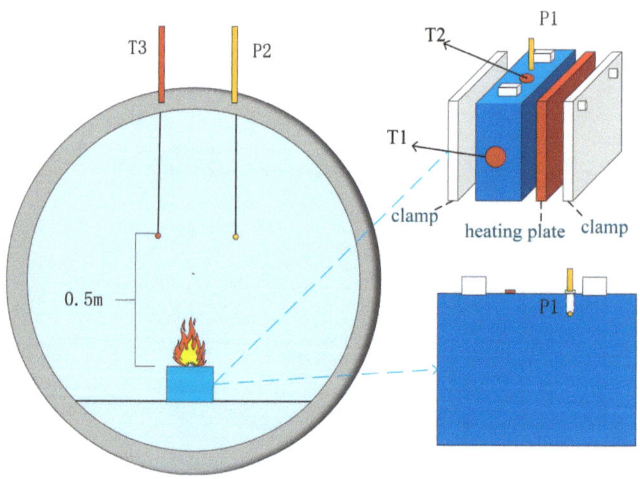

Figure 3. TR experimental diagram.

In this experiment, a front heating method was used to trigger TR of the battery. As shown in Figure 3, the heating plate is fixed to the large side of the battery using a fixture. The power of the heater is 2000 W. The heating plate and the battery are in close contact and are fixed in the chamber with a fixed splint. Then, the gas replacement of the chamber is carried out. Firstly, the vacuum pump is used to replace nitrogen to reach 40 kPa, 60 kPa, 80 kPa, and 101 kPa, respectively, to simulate the plateau environment. After holding for 1 h, the battery TR experiment can be started. The experimental data are automatically collected by the data acquisition system. The gas mixture in the chamber is collected by the airbag for GC analysis.

3. Thermal-Venting Characteristics at 40 kPa

3.1. Feature Temperature

At 40 kPa, the temperature characteristics of the safety valve (T_v), center temperature of battery side (T_s), and the center temperature of chamber (T_c) are obtained, as shown in Figure 4. Figure 4a–d show the temperature change curves of the T_v and the temperature rise rate of valves T_s and T_c, respectively.

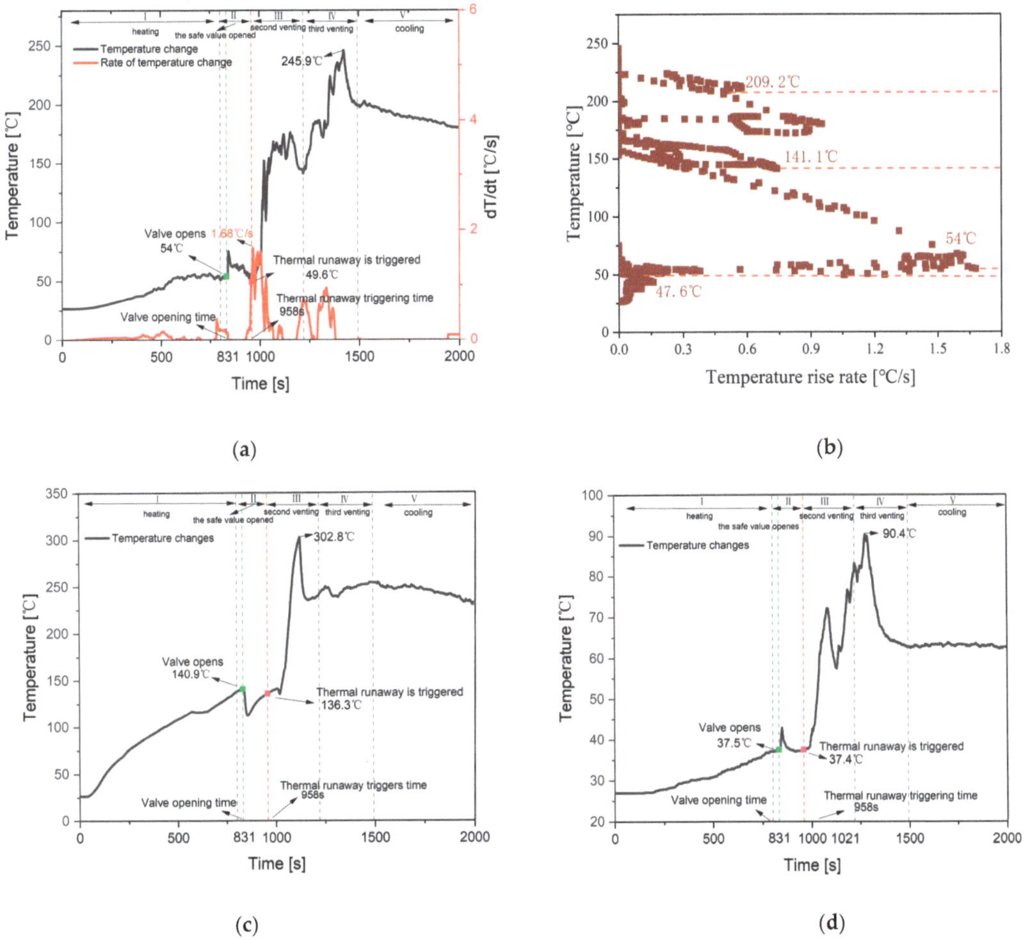

Figure 4. Feature temperatures at 40 kPa. (**a**) Battery safety valve temperature; (**b**) temperature rise rate of valve; (**c**) center temperature of battery side; (**d**) center temperature of chamber.

As shown in Figure 4a,b, at 40 kPa, the T_v shows three rapid increases. The T_c also shows a similar pattern of shift, while the T_s shows two rapid rises. Based on the temperature change pattern of the valve, the TR of the battery is divided into five stages as shown in Table 2, with the time range and typical characteristics of each stage listed in Table 2.

trode material to produce O_2 and react with the solvent to generate CO and CO_2, as shown in Equations (9)–(11).

$$Li_{1-x}FePO_4 \rightarrow (1-x)\ LiFePO_4 + x/2\ Fe_2P_2O_7 + x/4O_2 \tag{9}$$

$$5/2O_2 + C_3H_4O_3\ (EC) \rightarrow 3CO_2 + 2H_2O \tag{10}$$

$$O_2 + C_3H_4O_3\ (EC) \rightarrow 3CO + 2H_2O \tag{11}$$

Stage V: Chemical reactions weaken, and the battery enters the cooling stage.

3.2. Internal and Chamber Pressure

The internal pressure changes of the battery, the pressure inside the chamber, and the temperature and pressure difference changes of the safety valve at 40 kPa are shown in Figure 5.

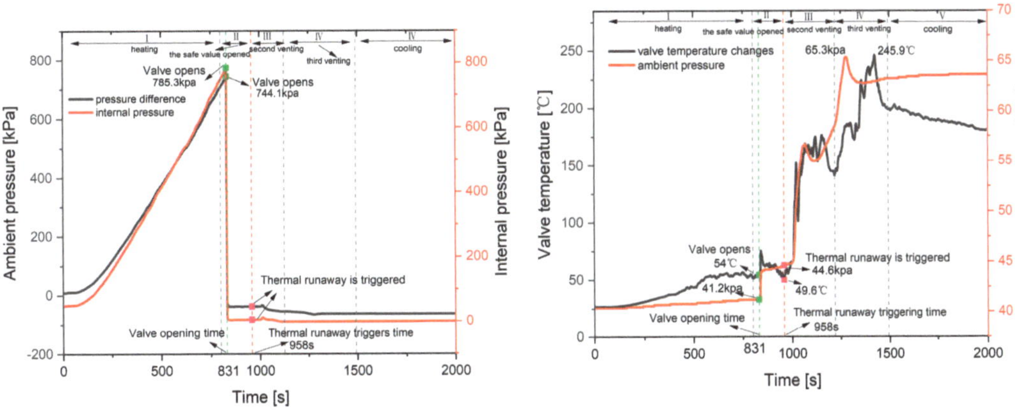

(**a**) (**b**)

Figure 5. Internal and chamber pressure. (**a**) Battery internal pressure and pressure difference; (**b**) chamber pressure and valve temperature.

From Figure 5a, it can be seen that the internal pressure of the battery exhibits a rapid increase, followed by a rapid decrease, and then stabilizes. From Figure 5b, it can be observed that the trend of pressure change in the chamber shows three rapid increases, similar to the three peak-like increases in the T_v.

Stage I: The internal pressure range of the battery is 49.4–747.7 kPa. From 0 s to 66 s, the internal pressure remains basically unchanged, and the T_v fluctuates around 26.5 °C. During this time range, the temperature is far below the decomposition temperature of the SEI membrane and the evaporation temperature of the electrolyte, indicating that the electrolyte is in a relatively stable state. From 66 s to 831 s, the internal pressure of the battery increases linearly, and this curve can be fitted into a pressure/time function expression as shown in Equation (12). Correspondingly, the range of T_s is 31.2–140.9 °C, and the SEI membrane begins to undergo a decomposition reaction.

$$y = -78.97 + 1.005x \tag{12}$$

Stage II: At 831 s, the critical pressure inside the battery before the safety valve opens is 747.7 kPa. After the valve opens, the interior of the battery is connected to the chamber, and the pressure inside the chamber is 44.1 kPa. Subsequently, the pressure slowly increases.

Stage III: At 958 s, TR is triggered, leading to a violent internal chemical reaction. The electrolyte boils, the separator melts, and the positive and negative electrodes come into contact. As the temperature rises, the internal energy of the gas increases, causing the pressure inside the chamber to rapidly rise during this stage. At 1068 s, the pressure inside the chamber reaches 56.7 kPa.

Stage IV: The pressure inside the chamber continues to rise. At 1277 s, it reaches the peak of the entire TR process, at 65.3 kPa.

Stage V: The internal chemical reactions of the battery gradually weaken, leading to a slower gas content rate. The pressure inside the chamber slowly increases. Eventually, due to the accumulated gas not being released, the pressure inside the chamber stabilizes at 63.6 kPa.

3.3. Venting Flow Characteristics

3.3.1. Component Identification

Using GC for gas component analysis, the gas composition of the 40 kPa is shown in Figure 6. From Figure 6, it can be seen that the main components of the gas produced by the battery under low pressure are H_2, CO_2, and CH_4, accounting for 46%, 26%, and 12%, respectively. Among them, hydrogen has the highest proportion.

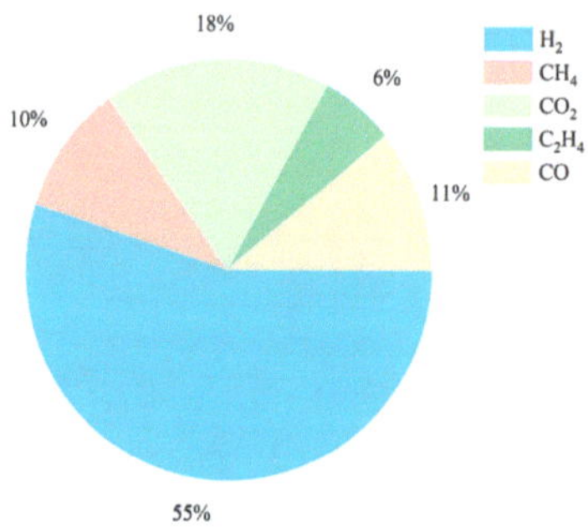

Figure 6. Gas composition distribution at 40 kPa.

3.3.2. Flammability Analysis

In order to evaluate the safety of gas generation during TR of batteries, because the mixture gas contains incombustible CO_2, the improved Le Chatelier formula is used to calculate the flammability limits of the gas mixture. The lowest and highest concentrations at which a flammable gas mixture can ignite with air are referred to as *LEL* and *UEL*, respectively [34]. The flammability limit range is obtained by subtracting the *LEL* from the *UEL*, as shown in Equations (13)–(17).

$$L_m = \frac{100}{\frac{V_1}{L_{m1}} + \frac{V_2}{L_{m2}} + \frac{V_3}{L_{m3}} + \frac{V_4}{L_{m4}}} \tag{13}$$

$$LEL = \frac{L_m\left(1 + \frac{\varnothing_D}{100-\varnothing_D}\right) \times 100}{100 + L_m\left(\frac{\varnothing_D}{100-\varnothing_D}\right)} \times 100\% \tag{14}$$

$$U_m = \frac{100}{\frac{V_1}{L_{u1}} + \frac{V_2}{L_{u2}} + \frac{V_3}{L_{u3}} + \frac{V_4}{L_{u4}}} \tag{15}$$

$$UEL = \frac{U_m\left(1 + \frac{\varnothing_D}{100-\varnothing_D}\right) \times 100}{100 + U_m\left(\frac{\varnothing_D}{100-\varnothing_D}\right)} \times 100\% \tag{16}$$

$$\text{Range} = UEL - LEL \tag{17}$$

In the formula, L_m is the lower limit of flammability without inert gases, V_{number} represents the volume proportion of different gases, L_{number} represents the lower flammability limit for different gases, and $\varnothing_D$ is the volume proportion of inert gases. U_m is the upper flammability limit excluding inert gases; U_{number} represents the upper flammability limit of different gases.

The text describes the flammability limits of various components in a mixed gas. It mentions the *LEL* as a percentage of the volume fraction of each component in the mixed gas, and the *UEL* as a percentage of the volume fraction of each component. It also introduces R as representing the flammability limit range.

The explosion upper and lower limits of each component gas are shown in Table 3. By substituting the *LEL* of each gas into Equation (14), *LEL* of the mixed gas can be obtained. By substituting into Equation (16), the *UEL* can be obtained. The calculation yields a LEL of 6.42% for the mixed gas and a *UEL* of 63%. The range is 56.6%.

Table 3. Gas flammability limits table.

Parameter	*LEL* (%)	*UEL* (%)
H_2	4	75
CO	12.5	74
CH_4	5	15
C_2H_4	2.7	36
Mixed gas	6.42	63
Flammability limit range	56.6	

3.3.3. Gas Content Calculation

In this experiment, it is assumed that the generated gas conforms to the ideal gas assumption. The mass flow rate, volume flow rate, and mass change characteristics of the gas produced by the battery can be calculated based on the differential form of the ideal gas state equation. Before the safety valve opens, the change in gas content inside the battery can be calculated based on the internal gas pressure characteristics of the battery, as in Equation (18). After the safety valve opens, the change in gas from the battery venting can be calculated based on the pressure inside the chamber, as in Equation (19).

$$\frac{dn_1}{dt} = \frac{\frac{dP_a}{dt}V_1}{R\frac{dT_a}{dt}} \tag{18}$$

$$\frac{dn_2}{dt} = \frac{\frac{dP_b}{dt}V_2}{R\frac{dT_b}{dt}} \tag{19}$$

The total gas content is obtained by integrating the rate of change of gas content after the valve is opened. During the time period t_1, the gas produced internally is of equal mass.

$$n = \int_{t_1}^{t_2} \frac{dn_2}{dt} dt \tag{20}$$

In the equation, n_1 is the gas content before valve opening, n_2 is the gas content after valve opening, T_a is the temperature before valve opening, T_b is the temperature after valve opening, P_a is the internal gas pressure of the battery, P_b is the gas pressure inside the flammability chamber, V_1 is the internal volume of the battery, V_2 is the volume of the chamber, and R is the gas constant, which is 8.314 J/mol·K [35].

After calculating the gas generation, the quality can be calculated using the following Formulas (21)–(24).

$$\frac{dm_1}{dt} = \frac{dn_1}{dt} M_0 \tag{21}$$

$$\frac{dm_2}{dt} = \frac{dn_2}{dt} M_0 \tag{22}$$

$$m = \int_{t_1}^{t_2} \frac{dm_2}{dt} dt \tag{23}$$

$$M_0 = p_1 * M_1 + p_2 * M_2 + p_n * M_n \tag{24}$$

In the equation: m_1 is the mass before the valve is opened, m_2 is the mass after the valve is opened, M_0 is the molar mass equivalent, and the calculated value of M_0 is 10.6 g/mol. p% represents the percentage composition of each gas and M_{number} represents the molar mass of a specific gas.

The volume can be calculated using the following formula.

$$V = \frac{\int_{t_1}^{t_2} \frac{dV_2}{dt} dt}{\rho_0} \tag{25}$$

$$\rho_0 = p_1 * \rho_1 + p_2 * \rho_2 + \ldots p_n * \rho_n \tag{26}$$

In the equation, the equivalent density can be calculated as 0.439 g/m^3, where p% represents the proportion of each gas (as shown in Figure 6), and is the density of a certain gas.

3.3.4. Venting Flow Characteristics

According to theoretical calculations, the cumulative gas content, cumulative mass, and cumulative volume of the battery over time at 40 kPa are shown in Figure 7. The blue area in the figure represents the gas content calculated using the internal pressure, while the red area represents the gas content calculated using the internal pressure of the chamber after the safety valve is opened.

From Figure 7, it can be seen that the gas content and the mass volume both show a trend of rapid increase and fluctuation. The reaction rate has three peaks corresponding to the three stages of TR.

Stage I: The battery is in the heating stage, maintaining a relatively stable state. At this time, the safety valve is not open and the gas content is minimal. At 800 s, the accumulated gas content is 0.067 mol, the accumulated mass is 0.79 g, and the accumulated volume is 1.62 L.

Stage II: At 831 s, the safety valve opens, and high-temperature and high-pressure gas inside the battery is ejected into the chamber. The rate of change of gas content reaches the first peak, with a rate of 0.09 mol/s, and the cumulative gas content is 0.93 mol. The cumulative gas volume and cumulative gas mass change trends are the same as the gas content, at 10.9 g and 22.4 L, respectively.

Stage III: At 958 s, the battery experienced TR, triggering a vigorous internal chemical reaction and a rapid increase in gas content. At 1008 s, the rate of change in gas content reached a second peak of 1.87 L/s. At this point, the cumulative gas content was 4.07 moles, with a cumulative gas mass of 48.06 g and a cumulative gas volume of 98.5 L. Subsequently, the cumulative gas content fluctuated around 4 moles due to convective heat exchange between gases at different temperatures inside the chamber, causing fluctuations in temperature and pressure, which in turn affected the cumulative gas content. As the convective heat exchange between hot and cold air inside the chamber ceased, the noticeable fluctuations disappeared, and the gas content continued to increase.

Stage IV: As the TR reaction of the battery continues, the rate of change in cumulative gas content reaches a third peak at 1224 s. At this time, the cumulative gas content is 6.61 mol, with a mass of 78.17 g and a volume of 160 L.

Stage V: The battery enters the cooling stage, and the chemical reaction basically stops. The gas content behavior of the battery stops at 1493 s. The final cumulative gas content is 6.76 mol, with a cumulative gas mass of 79.8 g and a cumulative gas volume of 163 L.

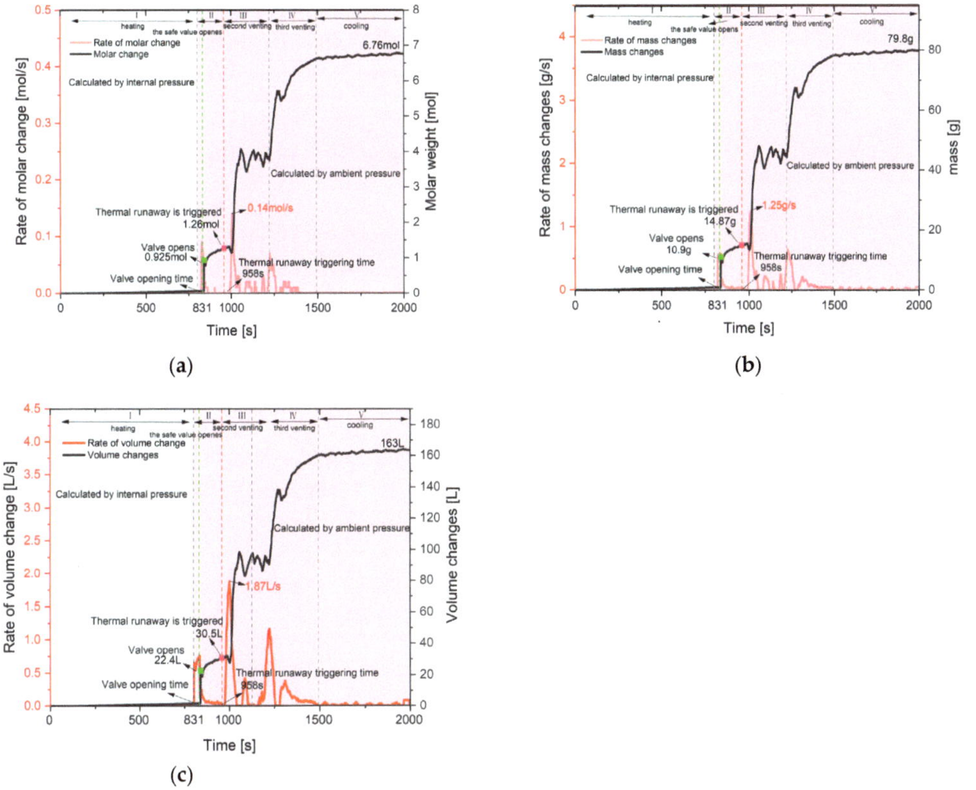

Figure 7. Gas venting characteristics at 40 kPa. (**a**) Molar of gas; (**b**) mass of gas; (**c**) volume of gas.

4. Pressure Effect on Thermal Behavior

4.1. Feature Temperature

The paper explores the influence mechanism of different chamber pressures on the failure characteristics of battery. The inert gas pressure inside the chamber is controlled at 40 kPa, 60 kPa, 80 kPa, and 101 kPa, respectively. The T_s and the peak temperature under different pressure environments are shown in Figure 8.

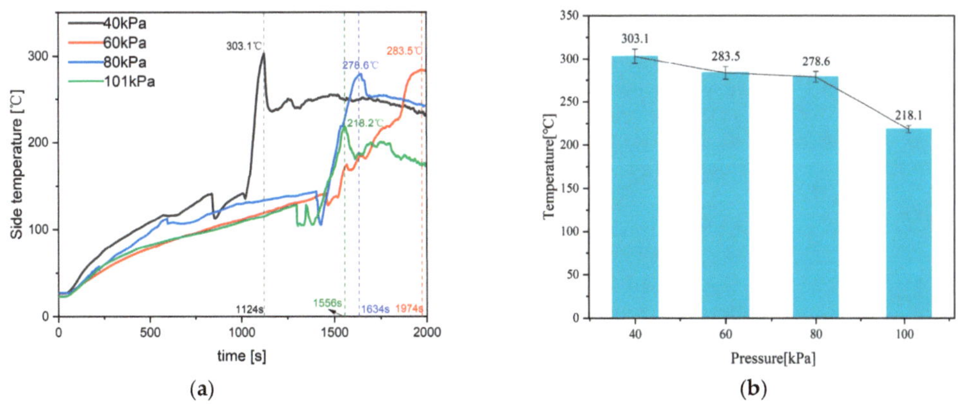

Figure 8. Side temperature comparison. (**a**) Side center temperature; (**b**) peak temperature.

As shown in Figure 8a, at 40 kPa, the peak temperature of T_s reached 303.1 °C at 1124 s. At 101 kPa, the peak of T_s reached 218.1 °C at 1556 s. At 80 kPa, the peak of T_s reached 278.6 °C at 1634 s. At 60 kPa, the peak of T_s reached 283.5 °C at 1974 s. The time to reach peak temperature is in the order of 40 kPa < 101 kPa < 80 kPa < 60 kPa, and the peak temperature is in the order of 40 kPa > 60 kPa > 80 kPa > 101 kPa. The peak temperatures for the four pressures are shown in Figure 8b, and the peak temperature decreases with increasing chamber pressure.

At 40 kPa, T_s reaches its peak faster compared to the other three pressures. TR is triggered earliest at 40 kPa, leading to a quicker generation of a large amount of heat. T_s increases as the pressure decreases because of the increase in pressure inside the chamber, where more inert gas is erupting into the chamber, effectively diluting the concentration of reaction gas, suppressing TR, and correspondingly exhibiting the suppression of T_s rise.

The T_c under four different pressures is shown in Figure 9. At 40 kPa, T_c first reaches its peak at 1276 s, reaching 90.4 °C. At 101 kPa, T_c first reaches its peak at 1392 s, reaching 103.6 °C. At 80 kPa, T_c first reaches its peak at 1567 s, reaching 103.1 °C. At the 60 kPa, T_c first reaches its peak at 1862 s, reaching 92.5 °C. The order of peak temperature attainment time is 40 kPa < 101 kPa < 80 kPa < 60 kPa, and the order of peak temperature from low to high is 40 kPa < 60 kPa < 80 kPa < 101 kPa. The peak temperatures at four pressures are shown in Figure 9b. As the pressure increases, the maximum T_c continues to increase.

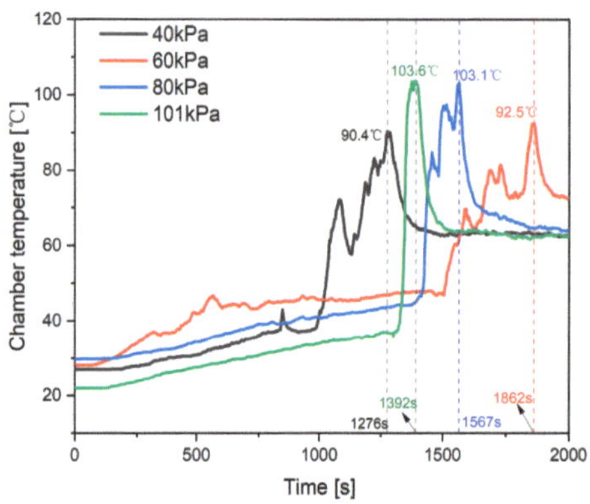

(a)

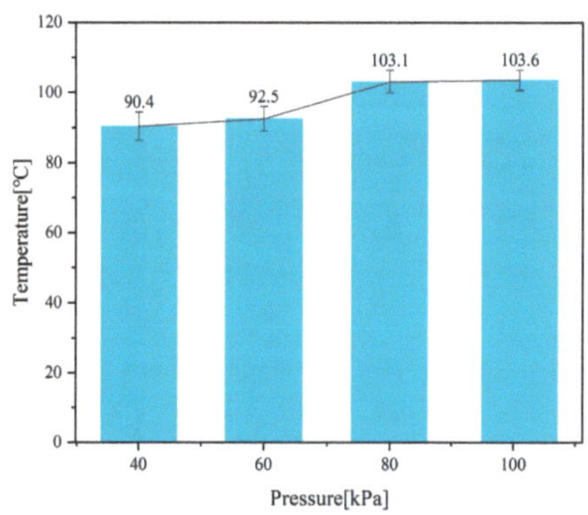

(b)

Figure 9. Comparison of the central temperature in chamber. (**a**) Temperature in center of chamber; (**b**) peak temperature.

The temperature trends under four different pressures are the same, and the peak of T_c increases at higher pressure. However, the T_s changes shown in Figure 8 decrease with increasing pressure. The reason for this difference is that with higher pressure, the gas heat transfer coefficient is higher, making it easier for the temperature inside the chamber to rise.

4.2. Internal and Chamber Pressure

This study replaces the air with nitrogen and maintains 40 kPa, 60 kPa, 80 kPa, and 101 kPa to conduct TR experiments. The curves of the internal pressure change of the battery are shown in Figure 10.

As shown in Figure 10a, the curve represents the internal pressure changes in the battery. Due to the opening of the safety valve, the trend is an initial increase followed by a decrease. At 40 kPa, the safety valve opens at 831 s, corresponding to a peak pressure of 725.3 kPa. At 101 kPa, the safety valve opens at 1301 s, with a peak pressure of 726.6 kPa. At 80 kPa, the safety valve opens at 1402 s, with a peak pressure of 739.2 kPa. At 60 kPa, the safety valve opens at 1457 s, with a peak pressure of 729.3 kPa. The peak pressures are ranked from high to low as 80 kPa > 60 kPa > 101 kPa > 40 kPa. It is worth noting that at 40 kPa, the internal pressure of the battery increases the fastest.

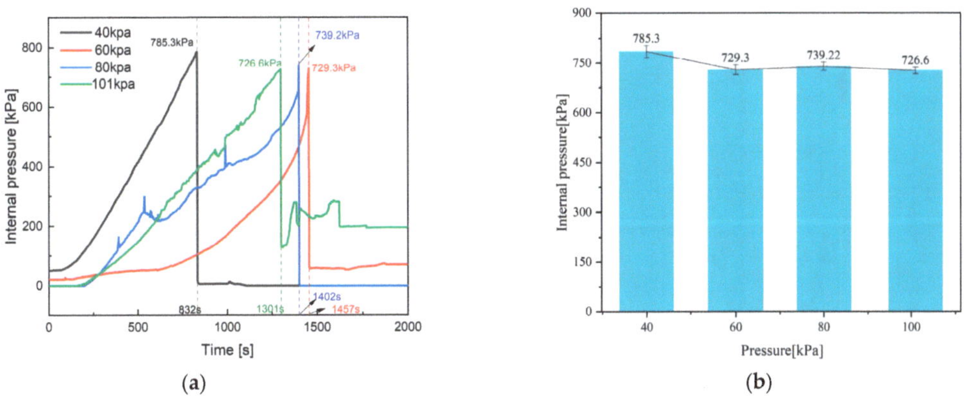

Figure 10. The internal pressure. (**a**) Internal pressure change; (**b**) peak internal pressure.

As shown in Figure 10b, the peak internal pressure of the battery under four different chamber pressures is similar. The difference is only 0.8%.

The pressure change curve inside the chamber is shown in Figure 11a. At the moment when the battery safety valve is opened, a large amount of gas enters the chamber, causing a rapid increase in pressure inside the chamber. At 40 kPa, the peak pressure is 65.3 kPa, at 60 kPa, the peak pressure is 88 kPa. At 80 kPa, the peak pressure is 116.9 kPa, and at 101 kPa, the peak pressure is 147.4 kPa. The peak pressures are sorted in ascending order as 40 kPa < 60 kPa < 80 kPa < 101 kPa.

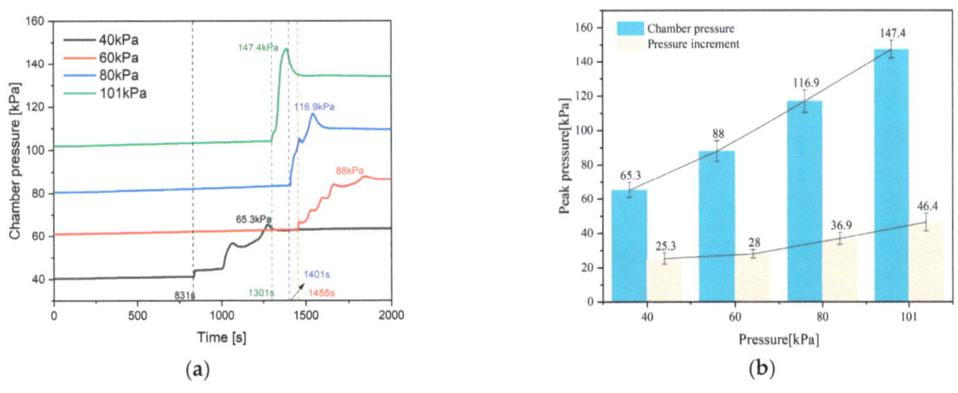

Figure 11. The chamber pressure. (**a**) Pressure changes in chamber; (**b**) peak pressure changes.

As shown in Figure 11b, the internal pressure and the pressure increment inside the chamber increase linearly with the increase in chamber pressure. The main reason for this is that the temperature inside the chamber increases with the increase in pressure, causing the gas inside the chamber to expand and the pressure to increase.

The change in pressure difference is shown in Figure 12a, and the variation is similar to that in Figure 10a. As shown in Figure 12b, as the pressure increases, the maximum pressure difference decreases slightly because the peak pressure inside the battery at different pressures is not significantly different. Subtracting the four increasing chamber pressure, the calculated maximum pressure difference shows a slightly decreasing trend.

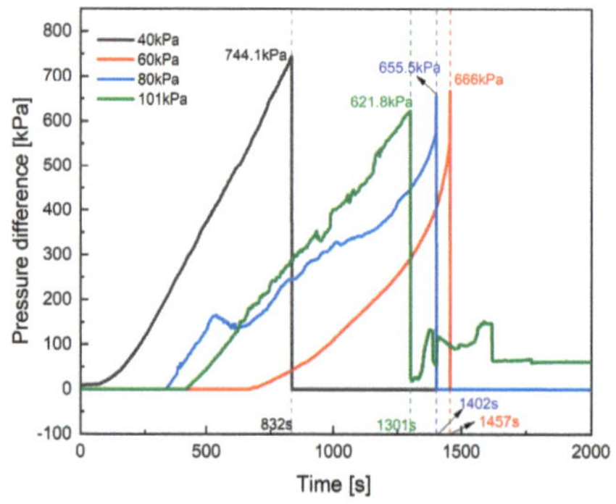

(**a**)

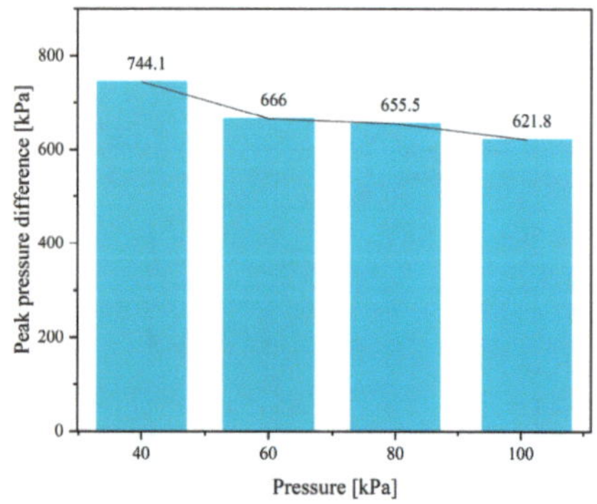

(**b**)

Figure 12. The pressure difference. (**a**) Pressure difference; (**b**) Peak pressure difference.

In summary, at four pressures, the internal pressure and maximum pressure difference in the battery do not differ significantly, which is similar to the research conclusion of Wang [36] and others. However, the pressure inside the chamber increases with the increase in pressure due to the rise in temperature.

4.3. Feature Time

The key time points for characterizing battery failure behavior include t_{TR}, t_{open}, and Δt. The impacts of these key times at four different pressures are shown in Figure 13.

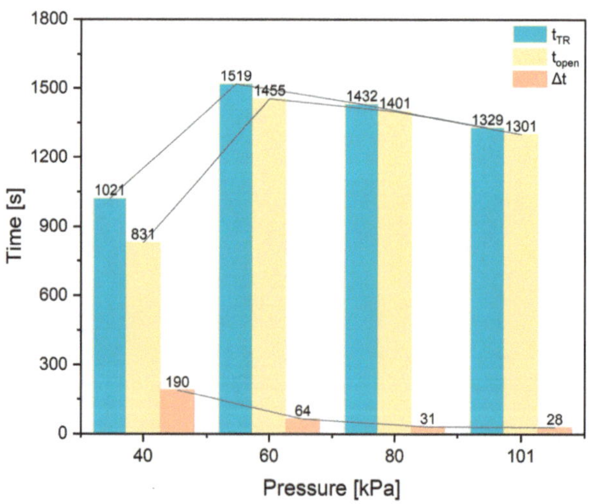

Figure 13. The key time of TR.

From the above Figure 13, at 40 kPa, t_{open} is 831 s, t_{TR} is 1021 s, and the difference is 190 s. At 60 kPa, t_{open} is 1455 s, t_{TR} is 1519 s, and the difference is 64 s. At 80 kPa, t_{open} is 1401 s, t_{TR} is 1432 s, and the difference is 31 s. At 101 kPa, t_{open} is 1301 s, t_{TR} is 1329 s, and the difference is 28 s. Sorting t_{TR} from low to high, we have 40 kPa < 101 kPa < 80 kPa < 60 kPa, and sorting t_{open} from low to high is 40 kPa < 101 kPa < 80 kPa < 60 kPa. Sorting Δt from low to high is 101 kPa < 80 kPa < 60 kPa < 40 kPa.

In comparison to 40 kPa, t_{open} and t_{TR} show a big increase, while the differences are not significant at the other three pressures. This indicates that t_{TR} and t_{open} will significantly advance when the chamber pressure is below a certain critical value because the chamber pressure of the battery after modification is 101 kPa, and the initial pressure difference in the 40 kPa environment is 61 kPa, both of which are greater than the initial pressure differences under the other three conditions. Therefore, reaching the safety valve pressure difference early and opening the valve earlier under 40 kPa. The TR after the electrolyte and gas discharge is triggered earlier.

The time difference decreases with the increase in pressure, and the time difference is particularly large at 40 kPa. There are two main reasons for this. Firstly, at low pressure, the safety valve opens early, leading to incomplete chemical reactions, requiring a longer time to reach TR. Secondly, at high pressure, it helps molecules to overcome activation energy, accelerating the rate of chemical reactions, and reaching the TR state faster after the valve opens.

5. Pressure Effect on Venting Behavior

5.1. Gas Content

Figure 14 shows the variation rate of gas generation during battery TR at different pressures, the accumulation characteristics over time, and the maximum gas generation.

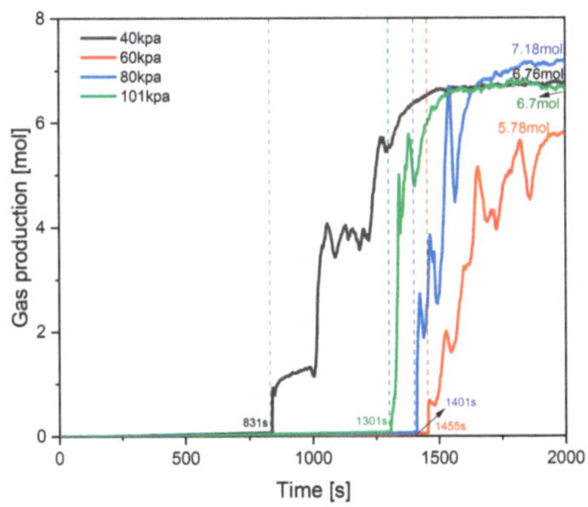

(**a**) Gas content changes.

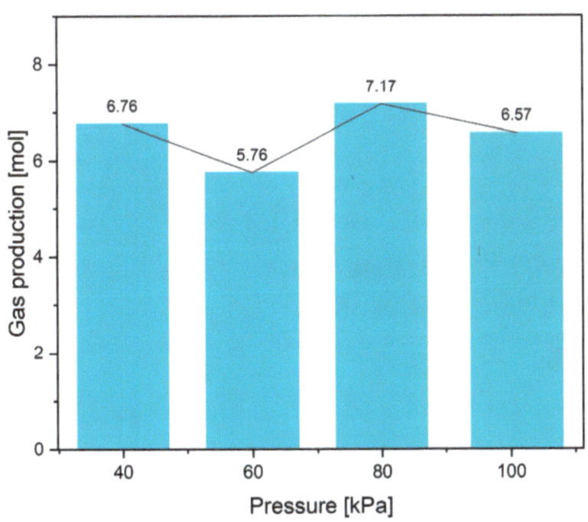

(**b**) Gas content.

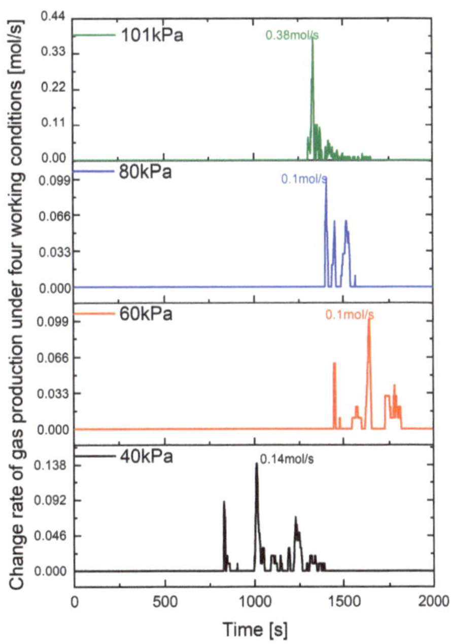

(**c**) Gas content.

Figure 14. Comparison of gas content.

According to Figure 14a, it can be seen that the cumulative gas content at 40 kPa, 60 kPa, and 80 kPa shows a trend of rapid increase three times, while at 101 kPa, it shows a rapid increase only once, which is significantly different from the other three pressures. The main reason is that at 101 kPa, the time taken for the safety valve to open and trigger TR is the shortest, and the chemical reaction rate is faster, leading to only one rapid increase. For the other three pressures, the first rapid increase in cumulative gas content occurs when the safety valve opens, and the remaining two occur after TR.

According to Figure 14b, the cumulative gas content at 40 kPa is 6.76 mol, at 60 kPa is 5.76 mol, at 80 kPa is 7.17 mol, and at 101 kPa is 6.57 mol. The cumulative gas content in ascending order is 60 kPa < 101 kPa < 40 kPa < 80 kPa. Pressure has little effect on gas production during TR.

From Figure 14c, it can be seen that there are three distinct peaks in the images of the three pressures, except for 101 kPa, which only has one distinct peak. This corresponds to the three rapid increases shown in Figure 14a. Table 4 shows the numerical values and times of the peak rate of change for each pressure.

Table 4. Change rate of the gas content.

Parameter		40	60	80	101
The first peak	Value (mol/s)	0.09	0.06	0.1	0.38
	Time (s)	831	1456	1411	1336
The second peak	Value (mol/s)	0.14	0.1	0.06	\
	Time (s)	1012	1638	1455	\
The third peak	Value (mol/s)	0.07	0.04	0.06	\
	Time (s)	1228	1778	1520	\

According to Table 4, the cumulative gas content rate exhibits a peak sequence of 101 kPa > 40 kPa > 80 kPa = 60 kPa. The highest gas content rate is observed at 101 kPa, and is primarily attributed to the enhanced likelihood of molecular overcoming of activation energy under high pressure, resulting in an accelerated chemical reaction rate.

5.2. Gas Composition and Proportion

In this study, we conducted four experiments, collected gas samples using gas sampling bags, and analyzed the components using GC. The gas composition and ratios under four different pressures are shown in Figure 15.

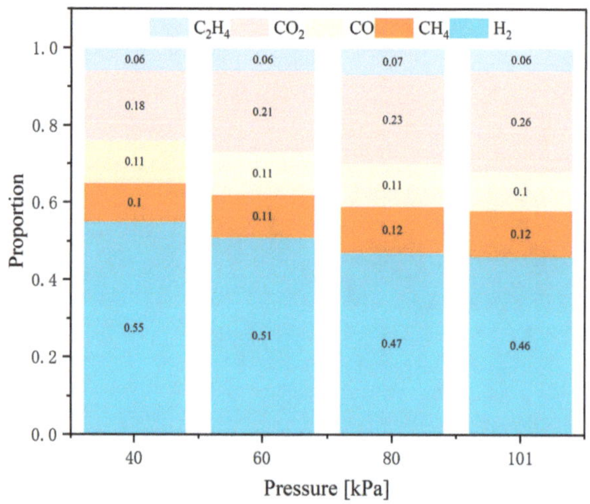

Figure 15. The proportion of gases at different pressures.

According to Figure 15, at 40 kPa, the main components of the mixed gas are hydrogen, methane, carbon monoxide, carbon dioxide, and ethylene, with proportions of 55:10:11:18:6. At 60 kPa, the proportions are 51:11:11:21:6. At 80 kPa, the proportions are 47:12:11:23:7. At 101 kPa, the proportions are 23:6:5:13:3.

At 40 kPa, the proportion of hydrogen in the gas mixture is 0.55. At 60 kPa, 80 kPa, and 101 kPa, the proportions of hydrogen are 0.51, 0.47, and 0.46, respectively. The proportion of hydrogen in the gas mixture is significantly higher at 40 kPa compared to the other three pressures, while the differences in hydrogen proportions are not significant among the remaining three pressures. The main source of hydrogen is the reaction between binder and lithium. At high temperatures, the graphite particles on the anode fall off, bringing the lithium metal into direct contact with the binder, producing hydrogen. Additionally, the binder material is polyvinylidene fluoride (PVDF) and carboxy methyl cellulose (CMC), as shown in Equations (27) and (28).

$$\mathrm{PVDF} + \mathrm{Li} \rightarrow \mathrm{LiF} + -\mathrm{CH} = \mathrm{CF} - + \frac{1}{2}\mathrm{H_2} \tag{27}$$

$$\mathrm{CMC} + \mathrm{Li} \rightarrow \mathrm{CMC} - \mathrm{OLi} + \frac{1}{2}\mathrm{H_2} \tag{28}$$

There is little difference in methane production under different pressures. Methane is generated as DMC is reduced by hydrogen to produce methane, as shown in Equations (29) and (30).

$$\mathrm{DMC} + \mathrm{Li^+} + \mathrm{e^-} + \frac{1}{2}\mathrm{H_2} \rightarrow \mathrm{CH_3OCO_2Li} + 2\mathrm{CH_4} \tag{29}$$

$$\mathrm{DMC} + 2\mathrm{Li^+} + 2\mathrm{e^-} + \mathrm{H_2} \rightarrow \mathrm{Li_2CO_3} + 2\mathrm{CH_4} \tag{30}$$

The proportion of carbon dioxide increases with increasing pressure, and the main source of carbon dioxide is the reaction of oxygen released from the electrolyte and cathode material, as shown in Equations (29)–(31).

$$2.5\mathrm{O_2} + \mathrm{C_3H_4O_3} \rightarrow 3\mathrm{CO_2} + 2\mathrm{H_2O} \tag{31}$$

$$3\mathrm{O_2} + \mathrm{C_3H_6O_3} \rightarrow 3\mathrm{CO_2} + 3\mathrm{H_2O} \tag{32}$$

$$5\mathrm{O_2} + \mathrm{C_4H_8O_2} \rightarrow 4\mathrm{CO_2} + 4\mathrm{H_2O} \tag{33}$$

5.3. Flammability Characteristic

LEL refers to the lowest concentration at which a combustible mixture can explode. Due to the insufficient concentration of flammability, the cooling effect of excess air prevents the spread of the flame, so it neither explodes nor catches fire at levels below LEL. Therefore, the lower the LEL, the more dangerous, and the larger the flammability limit range, the more dangerous [37]. Figure 16 shows the changes in the LEL, UEL, and flammability limit range under four pressures.

According to Figure 16a, the LEL at 40kPa is 6.42% and the UEL is 63%. At 60kPa, the LEL is 6.94%, and the UEL is 63.5%. At 80kPa, the LEL is 7.28% and the UEL is 63.6%. At 101 kPa, the LEL is 7.85% and the UEL is 64.5%. The UEL values are sorted in ascending order as 40 kPa < 60 kPa < 80 kPa < 101 kPa, and the LEL values are sorted in ascending order as 40 kPa < 60 kPa < 80 kPa < 101 kPa. As shown in Figure 16b, at different chamber pressures, the flammability limit range basically does not change.

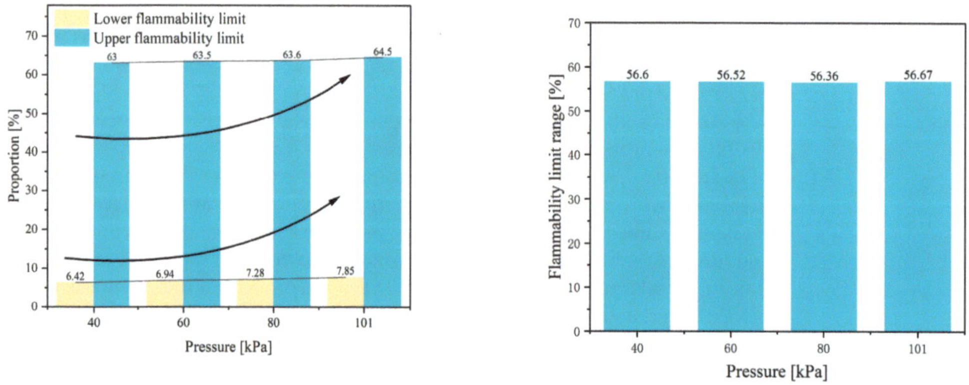

(a) (b)

Figure 16. Flammability limit of mixture. (**a**) UEL and LEL; (**b**) flammability limit range.

6. Safety Assessment

6.1. Methdology

The above research shows that the main parameters for characterizing battery safety include T_s, T_c, internal pressure of the battery, gas content, LEL, and TR trigger time. In order to comprehensively evaluate the risk of battery TR under four pressures, this paper refers to the maximum benchmark method to evaluate each characteristic. The evaluation method is shown in Equation (32). The scores of a characteristic under a certain pressure condition are calculated by dividing the value of the characteristic at this operating condition by the maximum value of the characteristic under the four operating conditions, and then multiplying by 100.

$$\text{Scores} = (\text{Value}/\text{Max value}) \times 100 \tag{34}$$

In terms of the triggering time for TR, a shorter triggering time indicates a higher level of danger. Therefore, the method described in Equation (33) is introduced for quantification.

$$\text{Time Scores} = 100 - [(\text{Value} - \text{Min value})/\text{Min value} \times 100] \tag{35}$$

6.2. Six Dimensional Radar Evaluation Results

As shown in Table 5, the allocation of six characteristic values under different pressure conditions is presented. The most dangerous characteristic is assigned a value of 100, with smaller numbers indicating relative safety. The average of the assigned values for the six characteristics is used to evaluate the risk of TR in a specific pressure for the battery.

Table 5. Scores of each characteristic quantity.

Parameter	40 kPa	60 kPa	80 kPa	101 kPa
Side temperature [°C]	100	93.5	91.9	72
Chamber temperature [°C]	87.3	89.3	99.5	100
Internal pressure [kPa]	100	92.9	94.1	92.5
Gas content [mol]	94.3	80.3	100	91.6
Lower flammable limit [%]	100	92	86.6	77.7
TR triggers time [s]	100	51.2	59.7	69.8
Average scores	96.9	83.2	88.6	83.9

To visually compare the TR risk under four different pressures, a hexagonal radar chart, as shown in Figure 17, is plotted for better illustration.

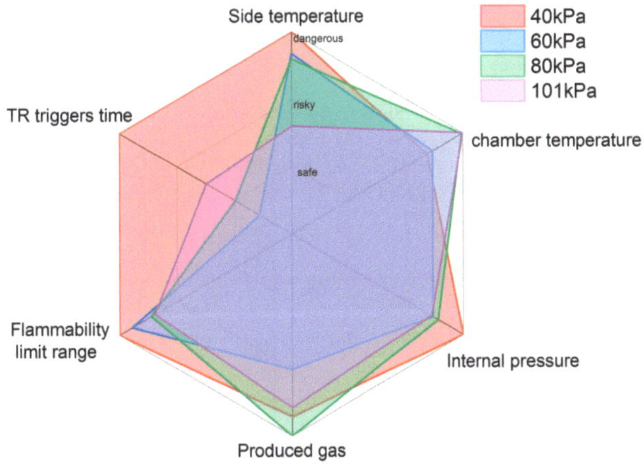

Figure 17. Comparison of radar charts in different pressures.

6.3. Sequence Analysis

Combining Table 5 and Figure 17, it can be seen that at 40 kPa, the area in Figure 17 is the largest. At the TR triggering time, internal pressure, *LEL*, and T_s are more dangerous than the other three chamber pressures, and the TR triggering time is far more dangerous than the other three chamber pressures. T_c at 80 kPa and 101 kPa is more dangerous than the other two pressures. The gas content is the highest at 80 kPa and the lowest at 60 kPa.

According to Table 5, the average scores in descending order are 40 kPa > 80 kPa > 101 kPa > 60 kPa. Therefore, it is considered that TR is most dangerous at 40 kPa, with an evaluation score far higher than those of the other three chamber pressures. Among the six characteristic parameters, four of them scored 100, indicating the most dangerous conditions. Next is the 80 kPa and 101 kPa, with one characteristic parameter scoring 100. Finally, for 60 kPa, none of the characteristic parameters were assigned a value of 100, indicating they are the safest.

7. Conclusions

This article compares the TR characteristics of batteries under different chamber pressures from the aspects of T_s, T_c, gas generation, t_{open}, t_{TR}, and gas composition, and draws the following conclusions.

1. The paper measures the TR characteristics at 40 kPa using temperature and pressure sensors. The results indicate that venting occurs twice during the TR process, with corresponding T_v of 163.3 °C and 245.9 °C, and corresponding chamber pressures of 55.3 kPa and 65.3 kPa.
2. At 40 kPa, gas chromatography analysis reveals that the highest proportion of H_2 in the mixed gas is 55%, followed by CO_2 at 18%, and then CO and CH_4 at 11% and 10%, respectively. The *LEL* and *UEL* are calculated using the Le Chatelier formula and are found to be 6.42% and 63%, respectively.
3. The paper further investigates the effects of 40 kPa, 60 kPa, 80 kPa, and 101 kPa on the TR characteristics of the battery. It is found that chamber pressure significantly affects the peak of T_s, T_c, t_{open}, t_{TR}, Δt, gas composition, *LEL*, and *UEL*. Conversely, pressure has minimal impact on the internal pressure and gas generation rate of the battery.
4. A six-dimensional radar chart analysis method is proposed to evaluate the danger of TR under different pressures. The results show that the most dangerous pressure is at 40 kPa, followed by 80 kPa and 101 kPa, while the 60 kPa is relatively safer.
5. At low pressure, it is found that t_{open} and t_{TR} are much shorter than that of the battery at normal pressure. Δt is significantly longer than that of the battery under normal

pressure. It is recommended that designers use the opening sound of the safety valve or the gas concentration to catch the opening action quickly and take action at longer time intervals.

This study proposes a testing method with embedded pressure sensors and an evaluation method based on a six-dimensional radar chart. It analyzes the characteristics and mechanisms of TR under different pressure conditions, providing theoretical references for the safe use and TR warning of lithium-ion batteries in the high-altitude energy storage industry.

Author Contributions: Writing—original draft, Y.W. (Yu Wang); Methodology, Y.W. (Yan Wang); Investigation, J.Z., H.L. and F.D.; Conceptualization, C.X.; Formal analysis, Y.L.; Validation, H.W. and R.Y.; Data curation, L.L. and F.Q. All authors have read and agreed to the published version of the manuscript.

Funding: This research was funded by [National Natural Science Foundation of China] grant number [No. 52207240], [Shandong Province Science and Technology Foundation] grant number [No. ZR2022QE099], and [Shandong Province Excellent youth & innovation team Foundation] grant number [No. 2023KJ323]. And The APC was funded by [Tsinghua University].

Data Availability Statement: Data available on request due to restrictions eg privacy or ethical. The data presented in this study are available on request from the corresponding author. The data are not publicly available due to [due to our lab's policies or confidentiality agreements, we cannot provide raw data. If editors and reviewers have questions about specific data, we will try to provide more detailed explanations and explanations].

Conflicts of Interest: As author Feng Dai was employed by Sichuan New Energy Vehicle Innovation Center Co., Ltd., the remaining authors declare that the research was conducted in the absence of any commercial or financial relationships that could be construed as a potential conflict of interest.

References

1. Wang, Q.; Mao, B.; Stoliarov, S.I.; Sun, J. A review of lithium ion battery failure mechanisms and fire prevention strategies. *Prog. Energy Combust. Sci.* **2019**, *73*, 95–131. [CrossRef]
2. Jia, Z.; Wang, S.; Qin, P.; Li, C.; Song, L.; Cheng, Z.; Jin, K.; Sun, J.; Wang, Q. Comparative investigation of the thermal runaway and gas venting behaviors of large-format LiFePO4 batteries caused by overcharging and overheating. *J. Energy Storage* **2023**, *61*, 106791. [CrossRef]
3. Peng, Y.; Yang, L.; Ju, X.; Liao, B.; Ye, K.; Li, L.; Cao, B.; Ni, Y. A comprehensive investigation on the thermal and toxic hazards of large format lithium-ion batteries with LiFePO4 cathode. *J. Hazard. Mater.* **2020**, *381*, 120916. [CrossRef] [PubMed]
4. Gyatso, N.; Li, Y.; Gao, Z.; Wang, Q.; Li, S.; Yin, Q.; Chen, J.; Jin, P.; Liu, Z.; Ma, Z.; et al. Wind power performance assessment at high plateau region: A case study of the wind farm field test on the Qinghai-Tibet plateau. *Appl. Energy* **2023**, *336*, 120789. [CrossRef]
5. Ren, D.; Feng, X.; Liu, L.; Hsu, H.; Lu, L.; Wang, L.; He, X.; Ouyang, M. Investigating the relationship between internal short circuit and thermal runaway of lithium-ion batteries under thermal abuse condition. *Energy Storage Mater.* **2021**, *34*, 563–573. [CrossRef]
6. Zhu, N.; Wang, X.; Chen, M.; Huang, Q.; Ding, C.; Wang, J. Study on the Combustion Behaviors and Thermal Stability of Aging Lithium-Ion Batteries with Different States of Charge at Low Pressure. *PROCESS Saf. Environ. Prot.* **2023**, *174*, 391–402. [CrossRef]
7. Zhang, Q.; Liu, T.; Wang, Q. Experimental study on the influence of different heating methods on thermal runaway of lithium-ion battery. *J. Energy Storage* **2021**, *42*, 103063. [CrossRef]
8. Feng, X.; Ouyang, M.; Liu, X.; Lu, L.; Xia, Y.; He, X. Thermal runaway mechanism of lithium ion battery for electric vehicles: A review. *Energy Storage Mater.* **2018**, *10*, 246–267. [CrossRef]
9. Tran, M.-K.; Mevawalla, A.; Aziz, A.; Panchal, S.; Xie, Y.; Fowler, M. A Review of Lithium-Ion Battery Thermal Runaway Modeling and Diagnosis Approaches. *Processes* **2022**, *10*, 1192. [CrossRef]
10. Li, Y.; Feng, X.; Ren, D.; Ouyang, M.; Lu, L.; Han, X. Thermal Runaway Triggered by Plated Lithium on the Anode after Fast Charging. *ACS Appl. Mater. Interfaces* **2019**, *11*, 46839–46850. [CrossRef]
11. Guo, Q.; Zhang, J.; Zhou, C.; Huang, Z.; Han, D. Thermal Runaway Behaviors and Kinetics of NCM Lithium-Ion Batteries at Different Heat Dissipation Conditions. *J. Electrochem. Soc.* **2023**, *170*, 080507. [CrossRef]
12. Xu, X.; Sun, X.; Zhao, L.; Li, R.; Tang, W. Research on thermal runaway characteristics of NCM lithium-ion battery under thermal-electrical coupling abuse. *Ionics* **2022**, *28*, 5449–5467. [CrossRef]
13. Pastor, J.V.; García, A.; Monsalve-Serrano, J.; Golke, D. Analysis of the aging effects on the thermal runaway characteristics of Lithium-Ion cells through stepwise reactions. *Appl. Therm. Eng.* **2023**, *230*, 120685. [CrossRef]

14. Li, Q.; Yang, C.; Santhanagopalan, S.; Smith, K.; Lamb, J.; Steele, L.A.; Torres-Castro, L. Numerical investigation of thermal runaway mitigation through a passive thermal management system. *J. Power Sources* **2019**, *429*, 80–88. [CrossRef]
15. Qian, F.; Wang, H.; Li, M.; Li, C.; Shen, H.; Wang, J.; Li, Y.; Ouyang, M. Thermal Runaway Vent Gases from High-Capacity Energy Storage $LiFePO_4$ Lithium Iron. *Energies* **2023**, *16*, 3485. [CrossRef]
16. Chen, S.; Wang, Z.; Wang, J.; Tong, X.; Yan, W. Lower explosion limit of the vented gases from Li-ion batteries thermal runaway in high temperature condition. *J. Loss Prev. Process. Ind.* **2020**, *63*, 103992. [CrossRef]
17. Zhang, Q.; Niu, J.; Yang, J.; Liu, T.; Bao, F.; Wang, Q. In-situ explosion limit analysis and hazards research of vent gas from lithium-ion battery thermal runaway. *J. Energy Storage* **2022**, *56*, 106146. [CrossRef]
18. Baird, A.R.; Archibald, E.J.; Marr, K.C.; Ezekoye, O.A. Explosion hazards from lithium-ion battery vent gas. *J. Power Sources* **2020**, *446*, 227257. [CrossRef]
19. Zhang, F.; Feng, X.; Xu, C.; Jiang, F.; Ouyang, M. Thermal runaway front in failure propagation of long-shape lithium-ion battery. *Int. J. Heat Mass Transf.* **2022**, *182*, 121928. [CrossRef]
20. Huang, Q.; Weng, J.; Ouyang, D.; Chen, M.; Wang, X.; Wang, J. Comparative studies on the combustion characteristics of electrolytes and carbonate mixed solvents with flame retardant additives under low pressures. *Case Stud. Therm. Eng.* **2023**, *43*, 102810. [CrossRef]
21. Zou, K.; Chen, X.; Ding, Z.; Gu, J.; Lu, S. Jet behavior of prismatic lithium-ion batteries during thermal runaway. *Appl. Therm. Eng.* **2020**, *179*, 115745. [CrossRef]
22. Kang, R.; Jia, C.; Zhao, J.; Zhao, L.; Zhang, J. Effects of capacity on the thermal runaway and gas venting behaviors of large-format lithium iron phosphate batteries induced by overcharge. *J. Energy Storage* **2024**, *87*, 111523. [CrossRef]
23. Wang, Z.; Jiang, X.; Ke, W.; Wang, W.; Zhang, S.; Zhou, B. Effect of lithium-ion battery diameter on thermal runaway propagation rate under one-dimensional linear arrangement. *Therm. Sci. Eng. Prog.* **2022**, *31*, 101301. [CrossRef]
24. Liu, Y.; Niu, H.; Liu, J.; Huang, X. Layer-to-layer thermal runaway propagation of open-circuit cylindrical li-ion batteries: Effect of ambient pressure. *J. Energy Storage* **2022**, *55*, 105709. [CrossRef]
25. Li, Y.; Jiang, L.; Huang, Z.; Jia, Z.; Qin, P.; Wang, Q. Pressure Effect on the Thermal Runaway Behaviors of Lithium-Ion Battery in Confined Space. *Fire Technol.* **2023**, *59*, 1137–1155. [CrossRef]
26. Liu, Q.; Zhu, Q.; Zhu, W.; Yi, X.; Han, X. Thermal Runaway Characteristics of 18650 NCM Lithium-ion Batteries under the Different Initial Pressures. *Electrochemistry* **2022**, *90*, 087004. [CrossRef]
27. Ding, C.; Zhu, N.; Yu, J.; Li, Y.; Sun, X.; Liu, C.; Huang, Q.; Wang, J. Experimental investigation of environmental pressure effects on thermal runaway properties of 21700 lithium-ion batteries with high energy density. *Case Stud. Therm. Eng.* **2022**, *38*, 102349. [CrossRef]
28. Sun, Q.; Liu, H.; Zhi, M.; Chen, X.; Lv, P.; He, Y. Thermal characteristics of thermal runaway for pouch lithium-ion battery with different state of charges under various ambient pressures. *J. Power Sources* **2022**, *527*, 231175. [CrossRef]
29. Zhao, J.; Feng, X.; Tran, M.-K.; Fowler, M.; Ouyang, M.; Burke, A.F. Battery safety: Fault diagnosis from laboratory to real world. *J. Power Sources* **2024**, *598*, 234111. [CrossRef]
30. Qiu, M.; Liu, J.; Cong, B.; Cui, Y. Research Progress in Thermal Runaway Vent Gas Characteristics of Li-Ion Battery. *Batteries* **2023**, *9*, 411. [CrossRef]
31. Zhang, Y.; Wang, H.; Wang, Y.; Li, C.; Liu, Y.; Ouyang, M. Thermal abusive experimental research on the large-format lithium-ion battery using a buried dual-sensor. *J. Energy Storage* **2021**, *33*, 102156. [CrossRef]
32. Song, L.; Zheng, Y.; Xiao, Z.; Wang, C.; Long, T. Review on Thermal Runaway of Lithium-Ion Batteries for Electric Vehicles. *J. Electron. Mater.* **2022**, *51*, 30–46. [CrossRef]
33. Sun, Y.; Jin, Y.; Jiang, Z.; Li, L. A review of mitigation strategies for li-ion battery thermal runaway. *Eng. Fail. Anal.* **2023**, *149*, 107259. [CrossRef]
34. Shen, H.; Wang, H.; Li, M.; Li, C.; Zhang, Y.; Li, Y.; Yang, X.; Feng, X.; Ouyang, M. Thermal Runaway Characteristics and Gas Composition Analysis of Lithium-Ion Batteries with Different LFP and NCM Cathode Materials under Inert Atmosphere. *Electronics* **2023**, *12*, 1603. [CrossRef]
35. Ohsaki, T.; Kishi, T.; Kuboki, T.; Takami, N.; Shimura, N.; Sato, Y.; Sekino, M.; Satoh, A. Overcharge reaction of lithium-ion batteries. *J. Power Sources* **2005**, *146*, 97–100. [CrossRef]
36. Wang, H.; Du, Z.; Liu, L.; Zhang, Z.; Hao, J.; Wang, Q.; Wang, S. Study on the Thermal Runaway and Its Propagation of Lithium-Ion Batteries Under Low Pressure. *Fire Technol.* **2020**, *56*, 2427–2440. [CrossRef]
37. Yu, R.; Liu, J.; Liang, W.; Law, C.K.; Wang, H.; Ouyang, M. On flammability limits of battery vent gas: Role of diffusion, radiation and chemical kinetics. *Combust. Flame* **2023**, *249*, 112631. [CrossRef]

batteries

MDPI

Article

Advanced Thermal Management of Cylindrical Lithium-Ion Battery Packs in Electric Vehicles: A Comparative CFD Study of Vertical, Horizontal, and Optimised Liquid Cooling Designs

Michael Murphy and Mohammad Akrami *

Department of Engineering, University of Exeter, Exeter EX4 4QF, UK; mcm218@exeter.ac.uk
* Correspondence: m.akrami@exeter.ac.uk

Abstract: Battery packs found in electric vehicles (EVs) require thermal management systems to maintain safe operating temperatures in order to improve device performance and alleviate irregular temperatures that can cause irreversible damage to the cells. Cylindrical lithium-ion batteries are widely used in the electric vehicle industry due to their high energy density and extended life cycle. This report investigates the thermal performance of three liquid cooling designs for a six-cell battery pack using computational fluid dynamics (CFD). The first two designs, vertical flow design (VFD) and horizontal flow design (HFD), are influenced by existing linear and wavy channel structures. They went through multiple geometry optimisations, where parameters such as inlet velocity, the number of channels, and channel diameter were tested before being combined into the third and final optimal design (OD). All designs successfully maintained the maximum temperature of the cells below 306.5 K at an inlet velocity of 0.5 ms^{-1}, meeting the predefined performance thresholds derived from the literature. The HFD design was the only one that failed to meet the temperature uniformity goal of 5 K. The optimal design achieved a maximum temperature of 301.311 K, which was 2.223 K lower than the VFD, and 4.707 K lower than the HFD. Furthermore, it produced a cell temperature difference of 1.144 K, outperforming the next-best design by 1.647 K, thus demonstrating superior temperature regulation. The OD design can manage temperatures by using lower inlet velocities and reducing power consumption. However, the increased cooling efficiency comes at the cost of an increase in weight for the system. This prompts the decision on whether to accommodate the added weight for improved safety or to allocate it to the addition of more batteries to enhance the vehicle's power output.

Keywords: thermal management; liquid cooling; wavy channel; linear channel; parameter optimisation; lithium-ion battery pack

Citation: Murphy, M.; Akrami, M. Advanced Thermal Management of Cylindrical Lithium-Ion Battery Packs in Electric Vehicles: A Comparative CFD Study of Vertical, Horizontal, and Optimised Liquid Cooling Designs. *Batteries* **2024**, *10*, 264. https://doi.org/10.3390/batteries10080264

Academic Editor: Mingyi Chen

Received: 20 June 2024
Revised: 16 July 2024
Accepted: 19 July 2024
Published: 25 July 2024

1. Introduction

1.1. Background

With increasing concerns about global warming, the need to transition to renewable energy sources and reduce greenhouse gas emissions is paramount for preserving the future of the planet. The transportation sector alone contributes to 24% of the UK's total emissions, with cars accounting for 52% of the sector's emissions [1]. The demand for electric vehicles (EVs) is expected to surge in the coming years due to the sustainability goals pursued by both the public and the government. Batteries are the primary power source for EVs, and continuous developments in battery technology have significantly improved EV performance, including charging speed, life cycle, and range. However, the absence of effective battery thermal management systems (BTMSs) can severely impact vehicle performance. For instance, abnormal temperatures can degrade the battery capacity, shorten the EV's lifespan, and, in extreme cases, lead to thermal runaway, resulting in combustion and explosion [2]. Due to the critical role of battery thermal systems in EVs, this report will examine and present the development of three cooling structures, the vertical flow design (VFD), horizontal flow design (HFD), and optimal design (OD), and assess their effectiveness in regulating temperature within the battery pack.

1.2. Literature Review

The following literature review aims to provide a comprehensive summary of the key areas influencing EV BTMSs, focusing on battery types and different cooling methods.

1.2.1. Batteries

As of 2015, lithium-ion batteries accounted for 85.6% of deployed energy storage systems [3]. Their widespread use in EVs is attributed to several advantages over alternative rechargeable batteries, such as nickel-, sodium-, and lead-based batteries. These advantages include a higher operating voltage of up to 3.7 V, making them ideal for high-power applications, and exhibit a low self-discharge rate, enabling them to store charge for longer periods and therefore extending their lifespan [4]. EV manufacturers primarily utilise three cell shapes: cylindrical, prismatic, and pouch cells. Cylindrical cells are available in standardised dimensions of 18,650, 21,700, and 46,800; for example, all units are in millimetres, with 18 and 65 representing the cell diameter and height, respectively [5]. Prismatic cells do not have a universal size but are rectangular in shape, and compared to cylindrical cells, they feature lower voltage and fewer connections, resulting in slower discharge rates. However, their fewer connections allow for larger energy capacities [6]. Cylindrical and prismatic cells are the preferred choice for EV manufacturers; conversely, pouch batteries are an emerging technology, inducing high production costs, but most importantly, their low mechanical resistance makes them a risk for installation in vehicles [5]. The significant hazards associated with pouch cells make them an unpopular choice for designing battery packs.

Excessive heat generation within batteries occurs during the charging and discharging process because of changes in enthalpy, electrochemical polarisation, and resistive heating [7]. Additional thermal issues, such as uneven temperature distribution, can arise from capacity fading, self-discharge, and electrical imbalance. Liu et al. [8] found that when the temperature of the battery exceeds 353.15 K, the chemical reaction rate inside the cells increases, causing excessive internal heating and initiating the thermal runaway process. According to Ji et al. [9], the maximum temperature range before performance degradation for a lithium-ion battery is 253.15 K to 333.15 K, while the optimal operating range is between 298 K and 313 K. Uneven temperature distribution can also affect the battery lifespan, so the difference in maximum and minimum temperatures in a battery pack should be less than 5 K to achieve the best performance [10]. Thermal problems can affect the lifespan of the battery pack, which is why cooling methods are implemented to dissipate heat, manage heat flow, and ensure the battery packs are operating within the optimal temperature range with minimal temperature variation. The literature has presented the three main types of cell types, and the focus of this study will be on cylindrical cells, given their widespread use.

1.2.2. Cooling Methods

This section explores the different cooling methods that can be employed in BTMSs in order to alleviate abnormal temperatures. Air cooling and liquid cooling are universally the most used methods of cooling and will therefore be discussed to guide the development of the cooling structures showcased in this report.

Air Cooling

Despite its poor thermal conductivity, air cooling is a preferred choice for certain Chinese and Japanese EV manufacturers, such as BYD, Toyota, and Nissan [11], due to its simplicity and low costs. Air cooling can be classified into forced and passive air cooling. Passive cooling relies on the movement of the vehicle to naturally send air through the system but does not provide a sufficient cooling capacity compared to forced air cooling, which makes use of fans to blow air across the surface of the cells. Although liquid cooling contains a more complex geometry and heavier design, it can be 2–3 times more efficient than air cooling. The average liquid-cooled structure lowers the cell temperature by 3 K, resulting in energy savings of up to 40% while also making the battery pack more compact [11].

Liquid Cooling

Liquid cooling is held at a higher degree of importance to researchers as it is the most efficient cooling method, being employed by many EV manufacturers such as Tesla, BMW, and Chevrolet [12]. Liquid cooling can be classified into direct and indirect cooling, where direct cooling involves immersing the cells directly into a fluid, offering advantages like maximum contact area and enhanced thermal uniformity [13]. Despite these advantages, direct cooling is rarely employed in BTMSs because of the potential short circuits and electrochemical reactions that can arise from the difficulty of producing completely water-resistant battery packs [13]. Indirect cooling utilises metal cooling plates, tubes, or jackets containing embedded miniature coolant channels [12]. These metal structures contact the surface of the cell, allowing heat generated from the cells to be transferred through the plates to the flowing coolant, which carries the heat out of the system. This approach eliminates the direct contact between the battery and coolant, minimising the risk of electrical short-circuiting. Tang et al. [14] investigated the effect of the coolant velocity at the inlet for a cold plate prismatic cell pack. A discharge of 2C was applied to the cells, and velocities ranging from 0.1 ms^{-1} to 1 ms^{-1} were evaluated. Greater temperature reductions and thermal uniformity were observed at velocities above 0.5 ms^{-1}. This was only tested with one coolant channel applied along the edge of the cell, opening the potential for investigation into more channels and their placements. Zhao et al. [15] investigated the common wavy/serpentine channel design, referencing a Tesla Model S battery pack consisting of 71, 18,650 Li-ion cells. The structure was made of an aluminium alloy with a width of 2 mm and a height of 65 mm. It consisted of a singular coolant channel spanning the length of the structure. Channel angles of 10°, 20°, 30°, and 40° were used, and the thermal performance improved as the angles increased. This improvement was attributed to the larger surface contact area achieved with larger channel angles, resulting in greater thermal transfer and a lower cell temperature. The study by Xu et al. [16] confirmed this effect and tested their own linear U-shaped structure with grooves to slot the cells into place. It was found to be inferior to the wavy channel design, achieving a lower maximum temperature of 0.37 K. A linear flowing channel design (LCD) constructed by Lloyd et al. [17] features a uniform arrangement with four channels and contact points per cell, providing 75% contact with the surface of each battery. This structure was able to maintain the cell temperature within the optimal operating range and achieve the recommended thermal uniformity proposed by Ji et al. [9]. A low mass flow rate of 7.5×10^5 kgs^{-1} was used to save energy consumption, and channels with diameters of 3 mm, 6 mm, and 9 mm were assessed. The results revealed that larger diameters only marginally reduced maximum temperatures. A mini channel-cooled cylinder (MCC) design by Yates et al. [18] consists of a 3 mm thick cylinder structure with multiple microchannels. This design fully encapsulates the cells and is therefore expected to provide the greatest heat transfer. Comparing the MCC design to the LCD at a cell operating at a 5C discharge rate, the maximum temperature of the MCC reached 309.308 K, while the LCD reached 310.26 K. This confirms that surface area contact plays an important role in determining the cooling effectiveness of designs.

The literature review has highlighted the importance of liquid cooling and the many types of cooling structures. The wavy channel design is prevalent, and this report will explore further angle variations, whilst examining various numbers of coolant channels to further optimise the design. Refining the linear channel design can potentially improve thermal capabilities, as it utilises less material while still providing commendable thermal regulation. Therefore, my iteration of the LCD will incorporate a slightly modified shape, and the coolant channel diameter will be adjusted accordingly. To further enhance the scientific relevance of the study, a third and unique design will also be included.

2. Materials and Methods

The objective of the project was to develop and evaluate the effectiveness of liquid cooling structures for thermal management within a battery pack. As identified in the literature, liquid cooling surpassed air cooling in terms of heat capacity and heat transfer efficiency, making it the chosen method for the investigation. Cylindrical cells are the most common cell geometries for battery packs; hence, the cooling system will be tailored to these specifications, as it offers more value and practicality.

2.1. Battery Chemistry

Lithium-ion batteries are the most popular type of battery, specifically those consisting of a LiFePO4 chemistry. In cylindrical cells, the internal components include the cathode, anode, separator, and current collectors, which are spirally wound [19]. The electrochemical reactions that occur during charging (←) and discharging (→) at the electrode/electrolyte interfaces are presented by Equations (1) and (2) [19]:

$$\text{Negative electrode}: \ Li_xC_6 \leftrightarrows C_6 + xLi^+ + ze^- \tag{1}$$

$$\text{Positive electrode}: \ Li_{1-x}FePO_4 + xLi^+ + ze^- \leftrightarrows LiFePO_4 \tag{2}$$

2.2. Battery Material and Properties

Modelling the physical and chemical reactions occurring within the cell can be challenging. Therefore, to reduce computational complexity, the internal battery components were simplified to a solid cylinder with uniform heat generation and homogeneous material properties. This simplification ensured a steady transfer of heat to the surroundings and reduced the computational requirements. The properties of the simplified cell model, sourced from Zhao et al. [20], are presented in Table 1, alongside other experimental materials, utilising the Ansys material database.

Table 1. Material properties.

Materials	Density (kg m^{-3})	Specific Heat Capacity (Jkg^{-1} K^{-1})	Thermal Conductivity (Wm^{-1} K^{-1})	Viscosity (kgm^{-1} s^{-1})
Liquid water	309.219	4182	0.6	1.003×10^{-3}
Aluminium	305.705	871	202.4	-
Li-ion battery	304.842	1108	3.91	-

A battery undergoing a 5C discharge rate experiences significant strain and shock, which can increase the chance of chain reactions. Wang et al. [21] studied the effects of a single cell operating at this discharge rate and found that the heat generation increased linearly, reaching a maximum temperature of 351 K after running transiently for 720 s. These results were validated by Lloyd et al. [17], who calculated the equivalent heat generation rate of the cell to be 138,000 Wm^{-3}. These findings served as the basis for establishing the heat generation of the battery pack in my own investigation.

Three models were created for the study. The first model featured a new partial jacket structure, resembling a four-point star configuration, which provides more contact with the cell. The second model focused on adapting the traditional wavy channel, where modifications were made to explore different channel angles and the number of coolant channels to assess variables not explored in the literature. Finally, the third and novel model integrated both the wavy channel design and the partial jacket design with the optimised parameters from the first two designs. One unique aspect of this model is the bidirectional flow of coolant, which has not been explored in the literature. Many parameters will need to be modified for each design, so utilising CAD and CFD software (Ansys 2024 R1) for modelling and experimentation will allow for an easy adjustment of the geometry and parameters. This approach will require significantly less time and resources as opposed to physical setup and model manufacture. SolidWorks 2022 will be used for model creation, and once these models are created, they will be downloaded as a STEP file and imported into Ansys 2024 R1, along with its fluent module, where meshing and simulations will take place to analyse the heat generation of the cell and the extent to which the cooling structures can manage the cell temperatures.

2.3. Numerical Model

The governing equations presented below explain the thermodynamics and coolant dynamic principles that are fundamental to the numerical investigation. The temperature (T_b) of a single lithium-ion cell can be modelled by the general conservation energy Equation (3) [19]:

$$\frac{\delta}{\delta t}\left(\rho_b * c_{p,b} * T_b\right) = \nabla * (k_b * \nabla T_b) + q_{gen} \tag{3}$$

where ρ_b denotes the generalised density of the battery, $c_{p,b}$ is the specific heat capacity of the battery, and k_b signifies the generalised thermal conductivity of the battery. q_{gen} is the total heat generation of the battery and can be determined by the simplified Bernardi Equation (4) [22]:

$$q_{gen} = \frac{1}{V_b}\left(I_b^2 R_b + I_b T_b \frac{dU_b}{dT}\right) \tag{4}$$

where V_b is the volume of the battery, I_b is the internal current of the battery, and R_b is the internal resistance of the battery. The term $\frac{dU_b}{dT}$ represents the temperature coefficient, a variable related to the electrochemical reaction of a cell and derived from the open-circuit voltage of the battery, U_b. Liquid water was the coolant used in the research, maintaining a constant mass throughout the system. According to Zhou et al. [23], the coolant flow and heat transfer can be expressed by continuity Equation (5) [23], momentum conservation Equation (6) [23], and energy conservation Equation (7) [23]:

$$\frac{\delta \rho_w}{\delta t} + \nabla * \left(\rho_w * \vec{v}_w\right) = 0 \tag{5}$$

$$\frac{\delta \vec{v}_w}{\delta t} + \left(\vec{v}_w * \nabla\right)\vec{v}_w = -\frac{1}{\rho_w}\nabla * P + \mu_w * \nabla^2\left(\vec{v}_w\right) \tag{6}$$

$$\frac{\delta}{\delta t}(\rho_w * c_{p,w} * T_w) + \nabla * \left(\rho_w * c_{p,w} * \vec{v}_w * T_w\right) = \nabla * (k_w * \nabla T_w) \tag{7}$$

where t represents time, $\vec{v}_w$ is the velocity of water, μ_w indicates the dynamic viscosity of water, and P represents pressure. The subscript 'w' is used to denote variables associated with water that have been previously identified. The heat transfer between the coolant, cooling channels, and the battery can be modelled by the heat flow Equations (8) and (9) [17]:

$$Q = \dot{m} * c_p * \nabla T \tag{8}$$

$$Q = h * A * \nabla T \tag{9}$$

where Q is the thermal energy, $\dot{m}$ is the mass flow rate, h is the heat transfer coefficient, and A is the surface area for heat transfer. The Reynold number (Re) (8) [16] can be used to calculate the coolant flow characteristics:

$$Re = \frac{\rho_w \vec{v}_w d}{\mu_w} \tag{10}$$

where d represents the characteristic length. The Reynolds number is used to determine the coolant flow characteristics, which can be laminar, transitional, or turbulent.

2.4. Battery Pack Development

The first model, the vertical flow design (VFD), draws inspiration from the structure created by Lloyd et al. [17]. The second model, the horizontal flow design (HFD), is based on the serpentine design by Tang et al. [24]. The third and final model combines the features of the VFD and HFD. The cooling systems were developed in SolidWorks and configured to regulate the temperature of six 18,650 cylindrical cells. These cells were uniformly arranged, with each cell positioned 20 mm apart from the adjacent cell.

2.4.1. Vertical Flow Design (VFD)

Typical linear liquid cooling structures contain a coolant domain located above and below the cells with one inlet and one outlet, as shown in Figure 1a. Both this model and a simplified version, as shown in Figure 1b, were tested with a set inlet velocity and cell heat generation. The results revealed only a minimal temperature difference of 0.02 K between the two designs. Therefore, the simplified version was adopted as it significantly reduced model complexity and simulation time. The four-point star-shaped structure, which is 65 mm in height, interlocks the cylindrical cell surfaces, as illustrated in the top view drawing in Figure 1c. This figure also depicts the variation in channel diameters, which will be evaluated in the simulations. Additionally, the inlet flow velocity will be tested at 0.01 ms^{-1}, 0.05 ms^{-1}, 0.1 ms^{-1}, 0.5 ms^{-1}, and 1 ms^{-1}. These parameters were selected to explore untested variations in the cooling design and to validate against the existing literature.

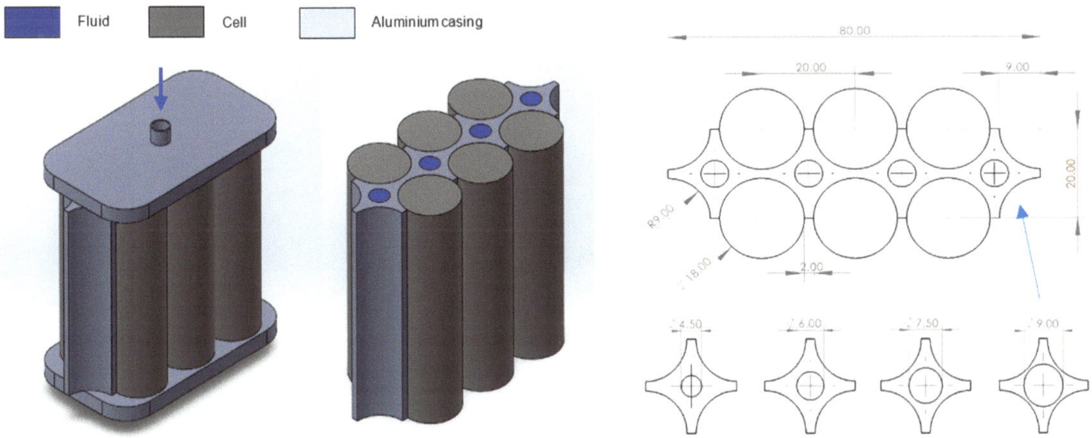

(a) (b) (c)

Figure 1. VFD views and dimensions (mm): (**a**) isometric view of VFD with fluid domain and inlet; (**b**) isometric view of simplified VFD; (**c**) battery pack top view and arrow indicating the section of the design that will have channels of varying diameters.

2.4.2. Horizontal Flow Design (HFD)

The second model created is illustrated in Figure 2a. Figure 2b shows the top view of the pack, depicting the cooling plate thickness and channel angles of 45 °, 55 °, 65 °, and 75° selected for the investigation. Then, 1, 2, 3, and 5 coolant channels will be examined, with the width of these channels remaining constant. The varying height dimensions of the coolant domain are illustrated in a front view in Figure 2c.

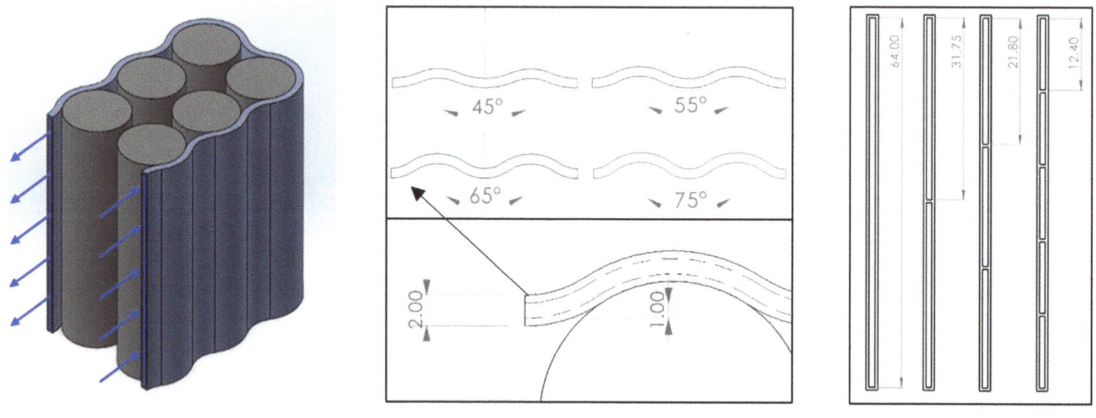

(**a**) (**b**) (**c**)

Figure 2. HFD views and dimensions (mm): (**a**) HFD isometric view; (**b**) top view and dimensions, with an arrow showing the detailed view; (**c**) front view dimensions of the inlet/outlet fluid domains.

2.4.3. Optimal Design (OD)

Illustrated in Figure 3 is the OD, the third and final cooling structure. This figure also depicts the design methodology of the OD which combines components from both the HFD and VFD. The OD incorporates an outer wavy channel and a star-shaped configuration into a single design. Two horizontal channels with an angle of 65° were used, determined through testing of the thermal efficiency based on simulations of the VFD and HFD.

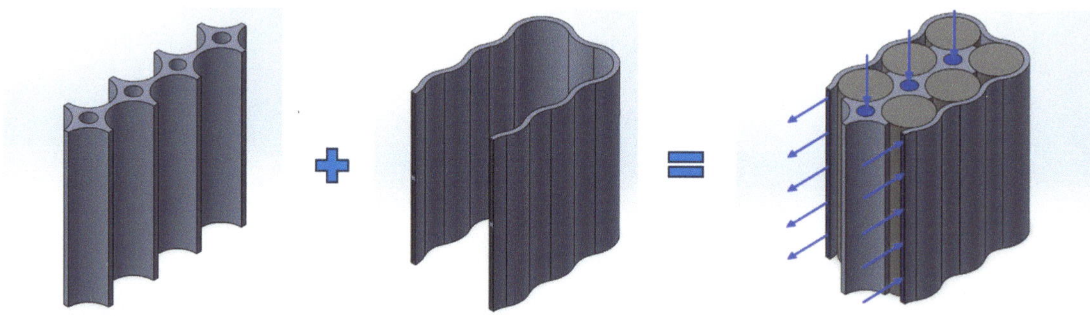

Figure 3. Construction of OD model with coolant inlet and outlet directions.

2.5. CFD Configuration and Boundary Conditions

Once the CAD model's geometries were created, they were imported into Ansys 2024 R1, and a tetrahedral-shaped mesh was selected due to its ease of generation in complex geometry cases. A mesh refinement study was conducted on the VFD to determine the optimal mesh density, using the maximum temperature as a basis to ensure accurate solutions. The results are presented in Figure 4 and show that the maximum temperature converges when more than 400,000 elements are produced. The final mesh utilises a 1 mm element size, producing 781,479 elements. As this mesh size produced satisfactory convergence, it was applied to the other designs as they share similar geometry. This mesh was further refined using inflation layers over walls, flow channels, and cells where heat transfer and thermal gradients are expected. The total number of mesh elements for each design and parameter change was maintained at a similar range where possible. Specifically, 781,479 elements were produced for the VFD, and 712,904 elements were produced for the HFD, as shown in Figure 5a,b, respectively.

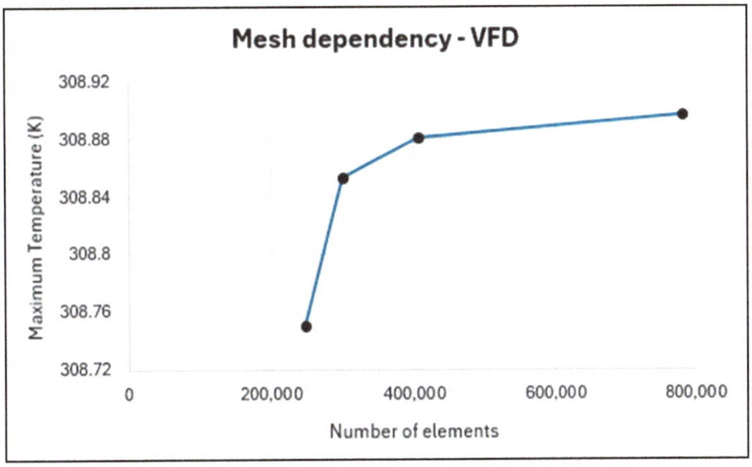

Figure 4. Mesh dependency study on VFD.

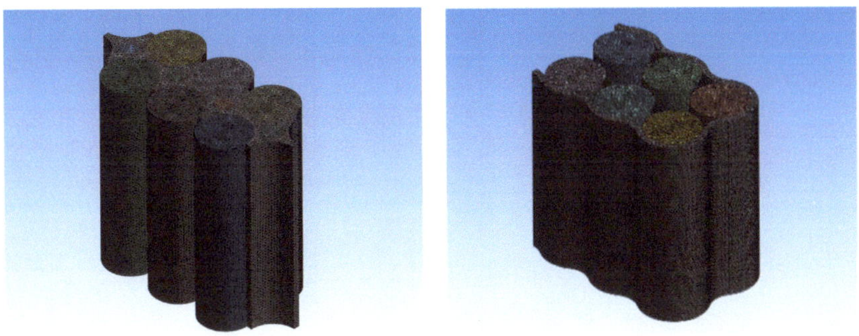

(**a**) (**b**)

Figure 5. Mesh generation for (**a**) VFD and (**b**) HFD.

In the fluent module, the double precision solver option was selected along with four solver processors. A pressure-based solver and absolute velocity formulations with a transient time step were chosen. The fluent energy model was activated to enable heat transfer, utilising energy conservation Equation (7). The viscous standard k-ε model with enhanced wall treatment was selected, as the flow characteristics were expected to range from laminar to turbulent flow, depending on variations in channel diameter and inlet velocities as per Equation (8). As recommended by Yates et al. [18], the ambient temperature of the environment is set to 300 K, assuming an adiabatic environment for the battery packs. Therefore, the outer surfaces of the battery packs were designated as thermally insulating boundaries, considering their confined environments with restricted airflow and minimal thermal influence. Material properties used in this investigation are shown in Table 1, with aluminium being assigned as the cooling structure material, custom battery material for the cells, and liquid water for the coolant. Wall surfaces within the channel were set as stationary with a no-slip condition. The solid and coolant cell zones were set as coupled to allow for natural convection in the system.

A heat generation rate of 138,000 Wm^{-3} was applied to each cell, highlighted in green in Figure 6, to replicate heat generation in a cell undergoing a 5C charge/discharge rate. The temperature of the coolant at the inlet was set to 300 K, and the velocities were adjusted accordingly to the project aims. The inlet location and coolant path of both designs are depicted in Figure 6 by the blue arrows and blue lines.

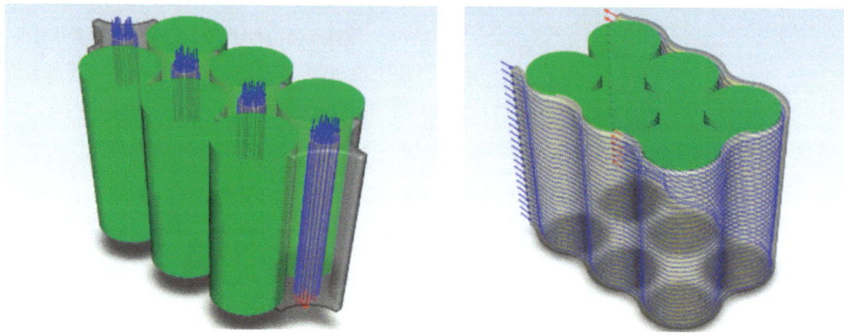

Figure 6. Boundary condition selection for (**a**) VFD and (**b**) HFD.The calculation initialisation was computed from all zones, and the calculation was set to run for 720 s to replicate the time it takes a cell to reach maximum temperature at 5C.

The timestep was set to 1 s for boosted accuracy, with a maximum number of iterations per time step set to 400 to increase the chances of convergence. Although, the maximum number of iterations it took for the time step to converge was 20. Following the completion of each design and variable change in the investigation, the boundary conditions, solution activities, and monitors to record and plot the maximum and minimum temperatures for the cell volume were all transferred to the subsequent stages. The simplified CFD configuration process can be summarised by the flow chart from Figure 7.

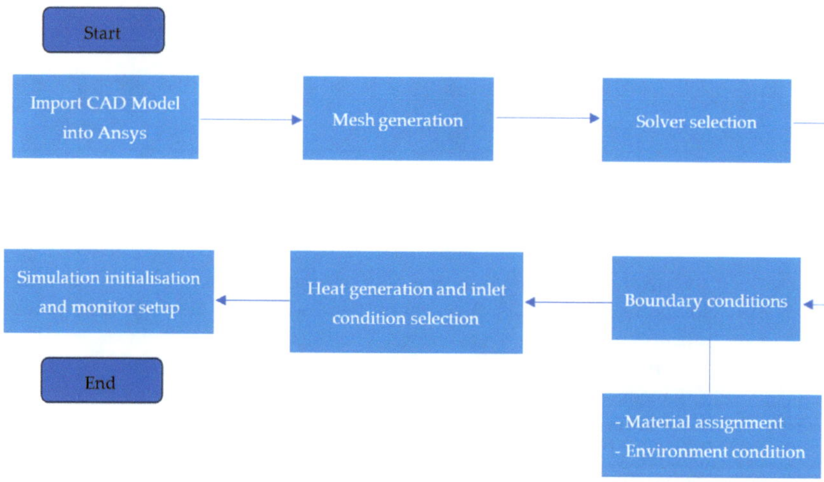

Figure 7. Flow diagram outlining CFD configuration of each design.

3. Results

3.1. Vertical Flow Design Results and Analysis

3.1.1. Influence of Inlet Velocity on VFD

The first parameter optimisation study explored the impact of inlet velocity on the maximum temperature (T_{max}) and temperature difference between the maximum and minimum (ΔT) of the battery pack. For this study, 6 mm channels were used. Figure 8a presents the results for the effect of the five different velocity values on the T_{max} of the cell. A clear trend is illustrated where increasing the velocity reduces the T_{max} of the cells. This trend aligns with heat flow Equations (8) and (9), where the rate of heat flow is directly proportional to the mass flow rate and hence the velocity. Initially, the temperature rises rapidly for all velocities within the first 60 s of cell discharge, after which the temperatures begin to diverge. Beyond 320 s, the T_{max} of the cell is reached and levels out for all velocities except for 0.01 ms^{-1}. Table 2 provides the maximum temperature and uniformity achieved by the cell after 720 s for different velocities. Comparing the results reveals that the difference in final maximum temperatures between 0.5 ms^{-1} and 1 ms^{-1} is only 0.327 K, while the difference between 0.1 ms^{-1} and 0.5 ms^{-1} is 1.308 K, indicating the diminishing effectiveness of higher flow rates beyond 0.5 ms^{-1}. This study can be validated against the work of Lloyd et al. [17], where their LCD of a six-cell lithium-ion battery pack with dimensions, material, and shape produced a T_{max} of 309.308 K, which is 0.029% different to the T_{max} from the work of this cooling structure.

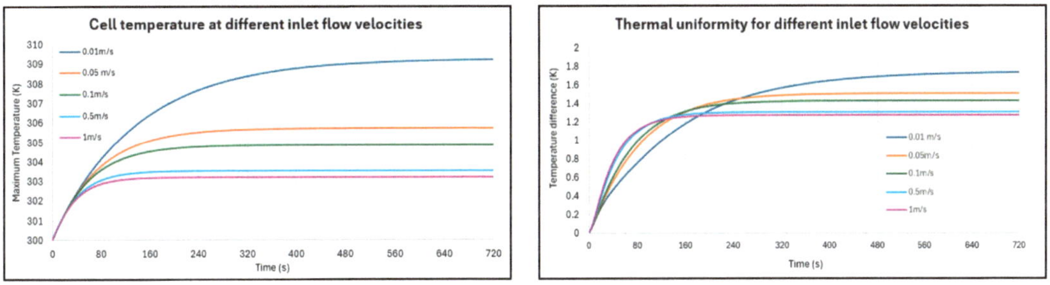

(**a**) (**b**)

Figure 8. Effects of flow velocity on (**a**) maximum cell temperature and (**b**) cell temperature difference.

Table 2. Maximum temperature and uniformity achieved by the cells at different inlet velocities.

Velocity (ms^{-1})	T_{max} (K) at 720 s	ΔT (K) at 720 s
0.01	309.219	1.733
0.05	305.705	1.503
0.1	304.842	1.426
0.5	303.534	1.306
1	303.207	1.271

Figure 8b shows the relationship between increasing inlet velocity and thermal uniformity. The first 80 s of the simulation feature a steep rise in ΔT before it stabilises as the lithium-ion core cools at the same rate as the surface. A temperature differential is present because the VFD comes into contact with half of each cell, leaving the outer part exposed to the air. This makes it difficult for the heat generated to dissipate, resulting in higher temperatures than the rest of the pack. Table 2 reveals that the ΔT is only 0.462 K between the largest and smallest velocities tested. Despite the lowest velocity producing a ΔT of 1.733 K, it is still within the maximum ΔT goal of 5 K. These results demonstrate that the inlet velocities have a greater effect on the maximum temperature of the cell than on temperature uniformity. All examined velocities satisfy the desired operating temperatures of 298 K < T_{max} < 313 K and ΔT < 5 K. As the results of the 0.5 ms^{-1} were very similar to and within 0.12% of the T_{max} observed at a velocity of 1 ms^{-1}, it will be used as the velocity for all remaining experiments as it balances effective thermal regulation and power consumption.

3.1.2. Influence of Channel Diameter on VFD

Next, the channel diameter for the VFD was varied while using a constant inlet velocity of 0.5 ms^{-1}. Figure 9a shows the relationship between T_{max} and channel diameter, indicating a reduction in T_{max} as the channel diameter increases. This is explained by the heat flow Equations (8) and (9), which suggest that increasing the channel diameter increases the heat transfer area between the coolant and channel interior. Table 3 provides the T_{max} and ΔT values recorded at the end of the battery discharge for the tested channel diameters. Notably, the T_{max} of the 4.5 mm and 9 mm diameter channels measured

303.750 K and 303.297 K, respectively, only showing a marginal difference of 0.453 K. This can be attributed to the velocity in the channel decreasing as channel diameter increases, reducing the heat transfer coefficient, and counteracting the cooling capability.

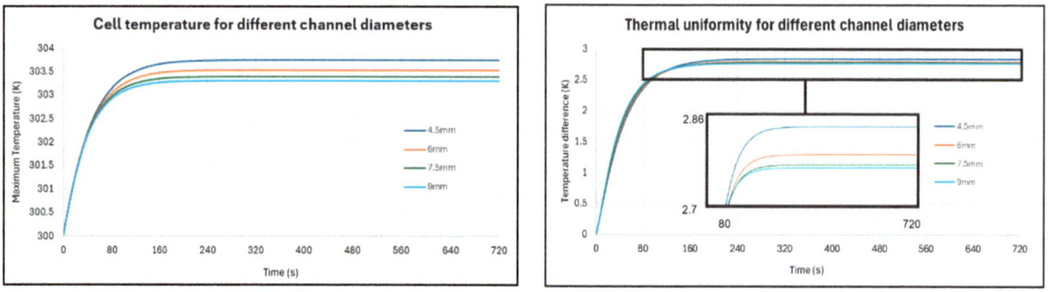

(a) (b)

Figure 9. Effects of channel diameter on (**a**) maximum cell temperature and (**b**) cell temperature difference with enlarged graph plot between ΔT of 2.7 K and 2.86 K and time step of 80 to 720.

Table 3. Maximum temperature and uniformity achieved by the cells at different channel diameters.

Channel Diameter (mm)	T_{max} (K) at 720 s	ΔT (K) at 720 s
4.5	303.750	2.840
6	303.534	2.791
7.5	303.388	2.772
9	303.297	2.767

The thermal uniformity for different diameters is shown in Figure 9b. While initial observations suggest a negligible difference between the variables, a closer examination within a time of 80–720 s and a ΔT of 2.7–2.86 reveals some variation. The 9 mm diameter exhibited a lower ΔT compared to the 4.5 mm at 720 s, but this was marginal; for the 4.5 mm channel, it was 2.840 K, and the 9 mm channel was 2.767 K, showing a difference of 0.0073 K. Although the channel diameter shows improvement in thermal uniformity, the impacts are minimal when compared to changing the velocity.

3.2. Horizontal Flow Design Results and Analysis

3.2.1. Influence of Channel Angle on HFD

For the HFD, channel angles of 45°, 55°, 65°, and 75° were tested with one flow channel. The relationship between T_{max} and ΔT is presented in Figure 10a,b, and the final values achieved at the end of the battery discharge are presented in Table 4. These results indicate that using a lower channel angle worsens the thermal performance, with a difference of 0.802 K and 0.858 K between the largest and smallest angles for the T_{max} and ΔT, respectively. The effect of channel angle has more influence on the thermal uniformity, with a 16.5% change in ΔT between the angles of 45° and 55°, compared to a 0.26% change in T_{max}. The improved performance of higher contact angles results from geometry changes during each design test, as the curves of the aluminium structure wedge themselves into the gaps between the cells, creating more surface contact area. According to Equation (9), this allows for more heat transfer from the lithium-ion cell to the aluminium structure, thereby lowering the temperature of the cell and being the cause of the significant difference in thermal uniformity. The T_{max} recorded was 306.250 K, meeting the condition of 298 K < T_{max} < 313 K. However, none of the angles in the simulation satisfied the ΔT < 5 K thermal uniformity goal, with the best-performing angle (75°) only producing a temperature difference of 5.197 K at the end of the 720 s. While the contact angle of 75° showed better performance than smaller angles, this design uses more material and hence weight. The sharper angles may introduce slight pressure drops in the system.

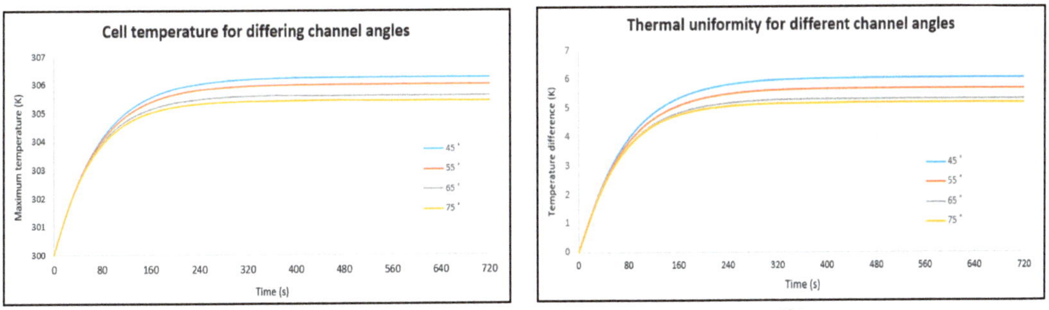

(a) (b)

Figure 10. Effects of channel angle on (**a**) maximum cell temperature and (**b**) cell temperature difference.

Table 4. Maximum temperature and uniformity achieved by the cells at different channel angles.

Channel Angle (°)	T_{max} (K) at 720 s	ΔT (K) at 720 s
45	306.250	6.055
55	306.018	5.710
65	305.620	5.326
75	305.448	5.197

3.2.2. Influence of Number of Coolant Channels on HFD

The graphs in Figure 11 present the results on how T_{max} and ΔT of the cell vary when one, two, three, and five coolant plate channels are present in the HFD. As seen in the first 80 s, both T_{max} and ΔT increase sharply, before diverging. Increasing the number of cooling channels leads to a decrease in the final T_{max} and ΔT within the battery pack. However, beyond one channel, the influence becomes minimal, as indicated by Table 5, which shows a T_{max} difference of 0.544 K between one and two channels, while the difference between T_{max} of channels two and five is only 0.121 K. In addition, the temperature uniformity showed a similar relationship as the number of channels increased.

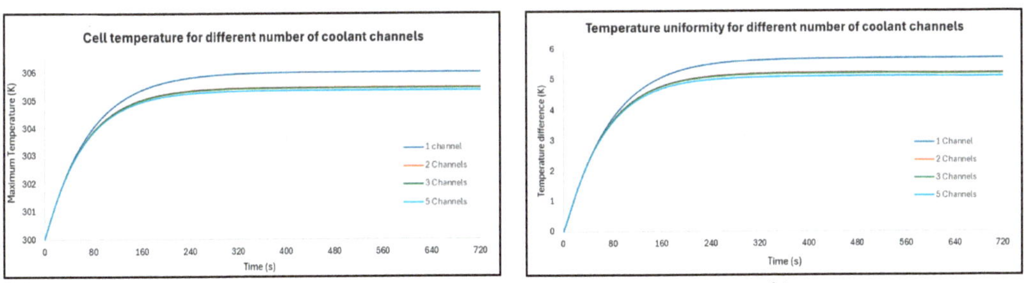

(a) (b)

Figure 11. Effects of number of coolant channels on (**a**) maximum cell temperature and (**b**) cell temperature difference.

Table 5. Maximum temperature and uniformity achieved by the cells at different numbers of flow channels.

Number of Channels	T_{max} (K) at 720 s	ΔT (K) at 720 s
1	306.018	5.710
2	305.474	5.223
3	305.451	5.204
5	305.353	5.111

For both the channel angle and number of channel optimisations for the HFD, the temperature difference ΔT exceeds 5 K, which is greater than the desired condition of ΔT < 5 K. This phenomenon is caused by the design itself, where the aluminium structures only wrap around the outside of the cell arrangement, leaving the inner cells unattended. Additionally, as the channels loop around, they provide more cover for the two outer cells compared to the cells located closest to the inlet and outlet, causing excessive thermal gradients which are evident in the final temperature contour of the HFD, as shown in Figure 12a. The results for the HFD can be validated against the work of Tang et al. [24] who found the ΔT to be approximately 5.29 K, which is close to the values I obtained.

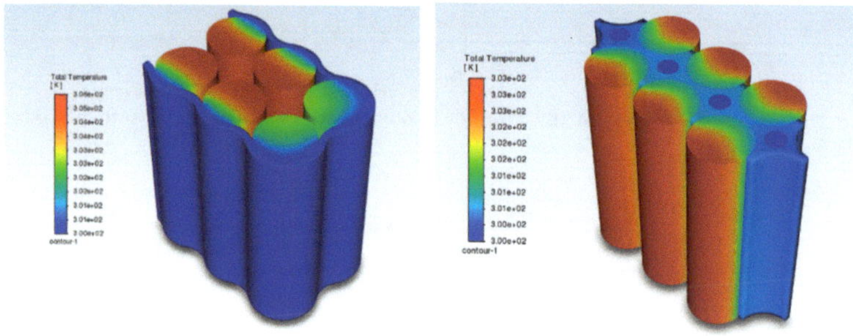

(**a**) (**b**)

Figure 12. Battery pack temperature contour after 720 s at 0.5 ms^{-1} inlet velocity: (**a**) HFD; (**b**) VFD.

3.3. *Optimal Design Results and Analysis*

The final optimal design (OD) for the six-cell 18,650 battery pack combines the VFD and HFD, featuring three vertical 6 mm diameter channels and two 31.75 mm × 1 mm horizontal channels for coolant flow, as shown in Figure 3. The thermal performances of the VFD, HFD, and OD are plotted in Figure 13, with the final T_{max} and ΔT values presented in Table 6. The OD design excelled in the thermal management of the pack compared to the other designs, efficiently dissipating the heat generated by the cells after 50 s and maintaining T_{max} at an extremely low temperature of 301.311 K for the remaining time steps. In contrast, the VFD took 160 s to reach a T_{max} of 303.534 K, while the HFD showed the poorest cooling performance, reaching a T_{max} of 306.018 K after 320 s. The superior performance of the OD is attributed to its greater contact surface area with the cells compared to the other designs; the cells are almost entirely encapsulated, whereas the HFD and VFD leave 50% of the cells without contact. The temperature contour of the OD is illustrated in Figure 14a. Furthermore, as explained by Equation (9), more heat generated by the battery can be transferred to the cooling structure and expelled from the system. All three designs comfortably met the 298 K < T_{max} < 313 K requirement, with only the OD and VFD satisfying the ΔT < 5 K requirement.

The OD exceeded thermal management expectations and was subsequently assessed with a smaller inlet velocity of 0.01 ms^{-1}. The temperature contour for both the front and back view of the structure at this lower velocity are shown in Figure 14b,c, respectively. The results indicate that the T_{max} at the end of the battery discharge cycle reached 304 K, well within the desired range of 298 K < T_{max} < 313 K, suggesting that the structure can effectively conserve energy, by using lower flow rates while maintaining efficient thermal management. However, the thermal gradients become more pronounced, with a ΔT of 4.05 K, approaching the 5 K thermal uniformity limit. The observed temperature variation within the cells reveals that those situated closest to the inlet are cooler, while those closer to the outlet exhibit higher temperatures. This phenomenon arises from the slower velocity, which limits the heat transfer rate between the coolant and cold plates. By the time the coolant circulates through the structures, it heats up, resulting in a slower heat dissipation process, as explained by heat flow Equation (9). This effect is also evident in the wavy aluminium structure, which experiences significant heating due to the coolant's reduced capacity to extract heat from the system at lower velocities. In contrast, the wavy structure from Figure 14a, remains closer to ambient temperature, due to the higher velocity of the coolant moving through the system, and providing a more even temperature distribution.

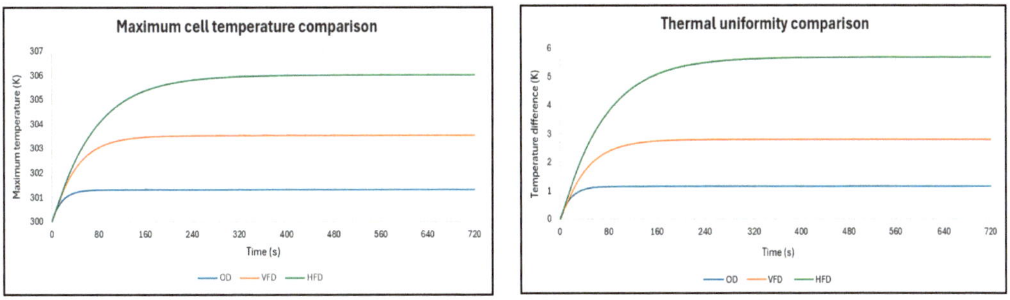

(a) (b)

Figure 13. VFD, HFD, and OD thermal comparison: (**a**) maximum cell temperature; (**b**) cell temperature difference.

Table 6. Maximum temperature and uniformity of the OD, VFD and HFD.

Cooling Structure	Tmax (K) at 720 s	ΔT (K) at 720 s
OD	301.311	1.144
VFD	303.534	2.791
HFD	306.018	5.690

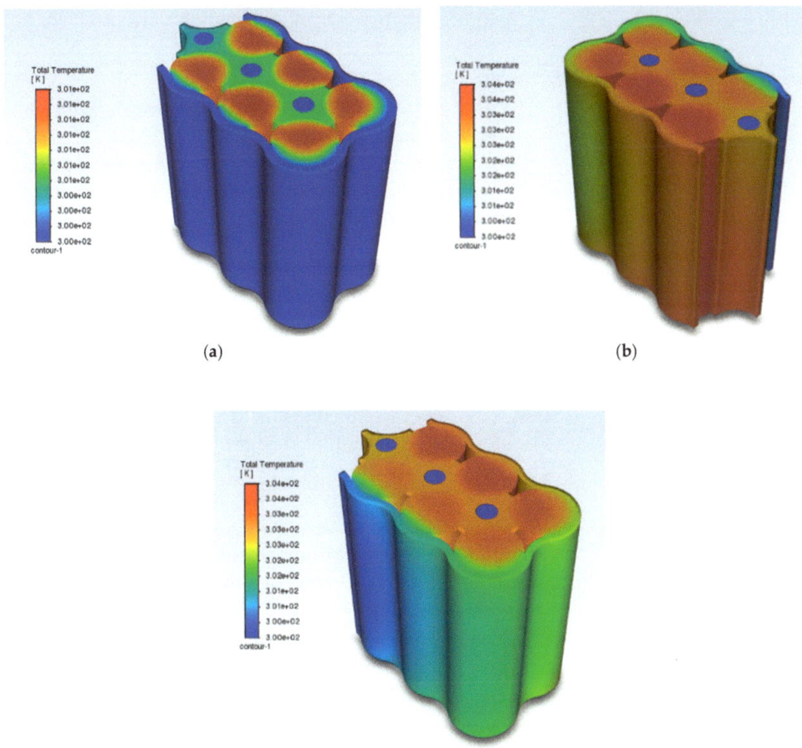

(a) (b) (c)

Figure 14. OD thermal contours (**a**) at 0.5 ms^{-1}; (**b**) at 0.01 ms^{-1}, view 1; and (**c**) at 0.01 ms^{-1}, view 2.

3.4. Summary of Key Findings

3.4.1. VFD—Influence of Inlet Velocity

- Increasing the inlet velocity reduces the T_{max} of the cells.
- The effectiveness of higher flow rates diminishes beyond 0.5 ms^{-1}.
- All tested velocities met the desired operating temperatures (298 K < T_{max} < 313 K) and thermal uniformity (ΔT < 5 K).
- A value of 0.5 ms^{-1} was chosen for further experiments as it balanced thermal regulation and power consumption.

3.4.2. VFD—Influence of Channel Diameter

- Increasing the channel diameter reduces T_{max}, but this effect is only minimal. The smallest diameter (4.5 mm) produced a T_{max} of 0.453 K higher than the largest diameter (9 mm).
- Thermal uniformity improved slightly with larger diameters.

3.4.3. HFD—Influence of Channel Angle

- Larger channel angles displayed better thermal performance and uniformity than lower angles.
- All angles met the thermal range; however, none of the angles tested met the thermal uniformity requirement.

3.4.4. HFD—Influence of Number of Coolant Channels

- Increasing the number of coolant channels reduced T_{max}, but this effect is minimal beyond one channel.
- The thermal uniformity improved when more than one channel is used but the effect is minimal when additional channels are used.

3.4.5. OD

- Superior thermal management was shown, maintaining a very low T_{max} (301.311 K) and excellent thermal uniformity (1.144 K).
- OD met all thermal requirements and outperformed both the VFD and HFD.
- Lowering the inlet velocity to 0.01 ms^{-1} for OD still kept T_{max} within the desired range, but thermal uniformity approached the 5 K limit (4.05 K).

4. Discussion and Conclusions

This investigative project evaluated two liquid cooling designs: one with water flowing in channels parallel to the cells (VFD), and the other with coolant channels placed perpendicular to the cells (HFD). These designs were investigated using CFD to assess their effectiveness in battery thermal management. Following the evaluation of the VFD and HFD, they were combined to create a novel cooling design, incorporating the most effective variables from each initial design. The simulations used a cylindrical 18,650 LiFePO4 battery model. The thermal generation within the cell was set to replicate a battery undergoing a 5C charge/discharge rate. The coolant and ambient temperatures were assumed to be 300 K. The influence of inlet velocity and channel diameter was investigated for the VFD model, while the influence of the wavy channel angle and number of channels was investigated for the HFD model. The performance of the liquid cooling structures was evaluated based on the maximum temperature reached and the temperature difference between the maximum and minimum temperatures within the cell. The main conclusions are as follows:

- Increasing the channel diameter in the VFD reduced the maximum temperature, and the thermal uniformity also improved due to the relationship between the surface contact area and heat transfer rate. These thermal effects were minimal in comparison to the effects of inlet velocity.

- As the inlet flow velocity increases, the maximum temperature and temperature difference decrease due to the relationship between flow velocity and heat transfer. The flow velocity was found to have a greater influence on temperature reduction compared to the other variables evaluated in the report. A value of 0.5 ms^{-1} was the most efficient choice.
- Increasing the number of channels in the HFD decreases the maximum temperature of the cells and improves thermal uniformity, but the effects are almost negligible as the number of channels increases. Two channels provide sufficient cooling and ease of manufacture.
- The maximum cell temperature increased as greater channel angles were used for the HFD. The angle increase also resulted in more of the casing coming into contact with the cell, providing greater temperature reduction and uniformity.
- The VFD produced a T_{max} of 303.534 K and a ΔT of 2.791 K, meeting the thermal objectives of 298 K < T_{max} < 313 K and ΔT < 5 K, using an inlet velocity of 0.5 ms^{-1} and 6 mm diameter channels.
- The HFD produced a T_{max} of 306.018 K and a ΔT of 5.690 K, meeting the thermal objective of 298 K < T_{max} < 313 K but failing to meet ΔT < 5 K, using two channels and an inlet velocity of 0.5 ms^{-1}.
- The OD combines the HFD and VFD and produces a T_{max} of 301.311 K and a ΔT of 1.144 K, comfortably within the thermal objective of 298 K < T_{max} < 313 K and ΔT < 5 K. It should be noted, in theory, that increasing the velocity in the HFD and VFD designs would help meet the thermal targets, but at the cost of extra energy consumption.

The OD demonstrated exceptional thermal management for the battery pack; however, this is outweighed by the larger structure, more components, and a greater volume of coolant flowing through the system, resulting in increased overall weight. Consequently, the additional weight may require the vehicle to expend more energy for transportation, and the study by Carlson et al. [25] investigated the impact of additional weight and found that a 10% mass increase can result in a 3–4% energy consumption increase for electric vehicles. Taking only the mass of the aluminium structure for a six-cell pack, we find that the OD has a mass of 89.93 g while the HFD, a design that is commonly found in modern EVs, has a mass of 28.73 g. The added weight may impact vehicle driving range, and the additional components, such as cooling channels and flow directions, increase the complexity of the battery pack, and its compatibility in existing EVs. Furthermore, there is a greater risk of leakage due to the greater potential for seal failures. The other designs, despite utilising half the structure volume of the OD, managed to produce satisfactory thermal management and would be seen as the most preferable options for implementation in EVs, but the OD would be best suited for high-temperature and safety-critical applications. The excellent thermal management of the OD allows for the vehicle to produce higher discharge rates which in turn produces a greater power output, without the risk of exceeding the thermal limit. Therefore, the OD would be best suited for high-temperature, power-intensive and safety-critical applications.

The findings provide valuable insights into the influential parameters and effectiveness of novel liquid cooling designs, but there are certain limitations to the report such as the simplification of the battery pack, using only six cells to mitigate computational requirements, and ignoring components like busbars and casing, which can further influence the temperature of the entire system. Additionally, the cell arrangement used in the simulations is less compact than other battery pack configurations which poses a slight constraint on its suitability for EVs. The material properties were assumed to be homogenous throughout, and the heat generation model of the cell was assumed to be uniform across the entire volume.

Future work could analyse these models using a larger battery pack and include additional components to more accurately simulate a real EV battery pack.

Furthermore, exploring a wider range of innovative and lightweight materials for the structure, such as phase change materials, could be beneficial. Investigating emerging coolant options such as coolants with nanometal additives or liquid metals may also further refine these designs and enhance their performance.

Author Contributions: Conceptualisation, M.M. and M.A.; methodology, M.M. and M.A.; software, M.M. and M.A.; validation, M.M. and M.A.; formal analysis, M.M. and M.A.; investigation, M.A.; resources, M.A.; data curation, M.M. and M.A.; writing—original draft preparation, M.M.; writing—review and editing, M.M. and M.A.; visualisation, M.M.; supervision, M.A.; project administration, M.A.; funding acquisition, M.A. All authors have read and agreed to the published version of the manuscript.

Funding: This research received no external funding.

Data Availability Statement: Data is available upon request.

Conflicts of Interest: The authors declare no conflicts of interest.

Nomenclature

Variables	
A	contact area (m^2)
c	specific heat capacity ($Jkg^{-1} K^{-1}$)
d	channel diameter (mm)
h	heat transfer coefficient ($Wm^{-2} K^{-1}$)
I	current (A)
k	thermal conductivity ($Wm^{-1} K^{-1}$)
L	characteristic length (m)
m	mass (kg)
$\dot{m}$	mass flow rate (kgs^{-1})
P	pressure (Pa)
q	heat generation (Wm^{-3})
Q	rate of heat flow (W)
R	resistance (Ω)
Re	Reynolds number
T	temperature (K)
t	time (s)
U	open-circuit voltage (V)
V	volume (m^3)
v	velocity (ms^{-1})
Acronyms	
CAD	computer-aided design
CFD	computational fluid dynamics
EV	electrical vehicle
HFD	horizontal flow design
VFD	vertical flow design
PCM	phase change material
Greek letters	
Δ	difference
μ	dynamic viscosity ($kgm^{-1} s^{-1}$)
ρ	mass density (kgm^{-3})
Subscripts	
b	battery
gen	generation
max	maximum
p	constant pressure process
w	water

References

1. Transport and Environment Statistics 2022. Available online: https://www.gov.uk/government/statistics/transport-and-environment-statistics-2022/transport-and-environment-statistics-2022 (accessed on 10 November 2023).
2. Liu, W.; Placke, T.; Chau, K.T. Overview of Batteries and Battery Management for Electric Vehicles. *Energy Rep.* **2022**, *8*, 4058–4084. [CrossRef]
3. Assad, M.; Rosen, M.A. *Design and Performance Optimization of Renewable Energy Systems*; Academic Press: Cambridge, MA, USA, 2021; pp. 205–219.
4. Plett, G.L. *Battery Management Systems. Volume I, Battery Modeling*, 1st ed.; Artech House: Norwood, MA, USA, 2015.
5. Comparison of Different Types of Electric Vehicle Battery Cells. Available online: https://www.keyence.co.uk/products/marker/laser-marker/resources/laser-marking-resources/comparison-of-different-types-of-electric-vehicle-battery-cells.jsp (accessed on 11 November 2023).
6. Pros and Cons of Lithium Prismatic Cells vs. Cylindrical Cells. Available online: https://blog.epectec.com/pros-and-cons-of-lithium-prismatic-cells-vs-cylindrical-cells (accessed on 15 November 2023).
7. Zhai, N.; Li, M.; Wang, L.; Majid, N.; Hafiz, S.; Arianto, S.; Kawano, S.; Nishimura, F.; Hazima, N.; Ismail, F.; et al. Simplified Heat Generation Model for Lithium Ion Battery Used in Electric Vehicle. *IOP Conf. Ser. Mater. Sci. Eng.* **2013**, *53*, 012014.
8. Liu, K.; Liu, Y.; Lin, D.; Pei, A.; Cui, Y. Materials for Lithium-Ion Battery Safety. *Sci. Adv.* **2018**, *4*, eaas9820. [CrossRef] [PubMed]
9. Ji, Y.; Zhang, Y.; Wang, C.-Y. Li-Ion Cell Operation at Low Temperatures. *J. Electrochem. Soc.* **2013**, *160*, A636–A649. [CrossRef]

10. Akbarzadeh, M.; Kalogiannis, T.; Jaguemont, J.; Jin, L.; Behi, H.; Karimi, D.; Beheshti, H.; Van Mierlo, J.; Berecibar, M. A Comparative Study between Air Cooling and Liquid Cooling Thermal Management Systems for a High-Energy Lithium-Ion Battery Module. *Appl. Therm. Eng.* **2021**, *198*, 117503. [CrossRef]
11. Xia, G.; Cao, L.; Bi, G. A Review on Battery Thermal Management in Electric Vehicle Application. *J. Power Sources* **2017**, *367*, 90–105. [CrossRef]
12. Liu, J.; Chen, H.; Huang, S.; Jiao, Y.; Chen, M. Recent Progress and Prospects in Liquid Cooling Thermal Management System for Lithium-Ion Batteries. *Batteries* **2023**, *9*, 400. [CrossRef]
13. Zhang, X.; Li, Z.; Luo, L.; Fan, Y.; Du, Z. A Review on Thermal Management of Lithium-Ion Batteries for Electric Vehicles. *Energy* **2022**, *238*, 121652. [CrossRef]
14. Tang, Z.; Liu, Z.; Li, J.; Cheng, J. A Lightweight Liquid Cooling Thermal Management Structure for Prismatic Batteries. *J. Energy Storage* **2021**, *42*, 103078. [CrossRef]
15. Zhao, C. Thermal Behaviour Study of Discharging/Charging Cylindrical Lithium-Ion Battery Module Cooled by Channelled Liquid Flow. *Int. J. Heat Mass Transf.* **2018**, *120*, 751–762. [CrossRef]
16. Xu, H.; Zhang, X.; Xiang, G.; Li, H. Optimization of Liquid Cooling and Heat Dissipation System of Lithium-Ion Battery Packs of Automobile. *Case Stud. Therm. Eng.* **2021**, *26*, 101012. [CrossRef]
17. Lloyd, R.; Akrami, M. A Critical Analysis of Helical and Linear Channel Liquid Cooling Designs for Lithium-Ion Battery Packs. *Batteries* **2022**, *8*, 236. [CrossRef]
18. Yates, M.; Akrami, M.; Javadi, A.A. Analysing the Performance of Liquid Cooling Designs in Cylindrical Lithium-Ion Batteries. *J. Energy Storage* **2021**, *33*, 100913. [CrossRef]
19. Amiribavandpour, P.; Shen, W.; Mu, D.; Kapoor, A. An Improved Theoretical Electrochemical-Thermal Modelling of Lithium-Ion Battery Packs in Electric Vehicles. *J. Power Sources* **2015**, *284*, 328–338. [CrossRef]
20. Zhao, J.; Rao, Z.; Huo, Y.; Liu, X.; Li, Y. Thermal Management of Cylindrical Power Battery Module for Extending the Life of New Energy Electric Vehicles. *Appl. Therm. Eng.* **2015**, *85*, 33–43. [CrossRef]
21. Wang, Z.; Li, X.; Zhang, G.; Lv, Y.; Wang, C.; He, F.; Yang, C.; Yang, C. Thermal Management Investigation for Lithium-Ion Battery Module with Different Phase Change Materials. *RSC Adv.* **2017**, *7*, 42909–42918. [CrossRef]
22. Bernardi, D.; Pawlikowski, E.; Newman, J. A General Energy Balance for Battery Systems. *J. Electrochem. Soc.* **1985**, *132*, 5–12. [CrossRef]
23. Zhou, H.; Zhou, F.; Zhang, Q.; Wang, Q.; Song, Z. Thermal Management of Cylindrical Lithium-Ion Battery Based on a Liquid Cooling Method with Half-Helical Duct. *Appl. Therm. Eng.* **2019**, *162*, 114257. [CrossRef]
24. Tang, Z.; Min, X.; Song, A.; Cheng, J. Thermal Management of a Cylindrical Lithium-Ion Battery Module Using a Multichannel Wavy Tube. *J. Energy Eng.* **2019**, *145*, 04018072. [CrossRef]
25. Carlson, R.B.; Lohse-Busch, H.; Diez, J.; Gibbs, J. The Measured Impact of Vehicle Mass on Road Load Forces and Energy Consumption for a BEV, HEV, and ICE Vehicle. *SAE Int. J. Altern. Powertrains* **2013**, *2*, 105–114. [CrossRef]

batteries

MDPI

Article

The Suppression Effect of Water Mist Released at Different Stages on Lithium-Ion Battery Flame Temperature, Heat Release, and Heat Radiation

Bin Miao [1,2,3], Jiangfeng Lv [1], Qingbiao Wang [1,2,4,*], Guanzhang Zhu [5], Changfang Guo [6], Guodong An [7] and Jianchun Ou [8]

1 College of Resources, Shandong University of Science and Technology, Tai'an 271019, China; miaobin@sdust.edu.cn (B.M.)
2 National Engineering Laboratory for Coalmine Backfilling Mining, Tai'an 271019, China
3 Zaozhuang Mining Group Co., Ltd, Zaozhuang 277000, China
4 School of Civil Engineering, Shandong Jianzhu University, Jinan 250101, China
5 Beijing Tianma Intelligent Control Technology Co., Ltd., Beijing 101399, China
6 Artificial Intelligence Research Institute, China University of Mining and Technology, Xuzhou 221116, China
7 Jining Haida Xingzhi School, Jining 272100, China
8 State Key Laboratory for Fine Exploration and Intelligent Development of Coal Resources, China University of Mining and Technology, Xuzhou 221116, China
* Correspondence: skd990748@sdust.edu.cn

Abstract: Thermal runaway (TR) is a serious thermal disaster that occurs in lithium-ion batteries (LIBs) under extreme conditions and has long been an obstacle to their further development. Water mist (WM) is considered to have excellent cooling capacity and is widely used in the field of fire protection. When used in TR suppression, WM also exhibits strong fire-extinguishing and anti-re-ignition abilities. Therefore, it has received widespread attention and research interest among scholars. However, most studies have focused on the cooling rate and suppression effect of TR propagation, and few have mentioned the effect of WM on flame heat transfer, which is a significant index in TR propagation suppression. This study has explored the suppression effect of WM released at different TR stages and has analyzed flame temperature, heat release, and heat radiation under WM conditions. Results show that the flame extinguishing duration for WM under different TR stages was different. WM could directly put out the flame within several seconds of being released when SV opened, 3 min after SV opening and when TR ended, and 3 min for WM when TR was triggered. Moreover, the heat radiation of the flame in relation to the battery Q_E could be calculated, and the case of WM released 3 min after SV opening exhibited the greatest proportion of heat radiation cooling η (with a value of 88.4%), which was same for the specific cooling capacity of WM Q_m with a value of 1.7×10^{-3} kJ/kg. This is expected to provide a novel focus for TR suppression in LIBs.

Keywords: lithium-ion battery; thermal runaway; water mist suppression; flame heat release; flame heat radiation

Citation: Miao, B.; Lv, J.; Wang, Q.; Zhu, G.; Guo, C.; An, G.; Ou, J. The Suppression Effect of Water Mist Released at Different Stages on Lithium-Ion Battery Flame Temperature, Heat Release, and Heat Radiation. *Batteries* **2024**, *10*, 232. https://doi.org/10.3390/batteries10070232

Academic Editor: Mingyi Chen

Received: 6 May 2024
Revised: 4 June 2024
Accepted: 7 June 2024
Published: 28 June 2024

1. Introduction

Lithium-ion batteries (LIBs) are widely used in electronic devices, electric vehicles, aerospace, and electrochemical energy storage, owing to their advantages of high energy density, a high-voltage platform, high charging and discharging efficiency, environment friendliness, a long lifespan, and wide applicability [1–5]. In practical appliance scenarios consisting of high energy densities, prismatic batteries stand out among other batteries, such as 18,650 batteries and pouch batteries, due to their large capacity [6]. However, a large capacity also constitutes a high thermal hazard in the process of thermal runaway (TR) [7]. The TR process is a severe exothermic disaster in LIBs, with the phenomena of combustible gases and jet flames [8]. Under extreme conditions, such as external heating, overcharging, penetration, and friction, the TR process may be triggered but not terminated [9,10]. In the process of TR, a series of pyrolysis side reactions are triggered continuously among the active materials inside the battery, exhibiting a high heat release rate, a high battery surface temperature, violent combustible gas release, and a serious risk of combustion and explosion [11–13]. Moreover, the process of TR in a single LIB will generally propagate to the adjacent batteries, ultimately resulting in a large-scale TR process in the LIB module, which will cause casualties and property damage [14,15].

The TR process in the LIB is both related to internal and external factors. The internal factors of commonly used LIBs mainly include the cathode material, the state of charge (SOC), the state of health (SOH), and the rated capacity [16,17]. The commonly used battery cathode materials are nickel cobalt manganese ternary lithium (NCM), nickel cobalt aluminum ternary lithium (NCA), and lithium iron phosphate (LFP). NCM and NCA batteries have a higher energy density but poorer thermal stability than LFP batteries, and they exhibit a fiercer TR process for a greater heat release rate and gas generation rate [18,19]. With the decline in battery SOC, the TR process becomes weaker and even disappears. In batteries with a high SOC, there are more lithium ions embedded in the anode material, which directly accelerates the side reactions of TR and reduces the TR triggering temperature [20]. The SOH does not directly affect the TR intensity, but it is more likely for lithium deposition and lithium dendrite growth to appear in the batteries with low SOH under extreme abuse conditions, causing a lower TR triggering temperature and charging security [21]. A battery with a higher-rated capacity will generate more heat and gases in its TR process due to a larger amount of active materials inside the battery [4]. The external factors mainly include the TR triggering method, the external oxygen content, and the ambient pressure. There are various thermal hazards of LIBs under external heating, overcharging, and penetration conditions. Most TR hazards occur under overcharging conditions, both in the single battery and the battery module [22,23]. When the oxygen content is low, the TR reactions inside the battery will be weakened, resulting in a lower surface temperature, mass loss, and heat release rate [24]. Additionally, under low or high ambient pressure levels, the TR hazard will be impaired, but the TR process is more likely to be triggered in low-pressure ambient environments [25].

In order to suppress the TR process and decrease the likelihood of a TR hazard, many scholars have tested different fire-extinguishing agents and suppression strategies [18,26,27]. The typical fire-extinguishing agents for LIBs are gaseous fire-extinguishing agents, dry powders, water-based fire-extinguishing agents, and aerosol fire-extinguishing agents. Zhang [28] has investigated the TR suppression effect of CO_2, HFC-227ea, and $C_6F_{12}O$ and found that $C_6F_{12}O$ could immediately extinguish the flame, CO_2 needed a longer duration for flame extinguishing, and HFC-227ea could not suppress the flame. Sun [29] has investigated the suppression effect of HFC-227ea and $C_6F_{12}O$ for TR propagation purposes and found that HFC-227ea could hardly suppress its propagation and $C_6F_{12}O$ could prolong the propagation time. Zhao [30] has investigated the TR suppression effect of ABC and BC ultrafine dry powders and found that they could not suppress TR propagation due to their poor cooling capacity. Davion [31] has tested aerosols for TR suppression purposes and found that they could immediately extinguish the flame, but a re-ignition phenomenon occurred. Tang [32] found that F500 had a good suppression effect for LIB TR. Liu [33] has investigated the water mist (WM) cooling strategy for TR propagation purposes and found that WM exhibited an excellent cooling capacity and could easily prevent TR propagation in the LIB module. Zhang [34] combined N_2, $C_6F_{12}O$, and WM in TR suppression and found that N_2-twin-$C_6F_{12}O$ mist could successfully inhibit the re-ignition of the battery flame and that the N_2-twin-H_2O mist synergistic technology could increase the cooling rate by over 20%. Li [35] added sodium dodecyl benzene sulfonate, sodium chloride and soy protein in WM and found with these additives, WM exhibited a greater cooling capacity and could cool down the LIB flame and surface temperature in a low duration. Zhang [36] has investigated the suppression effect of WM intermittent spray strategy for TR propagation and found this strategy could combine the advantages of low water consumption and high cooling capacity. Mei [37] has compared the suppression effect of three TR retardants based on parameters including flame height, and these novel retardants all expressed a greater TR suppression effect compared to paraffin.

Among various strategies for TR suppression, WM demonstrates outstanding cooling capacity and can effectively prevent battery flame re-ignition. However, existing studies predominantly focus on WM's cooling rate and its impact on TR propagation, largely overlooking its influence on flame heat transfer. When battery flames occur, the heat transferred from the flame to adjacent batteries similarly influences TR propagation dynamics. Therefore, investigating WM's cooling effects on LIB flames is crucial. This study has explored the suppression effect of WM released at different TR stages and has analyzed flame temperature, heat release, and heat radiation under WM conditions, aiming to introduce a new perspective on TR suppression in LIBs.

2. Experimental Settings

2.1. Battery Sample

The experimental batteries used were 100 Ah prismatic LFP batteries manufactured by Lishen (Qingdao). They consisted of a lithium iron phosphate ($LiFePO_4$) cathode, layered graphite (C) anode, aluminum (Al) positive current collector, copper (Cu) negative current collector, and an electrolyte mixture of ethylene carbonate (EC), propylene carbonate (PC), and dimethyl carbonate (DMC). The separator was polyethylene microporous membrane (PE), and the battery shell was aluminum (Al). Each battery was equipped with a safety valve (SV) to release gases during extreme conditions. The dimensions of each battery were 220 mm $\times$ 140 mm $\times$ 35 mm, with a mass of 2160 $\pm$ 2 g. Prior to experiments, all battery samples had their plastic covers removed. During charging and discharging, the maximum cut-off voltage was set at 3.6 V and the minimum at 2.5 V. Each battery underwent 3 cycles of charging/discharging, followed by a full charge to 100% State of Charge (SOC) using a Neware cycler and a 12-h rest period after each cycle. Each cycle included phases of constant current charging, constant voltage charging, and constant current discharge.

2.2. Experimental Setup

The schematic diagram of the experimental setup is shown in Figure 1. The experimental platform was constructed according to the standard of ISO 9705 [38]. The experiments of LIB TR and WM suppression were conducted in the combustion chamber where the observation window and WM nozzle were. The exhausting fume collecting hood was equipped at the top of the combustion chamber, and a fan was equipped at the end of the smoke exhaust duct with an exhaust volume of 0.13 m^3/s. A supply pipe for WM was installed in the combustion chamber and a WM pump was connected to the pipe ending. The WM-released pressure and mass flow were 6.5 MPa and 0.4 L/min, respectively.

The experimental module consisted of, a stainless steel module framework, two mica insulation plates, an LFP battery sample, and the heating plate. The dimension of the heating plate was the same as the battery sample, and the dimension of the mica insulation plate was slightly larger than the battery sample. The power of the heating plate was 500 W, and heating was stopped when TR was triggered or WM was released. One type K armored thermocouple was installed at the center of the battery surface and five were installed 10, 30, 50, 70, and 90 cm above the SV in order to monitor the temperature of the battery surface and flame. The measurement span, accuracy, and recording frequency of each type K armored thermocouple were -100–1200 °C, $\pm$1.5–5 °C, and 1 Hz, respectively. A disposable igniter was set in front of the experimental module. Once SV opened, the igniter was remotely activated, generating sparks to ignite the battery flame.

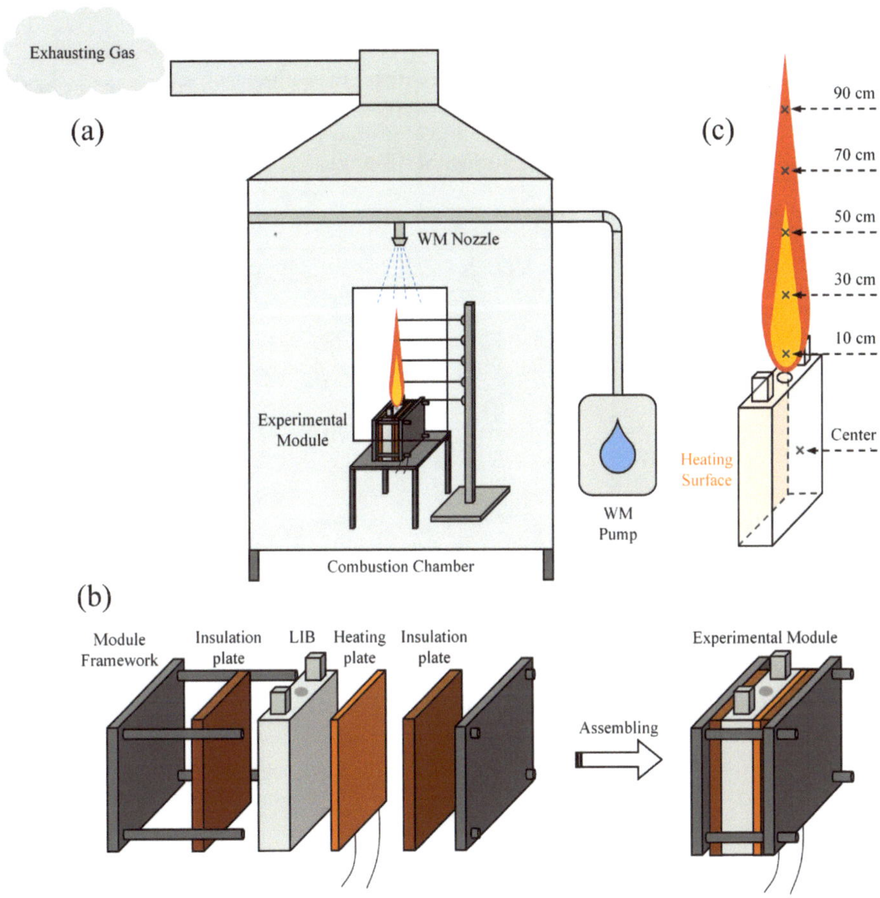

Figure 1. The schematic diagram of the experimental setup: (**a**) Experimental platform; (**b**) Experimental Module; (**c**) Thermocouple layout.

2.3. Case Setting

The experiment cases were set as Table 1 shown. In order to investigate the TR characteristic of the experimental battery sample, Case 1 was set to analyze the typical TR process and establish a reference for the extinguishing cases. Cases 2 to 5 were set to analyze the effect of WM on battery flame temperature and its heat release. In all the cases, the TR of the battery samples was triggered by external heating.

Table 1. Cases setting.

No.	WM Release Temperature	Case Description
Case 1	\	No WM extinguishing.
Case 2	108 °C	WM released at SV opening.
Case 3	116 °C	WM released 3 min after SV opening.
Case 4	140 °C	WM released at TR triggering.
Case 5	400 °C	WM released after TR ended.

3. Experimental Results and Analyses

3.1. Analysis of the Temperatures of Battery Surface and TR Flame

During thermal abuse conditions, the battery temperature generally rose due to heat transfer from an external heating plate, ultimately triggering the TR process accompanied by a vigorous jet flame. Figure 2 illustrates the curves of battery voltage, temperature, and temperature rise rate during the TR process.

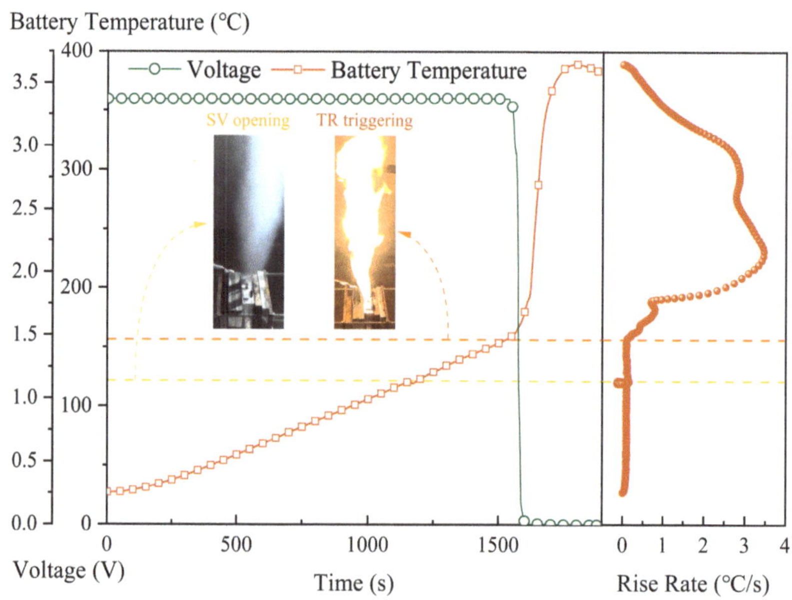

Figure 2. The curves of battery voltage, temperature, and its rise rate.

The battery temperature experienced a uniform rise, instantaneous descent, brief jumping ascent, and overall rapid rise, and they corresponded to the phenomena of external heating, SV opening, TR triggering, and TR peak. In the stage of external heating, with the internal temperature increase, the solid electrolyte interface (SEI) film was experiencing pyrolysis and slightly generated heat and combustible gases. When gases accumulated exceeded the threshold, SV was broken and a part of the heat was brought by the jet gas, resulting in the instantaneous descent of the temperature. The pyrolysis reactions of the SEI film are shown as follows [39]:

$$(CH_2OCO_2Li)_2 \rightarrow Li_2CO_3+C_2H_4+CO_2+1/2O_2 \quad (1)$$

$$2Li+(CH_2OCO_2Li)_2 \rightarrow 2Li_2CO_3+C_2H_4 \quad (2)$$

During the stage of TR triggering, the temperature rise rate initially increased and then decreased. Following the decomposition of the SEI film, lithium ions embedded in graphite came into direct contact with the organic electrolyte, releasing combustible gases and generating heat. Subsequently, lithium ions from the electrolyte deposited on graphite formed lithium dendrites that eventually bridged the cathode and anode, leading to internal short circuits and significant heat generation. This heat caused widespread melting of the separator, which absorbed some heat and contributed to a reduction in the temperature rise rate. Additionally, extensive internal short circuits occurring during this stage caused the battery voltage to plummet to 0 V. The pyrolysis reactions of the electrolyte are depicted as follows:

$$2Li + C_3H_4O_3(EC) \rightarrow Li_2CO_3+C_2H_4 \quad (3)$$

$$2Li + C_4H_6O_3(PC) \rightarrow Li_2CO_3{+}C_3H_6 \quad (4)$$

$$2Li + C_3H_6O_3(DMC) \rightarrow Li_2CO_3{+}C_2H_6 \quad (5)$$

With increasing heat accumulation within the battery, the pyrolysis reaction of the cathode material initiates, marking the peak of the TR process. During this phase, extensive pyrolysis reactions ensue among the battery's internal active materials, generating significant heat and gases and causing the vaporization of the organic electrolyte. Consequently, the battery temperature rises rapidly. The pyrolysis reactions involving the cathode material $LiFePO_4$ and the electrolyte can be illustrated as follows:

$$2Li_0FePO_4 \rightarrow Fe_2P_2O_7{+}1/2O_2 \quad (6)$$

$$5/2O_2{+}C_3H_4O_3(EC) \rightarrow 3CO_2{+}2H_2O \quad (7)$$

$$4O_2{+}C_4H_6O_3(PC) \rightarrow 4CO_2{+}3H_2O \quad (8)$$

$$3O_2{+}C_3H_6O_3(DMC) \rightarrow 3CO_2{+}3H_2O \quad (9)$$

The state of the battery flame was distinct from the TR process, and the curves of battery temperature and TR flame temperatures were 10, 30, 50, 70, and 90 cm above the SV, as shown in Figure 3.

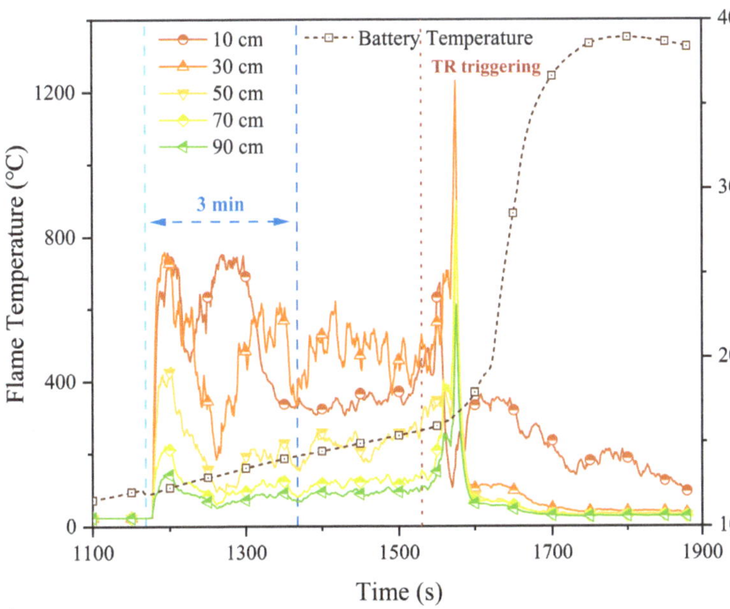

Figure 3. The curves of battery temperature and TR flame temperatures.

When the SV opened, combustible gases were expelled from the battery. Unlike the gases released after TR triggering, which predominantly contained organic electrolyte vapor, those released before TR were primarily inorganic gases like H_2 and CO, known for their higher calorific values, resulting in a higher flame temperature prior to TR triggering [40]. Approximately 3 min after SV opening, the battery flame temperature exhibited a regular fluctuation pattern, indicating stabilization thereafter, serving as a benchmark for WM release. During TR triggering, flame temperatures at various positions initially soared, then declined before stabilizing. At 30 cm, the maximum temperature exceeded 1200 °C. It was noted that while temperatures at other positions rose, the flame temperature at 10 cm dropped significantly due to flame extinction caused by high-speed gas jets at the flame base.

3.2. Analysis of the Effect of WM on TR Flame

The flame characteristics, including velocity and temperature, varied with the progression of the TR process. Consequently, the impact of WM on the battery flame differed across different stages of TR: WM swiftly extinguished the flame within seconds of release upon SV opening, after 3 min of SV opening, and at TR termination. However, it took approximately 3 min for WM to extinguish the flame once TR was triggered. Figure 4 illustrates the temperature curves of the battery flame under the influence of WM.

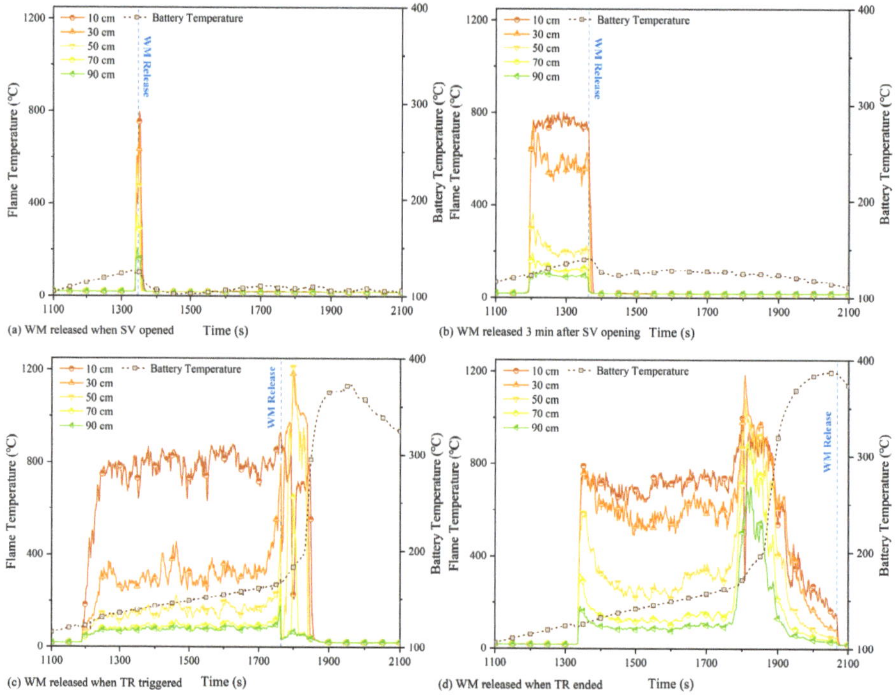

Figure 4. The curves of battery flame temperatures under the effect of WM: (**a**) WM released when SV opened; (**b**) WM released 3 min after SV opening; (**c**) WM released when TR triggered; (**d**) WM released when TR ended.

When WM was released before TR triggering, the battery flame could be extinguished immediately. During this phase, the gas generation rate was moderate, maintaining a stable flame that allowed WM droplets to effectively reach the flame's base and extinguish it. However, after TR triggering, the production of combustible gases accelerated significantly. Short-chain alkenes mixed with organic electrolyte vapor were expelled at high velocity, inducing a turbulent jet flame state. WM was less effective in suppressing this high-velocity flame and could only influence its upper portion, causing the flame to assume a conical shape. Before TR triggering, the flame temperature distribution showed lower temperatures at the top and higher temperatures at the bottom, with the region 10 cm above the SV exhibiting the highest temperature. However, due to the rapid jet gas velocity, the lower part of the flame experienced uneven oxygen mixing, leading to unstable combustion and resulting in higher temperatures observed at 30 and 50 cm above SV. Comparing the case of no WM release and release when TR triggered, the temperature at 30 and 50 cm were both greater in the case of WM released. This outcome occurred because WM suppressed the flame's upper region and concentrated the combustion into a smaller area, thereby elevating the core flame temperature. Upon TR cessation, the flame transitioned into an ember state, facilitating immediate extinguishment upon WM release.

3.3. Analysis of the Heat Cooling of TR Flame and Specific Cooling Capacity of WM

For disaster prevention and control, the heat release of battery flame is a key parameter. To calculate the heat release of battery flame Q, first, the heat flux density of battery flame q calculated by Equation (10) is needed [41].

$$q = h(T_{f,i} - T_a) \tag{10}$$

where $T_{f,i}$ and T_a are the flame temperature at i position and ambient temperature, respectively, and h is the convective heat transfer coefficient in the experimental environment which can be calculated by Equation (11). The Nusselt number Nu can be calculated by Equations (12)–(14) [42].

$$h = Nu\frac{\lambda}{d} \tag{11}$$

$$Nu = 0.27Re^{0.63}Pr^{0.36}(Pr_f/Pr_w)^{0.25} \tag{12}$$

$$Re = \frac{\rho v d}{\mu} \tag{13}$$

$$Pr = \frac{v}{\alpha} = \frac{v}{\lambda/\rho c} \tag{14}$$

where Re and Pr are the Reynolds number and Prandtl number of air in the experiment, respectively, λ is the thermal conductivity of air in the experiment, d is the combustion chamber diameter, ρ is the density of air in the experiment, v is the gas flow rate in the experiment, μ is the dynamic viscosity of air in the experiment, α is the thermal diffusion coefficient, c is the specific heat capacity of air in the experiment, and $(Pr_f/Pr_w)^{0.25}$ is the physical property correction factor with the value of 1.00 in this study. By substituting various coefficients into the calculation, the convective heat transfer coefficient h in the experiment was 70.4. Then, the heat release from battery flame Q could be calculated as Equation (15).

$$Q = qAt = Ah\sum_{i=1}^{5}\int_{t_{SV}}^{t_e}\left(T_{f,i} - T_a\right)dt \tag{15}$$

where A is the flame surface area assuming the flame shaping a cylinder, t_e and t_{SV} are the time of SV opening and TR ending, respectively. According to experimental videos, four time points of flame stabilizing after SV opening, TR triggering, flame peaking and TR ending were chosen to measure flame diameter and height. These were assumed to exhibit a linear variation between each two time points. The results for each case were 321.5 kJ in Case 1, 44.9 kJ in Case 2, 97.0 kJ in Case 3, 216.0 kJ in Case 4 and 284.4 kJ in Case 5. Using the time of SV opening, TR triggering and TR ending in Case 1 as the calculation reference, the differences in heat release between Case 1 and Cases 2 to 5 ΔQ were 11.6, 14.9, 54.4 and 37.1 kJ, respectively. Assuming the constant WM mass flow q_m, the specific cooling capacity of WM Q_m could be calculated by Equation (16). The results of each case are shown in Table 2.

$$Q_m = \frac{\Delta Q}{m_{WM}} = \frac{\Delta Q}{q_m \tau_{WM}} \tag{16}$$

where m_{WM} is the WM consumption for flame extinguishing and τ_{WM} is the release duration for WM to extinguish the flame. In order to clearly demonstrate the relationship between WM cooling and TR stages, Figure 5 is obtained by Q_m and battery temperature when WM is released.

Table 2. The heat release difference and specific cooling capacity of WM in Cases 2 to 5.

No.	ΔQ (kJ)	Q_m (kJ/kg)
Case 2	11.6	1.8×10^{-3}
Case 3	14.9	1.7×10^{-3}
Case 4	54.4	2.8×10^{-3}
Case 5	37.1	5.9×10^{-3}

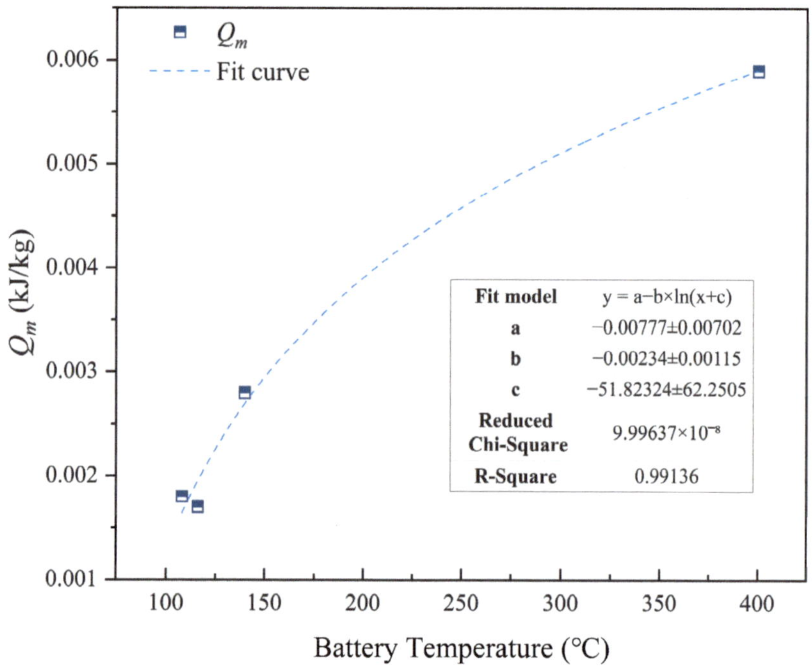

Figure 5. The scatter plot of Q_m and the battery temperature when WM is released.

As shown in Figure 5, the scatter plot could be well fitted as the logarithmic function in the figure, and the specific cooling capacity of WM Q_m was increased with the increase in battery temperature (or TR process). It was because when the battery temperature was high, the velocity of the jet flame became more violate and the flame became higher. And if WM was released under a high battery temperature, the upper part of the battery flame would be compressed and cooled to a lower flame temperature in this region. This would significantly decrease the heat release from the flame and exhibit a higher Q_m. According to the fitting curve, Q_m at each WM released temperature could be roughly estimated, and the cooling effect of WM on battery flame was better after TR triggering.

In the TR process, the battery's upper surface would be exposed to flame radiation continuously which would directly affect other batteries of the module in practical LIB applications. Thus, it is of great significance to consider the flame radiation to the battery. In this experiment, the heat radiation of flame to the battery could be simplified to the average of five temperature measurement points, and the heat radiation power of each measurement point on the surface of the battery $P_{E,i}$ could be calculated as Equation (17) [41].

$$P_{E,i} = A_{\text{LIB}}\varepsilon\frac{\sigma T_{F,i}^4}{4\pi L_i^2} \tag{17}$$

where A_{LIB} is the area of the battery's upper surface, ε is the surface emissivity of the aluminum battery surface which takes the value 0.05 [43], σ is the Stefan–Boltzmann constant, $T_{F,i}$ is the flame temperature at i position and L_i is the distance between i temperature measurement point and battery's upper surface. By integrating and averaging the $P_{E,i}$ of five measurement points, the heat radiation of flame to the battery's upper surface Q_E could be calculated as Equation (18).

$$Q_E = \frac{1}{5}\sum_{i=1}^{5}\int_{t_{SV}}^{t_e} P_{E,i}dt = \frac{1}{5}A_{LIB}\varepsilon\sigma\sum_{i=1}^{5}\frac{1}{4\pi L_i^2}\int_{t_{SV}}^{t_e} T_{F,i}^4 dt \quad (18)$$

The total flame heat radiation to the battery's upper surface in Case 1 was 117.33 kJ, and that of Cases 2 to 5 when WM released were 1.77, 0.56, 18.7 and 0.69 kJ, respectively. Also using the time of SV opening, TR triggering and TR ending in Case 1 as the calculation reference, the differences in flame heat radiation to the battery's upper surface between Case 1 and Cases 2 to 5 ΔQ_E were 5.83, 4.27, 26.31 and 1.94 kJ, respectively. Dividing ΔQ_E by the relative value of the reference case, the proportion of heat radiation cooling η in Cases 2 to 5 could be obtained. The calculation results of Cases 2 to 5 are shown in Table 3.

Table 3. The flame heat radiation to the battery's upper surface, its difference and cooling proportion in Cases 2 to 5.

No.	ΔQ_E (kJ)	η
Case 2	5.83	76.7%
Case 3	4.27	88.4%
Case 4	26.31	58.5%
Case 5	1.94	73.8%

As with the flame temperature above, after TR triggering, the battery flame was not directly extinguished and the temperatures of the flame bottom were still high. However, in the combustion process, the flame bottom was the main part affecting the heat radiation on the battery, resulting in a poorer suppression capacity for heat radiation. Although WM in Case 4 exhibited the poorest suppression effect, its η still reached 58.5%, implying a great suppression effect. For the whole TR process, WM released 3 min after SV opening exhibited the greatest suppression effect of flame heat radiation with the value of 88.4% which was the same for Q_m with the value of 1.7×10^{-3} kJ/kg.

4. Conclusions

During the TR process, a LIB undergoes distinct stages, including external heating, SV opening, TR triggering, and TR ending. The flame behavior varies significantly at each stage. After the SV opens, the battery flame exhibits a jet state, and its velocity intensifies following TR triggering. Prior to TR triggering, WM can effectively extinguish the battery flame, but its efficacy diminishes once TR is triggered. This study examined the impact of WM on battery flames at different TR stages, and the key findings are summarized as follows:

The flame extinguishing duration for WM under different TR stages was different. WM could directly put out the flame within several seconds released when SV opened, 3 min after SV opening and when TR ended, and it took about 3 min for WM to put out the flame released when TR triggered. When WM is released before TR triggering, the flame temperature exhibits a gradient, with lower temperatures at the top and higher temperatures at the bottom. Notably, the temperature at a position 10 cm above the SV reaches significantly higher values. Conversely, when WM is released after TR triggering, the temperatures at 30 cm and 50 cm are both higher. This observation correlates with uneven combustion at the base of the TR flame and compression at its apex.

The heat release of battery flame Q could be calculated by the flame temperatures, and Q in Cases 1 to 5 were 321.5, 44.9, 97.0, 216.0 and 284.4 kJ, respectively. Using the time of SV opening, TR triggering and TR ending in Case 1 as the calculation reference, the differences in heat release between Case 1 and Cases 2 to 5 ΔQ were 11.6, 14.9, 54.4 and 37.1 kJ, respectively, and the specific cooling capacity of WM Q_m in Cases 2 to 5 were 1.8×10^{-3}, 1.7×10^{-3}, 2.8×10^{-3} and 5.9×10^{-3} kJ/kg, respectively. The results for Q_m could be well fitted as the logarithmic function, according to which, Q_m at each WM released temperature could be roughly estimated. Moreover, the heat radiation of flame to the battery Q_E could be calculated, and the case of WM released 3 min after SV opening exhibited the greatest proportion of heat radiation cooling η with the value of 88.4% which was the same for Q_m with the value of 1.7×10^{-3} kJ/kg.

This is expected to provide a novel focus for TR suppression in the LIB and make contributions to prevent further expansion of disasters from the perspective of TR flames.

Author Contributions: Conceptualization, B.M. and J.L.; methodology, B.M.; software, J.L.; validation, J.L.; formal analysis, B.M.; investigation, Q.W.; resources, B.M. and Q.W.; data curation, G.Z.; writing—original draft preparation, B.M.; writing—review and editing, C.G.; visualization, G.A.; supervision, J.O.; project administration, B.M. and Q.W.; funding acquisition, B.M. and Q.W. All authors have read and agreed to the published version of the manuscript.

Funding: This work was supported by the National Natural Science Foundation of China (NSFC) (52278359), Natural Science Foundation of Jiangsu Province (No. BK20221130) and General Program of National Natural Science Foundation of China (51974305).

Data Availability Statement: Dataset available on request from the authors.

Acknowledgments: The authors are thankful to all those who have contributed to the article and the funding support.

Conflicts of Interest: Author Bin Miao was employed by the company Zaozhuang Mining Group Co., Ltd and Guanzhang Zhu was employed by the company Beijing Tianma Intelligent Control Technology Co., Ltd. The remaining authors declare that the research was conducted in the absence of any commercial or financial relationships that could be construed as a potential conflict of interest.

Nomenclature

A	Flame surface area, m^2
A_{LIB}	Area of battery upper surface, m^2
c	Specific heat capacity of air, J/(kg·K)
d	Combustion chamber diameter, m
h	Convective heat transfer coefficient, W/(m^2·K)
L_i	Distance between i temperature measurement point and battery's upper surface, m
m_{WM}	WM consumption for flame extinguishing, kg
Nu	Nusselt number
$P_{E,i}$	Heat radiation power of each measurement points on the surface of the battery, W
Pr	Prandtl number
$(Pr_f/Pr_w)^{0.25}$	Physical property correction factor
Q	Heat release of battery flame, J
Q_E	Heat radiation of flame to the battery, J
Q_m	Specific cooling capacity of WM, kJ/kg
ΔQ	Difference in heat release, J
ΔQ_E	Differences in flame heat radiation to battery's upper surface, J
q	Heat flux density of battery flame, W/m^2
q_m	WM mass flow, kg/s
Re	Reynolds number
T_a	Ambient temperature, °C
$T_{F,i}$	Flame temperature at i position, K
$T_{f,i}$	Flame temperature at i position, °C

t_e	Time of SV opening, s
t_{SV}	Time of TR ending, s
v	Gas flow rate, m/s
Abbreviations	
LIB	Lithium-ion Battery
LFP	Lithium Iron Phosphate
NCA	Nickel Cobalt Aluminum Ternary Lithium
NCM	Nickel Cobalt Manganese Ternary Lithium
SEI	Solid Electrolyte Interface
SOC	State of Charge
SOH	State of Health
SV	Safety Valve
TR	Thermal Runaway
WM	Water Mist
Greek	
λ	Thermal conductivity of air, W/(m·K)
ε	Surface emissivity of the aluminum battery surface
α	Thermal diffusion coefficient, m^2/s
η	Proportion of heat radiation cooling
μ	Dynamic viscosity of air, m^2/s
ρ	Density of air, kg/m^3
σ	Stefan–Boltzmann constant
τ_{WM}	Release duration for WM to extinguish the flame, s

References

1. Hu, X.Y.; Liu, T.; Zhu, G.Q.; Cui, S.Q.; Huang, J.H.; Dong, X.T.; Guo, X.Y. Study on temperature heterogeneity and flame confrontation of $LiFePO_4$ battery thermal runaway inhibition by water mist. *Appl. Therm. Eng.* **2024**, *244*, 122675. [CrossRef]
2. Li, Q.; Yu, J.S.; Liu, G.Z.; Ma, X.G.; Si, W.; Hu, X.Y.; Zhu, G.Q.; Liu, T. Study on the Effectiveness of Water Mist on Suppressing Thermal Runaway in $LiFePO_4$ Batteries. *Crystals* **2023**, *13*, 1346. [CrossRef]
3. Liu, J.L.; Duan, Q.L.; Qi, K.X.; Liu, Y.J.; Sun, J.H.; Wang, Z.R.; Wang, Q.S. Capacity fading mechanisms and state of health prediction of commercial lithium-ion battery in total lifespan. *J. Energy Storage* **2022**, *46*, 103910. [CrossRef]
4. Liu, T.; Huang, J.H.; Hu, X.Y.; Cui, S.Q.; Zhu, G.Q. Study on the variation of normalized heat and gas release of overcharge-induced thermal runaway in confined space. *Appl. Therm. Eng.* **2024**, *243*, 122636. [CrossRef]
5. Zheng, Y.S.; Che, Y.H.; Hu, X.S.; Sui, X.; Stroe, D.I.; Teodorescu, R. Thermal state monitoring of lithium-ion batteries: Progress, challenges, and opportunities. *Prog. Energy Combust. Sci.* **2024**, *100*, 101120. [CrossRef]
6. Al-Zareer, M.; Dincer, I.; Rosen, M.A. Comparative assessment of new liquid-to-vapor type battery cooling systems. *Energy* **2019**, *188*, 116010. [CrossRef]
7. Feng, X.N.; He, X.M.; Ouyang, M.G.; Wang, L.; Lu, L.G.; Ren, D.S.; Santhanagopalan, S. A Coupled Electrochemical-Thermal Failure Model for Predicting the Thermal Runaway Behavior of Lithium-Ion Batteries. *J. Electrochem. Soc.* **2018**, *165*, A3748–A3765. [CrossRef]
8. Wang, G.Q.; Kong, D.P.; Ping, P.; Wen, J.; He, X.Q.; Zhao, H.L.; He, X.; Peng, R.Q.; Zhang, Y.; Dai, X.Y. Revealing particle venting of lithium-ion batteries during thermal runaway: A multi-scale model toward multiphase process. *Etransportation* **2023**, *16*, 100237. [CrossRef]
9. Cao, Y.F.; Wang, K.; Wang, Z.R.; Wang, J.L.; Yang, Y.; Xu, X.Y. Utilization of liquid nitrogen as efficient inhibitor upon thermal runaway of 18650 lithium ion battery in open space. *Renew. Energy* **2023**, *206*, 1097–1105. [CrossRef]
10. Zhu, X.Q.; Wang, H.; Wang, X.; Gao, Y.F.; Allu, S.; Cakmak, E.; Wang, Z.P. Internal short circuit and failure mechanisms of lithium-ion pouch cells under mechanical indentation abuse conditions: An experimental study. *J. Power Sources* **2020**, *455*, 100237. [CrossRef]
11. Gao, T.F.; Wang, Z.R.; Chen, S.C.; Guo, L.S. Hazardous characteristics of charge and discharge of lithium-ion batteries under adiabatic environment and hot environment. *Int. J. Heat Mass Transf.* **2019**, *141*, 419–431. [CrossRef]
12. Mao, B.B.; Liu, C.Q.; Yang, K.; Li, S.; Liu, P.J.; Zhang, M.J.; Meng, X.D.; Gao, F.; Duan, Q.L.; Wang, Q.S.; et al. Thermal runaway and fire behaviors of a 300 Ah lithium ion battery with $LiFePO_4$ as cathode. *Renew. Sustain. Energy Rev.* **2021**, *139*, 110717. [CrossRef]
13. Wang, Z.; Ouyang, D.X.; Chen, M.Y.; Wang, X.H.; Zhang, Z.; Wang, J. Fire behavior of lithium-ion battery with different states of charge induced by high incident heat fluxes. *J. Therm. Anal. Calorim.* **2019**, *136*, 2239–2247. [CrossRef]
14. Han, Z.; Zhao, L.; Zhao, J.; Xu, G.; Liu, H.; Chen, M. An Experimental Study on the Thermal Runaway Propagation of Cycling Aged Lithium-Ion Battery Modules. *Fire* **2024**, *7*, 119. [CrossRef]
15. Wang, B.X.; Zhou, Z.Z.; Li, L.; Peng, Y.; Cao, J.D.; Yang, L.Z.; Cao, B. Experimental study on thermal runaway and its propagation of large format prismatic lithium-ion batteries. *J. Energy Storage* **2022**, *55*, 105550. [CrossRef]

16. Feng, X.N.; Ouyang, M.G.; Liu, X.; Lu, L.G.; Xia, Y.; He, X.M. Thermal runaway mechanism of lithium ion battery for electric vehicles: A review. *Energy Storage Mater.* **2018**, *10*, 246–267. [CrossRef]
17. Wang, Q.S.; Ping, P.; Zhao, X.J.; Chu, G.Q.; Sun, J.H.; Chen, C.H. Thermal runaway caused fire and explosion of lithium ion battery. *J. Power Sources* **2012**, *208*, 210–224. [CrossRef]
18. Ghiji, M.; Novozhilov, V.; Moinuddin, K.; Joseph, P.; Burch, I.; Suendermann, B.; Gamble, G. A Review of Lithium-Ion Battery Fire Suppression. *Energies* **2020**, *13*, 5117. [CrossRef]
19. Golubkov, A.W.; Fuchs, D.; Wagner, J.; Wiltsche, H.; Stangl, C.; Fauler, G.; Voitic, G.; Thaler, A.; Hacker, V. Thermal-runaway experiments on consumer Li-ion batteries with metal-oxide and olivin-type cathodes. *Rsc Adv.* **2014**, *4*, 3633–3642. [CrossRef]
20. Zhong, G.B.; Mao, B.B.; Wang, C.; Jiang, L.; Xu, K.Q.; Sun, J.H.; Wang, Q.S. Thermal runaway and fire behavior investigation of lithium ion batteries using modified cone calorimeter. *J. Therm. Anal. Calorim.* **2019**, *135*, 2879–2889. [CrossRef]
21. Feng, L.; Jiang, L.H.; Liu, J.L.; Wang, Z.Y.; Wei, Z.S.; Wang, Q.S. Dynamic overcharge investigations of lithium ion batteries with different state of health. *J. Power Sources* **2021**, *507*, 230262. [CrossRef]
22. Liu, P.J.; Li, S.; Jin, K.Q.; Fu, W.D.; Wang, C.D.; Jia, Z.Z.; Jiang, L.H.; Wang, Q.S. Thermal Runaway and Fire Behaviors of Lithium Iron Phosphate Battery Induced by Overheating and Overcharging. *Fire Technol.* **2023**, *59*, 1051–1072. [CrossRef]
23. Wei, D.; Zhang, M.Q.; Zhu, L.P.; Chen, H.; Huang, W.S.; Yao, J.; Yuan, Z.C.; Xu, C.S.; Feng, X.N. Study on Thermal Runaway Behavior of Li-Ion Batteries Using Different Abuse Methods. *Batteries* **2022**, *8*, 201. [CrossRef]
24. Tao, C.F.; Zhu, Y.H.; Liu, Z.Q.; Li, R.; Chen, Z.Y.; Gong, L.L.; Liu, J.H. The experimental investigation of thermal runaway characteristics of lithium battery under different nitrogen concentrations. *J. Therm. Anal. Calorim.* **2023**, *148*, 12097–12107. [CrossRef]
25. Li, Y.W.; Jiang, L.H.; Huang, Z.H.; Jia, Z.Z.; Qin, P.; Wang, Q.S. Pressure Effect on the Thermal Runaway Behaviors of Lithium-Ion Battery in Confined Space. *Fire Technol.* **2023**, *59*, 1137–1155. [CrossRef]
26. Qiu, Y.S.; Jiang, F.M. A review on passive and active strategies of enhancing the safety of lithium-ion batteries. *Int. J. Heat Mass Transf.* **2022**, *184*, 122288. [CrossRef]
27. Yuan, S.; Chang, C.Y.; Yan, S.S.; Zhou, P.; Qian, X.M.; Yuan, M.Q.; Liu, K. A review of fire-extinguishing agent on suppressing lithium-ion batteries fire. *J. Energy Chem.* **2021**, *62*, 262–280. [CrossRef]
28. Zhang, L.; Li, Y.Q.; Duan, Q.L.; Chen, M.; Xu, J.J.; Zhao, C.P.; Sun, J.H.; Wang, Q.S. Experimental study on the synergistic effect of gas extinguishing agents and water mist on suppressing lithium-ion battery fires. *J. Energy Storage* **2020**, *32*, 101801. [CrossRef]
29. Sun, H.L.; Zhang, L.; Duan, Q.L.; Wang, S.Y.; Sun, S.J.; Sun, J.H.; Wang, Q.S. Experimental study on suppressing thermal runaway propagation of lithium-ion batteries in confined space by various fire extinguishing agents. *Process Saf. Environ. Prot.* **2022**, *167*, 299–307. [CrossRef]
30. Zhao, J.C.; Xue, F.; Fu, Y.Y.; Cheng, Y.; Yang, H.; Lu, S. A comparative study on the thermal runaway inhibition of 18650 lithium-ion batteries by different fire extinguishing agents. *Iscience* **2021**, *24*, 102854. [CrossRef]
31. Hill, D. *Considerations for ESS Fire Safety*; DNVGL: Byrum, Norway, 2017.
32. Tang, W.; Yuan, L.M.; Thomas, R.; Soles, J. Comparison of Fire Suppression Techniques on Lithium-Ion Battery Pack Fires. *Min. Metall. Explor.* **2023**, *40*, 1081–1087. [CrossRef]
33. Liu, T.; Tao, C.F.; Wang, X.S. Cooling control effect of water mist on thermal runaway propagation in lithium ion battery modules. *Appl. Energy* **2020**, *267*, 115087. [CrossRef]
34. Zhang, T.W.; Liu, H.; Song, J.W.; Wang, B.; Wang, Y.; Shuai, X.C.; Guo, Z.D. Synergistic inhibition effect on lithium-ion batteries during thermal runaway by N_2-twin-fluid liquid mist. *Case Stud. Therm. Eng.* **2022**, *37*, 102269. [CrossRef]
35. Li, L.X.; Chen, Z.; Lu, Y.; Zang, P.J.; Zhan, W.; Cheng, Y.H. Study on the suppression of thermal runaway of lithium-ion battery by water mist with different additives. *Energy Sources Part A-Recovery Util. Environ. Eff.* **2023**, *45*, 11349–11362. [CrossRef]
36. Zhang, L.; Duan, Q.L.; Xu, J.J.; Meng, X.D.; Sun, J.H.; Wang, Q.S. Experimental investigation on suppression of thermal runaway propagation of lithium-ion battery by intermittent spray. *J. Energy Storage* **2023**, *58*, 106434. [CrossRef]
37. Mei, J.; Shi, G.Q.; Liu, H.; Wang, Z.; Chen, M.Y. Experimental study on the effect of passive retardation method for thermal runaway mitigation of lithium-ion battery. *Appl. Therm. Eng.* **2023**, *230*, 120861. [CrossRef]
38. Lönnermark, A. TOXFIRE-Fire Characteristics and Smoke Gas Analysis in under-Ventilated Large-Scale Combustion Experiments. Tests in the ISO 9705 Room. 1996. Available online: https://www.diva-portal.org/smash/get/diva2:962010/FULLTEXT01.pdf (accessed on 6 June 2024).
39. Diaz, F.; Wang, Y.; Weyhe, R.; Friedrich, B. Gas generation measurement and evaluation during mechanical processing and thermal treatment of spent Li-ion batteries. *Waste Manag.* **2019**, *84*, 102–111. [CrossRef]
40. Lee, J.-H.; Hong, S.-H.; Lee, H.-S.; Park, M.-W. Study on Gas-Generating Property of Lithium-Ion Batteries. *Fire Sci. Eng.* **2021**, *35*, 1–8. [CrossRef]
41. Incropera, F.P.; DeWitt, D.P.; Bergman, T.L.; Lavine, A.S. *Fundamentals of Heat and Mass Transfer*; Wiley: New York, NY, USA, 1996; Volume 6.

42. Han, R.; Tang, M.; Wang, D.; Zhang, S. Numerical analysis of the convective heat transfer coefficient effect on lithium battery thermal diffusion when considering temperature effect. *Sci. Technol. Rev.* **2023**, *41*, 104–112.
43. Hu, Z. *Experimental Investigation of Steel and Aluminum Alloy Surface Emissivity Characteristics*; Henan Normal University: Xinxiang, China, 2010.

batteries

MDPI

Article

Assessment of Run-Off Waters Resulting from Lithium-Ion Battery Fire-Fighting Operations

Arnaud Bordes, Arnaud Papin, Guy Marlair *, Théo Claude, Ahmad El-Masri, Thierry Durussel, Jean-Pierre Bertrand, Benjamin Truchot and Amandine Lecocq

Institut National de l'Environnement Industriel et des Risques (Ineris), Parc Technologique Alata, BP2, 60550 Verneuil-en-Halatte, France; arnaud.bordes@ineris.fr (A.B.); arnaud.papin@ineris.fr (A.P.); ahmad.el-masri@ineris.fr (A.E.-M.); benjamin.truchot@ineris.fr (B.T.)

* Correspondence: guy.marlair@ineris.fr; Tel.: +33-344556348

Abstract: As the use of Li-ion batteries is spreading, incidents in large energy storage systems (stationary storage containers, etc.) or in large-scale cell and battery storages (warehouses, recyclers, etc.), often leading to fire, are occurring on a regular basis. Water remains one of the most efficient fire extinguishing agents for tackling such battery incidents, and large quantities are usually necessary. Since batteries contain various potentially harmful components (metals and their oxides or salts, solvents, etc.) and thermal-runaway-induced battery incidents are accompanied by complex and potentially multistage fume emissions (containing both gas and particles), the potential impact of fire run-off waters on the environment should be considered and assessed carefully. The tests presented in this paper focus on analyzing the composition of run-off waters used to spray NMC Li-ion modules under thermal runaway. It highlights that waters used for firefighting are susceptible to containing many metals, including Ni, Mn, Co, Li and Al, mixed with other carbonaceous species (soot, tarballs) and sometimes undecomposed solvents used in the electrolyte. Extrapolation of pollutant concentrations compared with PNEC values showed that, for large-scale incidents, run-off water could be potentially hazardous to the environment.

Keywords: Li-ion battery; fire; safety; thermal runaway; toxicity; water; firefighting; pollutants

Citation: Bordes, A.; Papin, A.; Marlair, G.; Claude, T.; El-Masri, A.; Durussel, T.; Bertrand, J.-P.; Truchot, B.; Lecocq, A. Assessment of Run-Off Waters Resulting from Lithium-Ion Battery Fire-Fighting Operations. *Batteries* **2024**, *10*, 118. https://doi.org/10.3390/batteries10040118

Academic Editor: Mingyi Chen

Received: 6 February 2024
Revised: 22 March 2024
Accepted: 25 March 2024
Published: 31 March 2024

1. Introduction

The current development of Li-Ion batteries concerns numerous, application fields, and the thermal runaway hazard about those systems, often leading to fire and sometimes explosion events, remains a resilient issue. In parallel to the wide spread of Li-ion-powered consumer products in complex built environments, the increasing use of applications of LIB for e-mobility or large-scale battery energy storage systems (BESS), in the hundreds of MW power range, requires the urgent development of environmentally friendly strategies to fight lithium-ion battery fires. Considering that water remains one of the most efficient fire extinguishing agents to fight battery fires, and in many cases is the only extinguishing medium available in operational quantities to the fire-brigades, the potential impact of relating fire run-off waters to the environment should be considered and assessed carefully. Lessons of the past have primarily shown that uncontrolled release of toxic fire waters in rivers may lead to a dramatic consequence for water livestock, as primarily shown by major incidents involving large storage of toxic chemicals such as in Basle (Sandoz fire, Switzerland, 1986) [1,2] or in Tianjin (China, 2015) [3]. This is a prerequisite for establishing a clear and science-based firewater management doctrine [4]. In particular, the level of contamination of fire waters in terms of toxicity to aquatic ecosystems is needed to decide on the free release of extinguishing waters into the environment or into rainwater drain systems or on their containment in suitable systems for post-hazardous liquid waste management [5].

During the thermal runaway phenomenon—initiating stage of relating field failures—it is well-known that systems containing Li-ion batteries produce emissions or effluents which can range anywhere within the full spectrum of physical states, e.g., liquids (electrolyte leak or ejection), gases or vapors or solid aerosols [6–10], which adds complexity to both non-flaming and flaming conditions. These emissions may in turn interact with the environment and lead to pollution [11]. One of the contamination modes of both land and aquatic ecosystems is the aerosol sedimentation process arising during smoke plume dispersion, often at a stage where contaminant concentrations in the smoke plume are significantly diluted at a certain distance from the incident. Another possible and easier way of pollution is linked to the extinguishing agents used, typically water used by fire suppression systems or fire brigades, which can carry effluents emanating from the damaged battery. These various modes of contamination have been, unfortunately, largely confirmed in a significant number of fire records, as exemplified by Mc Namee et al. [12]. This shows, in particular, the diversity of influencing factors in terms of burning materials, size, and fire duration, potentially leading to environmental damage. Regarding batteries, the contaminants involved depend on the materials composing the system. These materials vary from one Li-ion battery chemistry/geometry to another and from one system to another, but the phenomena at stake and the resulting effects are close [13]. For small- or medium-isolated batteries (e.g., used for portable applications), the accidental contamination risk should be relatively low, but for more energy- or power-demanding applications leading to larger battery systems (containerized BESS, . . .) or large-scale cell and battery storages (warehouse, recyclers, . . .), the consequences might start to raise concerns in the absence—so far the usual case—of any fire water containment capacity. As a matter of fact, according to EPRI information, 64% of the BESS site owners are considering the implementation of water containment for the firefighting run-off waters [4]. Regarding fire extinguishing waters used to tackle car fires, if detailed studies [14,15] of fire water ecotoxicity had concluded that subsequent fire water run-off had a negligeable impact on the environment, as far as ICE cars are concerned, more caution is likely to be needed with EVs, given the significant differences applied from potential contaminants from the battery and the amount of water requirements.

Emissions during thermal events are directly linked to the materials constituting the battery. However, they will possibly be altered by reactions of thermal decomposition, electrolysis or even combustion that might drastically change the nature and properties of the ejected matter [8]. Carrying those substances by water will vary depending on the chosen extinguishing method. Three options are generally possible: (1) Direct watering of the batteries—when sprinklers or water fire hose are directed to the faulty system with direct contact with the batteries. (2) Fire plume watering for fire and smoke progress abatement—when water is not applied directly to the system but to its surroundings to prevent fire and subsequent damage propagation to adjacent elements and therefore minimize the impact of the root fire. (3) Water immersion—when the battery is immersed in a large volume of water, either after an incident to cool down the sample, or during an incident to try to limit it. In this last option, managing firefighting waters is relatively simple as water is already contained.

Water contamination in the smoke watering scenario (#2 firefighting option) was recently studied by EMPA [16,17] while analysis of immersion water (#3 firefighting option) has been performed both by EMPA and RIVM [18]. However, more globally, published information regarding contamination of fire waters used to tackle li-ion battery fires, regardless of the application, remains quite scarce. Therefore, further investigation is needed to confirm the early trends observed [19] and to address those issues in the entire value chain of LIBs.

In the present paper, the case of direct watering of the batteries is the only scenario studied. Commercially available NMC battery modules composed of two different cell formats (18,650 and prismatic) were chosen for the experimental approach selected in this study.

2. Materials and Methods

2.1. Description of the Samples

Two types of commercial Li-ion modules were used, both composed of NMC/graphite cells.

Module A comprises 16-metal-can prismatic cells (7.5 Ah) and has an electrical energy of 500 Wh. In addition to the electrochemical cells, the module also includes metallic (aluminum plates, cells connectors) and plastic (casing cover, wire insulation etc.) parts.

Module B is an assembly of 2 cell blocks, each one composed of 45-cylindrical 18,650 cells (2.4 Ah), circled with a metallic grid to ensure its mechanical integrity. The total energy of the battery assembly is 900 Wh. In addition to the electrochemical cells, the module also integrates a thin plastic film keeping the cells tightly together and connected.

The week before the abuse tests, the modules were fully charged using a constant current profile at C/5 using a cycling bay from FEV manufacturer.

2.2. Abuse Tests Set Up

Abuse tests on modules were performed in the Ineris 80 m^3 test chamber equipped with a smoke exhaust and treatment system remotely controlled to fully extract, measure, and eventually convey gases through the gas cleaning system of the facility before their rejection to the atmosphere. The room is also connected to a water-draining system to collect all liquid effluents produced by the fire or during the fire suppression process. In the testing room, the air entrance is located on one side, near the ground; extraction is placed in the center of the roof. All tests were performed under air with an extraction flow rate in the test chamber of approximately 2 500 m^3/h.

The sample was positioned in the center of the test chamber for each test, as represented in Figure 1. Modules were positioned on a metal grid, electrically insulated, using a small support made of inert material (calcium silicate).

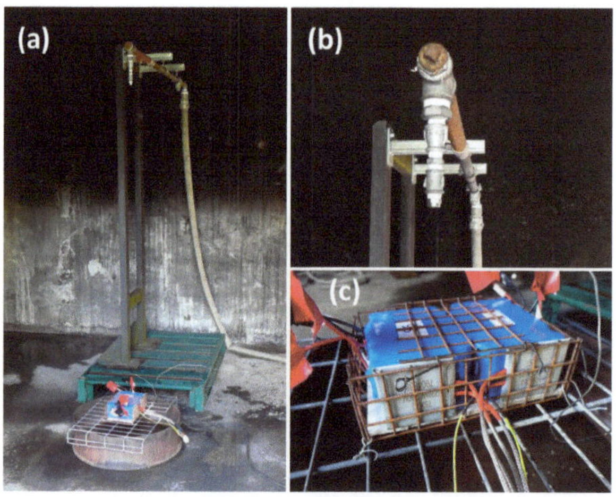

Figure 1. Pictures of the experimental set-up. (**a**) Overview, (**b**) sprinkler head, and (**c**) battery module B. White pads on the front faces of each battery block correspond to the heaters.

For module A, as the thermal pad failed to initiate a thermal runaway, a 20 kW gas burner was selected and positioned 30 cm from the sample and directed to the middle of the module. To prevent any interaction between the propane burner and the water used for firefighting, the burner was switched off as soon as thermal runaway was triggered.

For module B, two thermal pads with individual power of 220 W and a 50 cm^2 surface were put in contact with each cell block.

Since the objective of the tests was to evaluate water contamination in thermal runaway situations, the sprinkler activation was performed manually as soon as the thermal runaway was visually confirmed. As the modules were not equipped with thermocouples, the thermal runaway event was considered occurring when flames were escaping for the first time from the module. The application rate was set at 10 $L/m^2/min$. The basin surface was 0.25 m^2, and the volume of collected water was estimated by calculation using the water flow, the watering time, and the basin surface.

2.3. *Water Sampling*

After each test, 2 L of water was immediately sampled from the extinguishing water containment basin for chemical composition analysis. It is important to highlight that no filtration was made to keep all of the emissions in the analyzed samples, whatever the chemical or physical processes that were involved in the interactions of emissions from the battery module and extinguishing waters (condensation, dissolution, sedimentation etc.), since the objective of the test was to characterize the global composition of runoff water.

Before the test, the water receptacle was exposed to a direct flame to remove the potential traces of organic solvents. However, deposit remains possible, and a reference was then carried out by watering the same set-up, without any battery, in order to have a baseline of potential species inherently present in the water supply or due to receptacle component extraction during sampling.

2.4. *Water Analysis*

2.4.1. Inductively Couple Plasma Optical Emission Spectroscopy

Inductively Couple Plasma Optical Emission Spectroscopy (ICP-OES, Agilent 5110 equipment, Santa Clara, CA, USA) has been used for the analysis of major elements (Al, Fe, Li, Na, Ni, P).

2.4.2. Inductively Couple Plasma Mass Spectrometry

Inductively Couple Plasma Mass Spectrometry (ICP-MS, Agilent 7900 instrument, Santa Clara, CA, USA) has been used for the analysis of trace elements (Co, Cu, Mn). Instead of the ICP-OES used for major elements by measuring the light emitted from elements, ICP-MS uses a quadrupole to filter the ions according to their mass/charge ratio and counts each mass passed to the detector. The high sensitivity of the ICP-MS detector provides a much lower detection limit than ICP-OES.

2.4.3. Ion Chromatography

Chloride and fluoride species were measured by ion chromatography (Metrohm, 850 Professional IC, Herisau, Switzerland) with conductimetric detection. Ion chromatography is a method for separating ions (Cl^- and F^-) based upon their interactions with resin (stationary phase) and the eluent (mobile phase).

2.4.4. Liquid Chromatography

To extract polycyclic aromatic hydrocarbons (PAHs) from the water sample, a separation of the particle phase was carried out using glass wool. Aqueous phases were extracted using dichloromethane by liquid/liquid extraction and particulate phase was extracted using acetonitrile. Both extracts were evaporated and collected in 0.5 mL of acetonitrile each and recombined in the same vial before analysis.

Analysis of PAHs was performed on a liquid chromatography system, an ultimate 3000 from thermo coupled to a diode array detector (DAD) and fluorescence detector (FLD) detector. Molecules were separated on C18 column (Zorbax eclipse PAH 2.1 × 150 mm 1.8 micron from Agilent). All PAHs were quantified using the FLD detector except for Acenaphthylene that was quantified using the UV-DAD detector.

2.4.5. Gas Chromatography

Carbonates were analyzed using a gas chromatography system from Varian. Samples were diluted in methanol and 1µL was injected in split mode 1:10. Separation was performed on a capillary column from Agilent VF-5 ms 60 m, 0.25 mm internal diameter and 1 µm film thickness. A flame ionization detector was used to quantify the different compounds.

2.5. Particle Morphology Characterization

A particle size distribution analysis using the centrifugal disc method (by use of CPS Disc Centrifuge™ instrument, Tokyo, Japan) and further particle morphology study by Transmission Electron Microscopy (TEM, JEOL 1400 Plus instrument, Tokyo, Japan) were carried out on the sampled water. To enable this analysis, all particles larger than 2 µm in size were filtered beforehand.

To perform microscopic analysis of the particles, a droplet of the sample suspension was casted on a copper grid and dried at room temperature to be observed with a TEM (Transmission Electron Microscope, JEOL, 1400 Plus, Tokyo, Japan). A beam of electrons accelerated by a high voltage (120 kV) passes through a very thin sample, in this case a carbonized copper grid on which a microdrop of the sample to be analyzed was deposited. During the electron–matter interaction, the transmitted and diffracted electrons are used to form an image with high resolution in gray levels, and the X-ray photons allow for a micro or even a nano-volume of the sample to be chemically characterized.

3. Results

3.1. Test Conditions and TR Characteristics

Table 1 compiles the test conditions and reactions observed during the three successive test runs and Figure 2 gives details on the timeline of the experiments and presents pictures of the markers of significant events.

Table 1. Summary of test conditions and observations.

	Module Type	Module Energy	Heating Method	Reaction	Module State after Test	Sprinkler Flow	Amount of Water Delivered
Test 1	Prismatic cell assembly NMC (module A)	500 Wh	Gas burner	Venting + moderate fire	Upper plastic burnt Mechanical integrity conserved No module casing opening	10 $L/m^2/min$	7 L
Test 2	Prismatic cell assembly NMC (module A)	500 Wh	Gas burner	Jet fire + explosion	Module casing ejected All cells fully burnt with casing damaged	10 $L/m^2/min$	7 L
Test 3	Cylindrical cell assembly NMC (module B)	900 Wh	Thermal pad	Jet fire + explosion	All cells burnt Some jelly rolls visible	10 $L/m^2/min$	9 L

In the first experiment, the thermal runaway of module A was characterized by the emission of a large amount of white smoke followed by the appearance of flames. No jet fire was observed, but a rather moderate combustion process, as visible on the first line of Figure 2 was observed. Water was applied for 2 min 50 s, corresponding to a volume of collected water of 7 L. The flames stopped as soon as water was applied. After the test, no cells presented any side wall rupture, and their mechanical integrity was conserved.

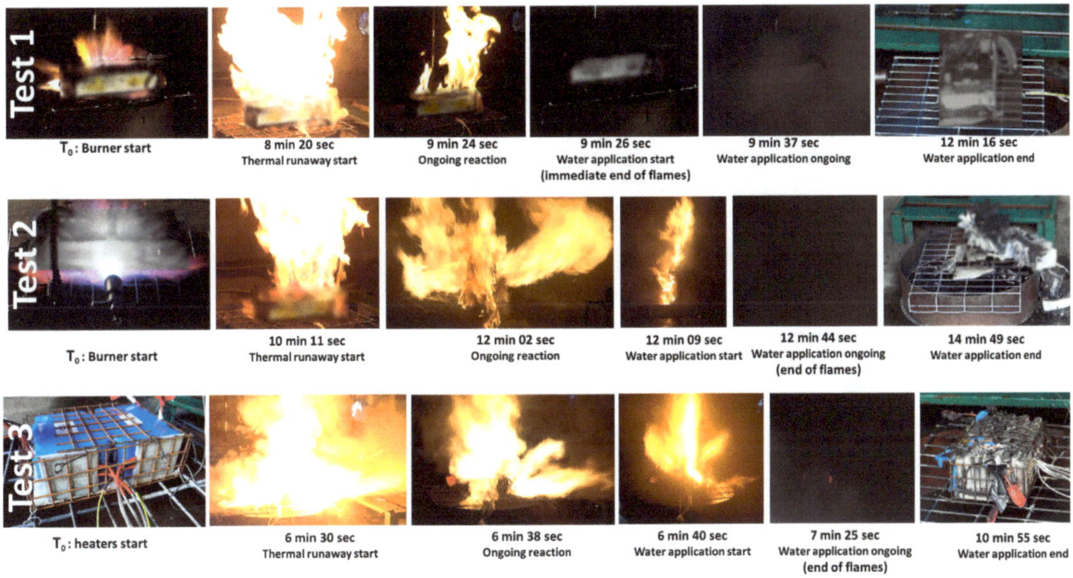

Figure 2. Extracts from the test video and timeline of the three experiments.

Test 2 was performed because module A was only moderately impacted by the first experiment. It was decided that the thermal runaway of module A should be further pushed and to restart the burner. After a few minutes of heating, the module entered again in the thermal runaway process. In this case, the reaction was much more violent since the jet fire was observed, the module casing was ejected, and all cells subsequently seemed damaged, some of them losing their mechanical integrity (casing opening). The second line of Figure 2 shows the reaction's visible effects just before water application (12 min 9 s). Water was applied for 2 min 40 s, leading to an additional volume of collected water of 7 L, i.e., a total of 12 L considering 5 L remains from test 1 (after that 2 L were sampled for analysis). Contamination levels indicated for test 2 are the values corresponding to the mix of the 5 L remaining from test 1 and the 7 L applied during test 2. The flames did not stop immediately upon water application, and an unknown portion of the water vaporized before reaching the receptacle. In the first approximation, this proportion of water vaporized was not considered for further calculation of the contaminant. The flames stopped 35 s after the application of water.

For module B, a single TR/fire water suppression step was carried out when the thermal runaway was reached. The third line of Figure 2 shows that the reaction was rather violent. All of the cells seemed damaged after the test and some of them lost their mechanical integrity (casing opening or jelly roll ejection). Water was applied for 3 min 30 s corresponding to a volume of collected water of 9 L, neglecting once more the vaporizer part. Flames stopped 45 s after water activation.

3.2. *Characterization of Water Contamination*

3.2.1. Halogens and Metals

Table 2 shows the results of the analyses for the presence of the two anions (F^- and Cl^-) as well as a selection of metal compounds. Those species have been chosen to reflect the foreseeable pollutants considering NMC Li-ion batteries composition [20].

As expected, the levels of fluorides and metals are found in large amounts, due to the composition of the cells. In module A, phosphorus and fluoride ions are the dominant species. In contrast, in module B, lithium is the more concentrated pollutant element compared to all other metallic elements and fluorides or chlorides. All these species are found in cell electrolytes or in the electrode for Li. Transition metals contained in the cathode (Ni, Mn, Co) are found mainly when the reaction was violent (tests 2 and 3). Their ratios, across different tests vary but in the three tests, Ni is overrepresented compared with Mn and Co, which is expected as stoichiometry of the current NMC cathode favors Ni. Their presence—in undetermined metal containing chemicals (oxides ? hydroxides ? metal complexes ?) [21]—is consistent, with composition of the selected cells. In order to better understand their respective amount, further studies on their chemical state and their solubility in water are necessary.

Table 2. Analysis of anions (F^- and Cl^-) and a selection of metals in the water before application and in the three samples after extinguishing. QL = quantification limit. Uncertainty values refer to expanded uncertainties (k = 2).

	QL	Uncertainty	Reference	Test 1 (Module A)	Test 2 (Module A)	Test 3 (Module B)
			Ions			
F^- (mg/L)	0.05	8%	0.25	142	91.6	93.7
Cl^- (mg/L)	0.01	3%	24.9	33.4	36.5	203
			Metals			
Al (mg/L)	0.17	15%	0.91	74.2	29.3	73.9
Co (mg/L)	0.03	10%	<LQ	0.42	12.8	7.07
Cu (mg/L)	0.03	10%	0.04	0.30	0.26	4.18
Fe (mg/L)	0.08	9%	0.30	5.92	4.59	0.30
Li (mg/L)	1.67	15%	<LQ	44.5	27.8	360
Mn (mg/L)	0.03	10%	<LQ	1.22	17.0	5.82
Na (mg/L)	1.67	14%	13.0	15.6	16.3	26.2
Ni (mg/L)	0.08	12%	<LQ	3.25	49.0	40.1
P (mg/L)	0.17	17%	<LQ	201	113	5.80

Aluminum, copper, and iron in pristine cells are present in sheets or bulk form and as particulate matter; therefore, they are expected to be less present in particulate emission. Aluminum is, however, found in noteworthy amounts probably because of its low melting point (660 °C). Iron and copper, which have higher melting points are found in relatively low amounts in the three tests.

By comparing the two extinguishing operations on the prismatic cells (test 1 and 2), it can be observed that when the thermal runaway thermal impact is characterized by a fully developed combustion process, the interacting water collected is much more concentrated in Polycyclic Aromatic Hydrocarbon (PAHs) and cathode metals (Ni, Mn, Co). On the other hand, the concentrations of elements essentially coming from the liquid electrolyte (typically Li, P, F) are present in higher quantities (1.6 to 1.8 factor) when the reaction is not fully developed, and where the electrolyte has a chance to be dragged in the water. This observation is coherent with the higher amount of organic carbonates found in test 1 presented in Section 3.2.3.

The comparison of the results between different cell geometries also confirms the importance of this parameter, influencing, in particular, the mechanical strength of the system and, therefore, the confinement of the species.

3.2.2. Poly Aromatic Hydrocarbons (PAHs)

Another important family of water contaminants in fire situations is PAHs. While the common specification for PAHs mentions 16 substances to be analyzed [22], 23 PAHs were analyzed; the results are reported in Table 3.

Table 3. Analysis of 23 PAHs in the water before application and in the three samples after extinguishing. QL = quantification limit. (Expanded: k = 2) Uncertainty of analysis for HAPs is 15% for all species.

PAH (Polycyclic Aromatic Hydrocarbons)					
	QL	**Reference**	**Test 1 (Module A)**	**Test 2 (Module A)**	**Test 3 (Module B)**
Naphtalène (ng/L)	10.0	<LQ	1279.2	2792.2	3114.6
Acénaphtylène (ng/L)	40.0	<LQ	2421.7	2405.1	1193.4
méthyl-1.naphtalène (ng/L)	10.0	<LQ	26.8	459.4	667.1
méthyl-2.naphtalène (ng/L)	10.0	<LQ	203.2	<LQ	2058.4
Acénaphtène (ng/L)	2.0	<LQ	34.1	110.6	275.7
Fluorene (ng/L)	2.0	<LQ	74.1	752.3	1055.0
Phénanthrène (ng/L)	4.0	5.7	360.9	3026.8	2581.6
Anthracène (ng/L)	2.0	<LQ	10.6	330.5	303.3
Fluoranthène (ng/L)	2.0	10.8	57.7	1280.9	349.8
Pyrène (ng/L)	2.0	7.2	45.1	1279.8	20.5
méthyl-2.fluoranthène (ng/L)	4.0	<LQ	7.3	45.1	21.3
B(a)A (ng/L)	2.0	<LQ	24.8	185.7	131.8
Chrysene (ng/L)	2.0	<LQ	32.5	212.3	40.8
Retene (ng/L)	2.0	<LQ	104.9	170.7	19.8
B(e)P (ng/L)	2.0	<LQ	7.5	306.3	50.4
B(j)F (ng/L)	20.0	<LQ	<LQ	106.3	<LQ
B(b)F (ng/L)	2.0	<LQ	34.6	259.6	5.8
B(k)F (ng/L)	2.0	<LQ	8.3	81.0	8.2
B(a)P (ng/L)	2.0	<LQ	13.0	163.9	20.8
D(a.h)A (ng/L)	2.0	<LQ	<LQ	36.7	4.5
benzo(ghi)P (ng/L)	2.0	<LQ	13.3	169.6	4.1
Indèno (ng/L)	4.0	<LQ	35.2	162.1	11.8
Coronene (ng/L)	2.0	<LQ	4.0	54.0	<LQ

It shows the presence of numerous PAHs including naphtalene and phenantrene, the most present, which typically indicates the combustion of hydrocarbon-based products. Specific attention should be paid to B(a)P as it is class 1 on the IARC scale (proven carcinogen). According to the potential ecotoxicological impact of those products, one should pay specific attention to the potential impact of runoff water.

3.2.3. Organic Carbonates

To complete the chemical characterization of the pollutants in the extinguishing waters, a selection of organic carbonates, classically used as electrolyte solvents or critical additives (VC, FEC), was quantified. Results are shown in Table 4.

Table 4. Analysis of 7 common carbonates used as electrolytes in the water before application and in the three samples after extinguishing. QL = quantification limit.

Species	**QL**	**Reference**	**Test 1 (Module A)**	**Test 2 (Module A)**	**Test 3 (Module B)**
DMC (μg/mL)	8.8	n/a	n/a	n/a	n/a
EMC (μg/mL)	8.3	n/a	138	59	n/a
VC (μg/mL)	9.4	n/a	n/a	n/a	n/a
DEC (μg/mL)	8.1	n/a	n/a	n/a	n/a
FEC (μg/mL)	10.2	n/a	n/a	n/a	n/a
EC (μg/mL)	7.7	n/a	1082	461	n/a
PC (μg/mL)	10	n/a	n/a	n/a	n/a

The presence of such compounds is found only in tests 1 and 2. This difference between the tests could be explained by the important combustion reaction observed during tests 3; it is most likely that the high temperature reached during this test led to the total evaporation and possible thermal decomposition of those volatile and easily flammable compounds before being dragged into the wastewaters. The boiling point for EC is typically 244 °C and 100 °C for EMC, which is significantly lower than the flame temperature. For the same reason, as the reaction in test 1 was less violent than in test 2, the quantity of carbonates found is higher for test 1 than for test 2. Species identified in the water are EMC and EC which are very commonly used as electrolyte solvents. Also, the boiling point difference might explain the difference between the quantity of EC and EMC found in the liquid phase, as EMC evaporates more easily. This also means that the massive use of water to cool down a whole system as a container could lead to a higher concentration of organic carbonates since part of the cells might, in such a case, be damaged but not burnt. Hydro solubility of those compounds may also play an important role (778 g/L for EC and 46.8 g/L at 20 °C for EMC) and explain the differences in the concentrations found. These compounds must be carefully monitored because they cannot easily be filtered out or left to settle.

3.2.4. Particle Size Analysis

To complete the chemical analysis of the water, particles sizes in the water were evaluated using the CPS method. Using Stokes' law, a hydrodynamic intensity-weighted particle-size distribution of diameters is obtained and transformed into a volume-weighted or number-weighted particle-size distribution, as presented in Figure 3.

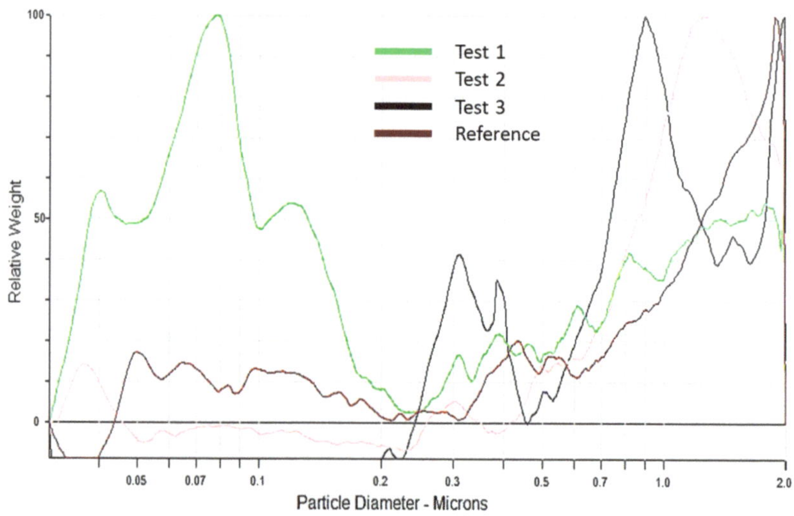

Figure 3. Number-weighted particle-size distribution of the particles presents in the water before application (reference) and in the three samples after extinguishing. Measurement was carried out by CPS.

This analysis leads to the conclusion that only extinguishing water from test 1 has a nanometric fraction, with particles around 70 nm in diameter. Other samples contain a majority of particles between 0.9 and 2 μm. This analysis confirms the possibility, mentioned in the literature [11], that extinguishing water might be loaded with nanoparticles, without being able to quantify them with the method used. Also, because nanoparticles are absent from tests where the reaction was the most developed, it can indicate that those particles might be dragged in the smoke plume before being dragged by water.

To get information on the nature of the nanometric particles in the extinguishing water of test 1, additional analysis by transmission electron microscopy were performed. Images are presented in Figure 4. Figure 4a shows a picture of a representative sample of what was observed over the entire grid. Several populations of particles of highly variable sizes are identified and presented on Figure 4b–d. The majority of particles are the finest and correspond to the smallest black dots in Figure 4a. According to Figure 4b, one can conclude that soot nanoparticles agglomerate and form nanostructured clusters. Spherical particles of intermediate size are then observed (Figure 4c) and are associated with tarballs, having a diameter around 100 nm. Finally, the largest particles (Figure 4d) have a characteristic size around one micron and are mainly metal particles, composed of iron and aluminum.

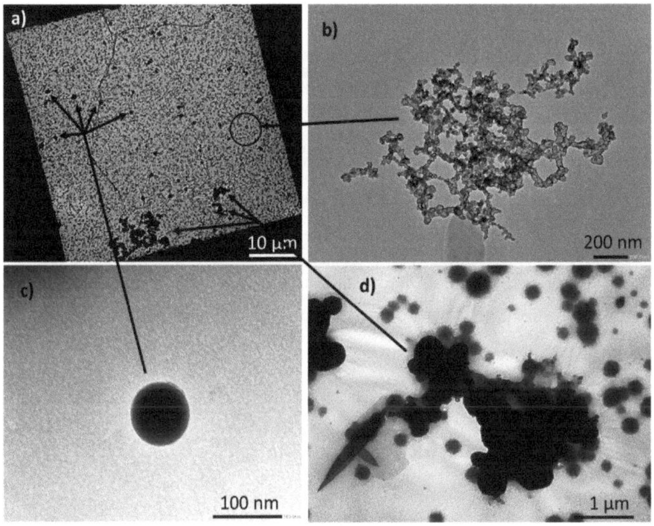

Figure 4. TEM analysis of the particles presents in the extinguished water of test 1. (**a**) Zoom-out of a representative area showing the relatives proportion of the three particles population encountered, (**b**) zoom-in of particles identified as soot agglomerate, (**c**) zoom-in of particles identified as tarballs and (**d**) zoom-in of particles identified as metallic.

Particles bellow 2.5 µm are inhalable and might pose a toxicological risk for humans [23]. In the case of this study, the particles are in water, making eco-toxicity the main risk identified. No size threshold is clearly defined in the literature nor in regulations. Some studies have nonetheless showed that particles with a size lower than 100 nm can enter the root system of higher plants and be translocated to aerial parts which demonstrate the possibility of trophic transfer [24]. In invertebrates (water flee), accumulation of several types of nanomaterials has been shown [25]. The interaction phenomenon between metallic oxide nano materials and freshwater micro-algae was also evidenced by Rivero et al. [26].

3.2.4.1. pH Measurement

Table 5 shows the pH measured in the sampled water immediately after each test.

Table 5. pH of the extinguishing water. (Expanded: k = 2) uncertainty of measurement is 1%.

	Test 1 (Module A)	Test 2 (Module A)	Test 3 (Module B)
pH	5.2	5.9	11

Depending on the test, the pH of extinguishing water is either acidic or basic. Values obtained in our tests would rate the corresponding water clearly outside recommended freshwater quality standards (6.5 < pH < 9) or limiting pH values for treatment in wastewater sewage systems (see Table 6).

Table 6. pH limit values in different local regulations.

pH Limit Values	Drinkable Water	Industrial Effluent Value for Discharge in Sewage Systems
EPA [27] (USA)	6.5 to 9	
Canada [28,29]	7 to 10.5	6–9
Switzerland [16,17]	6.8 to 8.2	6.5 to 9

Battery field failure incident reports as well as scarce pH values reported in the literature mostly report very basic contaminated water resulting from fire-fighting operations [16–18,30], although this is not always the case [31].

The basicity of the water is sometimes explained by the inner content of the cell that may contain soluble metal hydroxides. By contrast, the resulting acid fire water could be related to the interaction of the water with the acidic gases contained in the fire plume [31]. A difference in concentration in metallic species between the two tests might explain (see Table 2) the difference in pH observed. Depending on the environments in which the water will evolve (acidic or basic soils, etc.), it cannot be ruled out that these pHs are modified [32] before final pouring into aquatic ecosystems (surface of underground water resources).

4. Discussion

These tests were carried out at a small scale compared to what could occur, for example, in the event of an incident with a stationary storage container or storage warehouse. In such an event, the quantities of batteries involved, and the quantity of water used for extinction would be much higher. To estimate the orders of magnitude of water contamination values for a realistic situation (BESS container or storage warehouses), a simplistic extrapolation of the results obtained based on real incident data is proposed. In the Perles and Castelet (Ariège in France) battery stationary storage fire, which is well documented [33], and involving a stationary storage of 1500 kWh, the local authorities estimated that a volume of water of 180 m^3 was used by the firefighters, i.e., 0.12 L/Wh. This volume seems to be a good basis to extrapolate results as other feedback for other large-scale applications give similar values [34].

In the tests presented here, the volume of water used is coherent with other same-level studies [35] and, for test 1 on prismatic cells, 7 L were poured onto the 500 Wh battery (0.014 L/Wh) during test 2, and the total volume of water was 12 L (0.024 L/Wh). For the cylindrical cell, 3.9 L were poured onto the 900 Wh (0.01 L/Wh) battery. The values proposed in Table 7 correspond, for a selection of substances, to an extrapolation using a proportionality rule between the concentrations measured during the tests and the actual conditions reported during the Ariège incident (see Supplemental Material). This calculation also assumes that the normalized water flow rate (per watt-hour) does not significantly influence the mass transfer of pollutants in the run-off water.

In order to evaluate the potential environmental hazard of these wastewaters, the last column presents the "Predicted No-Effect Concentration"(PNEC) of the substance when available on the ECHA website [36]. Those values should be read with caution as they are given for a yearly average and are extracted from several sources, including industrial ones. The concentration in the wastewater was above PNEC values for all the substances studied when the data were available except for naphthalene, showing a potential environmental hazard. Two compounds show a particularly high hazardous potential: Co and EMC with concentrations, respectively, 2500 and 260 times greater than their PNEC. This means that, in a realistic scenario where two fire hoses are used to fight a fire using 1000 L/min, and the waste waters are flowing to a small river with a flow of 3 m^3/s, the concentrations of contaminants in the river are still above the PNEC for those compounds. It is also worth noting that some of the compounds' PNEC could not be found on the ECHA website but might be even more hazardous. For example, a PNEC as low as 0.0017 mg/L can be found for nickel [37] from sources other than ECHA. Another point to consider is the possible interaction between the contaminants. To assess this, the best method would be to test the particle mix directly. Few studies of this kind are available but, Yang et al. has recently shown [8] that particles from the NMC cell thermal runaway could cause inhibition of bacterial activities in the range of 25–200 mg/L and severe acute toxicity at 100 mg/L in 5 h [8] and Quant et al. showed the acute toxicity of the runoff water [19].

Table 7. Extrapolation of the experimental results to a real application and extinguishing. The last column presents the PNEC of the compound when available on ECHA website [36].

Substance	Test 1 (Module A)	Test 2 (Module A)	Test 3 (Module B)	PNEC Freshwater
Al (mg/L)	8.7	5.9	6.2	-
Co (mg/L)	0.05	2.6	0.6	0.00106
Cu (mg/L)	0.04	0.05	0.3	0.0063
Fe (mg/L)		Test 1 (module A)	Test 2 (module A)	Test 3 (module B)
Li (mg/L)	pH	5.2	5.9	11
Mn (mg/L)	0.1	3.4	0.5	0.034
Na (mg/L)	1.8	3.3	2.2	-
Ni (mg/L)	0.4	9.8	3.3	-
P (mg/L)	23.5	22.6	0.5	-
Fluorides (mg/L)	16.6	18.3	7.8	0.89
Chlorides (mg/L)	3.9	7.3	16.9	-
EMC (mg/L)	16.1	11.8	n/a	0.062
EC (mg/L)	126.2	92.2	n/a	5.9
Naphthalene (mg/L)	0.00015	0.00056	0.00026	0.0024

5. Conclusions

In the present work, the two battery modules were triggered in thermal runaway and subsequent degassing and fire. Water was applied to mock-up firefighting operations in order to analyze the composition of the extinguishing water.

The tests presented in this paper highlight that waters used for firefighting on NMC Li-ion batteries are susceptible to containing many metals, including Ni, Mn, Co, Li and Al. Those metals are mixed with other carbonaceous species (soots, tarballs). It is also important to note that particles present in the water can be nanometric or in the form of nanostructured clusters. In addition to the solid contaminants, liquid compounds can be present, especially organic carbonates coming from the electrolyte (EC and EMC in this case) and also gaseous species such as PAH. A comparison with PNEC values showed that this water could be potentially hazardous to the environment, depending on the actual situation encountered in the case of thermal runaway propagation with a Li-ion battery-based system.

These tests also make it possible to identify some trends concerning the reaction scenario. By comparing the two extinguishing operations on the prismatic cells, one can see that when the fire is developed, the water is much more concentrated in PAH and cathode metals (Ni, Mn, Co). On the other hand, the concentrations of elements coming from the liquid electrolyte (typically Li, P, F), more easily accessible, are present in equivalent quantities. Liquid organic carbonates are preferably found in the case of degassing without ignition. These low boiling point liquids are otherwise vaporized and found mainly in the gaseous phase. The comparison of the results between the prismatic cell module and the 18,650-cell module also confirms the importance of the cell and module geometry, influencing, in particular, the mechanical strength of the system and, therefore, the confinement of the inner materials.

As large Li-ion batteries are fast spreading (in so-called Battery Energy Storage Systems, BESS, for example), and only few data on the environmental impact of fires in those systems are available, it is crucial to further develop consolidated knowledge in this field. Several directions could be suggested for future tests like developing higher level (or full scale) testing to increase test representativity. Owing to field operational constraints in terms of emergency response following a fire, considering time between event initiation and water suppressant application as a parameter in futures studies also seems important. Other investigations worth being performed are, for instance, a detailed assessment of air, water and soil local impacts following Li-ion BESS significant incidents or in-depth environmental impact studies of key Li-ion substances like organic carbonate solvents (EC, EMC, etc.).

Supplementary Materials: The following supporting information can be downloaded at: https://www.mdpi.com/article/10.3390/batteries10040118/s1.

Author Contributions: A.B. was involved in conceptualization, investigation, formal analysis, in writing of the original manuscript and project administration; A.P. was involved in investigation and formal analysis; G.M. was involved in supervision, validation, review and editing; T.C. was involved in review; A.E.-M. was involved in investigation; T.D. was involved in investigation; J.-P.B. was involved in investigation; B.T. was involved in in conceptualization, investigation and review and A.L. was involved in supervision, validation and review. All authors have read and agreed to the published version of the manuscript.

Funding: This research was funded by the French Ministry for Ecological Transition (Program 181).

Data Availability Statement: The data presented in this study are available in article.

Conflicts of Interest: The authors declare no conflicts of interest.

References

1. Giger, W. The Rhine red, the fish dead—The 1986 Schweizerhalle disaster, a retrospect and long-term impact assessment. *Environ. Sci. Pollut. Res.* **2009**, *16*, 98–111. [CrossRef] [PubMed]
2. Schwabach, A. The Sandoz spill: The failure of international law to protect the Rhine from pollution. *Ecol. LQ* **1989**, *16*, 443.
3. Zhang, H.; Duan, H.; Zuo, J.; Song, M.; Zhang, Y.; Yang, B.; Niu, Y. Characterization of post-disaster environmental management for Hazardous Materials Incidents: Lessons learnt from the Tianjin warehouse explosion, China. *J. Environ. Manag.* **2017**, *199*, 21–30. [CrossRef] [PubMed]
4. EPRI. Battery Firewater Composition and Risk Assessment. 2020. Available online: https://www.epri.com/research/products/3002020017/ (accessed on 2 February 2024).
5. *ISO TR 26368*; Environmental Damage Limitation from Fire Fighting Water Run-Off (under Revision). ISO: Geneva, Switzerland, 2012.
6. Larsson, F.; Andersson, P.; Blomqvist, P.; Lorén, A.; Mellander, B.-E. Characteristics of lithium-ion batteries during fire tests. *J. Power Sources* **2014**, *271*, 414–420. [CrossRef]
7. Feng, X.; Ouyang, M.; Liu, X.; Lu, L.; Xia, Y.; He, X. Thermal runaway mechanism of lithium ion battery for electric vehicles: A review. *Energy Storage Mater.* **2018**, *10*, 246–267. [CrossRef]
8. Yang, Y.; Fang, D.; Maleki, A.; Kohzadi, S.; Liu, Y.; Chen, Y.; Liu, R.; Gao, G.; Zhi, J. Characterization of Thermal-Runaway Particles from Lithium Nickel Manganese Cobalt Oxide Batteries and Their Biotoxicity Analysis. *ACS Appl. Energy Mater.* **2021**, *4*, 10713–10720. [CrossRef]
9. Chen, S.; Wang, Z.; Yan, W. Identification and characteristic analysis of powder ejected from a lithium ion battery during thermal runaway at elevated temperatures. *J. Hazard. Mater.* **2020**, *400*, 123169. [CrossRef]
10. Bordes, A.; Marlair, G.; Zantman, A.; Herreyre, S.; Papin, A.; Desprez, P.; Lecocq, A. New insight on the risk profile pertaining to lithium-ion batteries under thermal runaway as affected by system modularity and subsequent oxidation regime. *J. Energy Storage* **2022**, *52*, 104790. [CrossRef]
11. Mrozik, W.; Rajaeifar, M.A.; Heidrich, O.; Christensen, P. Environmental impacts, pollution sources and pathways of spent lithium-ion batteries. *Energy Environ. Sci.* **2021**, *14*, 6099–6121. [CrossRef]
12. McNamee, M.; Marlair, G.; Truchot, B.; Meacham, B.J. *Research Roadmap: Environmental Impact of Fires in the Built Environment*; Fire Protection Research Foundation: Norfolk County, MA, USA, 2020.
13. Rappsilber, T.; Yusfi, N.; Krüger, S.; Hahn, S.-K.; Fellinger, T.-P.; von Nidda, J.K.; Tschirschwitz, R. Meta-analysis of heat release and smoke gas emission during thermal runaway of lithium-ion batteries. *J. Energy Storage* **2023**, *60*, 106579. [CrossRef]
14. Vilic, A. *Environmental Risk Assessment of Fire-Water Runoff from Vehicle Fire-Development of a Predictive Model Intended for the Fire-Rescue Service*; Fire Protection Research Foundation: Norfolk County, MA, USA, 2019.
15. Noiton, D.; Fowles, J.; Davies, H. The ecotoxicity of Fire Water Run-off, Part II: Analytical Results, New Zealand Fire Service Commission Research Report Number 18. 2001. ISBN 0-908920-61. Available online: https://www.researchgate.net/publication/272508692_Fire_Research_The_Ecotoxicity_of_Fire-Water_Runoff_Part_II_Analytical_Results_ESR#fullTextFileContent (accessed on 3 March 2024).

16. EMPA. Minimization of Fire Risks from Electric Vehicles in Underground Traffic Infrastructures. 2018. Available online: https://plus.empa.ch/images/2020-08-17_Brandversuch-Elektroauto/AGT_2018_006_EMob_RiskMin_Undergr_Infrastr_Final_Report_V1.0.pdf (accessed on 2 February 2024).
17. Held, M.; Tuchschmid, M.; Zennegg, M.; Figi, R.; Schreiner, C.; Mellert, L.D.; Welte, U.; Kompatscher, M.; Hermann, M.; Nachef, L. Thermal runaway and fire of electric vehicle lithium-ion battery and contamination of infrastructure facility. *Renew. Sustain. Energy Rev.* **2022**, *165*, 112474. [CrossRef]
18. RIVM. Risico's Van Rook Door Branden Van Li-Ionbatterijen. 2021. Available online: https://www.rivm.nl/bibliotheek/rapporten/2021-0019.pdf (accessed on 2 February 2024).
19. Quant, M.; Willstrand, O.; Mallin, T.; Hynynen, J. Ecotoxicity Evaluation of Fire-Extinguishing Water from Large-Scale Battery and Battery Electric Vehicle Fire Tests. *Environ. Sci. Technol.* **2023**, *57*, 4821–4830. [CrossRef] [PubMed]
20. Zhang, Y.; Wang, H.; Li, W.; Li, C.; Ouyang, M. Size distribution and elemental composition of vent particles from abused prismatic Ni-rich automotive lithium-ion batteries. *J. Energy Storage* **2019**, *26*, 100991. [CrossRef]
21. Dyer, J.A.; Scrivner, N.C.; Dentel, S.K. A practical guide for determining the solubility of metal hydroxides and oxides in water. *Environ. Prog.* **1998**, *17*, 1–8. [CrossRef]
22. Winberry, W.T.; Murphy, N.T.; Riggan, R. *Compendium of Methods for the Determination of Toxic Organic Compounds in Ambient Air*; EPA, National Service Center for Environmental Publications: Washington, DC, USA, 1988.
23. Geiser, M.; Kreyling, W.G. Deposition and biokinetics of inhaled nanoparticles. *Part. Fibre Toxicol.* **2010**, *7*, 2. [CrossRef] [PubMed]
24. Wang, X.; Xie, H.; Wang, P.; Yin, H. Nanoparticles in plants: Uptake, transport and physiological activity in leaf and root. *Materials* **2023**, *16*, 3097. [CrossRef] [PubMed]
25. Rivero Arze, A.; Mouneyrac, C.; Chatel, A.; Manier, N. Comparison of uptake and elimination kinetics of metallic oxide nanomaterials on the freshwater microcrustacean Daphnia magna. *Nanotoxicology* **2021**, *15*, 1168–1179. [CrossRef]
26. Rivero Arze, A.; Manier, N.; Chatel, A.; Mouneyrac, C. Characterization of the nano–bio interaction between metallic oxide nanomaterials and freshwater microalgae using flow cytometry. *Nanotoxicology* **2020**, *14*, 1082–1095. [CrossRef] [PubMed]
27. Available online: https://www.epa.gov/sdwa/drinking-water-regulations-and-contaminants (accessed on 2 February 2024).
28. Available online: https://www.canada.ca/fr/sante-canada/services/publications/vie-saine/recommandations-pour-qualite-eau-potable-canada-chlore-document-technique/page-3-recommandations-pour-qualite-eau-potable-canada-chlore-document-technique.html (accessed on 2 February 2024).
29. Ottawa City Service. Guide Discharging Waste Water Ind. *Facilities*. 2011. Available online: https://nchca.ca/wp-content/uploads/2017/06/Sewer-Use-Program_English_2011.pdf (accessed on 22 March 2024).
30. LCPP. Étude de l'Impact de Feux de Véhicules Electriques (RENAULT) sur les Intervenants des Services de Secours. 2015. Available online: http://iuv.sdis86.net/wp-content/uploads/2015/09/Rapport-LCPP-brulages-vehicules-electriques-Renault.pdf (accessed on 2 February 2024).
31. Alberto Project. Available online: https://alberoprojekt.de/index_htm_files/WP%201.4%20Contamination%20of%20extinguishing%20water%20after%20fires%20of%20Li-Ion%20Batteries.pdf (accessed on 2 February 2024).
32. Merseyside Fire & Rescue Service. SIGNIFICANT INCIDENT REPORT—018965—15092020—Orsted BESS, Carnegie Road, Liverpool, L137HY. 2021. Available online: https://planningregister.cherwell.gov.uk/Document/Download?module=PLA&recordNumber=154109&planId=1951104&imageId=30&isPlan=False&fileName=Appendix%202%20%20-%20Liverpool%20BESS%20Significant%20Investigation%20Report%20(1).pdf (accessed on 28 March 2024).
33. BEA-RI. Rapport d'Enquête Technique sur l'Incendie au Sein du Poste de Transformation RTE. de Perles et Castelet (09). 2021. Available online: https://www.igedd.developpement-durable.gouv.fr/IMG/pdf/rapportperlesvdif_cle286783.pdf (accessed on 2 February 2024).
34. Tesla. Model 3 Emergency Response Guide. 2018. Available online: https://www.tesla.com/sites/default/files/downloads/2017_Model_3_Emergency_Response_Guide_en.pdf (accessed on 2 February 2024).
35. Zhang, L.; Duan, Q.; Liu, Y.; Xu, J.; Sun, J.; Xiao, H.; Wang, Q. Experimental investigation of water spray on suppressing lithium-ion battery fires. *Fire Saf. J.* **2021**, *120*, 103117. [CrossRef]
36. Available online: https://echa.europa.eu/fr/home (accessed on 2 February 2024).
37. Available online: https://substances.ineris.fr/fr/substance/1301/3 (accessed on 2 February 2024).

batteries

MDPI

Article

Innovative Early Detection of High-Temperature Abuse of Prismatic Cells and Post-Abuse Degradation Analysis Using Pressure and External Fiber Bragg Grating Sensors

André Hebenbrock [1,2,*], Nury Orazov [1], Ralf Benger [1], Wolfgang Schade [1,3], Ines Hauer [1,4] and Thomas Turek [1,2]

1 Research Center Energy Storage Technologies, Clausthal University of Technology, Am Stollen 19A, 38640 Goslar, Germany
2 Institute of Chemical and Electrochemical Process Engineering, Clausthal University of Technology, Leibnizstraße 17, 38678 Clausthal-Zellerfeld, Germany
3 Fraunhofer Institute for Telecommunications, Heinrich Hertz Institute, Am Stollen 19H, 38640 Goslar, Germany
4 Institute of Electrical Power Engineering and Electrical Energy Engineering, Clausthal University of Technology, Leibnizstraße 28, 38678 Clausthal-Zellerfeld, Germany
* Correspondence: andre.hebenbrock@tu-clausthal.de

Abstract: The increasing adoption of lithium-ion battery cells in contemporary energy storage applications has raised concerns regarding their potential hazards. Ensuring the safety of compact and modern energy storage systems over their operational lifespans necessitates precise and dependable monitoring techniques. This research introduces a novel method for the cell-specific surveillance of prismatic lithium-ion cells, with a focus on detecting pressure increases through the surface application of a fiber Bragg grating (FBG) sensor on a rupture disc. Commercially available prismatic cells, commonly used in the automotive sector, are employed as test specimens and equipped with proven pressure and innovative FBG sensors. Encompassing the analysis capacity, internal resistance, and pressure (under elevated ambient temperatures of up to 120 °C), this investigation explores the thermal degradation effects. The applied FBG sensor on the rupture disc exhibits reversible and irreversible state changes in the cells, offering a highly sensitive and reliable monitoring solution for the early detection of abuse and post-abuse cell condition analysis. This innovative approach represents a practical implementation of fiber optic sensor technology that is designed for strain-based monitoring of prismatic lithium-ion cells, thereby enabling customized solutions through which to address safety challenges in prismatic cell applications. In alignment with the ongoing exploration of lithium-ion batteries, this research offers a customizable addition to battery monitoring and fault detection.

Keywords: lithium-ion battery; battery safety; prismatic cell; thermal fault; early detection; solid electrolyte interphase decomposition; fiber Bragg grating

Citation: Hebenbrock, A.; Orazov, N.; Benger, R.; Schade, W.; Hauer, I.; Turek, T. Innovative Early Detection of High-Temperature Abuse of Prismatic Cells and Post-Abuse Degradation Analysis Using Pressure and External Fiber Bragg Grating Sensors. *Batteries* **2024**, *10*, 92. https://doi.org/10.3390/batteries10030092

Academic Editor: Mingyi Chen

Received: 15 January 2024
Revised: 23 February 2024
Accepted: 29 February 2024
Published: 4 March 2024

1. Introduction

The safety of lithium-ion battery technology in the context of battery electric vehicles and large-scale battery storage has become increasingly crucial as the market for large lithium-ion battery systems continues to grow [1,2]. Foremost of the demands on present battery systems arise from heightened energy density and rapid charging capabilities, necessitating robust safety measures in response [3]. However, achieving these goals is a complex task, with cost pressures and the need for more rigorous thermal management (owing to the rising energy density) challenging the design of safe battery systems [4,5]. Recent advancements have led to substantial improvements in understanding and addressing thermal runaway (TR) events within lithium-ion batteries. These include breakthroughs in replicating TR events [6–8], early detection methods for battery cell TR [9–14], strategies to mitigate TR [15–17], and the prevention of thermal propagation [18]. From these ongoing advancements around the TR event ensue recommendations that translate to various regulatory levels [19,20]. Despite these improvements, there are still challenges in the field of early detection like the high computational effort required to extract accurate and relevant battery state data, as well as accounting for the development of relevant fault features due to the aging of the cell [5,14,21].

In recent years, various calorimetric studies have shed light on the reversible and irreversible degradation phenomena associated with thermal degradation in batteries leading up to a TR event, especially lithium-ion batteries using graphite-based anodes [22–24]. It has been proposed that the irreversible pressure increase observed during battery cell operation can be attributed to gas formation [25,26]. The nature and extent of gas-forming reactions in battery cells depend on factors such as temperature and the state of charge (SOC), which vary depending on the cell chemistry [18,27–30]. In addition to gas emission, the typical progression of TR inevitably leads to a significant increase in temperature [31]. The escalating cell pressure inexorably results in the opening of the battery cell, which is accompanied by a loud noise and visible smoke generation [32].

Therefore, leveraging the expansion of battery cells is one promising avenue for the early detection of battery degradation and critical battery states [33]. This volume increase can be utilized for cell state determination, early fault detection, and the identification of lithium plating in batteries [26,34,35]. Recent safety-related FBG application research in battery technology has particularly evolved based on strain measurements and internal sensing methods [35]. Complementary mechanical monitoring with fixed FBG sensors has proven to be particularly applicable in mechanically sensitive and lighter pouch cells [36], and it has also found applications in improved temperature sensing for inherently smaller cylindrical cells [37,38].

However, prismatic cells, commonly used as large format cells in various high-capacity systems, possess sturdy casings that make them less prone to expansion during operation compared to pouch cells, thereby limiting the usefulness of strain measurements. Additionally, lithium-ion battery cells are often externally braced in applications, further impeding the measurement of volume changes [34]. During operation, large format cells develop temperature fields that decrease in uniformity based on operational demand and cell size [39,40]. This is due to the small surface-to-volume ratio, which dampens cooling efficiency. Representative and system-wide temperature measurement is therefore challenging for larger cells.

This study introduces an innovative and non-invasive approach with FBG sensors for meaningful strain monitoring on commercial large format prismatic cells in a realistic braced application, thereby leveraging a mechanically fixed connection to the standard rupture disc in these cells for improved battery state determination. The experimental investigation targeting thermal abuse and internal gas evolution with parallel pressure measurement demonstrated the effectiveness of the FBG sensor application for sensitive strain measurements as an additional state parameter—complementary to voltage, current, and temperature—to ensure safe battery operation. Analysis of the underlying degradation reactions through phases of elevated temperatures of up to 120 °C revealed the potential for the early detection of critical pressures before a cell opening event. Thus, this work demonstrates a novel possibility for customizable and reliable cell-specific strain monitoring.

2. Materials and Methods

2.1. The Device under Test

The tested cell was an industry-standard, available prismatic cell of the PHEV2 type (according to the German Association of the Automotive Industry), and it was disassembled from a module with a nominal capacity of 37 Ah [41]. It was composed of nickel manganese cobalt oxide ($Li_x Ni_{0.33} Mn_{0.33} Co_{0.33} O_2$) (NMC111)/graphite intercalation compound (GIC) chemistry, as well as polypropylene (PP) and polyethylene (PE) separator combination, and it is usually employed in battery electric vehicle applications. Further relevant cell data from the manufacturer can be found in Table 1.

Table 1. Cell characteristics given by the manufacturer.

Property	Value
Mass/g	810
Length × width × thickness/mm	91 × 148 × 26.5
Upper cut-off voltage/V	4.2
Lower cut-off voltage/V	3.0
Maximum charge current/A	55
Maximum discharge current/A	123
Maximum storage temperature/°C	70
Minimum storage temperature/°C	−40
Maximum operating temperature/°C	60

Positioned in a symmetrically fastened arrangement, 8 mm thick aluminum plates clamped the cell on each side with a torque of 0.5 Nm. This applied torque resulted in a state-of-charge (SOC)-dependent pressure ranging from an approximate range of 0.074–0.11 MPa, which fell within the typical pressure range observed in automotive applications [42,43]. The cell was equipped with an overcharge safety device (OSD), which consisted of a pressure-sensitive aluminum diaphragm that causes an external short circuit when a certain pressure threshold is reached (Figure 1). The short-circuit current would cause the fuse to melt, which would interrupt the current flow [44]. Triggering the OSD would thus prevent further operation of the cell when an increased pressure is reached. To test the thermal degradation and to evaluate the effects on discharge capacity and internal resistance, the electrical operability must be preserved. The OSD was, therefore, deactivated beforehand by filling the membrane cavity with epoxy resin.

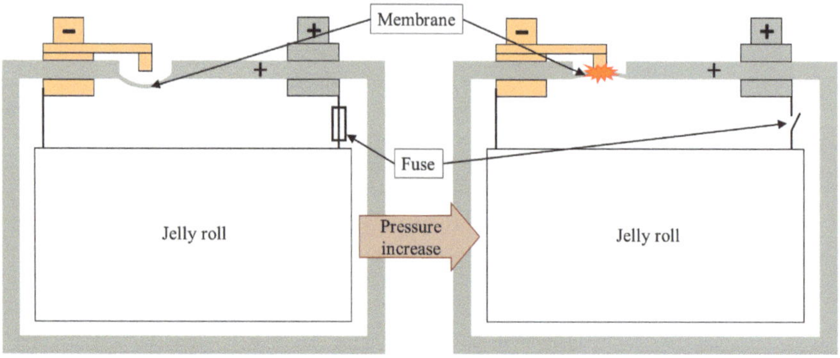

Figure 1. Schematic representation of the operation of the OSD in the prismatic cell showing the consequence of an external short circuit in the event of a pressure increase, as well as the subsequent triggering of the fuse. Based on ref. [44].

To allow for the cell opening to take place in a controlled way, the cell was equipped with a rupture disc as standard. The rupture disc was an approximately 0.15 mm thick aluminum sheet, and it was measured as 24 mm × 9 mm at its longest dimensions. It had a 12 mm-elongated perforation in the center. On both sides, two additional perforations protruded from it with a length of 4 mm at 135° (bottom left Figure 2).

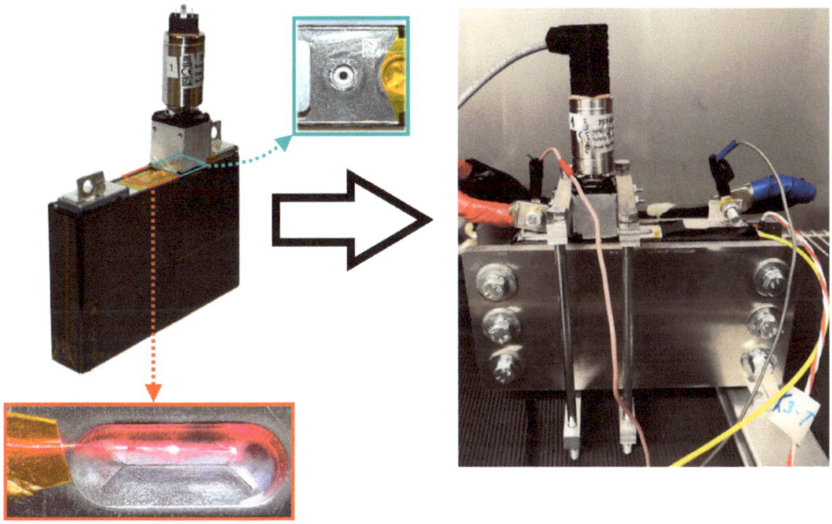

Figure 2. The cell was equipped with a pressure sensor (**top left**) and a fiber optic sensor (**bottom left**). The connection to the gas space was made by a penetration of the sealing pin. The sealing pin without the aluminum cover can be seen in the top middle breakout. The fiber optic sensor was positioned next to the elongated perforation and is marked by the red glowing dot in the middle. The clamped cell that was inside the climate chamber in the complete setup is shown on the right side.

2.2. Thermal Abuse Conditions

During the experiment, the cell was exposed to elevated ambient temperature phases in a 50–120 °C range. This high-temperature range was chosen to better quantify the less identifiable phenomena of cells overheating due to processes at the solid electrolyte interphase (SEI) layer of the GIC electrode. The test range, thus, started just below the maximum operating temperature, and it ended below the melting temperature of the PE layer in the separator, which would lead to pore closure, as well as drastic and thus easily identifiable irreversible degradation effects [31]. As for the used temperature range, we looked mostly at the reactions at the SEI layer on the graphite active material (Figure 3). The SEI layer is essential for the sustainable usage of lithium-ion batteries as a rechargeable energy storage, as the anode potential is otherwise outside the stability window of the electrolyte [45]. According to Peled et al. [46], the desired attributes are high electrical resistance, high selectivity, and low diffusion resistance for lithium-ion batteries, as well as low thickness, high mechanical stability against expansion, and a high tolerance for high temperatures and high potentials. The formation of the electrically insulating SEI layer started with the initial charging of a lithium-ion battery with a GIC anode to stabilize it against the potential window used. However, this SEI could not meet all the formulated requirements. With increased temperatures, electric cycling, and longer storage times, the layer thickness continued to increase [47].

The components of the initial SEI formation are composed of insoluble inorganic and partially soluble organic compounds like the reduction products of the electrolyte resulting from its reaction with lithium ions and electrons [46].

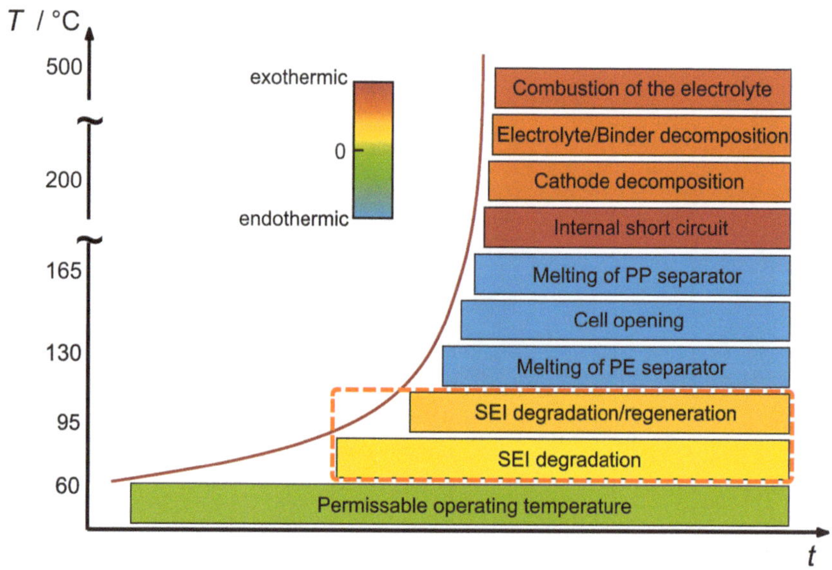

Figure 3. Progression of the cell events for the NMC/GIC chemistry and PE/PP separator with increasing temperature. Illustration style based on ref. [48] with insights from refs. [9,31,49].

The mix was strongly dependent on the solvent composition of the used additives and impurities [23]. The SEI formation reactions were irreversible and led to the binding of lithium ions, thereby contributing to capacity fade and Ohmic resistance increase (ORI) [22,47]. The exact electrolyte composition of the tested commercial cell is not known. Due to the similarity of the organic carbonate solvents commonly used in lithium-ion batteries, and for simplification in subsequent qualitative discussions, we assume the presence of EC. Consequently, $(CH_2OCO_2Li)_2$ emerges as the primary component in SEI formation, accompanied by the release of C_2H_4, as expressed in Equation (1) [50]:

$$2\,C_3H_4O_3 + 2\,Li^+ + 2\,e^- \longrightarrow (CH_2OCO_2Li)_2 + C_2H_4 \tag{1}$$

The decomposition of the main component of the SEI can be described by the global reaction form at which the metastable $(CH_2OCO_2Li)_2$ reacts to the stable inorganic Li_2CO_3, as expressed in Equation (2) [51]. Furthermore, the reaction was shown to be largely independent of the lithium content of the anode [52]. The onset temperature of the decomposition reaction depends on the electrolyte composition [23,53]. Additionally, the initial heating rate/reaction rate was found to be greatly dependent on the graphite surface area [31]. Therefore, with the use of calorimetry, the starting temperature of the SEI degradation was found to vary in a range of 57–80 °C [52,53].

$$(CH_2OCO_2Li)_2 \longrightarrow Li_2CO_3 + C_2H_4 + CO_2 + 0.5\,O_2 \tag{2}$$

With increasing temperature, an acceleration in the decomposition of metastable SEI components to stable components can be assumed [31,52,54,55]. The SEI decomposition has a peak at roughly 100 °C. As shown by Richard and Dahn, the self-heating rate slows down but increases again as new SEI is formed if lithium ions are available in the GIC [52]. As in the initial SEI formation, all reactants (as expressed in Equation (1)) have to come in contact, which can be at the graphite/electrolyte or SEI/electrolyte interface and can lead to the increasing growth of the SEI layer even after complete coverage of the graphite [46]. These necessary reaction conditions for the reforming process are mainly managed by two mechanisms.

- Cracking of the SEI layer due to the expansion of the graphite during charging leads to direct contact of lithiated graphite with the electrolyte and thus to the reformation of metastable SEI [46,54].
- Since the SEI layer is conductive for lithium ions, and with the tunneling effect for electrons, a new metastable SEI layer can be formed on already existing inorganic SEI [52].

A parallel occurrence between SEI decomposition and formation could also be observed from about 100 °C by the calorimetry experiments [52,56]. This then led to an equilibrium in the reformation of metastable organic species being set externally over the inorganic SEI layer, which increasingly reacted to form the stable inorganic layer (Figure 4).

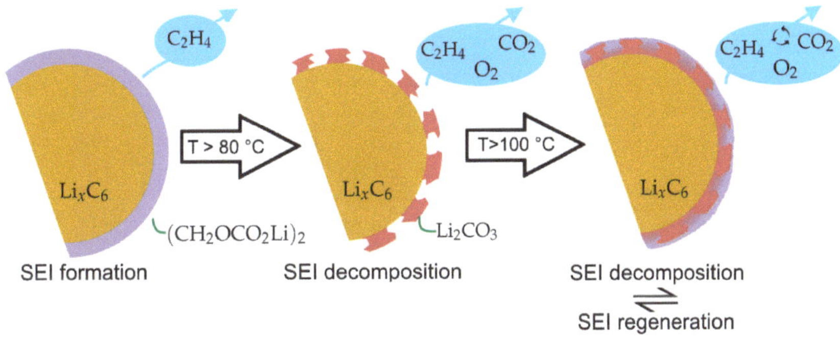

Figure 4. Schematic display of the temperature-dependent stages of the SEI composition and occurring reactions. Based on refs. [31,51].

At high temperatures, these reactions contribute to an irreversible growth of the stable SEI component (Li_2CO_3), which passivizes the active material and thus increases the cell's resistance, as well as a deactivation of active anode regions [57]. Both the formation (Equation (1)) and the decomposition (Equation (2)) of the SEI layer are exothermic reactions which bind of lithium ions and lead to gas evolution [31,51]. When the emission of gases is compared (Equations (1) and (2)), it is obvious that significantly more gaseous species are released during the decomposition of the SEI layer, which shows that the measured pressure increase strongly depends on the onset of the decomposition.

2.3. Sensors

Battery cell degradation due to SEI growth during use manifests itself in changes in internal resistance and in the decrease in capacity due to loss of lithium inventory (LLI) and loss of active material (LAM) [58,59]. Depending on the use case, this leads to an unknown internal state and uncertain thermal stability [18]. To prevent the uncertainty conceptually, the examined prismatic cell was equipped with a pressure sensor and an FBG sensor on the rupture disc as a basis for the early detection and monitoring of the degradation progress.

2.3.1. Pressure Sensor

In principle, a pressure sensor can be installed inside or outside the cell [60]. Here, an external cell sensor with a threaded connection was chosen for its simple installation. Despite its advantages, the sensor had some drawbacks, including dead volume and thermal mass, which were not considered in this work due to the large cell size (additional volume < 1% and mass < 0.5%). The selected piezoresistive ceramic sensor was suitable for use during the thermal loading of the cell, with a maximum storage temperature of 125 °C and a maximum media temperature of up to 135 °C. Exhibiting good media stability, the pressure sensor can measure pressures up to 6 bar relative to atmospheric pressure and provides an accuracy of ±0.25%. Between the positive pole and the rupture disc, a welded-on protective cap on top of the sealing pin can be seen on the cell. After removing the protective cap, there is an opening sealed with a plastic sealing pin, which is used during the production process to fill the cell with electrolyte (Figure 2). For attaching the sensor, an aluminum block with concentric inner threading was positioned over the sealing pin of the prismatic cell. Using this location for the measurement setup is considered advantageous due to the following reasons:

- It provides easy access to the opening without the need for drilling, which could introduce metal shavings into the cell and potentially cause an internal short circuit.
- Remaining uncovered, the rupture disc of the cell preserves its functionality, thus enabling the authentic behavior that is crucial to the FBG sensor.
- The electrolyte does not flow into the measurement block because of its position on top of the cell, thereby ensuring that the electrochemical properties of the cell remain unchanged.

For fixing the aluminum block, an adhesive silicon-based sealing paste (Acc Silicones–Part No.: 740010640) and epoxy resin (MG Chemicals–Part No.: 832HT-375ML) were utilized to bond the aluminum block to the cell housing. After hardening, the silicone-based adhesive sealing paste forms a resilient silicone elastomer and can withstand temperatures up to 300 °C. It was used as the outer ring sealant and the electric insulation between the aluminum block and the cell housing, which was connected to the positive pole.

An 8 mm-long tinned copper tube was fixed into the electrolyte pin opening with epoxy resin (maximum operating temperature of 200 °C) on the inside of the aluminum block. The chemical stability of the inner cured epoxy resin was tested by exposure to an electrolyte mixture of ethylene carbonate (EC):dimethyl carbonate (DMC) in a ratio of 1:1 with the conducting salt $LiPF_6$ in the concentration of 1 mol L^{-1} and had shown no visible changes over 2 years.

Before the cell was opened by puncturing the seal pin's final plastic layer, the cell was transferred into an argon-filled glove box. As the final steps, the pressure sensor was mounted on the aluminum block together with the silicone O-ring seal (max. operating temperature 200 °C) inside the glove box, and it was cast on the outside with epoxy resin to ensure complete sealing. The successfully retrofitted sensor was braced against the cell and can be seen on the right side of Figure 2.

2.3.2. Fiber Bragg Grating Sensor

The fiber optic sensors used were single-mode 800 polyimide-coated glass fibers. In addition to their mechanical robustness and immunity to electromagnetic interference, these sensors have a small form factor. Inside this fiber are uniform FBGs inscribed by a femtosecond laser. Depending on the grating period Λ and the effective refractive index n_{eff}, the FBG reflects a specific Bragg wavelength (λ_{B}), as can be seen in Equation (3) [61,62].

$$\lambda_{\mathrm{B}} = 2\Lambda n_{\mathrm{eff}} \tag{3}$$

These physical properties of the FBG grating are sensitive not only to mechanical strain, but also to temperature changes (Equation (4)) [61,62].

$$\frac{\Delta\lambda_{\mathrm{B}}}{\lambda_{\mathrm{B}}} = (1-\rho_{\mathrm{e}})\Delta\varepsilon + (\alpha+\eta)\Delta T \tag{4}$$

The stress-optic coefficient ρ_{e}, the thermo-optic coefficient η, and the thermal expansion α describe the effect of strain changes ($\Delta\varepsilon$), as well as the temperature changes ΔT on the Bragg wavelength when compared to the initial state of the measurement. The fiber optic sensor was mounted on the outside of the rupture disc parallel to the long perforation in the middle (bottom left Figure 2). This designed mechanical weakness as a safety feature is accessible, and it demonstrates the highest sensor sensitivity to state changes in the cell compared to other positions on the cell due to its thinness. Because of the orientation of the fiber on the rupture disc, the FBG was compressed with increasing swelling along the perforation as a result of the plastic deformation due to high cell pressure. After the irreversible deformation, the FBG thus relaxed to a compressed state at reference conditions and was directly dependent on the degree of the increasing curvature, which led to a reduction in the grating period Λ and therefore to a decrease in λ_B (Equation (3)). This compression made it necessary to apply the fiber under tension during the application; otherwise, no measurement was possible [63]. To test for the highest sensitivity, a configuration with the sensor glued on one side and free on the other side along the elongated perforation (one-sided adhesive), which had an even distribution of the adhesive on both sides of the elongated perforation (balanced adhesive) for a more symmetrical deformation behavior, were tested (Figure 5).

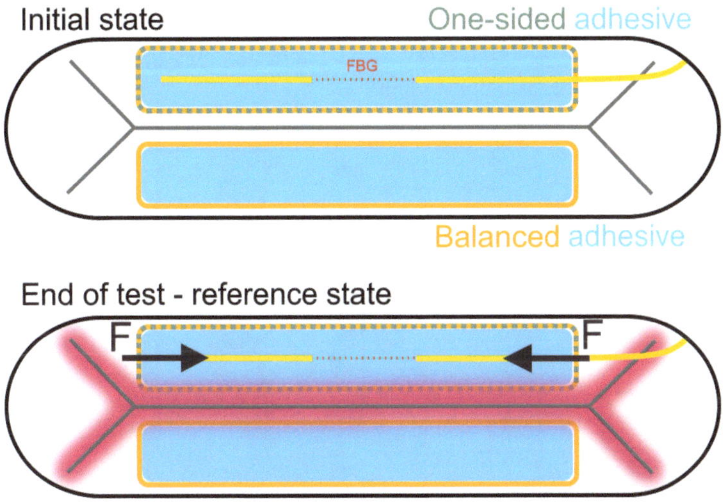

Figure 5. Scheme of transition from the initial state of the rupture disc to the state with swellings around the perforation at 25 °C after the thermal abuse test (red). In the one-sided configuration, only the area around the FBG sensor was glued to the rupture disc (gray box). In the balanced configuration (yellow boxes), both sides of the elongated perforation were glued for a more symmetric deformation.

To optimize the strain transfer, the fiber was bonded with a cyanoacrylate-based adhesive (Loctite 406) with a high Young modulus for optimal strain transfer (which was stable up to 120 °C and parallel to the center perforation with low adhesive thickness between the bursting disc and fiber). The bond length was chosen to be as long as the elongated perforation. To further improve the strain transfer of the FBG sensor, the polyimide coating was removed from the fiber [63,64]. Before the adhesive was used, the surface was roughened with 80-grit abrasive paper, which was then afterward cleaned with isopropyl alcohol. The bonding process took place in two steps. First, the tip of the sensor was glued next to the end of the elongated perforation that was midway to the wall of the rupture disc. This was followed by bonding the remaining fiber to the aluminum next to the perforation under tension. Using this two-step method allowed for the application of sufficient pretension to the fiber so as to allow the measurement of the increasing compression. For the balanced glue configuration, the other side of the long perforation was also coated with the same amount of adhesive to ensure the symmetrical behavior of the rupture disc's deformation. In addition to the rupture disc's surface-applied FBG sensor, there was an additional FBG sensor in a 3 cm distance along the fiber, which was attached loosely to the cell with a narrow adhesive strip that was perpendicular to the fiber course. This free FBG sensor served as a temperature reference to the bonded FBG sensor and was mechanically decoupled from the cell.

2.4. Experimental Method

In Figure 6a, it can be seen that the clamped cell in the climate chamber was electrically connected to an electric battery test rig (Scienlab–SL60/200/12BT4C, Bochum, Germany), which controls the electrical load; supports the galvanostatic electrochemical impedance spectroscopy (GEIS) that operates in a frequency range of 10 mHz to 5 kHz and a maximum current amplitude of 5 A, as well as logs the pressure sensor, the cell temperature by a Pt100 sensor on top of the cell, the chamber temperature, current, and voltage of the cell. Additionally, the fiber with the FBGs was monitored by a commercial FBG measurement system (FiSens–X400, Braunschweig, Germany). Each adhesive configuration on the rupture disc was tested two times (balanced: B1 and B2 and one-sided: O1 and O2). Overall, the measurement frequency was 1 Hz. The pressure and free FBG measurement was performed only for the B1 and O1 configurations due to the high similarity of the measurement results.

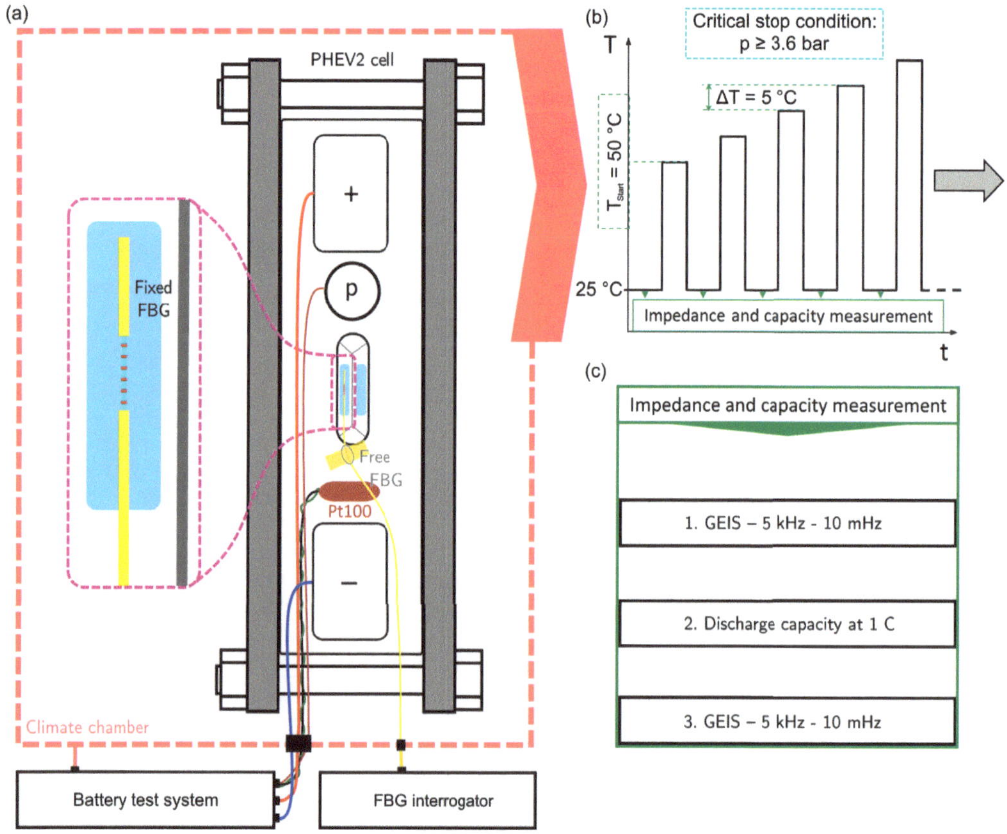

Figure 6. (**a**): Test setup of the prismatic PHEV2 cell used in the experiment with a pressure sensor and FBG sensor in the balanced configuration. (**b**): Gradual temperature increase in 5 °C steps from a range pf 50–120 °C or until a measured pressure of 3.6 bar was reached. (**c**): The electrical characterization procedure during the reference temperature phases after a 3 h relaxation at 3.95 V.

A climate chamber (Binder–MK720) with a heating rate of 4 K min^{-1} and a cooling rate of 4.5 K min^{-1} was chosen to minimize the transition times between each phase. To ensure the safety of the test, the experiments were carried out in a safe environment with a set critical pressure limit that was determined beforehand by heating an identical cell until the rupture disc burst. The critical pressure limit until test termination was set to 3.6 bar. To investigate the increasing thermal degradation effects of the cell, a sequence of reference and high-temperature phases going from a range of 50–120 °C in 5 °C increments were applied to the cell (Figure 6b). The test was finished when the 120 °C high-temperature phase was concluded, or it was ended prematurely when the critical pressure limit was reached.

After each 3 h high-temperature phase, the cell was electrically characterized during a 20 h-long reference temperature phase at 25 °C by a procedure comprising a GEIS before and after a capacity test to monitor the change in cell impedance after every high-temperature phase (Figure 6c). The complete electrical procedure from the start of the cooling from a high-temperature phase to the start of the next is listed below. There were two cells in the climate chamber for each experiment. Steps 2 and 3, as well as 6 and 7, were swapped for the cells in the O1 and O2 configuration to perform GEIS on only one cell at a time, thereby avoiding interference.

1. Hold voltage at 3.95 V for 4 h once the reference temperature of 25 °C is reached;
2. GEIS at 5 A from 5 kHz to 10 mHz for 10 sine periods;
3. Rest 1 h;
4. Capacity check at 1 C (2×)
   (a) Discharge at 1 C until 3 V;
   (b) Rest 1 h;
   (c) Charge at 1 C until 4.2 V;
   (d) Hold voltage 4.2 V until C-rate drops to 0.05 C;
   (e) Rest 1 h.
5. Discharge at 1 C to the desired 80% SOC;
6. Rest 1 h;
7. GEIS at 5 A from 5 kHz to 10 mHz for 10 sine periods;
8. Hold voltage at 3.95 V during the next high-temperature phase.

Before the start of the abuse test, the SOC of the cell was charged to 100% by a 1 C constant current (CC) charge and a 4.2 V constant voltage (CV) phase; this phase ended when the current dropped to 0.05 C, and it was then set to 80% SOC by a 1 C CC discharge phase. The cell was held at a constant 3.95 V (80% SOC) during the high-temperature phase to keep the boundary conditions for the SEI reactions constant. Thus, the experiments were conducted at a higher SOC close to the upper operation limit of commercially used cells [65]. Increasing the SOC any further would result in intensified gas emissions due to higher cell voltage, which are not related to the degradation phenomena under investigation [30]. After the end of the abuse test, the cells were cycled two times at 20 °C and in a CC phase at 0.05 C. These measured voltage responses were used for incremental capacity analysis (ICA) and were compared against cells without thermal abuse (Section 3.2).

3. Results and Discussion

All the temperature abuse tests were finished within a complete temperature range of 50–120 °C (Table 2). The overview of the capacity fade, ORI, maximum pressure increase (only B1 and O1), as well as the maximum FBG wavelength change compared to before the abuse, can be seen in Table 2. The capacity fade after the experiment was approximately 6% of the initial capacity. The resistance increases were around 25% except for the test O2, which had shown a problem with the GEIS measurement. A maximum pressure of 3.4 bar was reached during the 120 °C phase for the B1 and O1 tests. The start-to-end change in the wavelength change (at 1 Hz with an accuracy of ±2 pm) of each fixed FBG adhesive configuration was spread from a 0.74–1.20 nm range, thereby showing that there was potential for improving the quality of application methodology to reach more consistent results. The trends of all the tabulated characteristics over the thermal abuse can be seen in Appendix A.2.

Table 2. Degradation overview of the capacity fade, FBG wavelength change at reference conditions, pressure, and ORI after completion of the thermal abuse test. Additionally, the temperature-specific wavelength change of the fixed FBG sensors at 50 °C are displayed.

Property	Value			
ID	O1	O2	B1	B2
Max pressure/bar	3.40	-	3.44	-
FBG hub/nm	−1.02	−0.74	−0.86	−1.20
Capacity decrease/%	5.32	6.15	5.00	7.09
ORI/%	26.29	134.92	26.77	27.45
$\Delta\lambda_{fix}/\Delta T$/pm K^{-1}—before until end of 50 °C phase	21.2	19.1	13.2	18.4
$\Delta\lambda_{fix}/\Delta T$/pm K^{-1}—cycle after 50 °C phase	36.8	43.4	37.4	36.1

3.1. State Development during Thermal Degradation

The test performance of the conducted tests proceeded in a very similar fashion. The following describes the results of experiment B1. It can be seen in Figure 7a that, in each phase of increased temperature, the wavelength change of the free FBG sensor was found to be directly proportional to the cell temperature. By comparison, the reversibility of the Bragg wavelength of the fixed FBG was equal to the free FBG up to the maximum storage temperature of 70 °C (Figure 7b). Thus, up to this temperature, the deformations of the rupture disc were exclusively elastic. Exceeding this temperature led to a plastic bulging in the perforations on the rupture disc during the high-temperature phase, which was shown as an irreversible wavelength decrease due to the resulting compression of the fixed FBG in the following phase at reference temperature. However, like the free FBG, the relative signal change of the fixed FBG can be directly correlated with the temperature rise during the high-temperature phase, thereby allowing for temperature monitoring. While the sensitivity for the free FBG was based on the thermal part of Equation (4), and showed an effective sensitivity of 8 pm K^{-1}, the sensitivity of the fixed FBG during the high-temperature phase was three times higher (Table 2). As the complete fiber was assumed to be at an ambient temperature, the additional wavelength shift could be attributed to the thermal expansion of the host material. In contrast to the increasingly visible irreversible changes in the FBG that were fixed on the thin rupture disc, the expansion-induced higher temperature sensitivity could also be shown for other fixed FBG measuring points on the robust aluminum housing of the prismatic cell in the preliminary experiments.

Before starting the experiment, the pressure at the reference conditions inside the cell was at a slight over-pressure at a constant 160 mbar (Figure 7b). The pressure exhibited purely reversible pressure increases due to thermal expansions up to 75 °C. An irreversible pressure rise became apparent within the following high-temperature phases. In the closed cell, the increase in pressure at reference conditions could only be related to the gas evolution inside the cell due to the onset of the SEI degradation reactions. These irreversible contributions of both the pressure and the fixed FBG sensor were superimposed with the thermal expansion at elevated temperatures. As the deformation state was directly dependent on the cell pressure due to the design of the rupture disc, the measurable compression of the fixed FBG was found to be dependent upon the significantly larger pressure increase from the thermal expansion rather than the additional gas generation.

In addition to the increasing irreversible changes in the sensor signals, the GEIS performed also showed an ORI of the cell. The repeated GEIS measurements at 80% SOC and at the reference conditions showed a high reproducibility between the measurements before and after the determination of the discharge capacity. Therefore, only the first impedance spectrum after the high-temperature phases at 50, 85, and 120 °C can be seen in Figure 7c. The GEIS measurements showed an ORI leading to a shift along the real-axis in the characteristic impedance spectrum (at 50 °C) that saw increasing temperature.

In Figure 7d, the current ($I < 0$ is a discharge) of the capacity measurement with two discharge cycles at 1 C is shown at 50 °C. The second discharge $\Delta Q_{dis,2}$ was used as a reference for capacity determination. With the measurement after 50 °C taken as a reference measurement with 38.82 Ah, the discharge capacity decreased to 36.89 Ah after the 120 °C phase. On the secondary axis of this plot, the measured temperature of the Pt100 sensor on top of the cell showed a temperature increase of 1.5 °C at the end of the discharge phase.

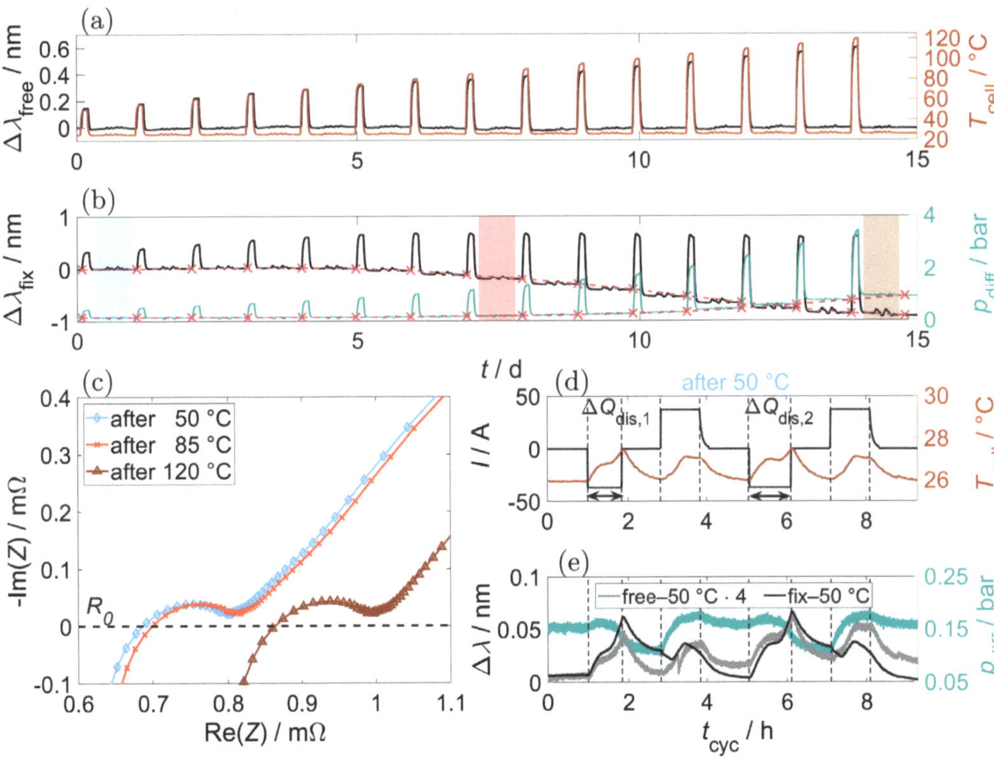

Figure 7. Overview of the results of experiment B1. (**a**): The free FBG wavelength change and temperature profile of the Pt100 sensor on top of the cell during test time. (**b**): Progression of the pressure and fixed FBG signals over the test duration and high-temperature phases that saw initial reversible changes up to 70 °C and which accumulated irreversible changes with increasing temperature. (**c**): Impedance spectroscopy spectra after 50, 85, and 120 °C temperature phases before capacity determination at 25 °C. (**d**): Current profile for the capacity determination and measured temperature at 25 °C after the 50 °C phase. (**e**): Sensor performance of the pressure, as well as the free and fixed FBG sensor during cycling after 50 °C. The wavelength change of the free FBG sensor was multiplied by four to create a comparable signal level to the fixed FBG sensor.

The cyclic load with changes in the SOC, which led to minimal cell temperature increase, was particularly interesting for the applied sensors to test their sensitivity during normal operation at the constant ambient temperature of 25 °C. Therefore, the optical wavelength shift of the fixed and free FBG sensor and the pressure, which were obtained during the same procedure, are displayed in Figure 7e. While dependent on the gas temperature inside the cell, the equilibrium pressure was directly proportional to the SOC of the cell, which can be seen during the relaxation phases in between the charge and discharge phases. This was in line with other results found in the literature, and it can be explained by the different states of the volumetric expansion of the electrodes depending on the SOC [66,67].

Looking at the FBGs, the results for the free FBG were corrected accordingly in the plot to allow for a comparison of trends (Figure 7e). The direct proportionality of the free FBG to the temperature profile became less accurate during cycling due to the low sensitivity. In addition, between hours 3 and 4, there was a steep decrease in λ_B. This irregular effect can be observed on average 2 to 3 times a day, and it was related to the mechanical disturbances outside the described test environment that were due to other activities in the lab environment to which the free FBGs sensors were susceptible.

The temperature profile was captured more precisely by the fixed FBG sensor as it displays a sensitivity over 35 pm K^{-1} (Table 2), which was averaged by the temperature difference at 150 mbar $\pm$ 10 mbar during the discharge between hours 1 and 2, as well as 5 and 6 (Figure 7e). By comparing the sensitivity of the fixed sensor during the high-temperature phase and cycling, it became evident that the pressure increase led to a compression in the FBG (Table 2). This became especially apparent when we compared the sensitivities to the temperature during cycling with small pressure changes of $\approx$70 mbar and the 50 °C phase with increased pressures at $\approx$270 mbar in Table 2. The apparent sensitivity to even minor pressure changes in the performed tests could be seen at the start of the charge phase for cell B1 (just after hours 3 and 7), where the charge led to a steep increase in pressure due to the parallel temperature and SOC increase. During the start of the charge, it could thus be seen that the fixed FBG signal decreased despite the increase in temperature. Later, during the charge, the pressure gradient decreased, which led to an increase in Bragg wavelength until the temperature reached a constant of 27 °C. Overall, the fixed FBG sensitivity to temperature predominated due to the thermal expansion of the material, while the SOC dependency was only noticeable under certain conditions due to the minimal changes in pressure in the SOC range.

The small elastic geometry changes of the rupture disc leading to the elastic compression of the fiber show an increasing wavelength change after the completion of the test due to an increase in the sensitivity of 30–80% at the last cycle depending on the test (Figure 7b). The sensitivity toward these small pressure changes seemed to vary between each test and was most likely highly susceptible to minor changes in the fiber positioning and adhesive geometry on the whole rupture disc.

A lower sensitivity toward pressure changes and a more direct proportionality of the fixed FBG to the temperatures during these cycling conditions can be seen in the analog figures of the experiments O1 (Figure A1), B2 (Figure A2), and O2 (Figure A3).

The development of the start-to-end normalized state changes of pressure, ORI, and capacity at reference conditions before the start of the next high-temperature phase showed an exponential deterioration over the intensified thermal abuse (Figure 8). The normalized measurement profiles of the irreversible wavelength changes in the first experimental test, which exemplified the differences in the fixed FBG adhesive configurations, are also depicted in this figure. These profiles did not exhibit an exponential dependence on the temperature level of the preceding phase.

After each high-temperature phase, there was an increasing change in the Ohmic resistance and discharge capacity (Figures A4 and A5). However, the irreversible rise of pressure only started after the high-temperature phase at 75 °C (Figure A6). Reaching the onset temperature of the SEI decomposition led to gas emissions that increased exponentially with increasing temperatures due to the decomposition reaction rate (Figure 8) [51]. This was also consistent with the measurable plastic deformation of the rupture disc by the fiber optic sensor system after the high-temperature step in the setup, which was conducted with a balanced distribution of adhesive on both sides of the middle perforation. However, this did not develop exponentially but correlated linearly in a range of 75–115 °C for B1 and a range of 85–115 °C for B2, which saw increasing temperatures in the high-temperature phases (Figure A7).

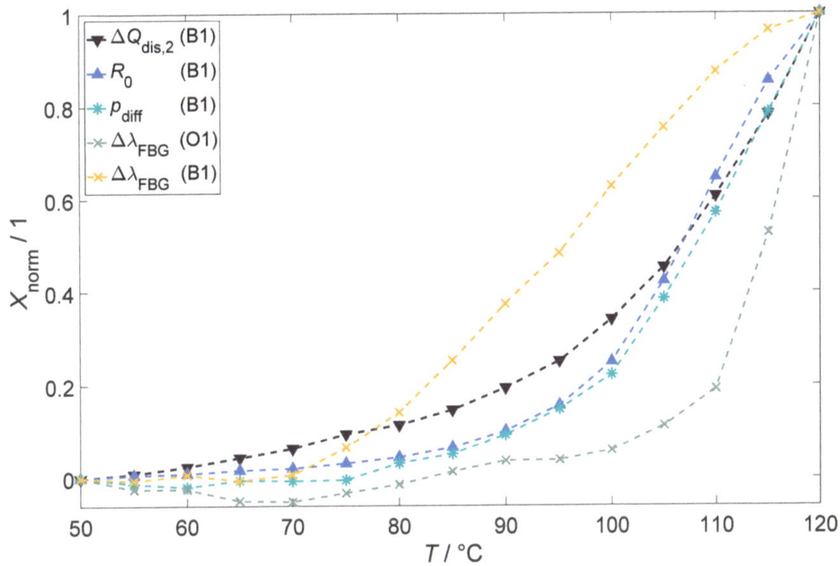

Figure 8. Progression start-end that was normalized in terms of the discharge capacity fade, ORI, pressure increase, and Bragg wavelength decrease in the fixed FBG, which was configured with an adhesive applied to one side (O), as well as both of the sides (B) of the middle rupture disc perforation after each high-temperature phase at 25 °C.

Because of the designed structural weakness to a plastic deformation of the aluminum rupture disc, the FBG sensor used here showed a clearer signal change compared to the pressure sensor. As the pressure approached the critical pressure limit, the linearity of the decline in the relaxed wavelength change decreased. This is an indication that the tensile strength of the aluminum of the bulging rupture disc had been reached and that the mechanical failure of the rupture disc had begun. Due to its direct dependence on the internal pressure, the deviation from linearity was a good indicator of the coming failure of the rupture disc, which thus enabled reactions to be made. Therefore, the early detection of critical rupture disc deformations due to critical pressure developments was directly enabled by monitoring the intentionally weakest point of the cell housing.

In contrast, the results of the one-sided configuration exhibited no reproducibility in the λ_B change during the high-temperature abuse. Particularly striking was, before the period of the expected decrease in the λ_B after 110 °C, the increase in wavelength in the O2 configuration. A positive wavelength change indicates an elongation. This can be explained by the non-parallel fixation of the FBG sensor near the center perforation, whose increasing curvature on the side of the FBG sensor explained the increase in λ_B from a range of 90–110 °C (Figure A7). When increasing the pressure, depending on the adhesive geometry on the rupture disc, there were variations in the extent of deformation. The use of a symmetric adhesive geometry resulted in a more homogeneous resistance behavior against deformation on the rupture disc. This led to a balanced configuration on the sensor side detecting (linear) changes significantly earlier. On the other hand, in the one-sided configuration, the adhesive-reinforced side, along with the sensor perforation, caused a higher mechanical resistance on the sensor side, thus resulting in lower sensitivity to pressure changes.

Thus, the sensor in the one-sided configuration effectively showed a less significant signal at slightly elevated temperatures, as well as drastic increases in the plastic deformation above 110 °C (Figure 8). The non-linearity and lower signal expression of this application alternative thus reduced the easily accessible informative value for state detection.

3.2. Incremental Capacity Analysis

Due to the good suitability of the ICA for the investigation of various significant degradation phenomena like the loss of mobile lithium inventory (LLI), loss of active material (LAM), and changes in ORI, the change in incremental capacity (dQ/dU) due to the thermal abuse test of the damaged cells was considered [59,68]. To evaluate the degradation effects, the results of the two cycles at a current rate of 0.05 C at a 20 °C ambient temperature on the cells after the high-temperature step test, as well as two undamaged cells, were compared to each other. Before the determination of the incremental capacity of each of these cells, the final charge and discharge curve of the voltage was smoothed with a Gaussian filter for a window of 400 s to improve the curve feature visibility. Figure 9 shows the dQ/dU curves of the undamaged cells, as well as the B1 and O1 cells as a function of the cell voltage for charge ($dQ/dU > 0$) and discharge ($dQ/dU < 0$).

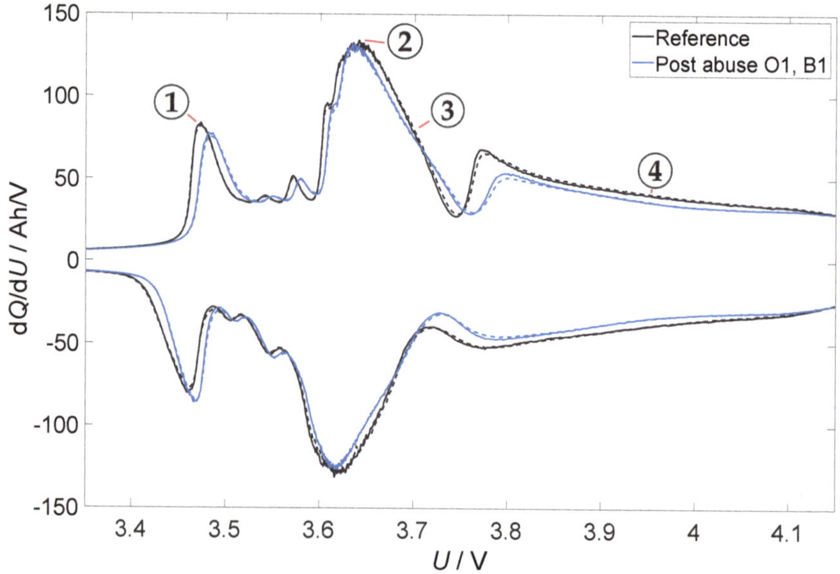

Figure 9. Incremental capacity analysis of the two reference cells without thermal degradation and the two cells after the thermal abuse test (B1 and O1) at a constant SOC of 80% during charge ($dQ/dU > 0$) and discharge ($dQ/dU < 0$). The features of interest during the charge phase are annotated by ① to ④.

The incremental capacity curves of the damaged cells and the reference cells showed good agreement, which indicates good cell quality and high reproducibility in the thermal step tests concerning the degradation state. Significant features of interest (FOI) for the NMC111/GIC cell chemistry are shown in Figure 9 by ① to ④. These FOI enabled assessments of the degradation phenomena [59]. Each feature was a combination of the electrochemical reactions at the positive and negative electrodes depending on the material [68]. The peaks identifiable in the NMC111/GIC results are largely dependent on the graphite [59,68,69]. There are further smaller peaks that are identifiable; these have been, however, disregarded due to their uncertain meaning [59]. Due to the oversized capacity for lithium ions after formation, the LAM was only detectable with a delay (incubation time) [58,59,70]. Any LAM may, therefore, be masked by the LLI accordingly [58,70].

For the NMC111/GIC chemistry, the effects can be interpreted according to Table 3, which lists the individual contributions of delithiated LAM (deNMC111 and deGIC), lithiated LAM (liNMC111 and liGIC), LLI, and ORI that have been studied by Dubarry et al. [59].

Table 3. Interpretation of the prominent feature shift of the NMC111/GIC cell chemistry due to different degradation effects. Multiple arrows correspond to the multiple positions that are dependent on the progression of degradation. The table is based on degradation maps for NMC111/GIC [59].

Property	①	②	③	④
LLI	→	↘	=	= ↘ ↗
$LAM_{deNMC111}$	← ↙	↙	↙ ↗	↓
$LAM_{liNMC111}$	↘	↓	=	↓
LAM_{deGIC}	=	=	↙ ↖	= ↘ ↗
LAM_{liGIC}	→	↘	=	= ↘ ↗
ORI	→	→	→	→

The increase in the ORI by about 25% led to no constant shifts in the peaks in the charging or discharging directions to higher or lower potentials, respectively (Figure 9), which quantitatively corresponded to the degradation map by Dubarry et al. [59]. The shift of the feature ① for the NMC111/GIC material system was found to be suitable for determining whether the positive electrode was limiting. In this case, the peak would shift to the left (←) until it disappeared (↙) [59]. A shift to ↘ was observed at ①, this corresponded to the LLI, which would include the loss of lithiated active material (LAM_{liGIC} and/or $LAM_{liNMC111}$) and could not be distinguished in this full cell ICA (Table 3). However, due to the continuous SEI formation, the LLI can be assumed in that case. Feature ③ (=) fits the degradation mode while ④ (↓) remains rather inconclusive as the position of this feature varies a great deal depending on the progression of the LLI [59]. Interestingly, feature ② showed only the slightest decrease in potential and incremental capacity, and it was found to be unusually independent of the combination of degradation phenomena compared to the other literature data [58,59]. Overall, the ICA supported the LLI and showed no detectable LAM that corresponded to the expected degradation phenomena in this temperature range (Figure 3).

3.3. Pre-Critical Thermal Abuse Correlations

When the onset of the irreversible pressure increase was examined, the exothermic decomposition of the SEI layer began after the 80 °C phase (Figure A6). The ensuing reformation reactions of the SEI layer contributed to an irreversible growth in the stable SEI components. Subsequent increments in temperature beyond 100 °C did not result in a further acceleration of the SEI layer decomposition and reformation, and this was evident through a linearization that commenced from this temperature phase during pressure rise, as depicted in Figure A6. This behavior, indicative of the kinetic limitations on the involved reactions, aligns with documented evidence in the literature [31]. This limited reaction equilibrium binds an increasing number of lithium ions and further passivates the GIC. The resulting SEI recomposition causes—in addition to the irreversible pressure increase due to gas evolution (as expressed by Equations (1) and (2))—increasing ORI (Figure 7b) and LLI, as shown by the ICA (Figure 9), which is also reflected in the increasing loss of capacity. As these effects are based on the same reactions at the tested temperature range, a strong correlation between them can be shown (Figure 8). The wavelength shift of the FBG sensor on the rupture disc at reference conditions was directly dependent on the plastic deformations on the rupture disc. In comparing the one-sided and balanced configuration, it was found that the development of the plastic deformations was strongly dependent on the glue distribution and correlates with the maximum pressure, which is a result of thermal expansion during the high-temperature phase. Therefore, it was not directly connected to the changes in the SEI layer composition. For the balanced configuration, the linear correlation to the reversible pressure increase was preserved due to the symmetric deformation behavior of the rupture disc. In the one-sided configuration, the plastic deformations of the glue-reinforced rupture disc side were only unambiguous at temperatures over 110 °C. The correlation of the degradation phenomena in the pre-critical thermal abuse can be summarized as shown in Figure 10.

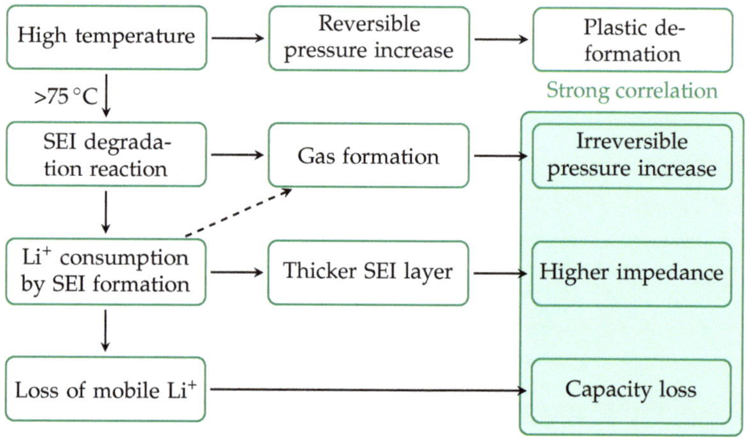

Figure 10. Correlation map of the degradation phenomena due to irreversible changes in measurable cell characteristics with an onset of SEI decomposition due to measurable gas formation over 75 °C.

4. Conclusions

Analysis of the experimental results showed that exceeding the maximum storage temperature during high-temperature phases contributes to the expected recomposition of the SEI layer, thus leading to increasing passivization (ORI), irreversible pressure increase, and capacity fade due to LLI, which was shown by ICA. The presented instrumentation of the standard rupture disc by fixed FBG sensors was shown to be reliable up to a temperature of 120 °C in terms of allowing temperature and rupture disc deformation surveillance.

A one-sided and balanced configuration of glue was tested for the FBG fixation. The balanced adhesive configuration displayed optimal conditions for a cell-specific, combined sensitive temperature sensor, as well as for reliable rupture disc monitoring due to the linear trend for plastic deformation that was shown. In contrast, the one-sided configuration failed to present a reproducible trend and unambiguous progression toward critical pressures. This showed the individual optimization potential for monitoring solutions, which depends on the exact location and application of the fixed FBG sensor.

The key features displayed by the fixed FBG sensor during these thermal abuse experiments were as follows:

- On the rupture disc, the fixed FBG sensors showed highly sensitive proportionality to temperature, as well as an inverse proportionality to the cell pressure.
- The fixed FBG application displayed regardless of the adhesive configuration sufficient sensitivity for cell temperature monitoring, and it also showed potential for the in operando detection of critical cell states.
- The direct correlation of the plastic deformations to the pressure via the mechanical properties of the rupture disc allowed for the determination of cell-specific critical states within a battery system, as well as ensured early warnings for the cell rupture event.

These sensor properties, thus, showed a robust baseline as a sensor in both normal operation and critical stress. Therefore, they also showed a feasible extension with favorable properties such as the small form factor and reliability in the monitoring of large format prismatic cells. However, it is important to investigate the potential for the early fault detection of developing local defects in large format prismatic cells with the proposed concept to establish a minimal severity in developing cell faults in future work.

Author Contributions: Conceptualization, A.H. and N.O.; methodology, A.H. and N.O.; software, A.H.; validation, A.H. and N.O.; formal analysis, A.H.; investigation, A.H. and N.O.; resources, T.T., R.B. and W.S.; data curation, A.H.; writing—original draft preparation, A.H.; writing—review and editing, A.H., N.O. and T.T.; visualization, A.H.; supervision, R.B., I.H. and T.T.; project administration, R.B.; funding acquisition, R.B. and T.T. All authors have read and agreed to the published version of the manuscript.

Funding: Parts of this research were conducted within the Research Training Group CircularLIB, which was supported by the Ministry of Science and Culture of Lower Saxony, with funds from the program zukunft.niedersachsen of the Volkswagen Foundation. Parts of this research were also funded by the Federal Ministry for Economic Affairs and Energy of Germany in the project RiskBatt (project number: 03EI3010A). Other parts were funded by the Federal Ministry of Education and Research of Germany in the project VentBatt (project number: 03XP0535A).

Data Availability Statement: The data supporting the findings of this study are openly available on Zenodo, a trusted open-access repository. The datasets associated with this research can be accessed at the following doi link: 10.5281/zenodo.10262033. The provided data include the raw optical, thermal, and electrical values for the abuse test, as well as ICA data, which allowed for the reproducibility and verification of the results presented in this manuscript. Researchers, readers, and interested parties are encouraged to access and utilize the provided datasets for further analysis or validation purposes. For any inquiries related to the data or additional information, please contact the corresponding author.

Acknowledgments: We would like to thank the Fraunhofer Heinrich Hertz Institute for providing the FBG sensors. We also acknowledge the support by the Open Access Publishing Fund of Clausthal University of Technology.

Conflicts of Interest: The authors declare no conflicts of interest. The founders had no part in the design of the study; in the collection, analyses, or interpretation of data; in the writing of the manuscript, or in the decision to publish the results.

Abbreviations

The following abbreviations are used in this manuscript:

CC	Constant Current
CV	Constant Voltage
DMC	Dimethyl Carbonate
EC	Ethylene Carbonate
FBG	Fiber Bragg Grating
FOI	Feature(s) Of Interest
ICA	Incremental Capacity Analysis
LAM	Loss of Active Material
LLI	Loss Lithium Inventory
ORI	Ohmic Resistance Increase
OSD	Over-charge Safety Device
PHEV	Plug-In Hybrid Electric Vehicle
SEI	Solid Electrolyte Interphase
SOC	State Of Charge
TR	Thermal Runaway

Appendix A

Appendix A.1. Experimental Results

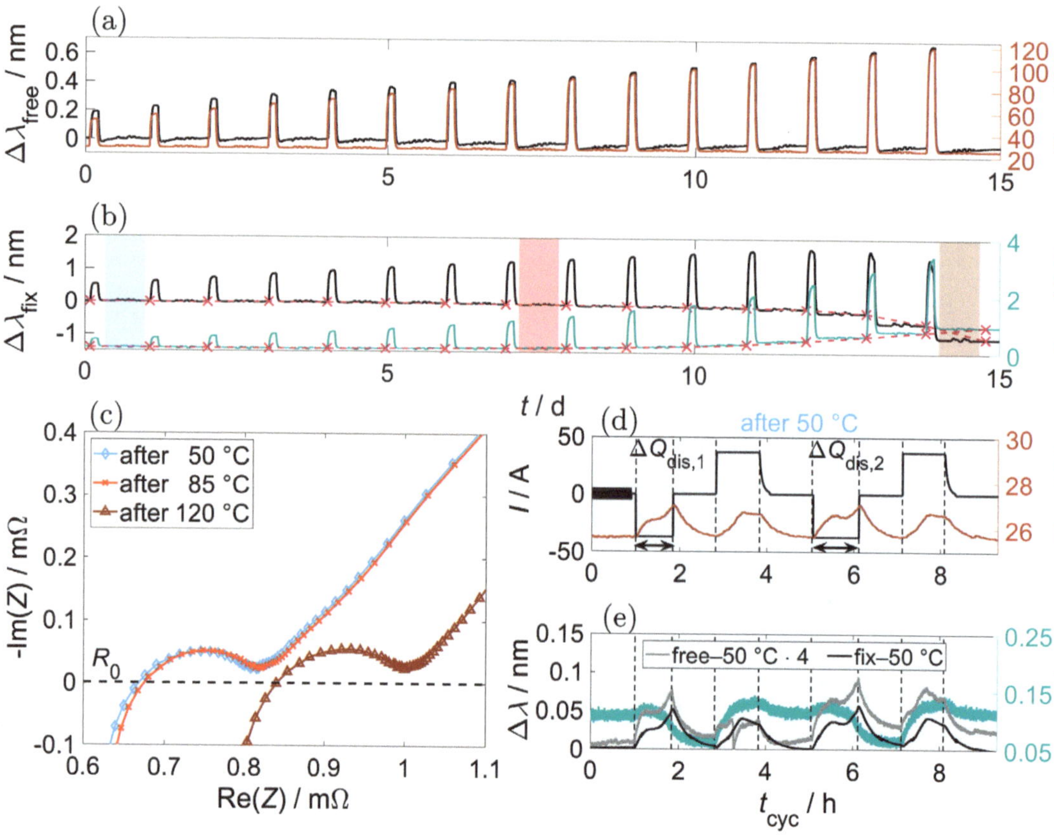

Figure A1. Overview of the results of experiment O1. (**a**): The free FBG wavelength change and temperature profile of the Pt100 sensor on top of the cell during test time. In comparison, the contact of the free FBG sensor with tape was realized worse than in the B1 experiment that led to less stable wavelength changes. (**b**): The progression of pressure and fixed FBG signals over the test duration and high-temperature phases with initial reversible changes to 70 °C and accumulating irreversible changes with increasing temperature. (**c**): Impedance spectroscopy spectra after 50, 85, and 120 °C temperature phases before a capacity determination at 25 °C. (**d**): Current profile for capacity determination and measured temperatures at 25 °C after the 50 °C phase. (**e**): Sensor performance of the pressure, as well as the free and fixed FBG sensor during cycling after 50 °C. The wavelength change of the free FBG sensor was multiplied by four to create a comparable signal level to the fixed FBG sensor.

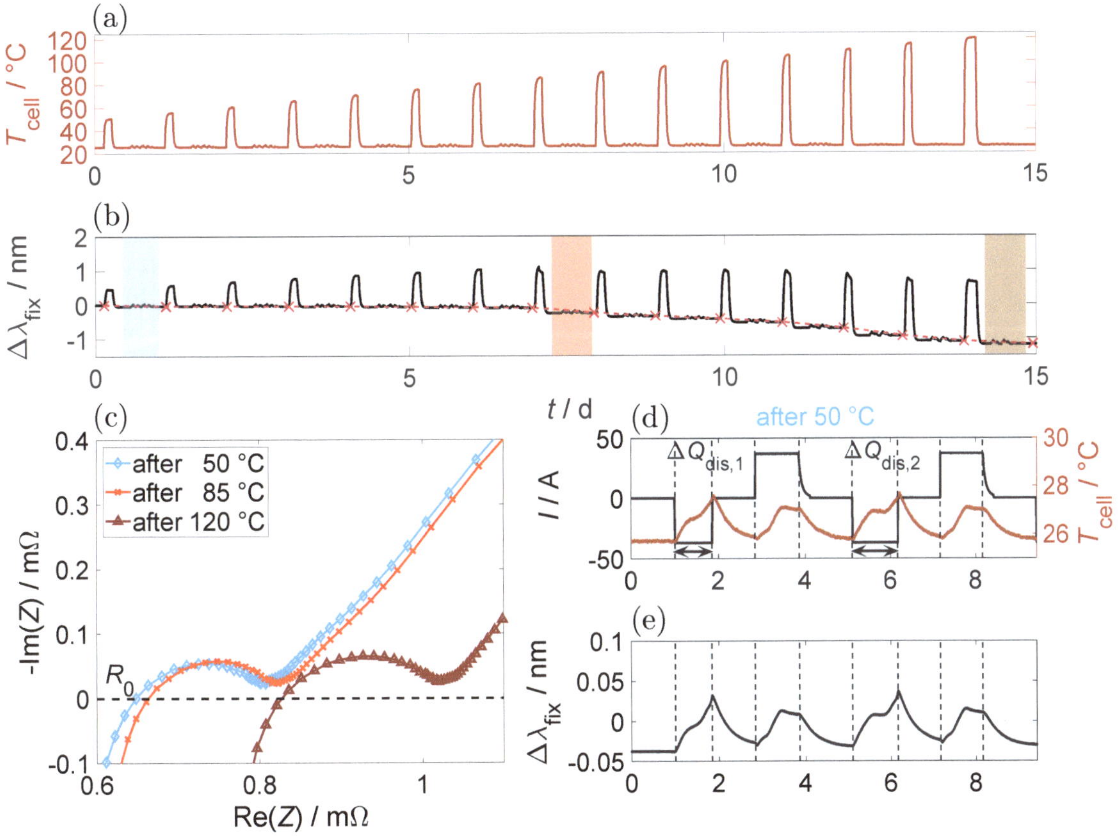

Figure A2. Overview of the results of experiment B2. (**a**): Temperature profile of the Pt100 sensor on top of the cell during test time. Due to an error, the 120 °C rest was prolonged from the usual duration of 3 h to 4.5 h, and the contact loosening of the Pt100 sensor on top of the cell after 110 °C, which resulted in an insufficient temperature measurement during cycling. (**b**): Progression of the fixed FBG signals over the test duration and high-temperature phases with initial reversible changes to 70 °C, as well as the accumulation of irreversible changes with increasing temperature. (**c**): Impedance spectroscopy spectra after 50, 85, and 120 °C temperature phases before capacity determination at 25 °C. (**d**): Current profile for capacity determination and the measured temperature at 25 °C after the 50 °C phase. (**e**): Sensor performance of the fixed FBG sensor during cycling.

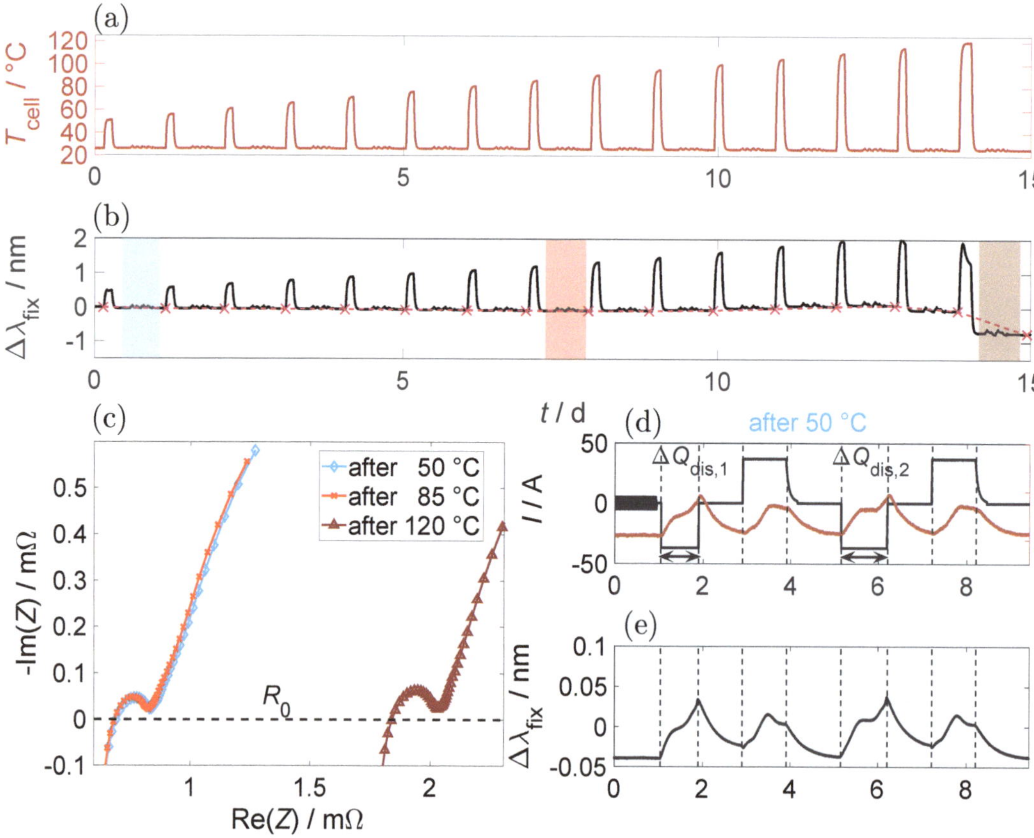

Figure A3. Overview of the results of experiment O2. (**a**): Temperature profile of the Pt100 sensor on top of the cell during test time. Due to an error, the 120 °C rest was prolonged from the usual duration of 3 h to 4.5 h. (**b**): Progression of the fixed FBG signal over the test duration and high-temperature phases with initial reversible changes to 70 °C, as well as the accumulation of irreversible changes with increasing temperature. (**c**): Impedance spectroscopy spectra after 50, 85, and 120 °C temperature phases before capacity determination at 25 °C. (**d**): Current profile for the capacity determination and measured temperature at 25 °C after the 50 °C phase. (**e**): Sensor performance of the fixed FBG sensor during cycling.

Appendix A.2. Characteristic Trends

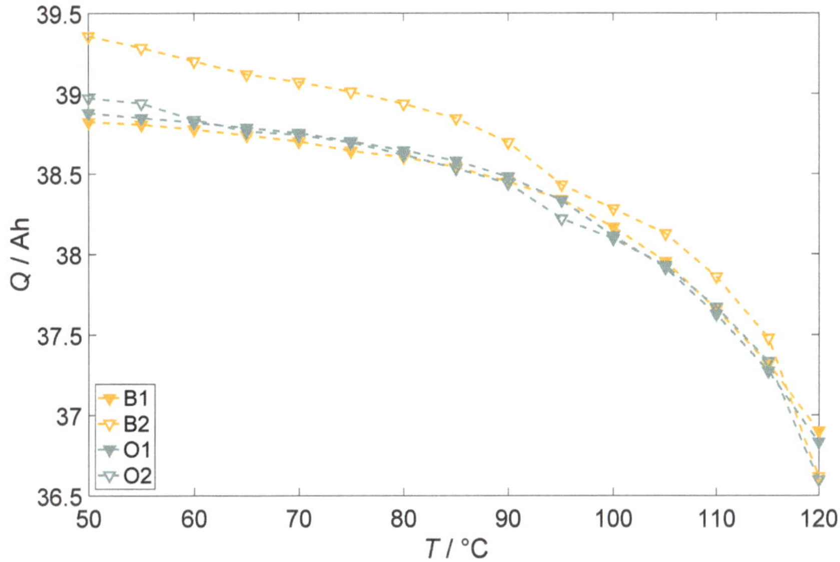

Figure A4. Decrease in the cell capacity at 25 °C after the high-temperature phase for each cell at the time before the start of the next high-temperature phase.

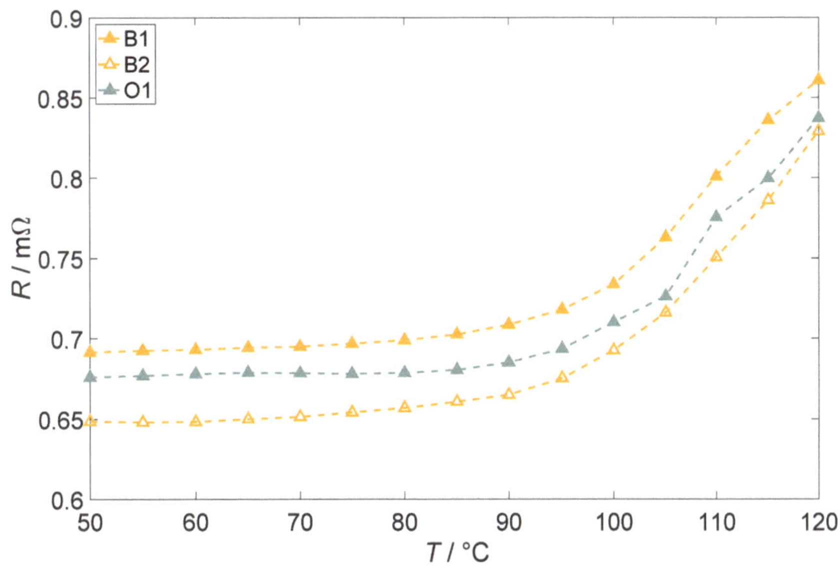

Figure A5. Increase in the Ohmic cell resistance at 25 °C after the high-temperature phase for each cell at the time before the start of the next high-temperature phase. The result of the O2 setup is not displayed due to an error during GEIS measurement.

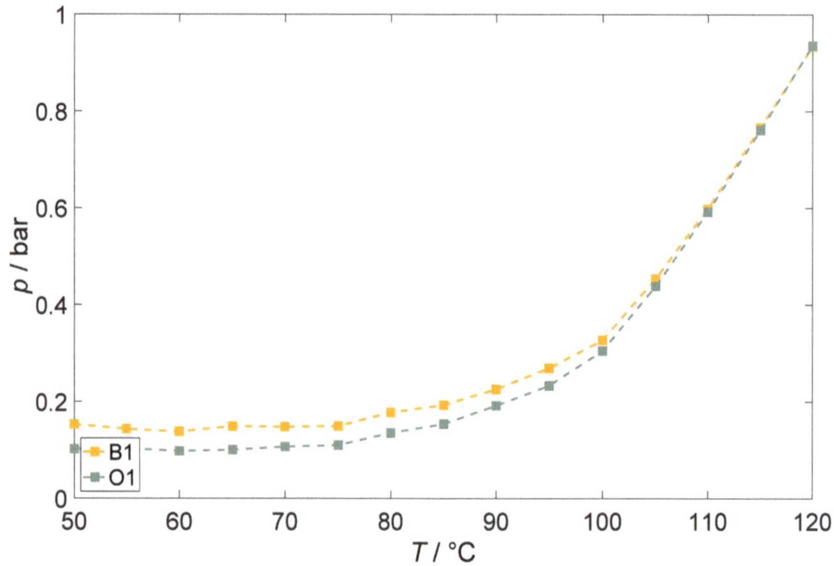

Figure A6. Increase in cell pressure at 25 °C after the high-temperature phase for each cell at rest before the start of the next high-temperature phase, as marked in Figures 7b and A1b.

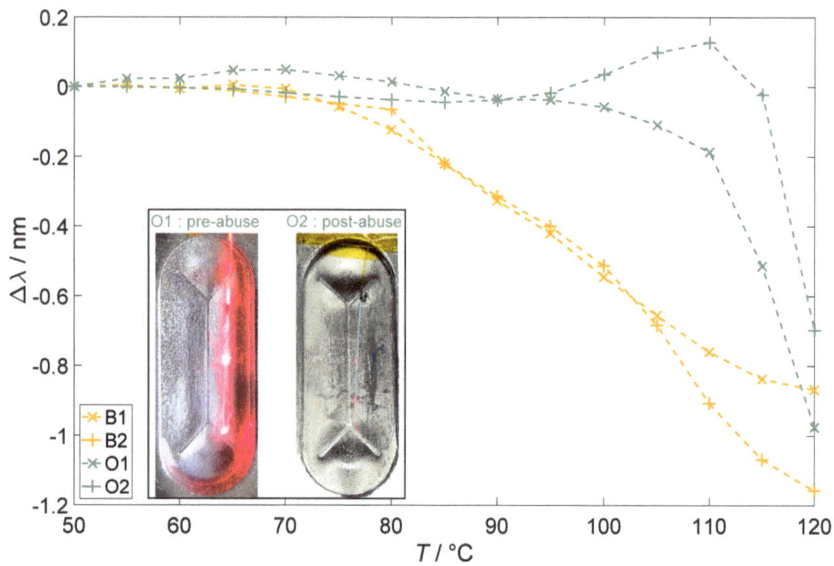

Figure A7. Change in the Bragg wavelength in the FBG sensor that was fixed on the rupture disc at 25 °C after the high-temperature phase for each cell at the time before the start of the next high-temperature phase relative to the measurement after the 50 °C phase. Inset picture of the fixed FBG position on the rupture disc for the experiments O1 (before abuse test) and O2 (after abuse test).

References

1. Bhandari, N.; Cai, A.; Yuzawa, K.; Zhang, J.; Joshi, V.; Fang, F.; Lee, G.; Harada, R.; Shin, S. Batteries: The Greenflation Challenge. Available online: https://www.goldmansachs.com/intelligence/pages/gs-research/batteries-the-greenflation-challenge/report.pdf (accessed on 11 January 2024).
2. Degen, F.; Winter, M.; Bendig, D.; Tübke, J. Energy consumption of current and future production of lithium-ion and post lithium-ion battery cells. *Nat. Energy* **2023**, *8*, 1284–1295. [CrossRef]

3. Liu, W.; Placke, T.; Chau, K.T. Overview of batteries and battery management for electric vehicles. *Energy Rep.* **2022**, *8*, 4058–4084. [CrossRef]
4. Finegan, D.P.; Darcy, E.; Keyser, M.; Tjaden, B.; Heenan, T.M.M.; Jervis, R.; Bailey, J.J.; Malik, R.; Vo, N.T.; Magdysyuk, O.V.; et al. Characterising thermal runaway within lithium-ion cells by inducing and monitoring internal short circuits. *Energy Environ. Sci.* **2017**, *10*, 1377–1388. [CrossRef]
5. Chen, Y.; Kang, Y.; Zhao, Y.; Wang, L.; Liu, J.; Li, Y.; Liang, Z.; He, X.; Li, X.; Tavajohi, N.; et al. A review of lithium-ion battery safety concerns: The issues, strategies, and testing standards. *J. Energy Chem.* **2021**, *59*, 83–99. [CrossRef]
6. Finegan, D.P.; Darst, J.; Walker, W.; Li, Q.; Yang, C.; Jervis, R.; Heenan, T.M.; Hack, J.; Thomas, J.C.; Rack, A.; et al. Modelling and experiments to identify high-risk failure scenarios for testing the safety of lithium-ion cells. *J. Power Sources* **2019**, *417*, 29–41. [CrossRef]
7. Essl, C.; Golubkov, A.W.; Fuchs, A. Comparing Different Thermal Runaway Triggers for Two Automotive Lithium-Ion Battery Cell Types. *J. Electrochem. Soc.* **2020**, *167*, 130542. [CrossRef]
8. Grabow, J.; Klink, J.; Benger, R.; Hauer, I.; Beck, H.P. Particle Contamination in Commercial Lithium-Ion Cells—Risk Assessment with Focus on Internal Short Circuits and Replication by Currently Discussed Trigger Methods. *Batteries* **2023**, *9*, 9. [CrossRef]
9. Li, H.; Duan, Q.; Zhao, C.; Huang, Z.; Wang, Q. Experimental investigation on the thermal runaway and its propagation in the large format battery module with Li(Ni1/3Co1/3Mn1/3)O2 as cathode. *J. Hazard. Mater.* **2019**, *375*, 241–254. [CrossRef]
10. Cai, T. Detection of Lithium-Ion Battery Failure and Thermal Runaway. Ph.D. Thesis, The University of Michigan, Ann Arbor, MI, USA, 2021. [CrossRef]
11. Klink, J.; Grabow, J.; Orazov, N.; Benger, R.; Börger, A.; Ahlberg Tidblad, A.; Wenzl, H.; Beck, H.P. Thermal fault detection by changes in electrical behaviour in lithium-ion cells. *J. Power Sources* **2021**, *490*, 229572. [CrossRef]
12. Tran, M.K.; Mevawalla, A.; Aziz, A.; Panchal, S.; Xie, Y.; Fowler, M. A Review of Lithium-Ion Battery Thermal Runaway Modeling and Diagnosis Approaches. *Processes* **2022**, *10*, 1192. [CrossRef]
13. Klink, J.; Hebenbrock, A.; Grabow, J.; Orazov, N.; Nylén, U.; Benger, R.; Beck, H.P. Comparison of Model-Based and Sensor-Based Detection of Thermal Runaway in Li-Ion Battery Modules for Automotive Application. *Batteries* **2022**, *8*, 34. [CrossRef]
14. Zhang, Y.; Jiang, M.; Zhou, Y.; Zhao, S.; Yuan, Y. Towards High-Safety Lithium-Ion Battery Diagnosis Methods. *Batteries* **2023**, *9*, 63. [CrossRef]
15. Feng, X.; Ren, D.; He, X.; Ouyang, M. Mitigating Thermal Runaway of Lithium-Ion Batteries. *Joule* **2020**, *4*, 743–770. [CrossRef]
16. Li, A.; Yuen, A.C.Y.; Wang, W.; de Cachinho Cordeiro, I.M.; Wang, C.; Chen, T.B.Y.; Zhang, J.; Chan, Q.N.; Yeoh, G.H. A Review on Lithium-Ion Battery Separators towards Enhanced Safety Performances and Modelling Approaches. *Molecules* **2021**, *26*, 478. [CrossRef]
17. Zhang, J.; Shao, D.; Jiang, L.; Zhang, G.; Wu, H.; Day, R.; Jiang, W. Advanced thermal management system driven by phase change materials for power lithium-ion batteries: A review. *Renew. Sustain. Energy Rev.* **2022**, *159*, 112207. [CrossRef]
18. Qiu, M.; Liu, J.; Cong, B.; Cui, Y. Research Progress in Thermal Runaway Vent Gas Characteristics of Li-Ion Battery. *Batteries* **2023**, *9*, 411. [CrossRef]
19. Ruiz, V.; Pfrang, A.; Kriston, A.; Omar, N.; van den Bossche, P.; Boon-Brett, L. A review of international abuse testing standards and regulations for lithium ion batteries in electric and hybrid electric vehicles. *Renew. Sustain. Energy Rev.* **2018**, *81*, 1427–1452. [CrossRef]
20. *Regulation (EU) 2023/1542*; Concerning Batteries and Waste Batteries, Amending Directive 2008/98/EC and Regulation (EU) 2019/1020 and Repealing Directive 2006/66/EC. European Parliament: Strasbourg, France, 2023.
21. Feng, X.; Ren, D.; Zhang, S.; He, X.; Wang, L.; Ouyang, M. Influence of aging paths on the thermal runaway features of lithium-ion batteries in accelerating rate calorimetry tests. *Int. J. Electrochem. Sci.* **2019**, *14*, 44–58. [CrossRef]
22. Yamaki, J. Thermal stability of graphite anode with electrolyte in lithium-ion cells. *Solid State Ion.* **2002**, *148*, 241–245. [CrossRef]
23. Spotnitz, R.; Franklin, J. Abuse behavior of high-power, lithium-ion cells. *J. Power Sources* **2003**, *113*, 81–100. [CrossRef]
24. Zheng, S.; Wang, L.; Feng, X.; He, X. Probing the heat sources during thermal runaway process by thermal analysis of different battery chemistries. *J. Power Sources* **2018**, *378*, 527–536. [CrossRef]
25. Schmidt, J.P. Verfahren zur Charakterisierung und Modellierung von Lithium-Ionen Zellen. Ph.D. Thesis, Karlsruhe Institute of Technology, Karlsruhe, Germany, 2013. [CrossRef]
26. Zhang, Y.; Wang, H.; Wang, Y.; Li, C.; Liu, Y.; Ouyang, M. Thermal abusive experimental research on the large-format lithium-ion battery using a buried dual-sensor. *J. Energy Storage* **2021**, *33*, 102156. [CrossRef]
27. Golubkov, A.W.; Fuchs, D.; Wagner, J.; Wiltsche, H.; Stangl, C.; Fauler, G.; Voitic, G.; Thaler, A.; Hacker, V. Thermal-runaway experiments on consumer Li-ion batteries with metal-oxide and olivin-type cathodes. *RSC Adv.* **2014**, *4*, 3633–3642. [CrossRef]
28. Wang, H.; Frisco, S.; Gottlieb, E.; Yuan, R.; Whitacre, J.F. Capacity degradation in commercial Li-ion cells: The effects of charge protocol and temperature. *J. Power Sources* **2019**, *426*, 67–73. [CrossRef]
29. Rowden, B.; Garcia-Araez, N. A review of gas evolution in lithium ion batteries. *Energy Rep.* **2020**, *6*, 10–18. [CrossRef]
30. Mattinen, U.; Klett, M.; Lindbergh, G.; Wreland Lindström, R. Gas evolution in commercial Li-ion battery cells measured by on-line mass spectrometry—Effects of C-rate and cell voltage. *J. Power Sources* **2020**, *477*, 228968. [CrossRef]
31. Feng, X.; Ouyang, M.; Liu, X.; Lu, L.; Xia, Y.; He, X. Thermal runaway mechanism of lithium ion battery for electric vehicles: A review. *Energy Storage Mater.* **2018**, *10*, 246–267. [CrossRef]

32. Zou, K.; Chen, X.; Ding, Z.; Gu, J.; Lu, S. Jet behavior of prismatic lithium-ion batteries during thermal runaway. *Appl. Therm. Eng.* **2020**, *179*, 115745. [CrossRef]
33. Cheng, A.; Xin, Y.; Wu, H.; Yang, L.; Deng, B. A Review of Sensor Applications in Electric Vehicle Thermal Management Systems. *Energies* **2023**, *16*, 5139. [CrossRef]
34. Popp, H.; Koller, M.; Jahn, M.; Bergmann, A. Mechanical methods for state determination of Lithium-Ion secondary batteries: A review. *J. Energy Storage* **2020**, *32*, 101859. [CrossRef]
35. Chen, D.; Zhao, Q.; Zheng, Y.; Xu, Y.; Chen, Y.; Ni, J.; Zhao, Y. Recent Progress in Lithium-Ion Battery Safety Monitoring Based on Fiber Bragg Grating Sensors. *Sensors* **2023**, *23*, 5609. [CrossRef]
36. Jia, T.; Zhang, Y.; Ma, C.; Li, S.; Yu, H.; Liu, G. The early warning for overcharge thermal runaway of lithium-ion batteries based on a composite parameter. *J. Power Sources* **2023**, *555*, 232393. [CrossRef]
37. Zhang, L.; Liu, X.; Li, K.; Du, D.; Zheng, M.; Niu, Q.; Yang, Y.; Zhou, Q.; Sun, T.; Grattan, K.T.V. Real-Time Battery Temperature Monitoring Using FBG Sensors: A Data-Driven Calibration Method. *IEEE Sens. J.* **2022**, *22*, 18639–18648. [CrossRef]
38. Liu, S.; Li, K. Thermal monitoring of lithium-ion batteries based on machine learning and fibre Bragg grating sensors. *Trans. Inst. Meas. Control.* **2023**, *45*, 1570–1578. [CrossRef]
39. Chang, L.; Chen, W.; Mao, Z.; Huang, X.; Ren, T.; Zhang, Y.; Cai, Z. Experimental study on the effect of ambient temperature and discharge rate on the temperature field of prismatic batteries. *J. Energy Storage* **2023**, *59*, 106577. [CrossRef]
40. Tardy, E.; Thivel, P.X.; Druart, F.; Kuntz, P.; Devaux, D.; Bultel, Y. Internal temperature distribution in lithium-ion battery cell and module based on a 3D electrothermal model: An investigation of real geometry, entropy change and thermal process. *J. Energy Storage* **2023**, *64*, 107090. [CrossRef]
41. *DIN 91252:2016-11*; Elektrische Straßenfahrzeuge–Batteriesysteme–Electrically Propelled Road Vehicles–Battery Systems–Design Specifications for Lithium-Ion Battery Cells. German Institute for Standardization: Berlin, Germany, 2016. [CrossRef]
42. Barai, A.; Tangirala, R.; Uddin, K.; Chevalier, J.; Guo, Y.; McGordon, A.; Jennings, P. The effect of external compressive loads on the cycle lifetime of lithium-ion pouch cells. *J. Energy Storage* **2017**, *13*, 211–219. [CrossRef]
43. Holland, A.A. The Effects of Compression on Lithium-Ion Batteries. Ph.D. Thesis, Imperial College London, London, UK, 2018. [CrossRef]
44. Kohlberger, M. Sicherheitselement für Batteriezelle. European Patent Office EP2779271A2, 21 January 2014.
45. Peled, E. The Electrochemical Behavior of Alkali and Alkaline Earth Metals in Nonaqueous Battery Systems—The Solid Electrolyte Interphase Model. *J. Electrochem. Soc.* **1979**, *126*, 2047–2051. [CrossRef]
46. Peled, E.; Menkin, S. Review—SEI: Past, Present and Future. *J. Electrochem. Soc.* **2017**, *164*, A1703–A1719. [CrossRef]
47. Pinson, M.B.; Bazant, M.Z. Theory of SEI Formation in Rechargeable Batteries: Capacity Fade, Accelerated Aging and Lifetime Prediction. *J. Electrochem. Soc.* **2013**, *160*, A243–A250. [CrossRef]
48. Son, K.; Hwang, S.M.; Woo, S.G.; Paik, M.; Song, E.H.; Kim, Y.J. Thermal and chemical characterization of the solid-electrolyte interphase in Li-ion batteries using a novel separator sampling method. *J. Power Sources* **2019**, *440*, 227083. [CrossRef]
49. Chen, S.; Gao, Z.; Sun, T. Safety challenges and safety measures of Li-ion batteries. *Energy Sci. Eng.* **2021**, *9*, 1647–1672. [CrossRef]
50. Aurbach, D.; Ein-Eli, Y.; Markovsky, B.; Zaban, A.; Luski, S.; Carmeli, Y.; Yamin, H. The Study of Electrolyte Solutions Based on Ethylene and Diethyl Carbonates for Rechargeable Li Batteries: II. Graphite Electrodes. *J. Electrochem. Soc.* **1995**, *142*, 2882–2890. [CrossRef]
51. Tanaka, N. Modeling and Simulation of Thermo-Electrochemistry of Thermal Runaway in Lithium-Ion Batteries. Ph.D. Thesis, University of Stuttgart, Stuttgart, Germany, 2015. [CrossRef]
52. Richard, M.N.; Dahn, J.R. Accelerating Rate Calorimetry Study on the Thermal Stability of Lithium Intercalated Graphite in Electrolyte. I. Experimental. *J. Electrochem. Soc.* **1999**, *146*, 2068–2077. [CrossRef]
53. Wang, Q.; Sun, J.; Yao, X.; Chen, C. Thermal stability of LiPF6/EC+DEC electrolyte with charged electrodes for lithium ion batteries. *Thermochim. Acta* **2005**, *437*, 12–16. [CrossRef]
54. Deshpande, R.D.; Bernardi, D.M. Modeling Solid-Electrolyte Interphase (SEI) Fracture: Coupled Mechanical/Chemical Degradation of the Lithium Ion Battery. *J. Electrochem. Soc.* **2017**, *164*, A461–A474. [CrossRef]
55. Palacín, M.R.; de Guibert, A. Why do batteries fail? *Science* **2016**, *351*, 1253292. [CrossRef]
56. Zhou, M.; Zhao, L.; Okada, S.; Yamaki, J.I. Quantitative Studies on the Influence of LiPF 6 on the Thermal Stability of Graphite with Electrolyte. *J. Electrochem. Soc.* **2011**, *159*, A44–A48. [CrossRef]
57. Agubra, V.A.; Fergus, J.W. The formation and stability of the solid electrolyte interface on the graphite anode. *J. Power Sources* **2014**, *268*, 153–162. [CrossRef]
58. Carter, R.; Kingston, T.A.; Atkinson, R.W.; Parmananda, M.; Dubarry, M.; Fear, C.; Mukherjee, P.P.; Love, C.T. Directionality of thermal gradients in lithium-ion batteries dictates diverging degradation modes. *Cell Rep. Phys. Sci.* **2021**, *2*, 100351. [CrossRef]
59. Dubarry, M.; Anseán, D. Best practices for incremental capacity analysis. *Front. Energy Res.* **2022**, *10*, 1023555. [CrossRef]
60. Schmitt, J.; Kraft, B.; Schmidt, J.P.; Meir, B.; Elian, K.; Ensling, D.; Keser, G.; Jossen, A. Measurement of gas pressure inside large-format prismatic lithium-ion cells during operation and cycle aging. *J. Power Sources* **2020**, *478*, 228661. [CrossRef]
61. Werneck, M.M.; Allil, R.C.S.B.; Ribeiro, B.A.; de Nazaré, F.V.B. A Guide to Fiber Bragg Grating Sensors. In *Current Trends in Short- and Long-Period Fiber Gratings*; Cuadrado-Laborde, C., Ed.; InTech: London, UK, 2013. [CrossRef]
62. Sahota, J.K.; Gupta, N.; Dhawan, D. Fiber Bragg grating sensors for monitoring of physical parameters: A comprehensive review. *Opt. Eng.* **2020**, *59*, 060901. [CrossRef]

63. Schlüter, V.G. Entwicklung Eines Experimentell Gestützten Bewertungsverfahrens zur Optimierung und Charakterisierung der Dehnungsübertragung Oberflächenapplizierter Faser-Bragg-Gitter-Sensoren. Ph.D. Thesis, Technische Universität Berlin, Berlin, Germany, 2009.
64. Wan, K.T.; Leung, C.K.Y.; Olson, N.G. Investigation of the strain transfer for surface-attached optical fiber strain sensors. *Smart Mater. Struct.* **2008**, *17*, 035037. [CrossRef]
65. Barai, A.; Uddin, K.; Dubarry, M.; Somerville, L.; McGordon, A.; Jennings, P.; Bloom, I. A comparison of methodologies for the non-invasive characterisation of commercial Li-ion cells. *Prog. Energy Combust. Sci.* **2019**, *72*, 1–31. [CrossRef]
66. Matasso, A.; Wetz, D.; Liu, F. The Effects of Internal Pressure Evolution on the Aging of Commercial Li-Ion Cells. *J. Electrochem. Soc.* **2015**, *162*, A92–A97. [CrossRef]
67. Schweidler, S.; de Biasi, L.; Schiele, A.; Hartmann, P.; Brezesinski, T.; Janek, J. Volume Changes of Graphite Anodes Revisited: A Combined Operando X-ray Diffraction and In Situ Pressure Analysis Study. *J. Phys. Chem. C* **2018**, *122*, 8829–8835. [CrossRef]
68. Ansean, D.; Gonzalez, M.; Blanco, C.; Viera, J.C.; Fernandez, Y.; Garcia, V.M. Lithium-ion battery degradation indicators via incremental capacity analysis. In Proceedings of the 2017 IEEE International Conference on Environment and Electrical Engineering and 2017 IEEE Industrial and Commercial Power Systems Europe (EEEIC/I&CPS Europe), Milan, Italy, 6–9 June 2017; pp. 1–6. [CrossRef]
69. Feng, X.; Sun, J.; Ouyang, M.; He, X.; Lu, L.; Han, X.; Fang, M.; Peng, H. Characterization of large format lithium ion battery exposed to extremely high temperature. *J. Power Sources* **2014**, *272*, 457–467. [CrossRef]
70. Birkl, C.R.; Roberts, M.R.; McTurk, E.; Bruce, P.G.; Howey, D.A. Degradation diagnostics for lithium ion cells. *J. Power Sources* **2017**, *341*, 373–386. [CrossRef]

batteries

MDPI

Article

Monitoring of Thermal Runaway in Commercial Prismatic High-Energy Lithium-Ion Battery Cells via Internal Temperature Sensing

Niklas Kisseler *, Fabian Hoheisel, Christian Offermanns, Moritz Frieges, Heiner Heimes and Achim Kampker

Chair for Production Engineering of E-Mobility Components, RWTH Aachen University, Bohr 12, 52072 Aachen, Germany; f.hoheisel@pem.rwth-aachen.de (F.H.); c.offermanns@pem.rwth-aachen.de (C.O.); m.frieges@pem.rwth-aachen.de (M.F.); h.heimes@pem.rwth-aachen.de (H.H.); a.kampker@pem.rwth-aachen.de (A.K.)

* Correspondence: n.kisseler@pem.rwth-aachen.de

Abstract: The temperature of a lithium-ion battery is a crucial parameter for understanding the internal processes during various operating and failure scenarios, including thermal runaway. However, the internal temperature is comparatively higher than the surface temperature. This particularly affects cells with a large cross-section, which is due to heat development within the cell and lower heat dissipation due to a poorer ratio of volume to surface area. This paper presents an approach that enables real-time monitoring of the behavior of a commercial prismatic high-energy battery cell (NMC811/C, 95 Ah, Contemporary Amperex Technology Co., Limited (Ningde, China)) in the event of thermal runaway induced by overcharging. The internal cell temperature is investigated by the subsequent integration of two hard sensors between the two jelly rolls and additional sensors on the surface of the aluminum housing of the battery cell. The sensor's signals show a significant increase in the temperature gradient between the temperature in the core of the cell and the cell casing surface until the onset of venting and thermal runaway of the battery. The data enable a detailed investigation of the behavior of the battery cell and the comparatively earlier detection of the point of no return in the event of thermal runaway.

Keywords: lithium-ion battery; thermal runaway; temperature monitoring; internal temperature monitoring; battery safety

Citation: Kisseler, N.; Hoheisel, F.; Offermanns, C.; Frieges, M.; Heimes, H.; Kampker, A. Monitoring of Thermal Runaway in Commercial Prismatic High-Energy Lithium-Ion Battery Cells via Internal Temperature Sensing. *Batteries* **2024**, *10*, 41. https://doi.org/10.3390/batteries10020041

Academic Editor: Mingyi Chen

Received: 21 December 2023
Revised: 15 January 2024
Accepted: 16 January 2024
Published: 23 January 2024

1. Introduction

The increasing electrification of passenger transportation is resulting in a significant increase in demand for lithium-ion batteries (LIBs) [1]. The demand in LIBs for battery electric vehicles (BEVs) for the Chinese, US and European markets currently amounts to 400 GWh and is expected to increase to 6800 GWh by 2030 [2]. However, there are still technological challenges that affect the broad acceptance of the technology for mobility applications in society. In addition to the technological challenges of increasing energy density, reducing charging time and reducing production cost, safety concerns are a critical factor [3]. A key to maximizing the safety of LIBs is a comprehensive understanding of the behavior before and during the occurrence of a failure [4].

The thermal runaway (TR) of a LIB as a failure mode is the key scientific problem in battery safety research, especially for batteries with Ni-rich cathode materials such as $LiNi_{0.8}Co_{0.1}Mn_{0.1}O_2$ (NCM811) due to their comparatively low thermal stability. Thermal runaway defines the uncontrolled increase in cell temperature, often resulting in fire, due to an exothermic chain reaction within the cell. The TR of a cell can be caused by mechanical abuse, electrical abuse, such as an external short circuit or overcharging, or thermal abuse [5,6].

In this context, the integration of sensors into the battery cell is a promising way of quantitatively monitoring internal cell behavior during operation and failure [7,8]. In recent years, increased efforts have been made to integrate various sensor types into different battery cell sizes and form factors to be able to directly monitor the internal battery cell temperature. Novais et al. use fiber Bragg grating (FBG) sensors to monitor the internal and external temperature of LIBs [9]. Nascimento et al. report that the hybrid sensor network, consisting of a Fabry–Perot (FP) and an FBG sensor embedded in the commercially available LFP pouch cell, proves to be effective, non-invasive and accurate [10]. Raghavan et al. integrate two FBG sensors into a pouch battery cell to monitor the state of the battery [11]. Lee et al. use a resistance temperature detector (RTD) to monitor the internal temperature of LFP/LTO coin cells [12]. A similar approach is taken by Zhu et al., who embed RTD thin film sensors in a pouch format LIB to monitor the internal cell temperature at multiple locations [13]. In another study by Fleming et al., NTC thermistors bonded to a flexible polyimide substrate are used as internal sensors for both a cylindrical 18650 cell and a 5.5 Ah pouch format battery cell [14]. In contrast, Mutyala et al. use a thin film K-type thermo-couple (TFTC) to monitor the temperature inside a 3 Ah pouch cell [15]. In various other approaches, thermocouples were also integrated into LIBs to quantify the temperature behavior within the cell [16–20].

However, most of the existing approaches are limited to investigating the internal cell temperature behavior of small-format pouch, cylindrical or coin cells during regular cycle tests. In contrast to this, Xu et al. present a study analyzing the behavior of a medium-size prismatic 37 Ah NMC111/C battery cell during thermal runaway with integrated temperature sensors in addition to a cylindrical 21700 cell and a pouch cell [21]. Parekh carries out a similar experiment with an LCO coin cell with integrated RTD and induces the thermal runaway by overcharging the cell [22]. In their work, Mei et al. present an approach to monitor the internal temperature and pressure during the thermal runaway of a commercially available 18650 LFP battery cell after applying external heat, using FBGs [23].

This work aims to investigate the temperature behavior of a large-format prismatic automotive grade LIB with a high energy density of 246 Wh/kg and NMC811/C electrode chemistry during thermal runaway using two redundant integrated thermocouples.

2. Experimental Set-Up and Methods

2.1. Cell Preparation and Sensor Integration

Prismatic lithium-ion cells from the manufacturer "Contemporary Amperex Technology Co., Limited" (Ningde, China) are used for the experiment. The specified nominal capacity is 95 Ah with a nominal voltage of 3.67 V, which results in a total energy of 350 Wh and a gravimetric energy density of 246 Wh/kg. The cell housing is made of aluminum and has the following dimensions: 35.7 mm × 180 mm × 102 mm. The wall thickness of the sides is 1.2 mm, whereas the wall thickness of the base is 1.4 mm. The internal cell structure consists of two jelly rolls connected in parallel. The jelly rolls each have a width of 178.1 mm, a height of 84 mm and a thickness of 16.65 mm. Graphite is used as the active material on the anode side and NMC811 on the cathode side. The aluminum foil of the cathode has a material thickness of 0.014 mm and is coated on both sides, whereby the total electrode thickness including coating is 0.12 mm. The copper foil of the anode has a comparatively lower material thickness of 0.012 mm and is also coated on both sides, whereby the total electrode thickness including coating is 0.165 mm. The separator has a thickness of 0.015 mm.

Type K thermocouples from the manufacturer "TC Mess-und Regeltechnik GmbH (Mönchengladbach, Germany)" with a measuring range of up to 1350 °C in short-time operation are used as temperature sensors to monitor the relevant temperature range despite the high temperatures during the thermal runaway of a battery cell.

A systematic process is used to enable safe and reliable integration of the thermocouples into the battery cell. During the integration process, the battery cell's internal structure must not be damaged, as this could otherwise lead to an irreversible loss of capacity and performance or to an internal short circuit. Before the integration process, the cells are preconditioned using a constant current–constant voltage (CC-CV) charging protocol over 2 full cycles at 0.25 C and 100% depth of discharge (DOD). The thermocouples are integrated after an additional relaxation phase of 2 h at approximately 0% state of charge (SOC).

The integration process is divided into six steps according to Figure 1 and is carried out in a conventional glovebox environment under a controlled argon atmosphere. To open the cell, the first step was to make a flat circumferential cut on the top of the cell, which separates the housing cover, including the electrode stacks, from the rest of the housing. To prevent an internal short circuit caused by aluminum chips penetrating the battery cell, the housing can is cut in a two-stage cutting process.

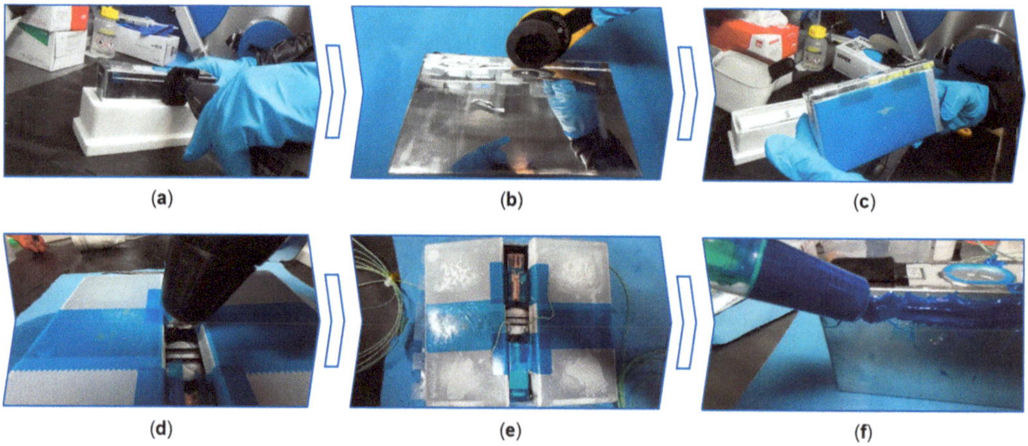

Figure 1. Systematic approach for integrating two thermocouples into a large-format prismatic cell; (**a**) milling a groove 5.0 mm below the housing cover; (**b**) cutting through the thin cut edge; (**c**) extraction of the jelly rolls and the housing lid from the housing can; (**d**) drilling two feed-throughs for the thermocouples; (**e**) integration and positioning of the thermocouples; (**f**) closing and sealing of the cell.

First, the upper layer of the cut edge is removed by a spiral cutter and the metal chips are carefully removed before a round blade cutter cuts through the remaining thin aluminum layer in the second step. In the third step, the two jelly rolls, including the housing cover, are extracted from the housing. In the fourth step, the jelly rolls are unfolded to drill two holes in the housing cover near the safety valve for the thermocouples to be integrated. Once the holes have been drilled and the swarf removed, in the fifth step, the two type K thermocouples are inserted into the cell and positioned on the separator surface of the side of the two jelly rolls facing inwards. The two measuring tips of the thermocouples are positioned centrally under the electrodes at the same height. No additional measures are required to keep the two thermocouples in place, as the stiffness of the thermocouple ensures that its position does not change during folding. In the sixth and final step, the two electrode stacks with the thermocouples are again folded and reinserted into the original housing. However, while working on the open cell, the solvent of the electrolyte evaporates at room temperature. To compensate for this loss of solvent and the expected loss of capacity, 10 mL of electrolyte containing $LiPF_6$ in EC:EMC (3:7, by wt.) is added before the cell is closed again.

To seal the reassembled cell, the 2-component structural adhesive type "BETAMATE 2090" from the manufacturer "Dow Europe GmbH (Horgen, Switzerland)" is applied to the circumferential cut edge and around the two holes for the thermocouples. The cell is left to rest for 72 h at ambient temperature to allow the adhesive to cure completely. After the curing time, the modified battery cell is tested in a vacuum leak test stand to prevent electrolyte leakage during handling and operation. In the following, the cell with integrated sensors is referred to as $Cell_{INT}$.

In addition to the two integrated thermocouples in the cell, the temperature is recorded on the cell surface and near the cell environment, as shown in Figure 2. The two thermocouples on the cell surface (1, 3) are located at the same respective positions as the two internal sensors (2, 4). Three additional thermocouples are used to monitor the temperature in the environment, in the center of the cell (5) and at the height of 10 mm above the safety vent (6) to monitor the temperature of the escaping flammable gases during thermal runaway.

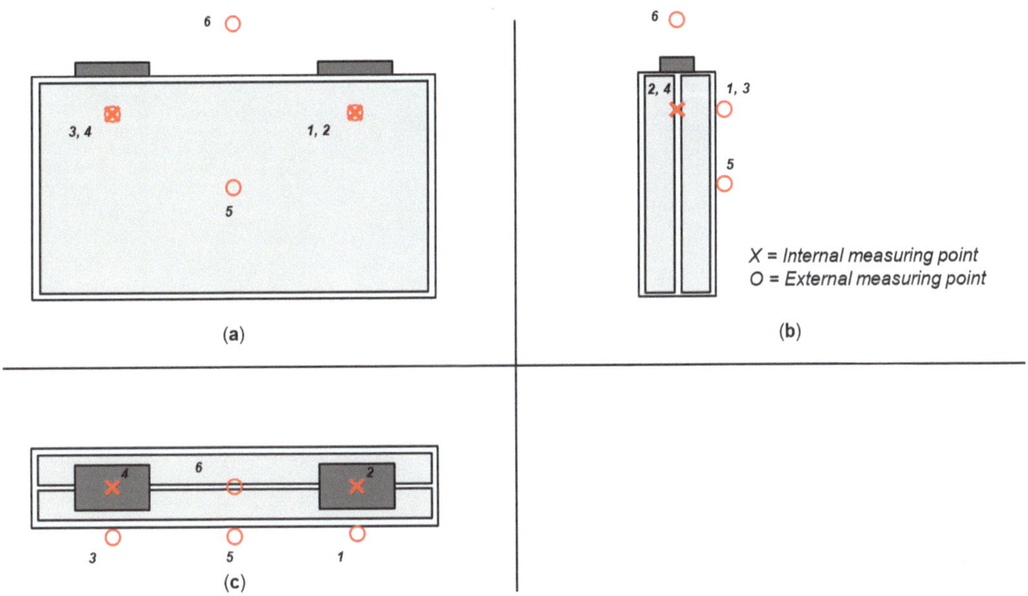

Figure 2. Projection view of the cell with thermocouple position; (**a**) front view; (**b**) side view; (**c**) top view.

2.2. Cyclization Pre-Tests

$Cell_{INT}$ is first tested within a comparably low number of charge–discharge cycles to exclude a significant cell defect due to the integration of the sensor technology in the first instance. The cells therefore undergo several CC-CV cycles with 24 A (0.25 C) charging current and 14 A (0.15 C) discharging current at 100% DOD with an additional 10 min break between each charging and discharging step. $Cell_{INT}$ is placed in a test chamber at a temperature of 22 °C during cycling. During cyclization of the modified cell, the temperature data of the internally and externally mounted temperature sensors are also recorded. This makes it possible to additionally evaluate the temperature gradient across the cell cross-section of the prismatic battery cell during normal operation. The cyclization pre-tests are carried out with a battery cell tester model "SI-9300R" from the manufacturer "Ametek, Inc. (Berwyn, PA, USA)". The temperature data are logged using a data acquisition system from the manufacturer "Gantner Instruments GmbH (Schruns, Austria)".

2.3. Thermal Runaway Tests

The thermal runaway tests take place in a test chamber suitable for abuse tests as displayed in Figure 3. The test chamber is equipped with active temperature control and a fume extraction and filtering system. The respective battery cell is connected to the battery cell tester and the temperature data recording unit presented in Section 2.3. The battery cell is placed upright. In case the cell cannot be brought into thermal runaway by overcharging, an external heating mechanism is installed as a backup thermal runaway trigger. For this purpose, the cell is fitted with tungsten wire heating elements encased in polyimide adhesive foil. To be able to investigate a possible influence of the integrated sensors on the thermal runaway behavior, the thermal runaway test is carried out both with $Cell_{INT}$ and a reference cell without integrated thermocouples, referred to below as $Cell_{REF}$.

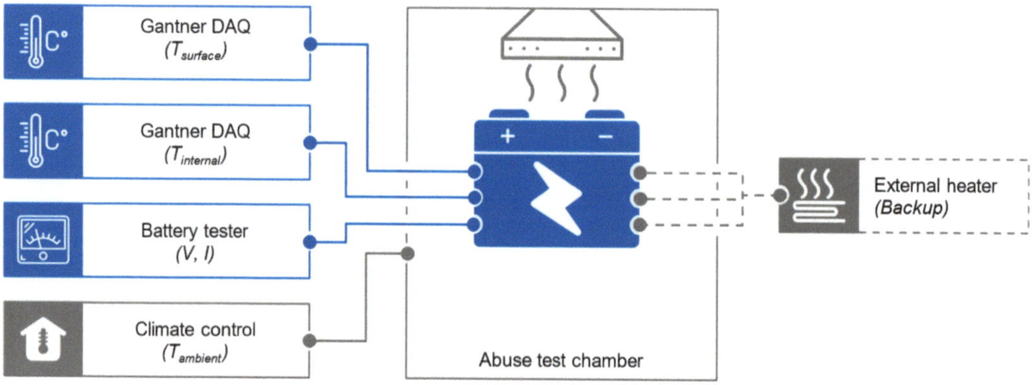

Figure 3. Schematic representation of the test setup in the abuse test chamber and the data interfaces.

The SOC of the cell is approximately 0% SOC before initiating the overcharging process. A constant electrical current of 26 A is continuously supplied for this purpose. The (over-)charging continues until thermal runaway occurs.

3. Results and Discussion

3.1. Cyclization with/without Integrated Thermocouples

To estimate the influence of the sensor insertion in the prismatic cell, the discharge capacity and quasi-open-circuit voltage (qOCV) curve at a low C rate of $Cell_{INT}$ before and after sensor integration are compared.

According to Schmalstieg et al., it should be noted that the qOCV measurement only approximates the actual values for the open-circuit voltage, as a small current always flows and the cell is not in a relaxed state. However, the actual deviation depends on the current and is therefore comparatively small at low C rates [24]. Experimental results of the qOCV curve and discharge capacity as well as the 1 kHz impedance before and after sensor integration are shown in Figure 4 and Table 1. The deviation of the qOCV after sensor integration fluctuates around the zero value and reaches its maximum at an SOC of 0% and is 0.16% compared to the qOCV before sensor integration. The value of the discharge capacity at 22 °C and 0.15 C is 1.22% above the initial value after sensor integration. This suggests that the amount of electrolyte added during the integration process to compensate for electrolyte evaporation has expanded the lithium-ion inventory in the cell and the evaporation effect has been overcompensated.

The correlation between electrolyte quantity and cell (discharge) capacity has already been described by Günter et al. in their study [25]. For future studies on integrated sensing, it is important to determine the actual amount of evaporated solvent as precisely as possible during integration to be able to carry out the compensation precisely.

Based on the results, it can be concluded that the insertion of the thermocouples did not lead to significant impairment of the electrochemical behavior of the battery cell. However, it cannot be determined to what extent the intervention in the internal structure of the battery cell has long-term effects regarding cell aging. This would require a reference analysis of the cyclical aging behavior of $Cell_{INT}$.

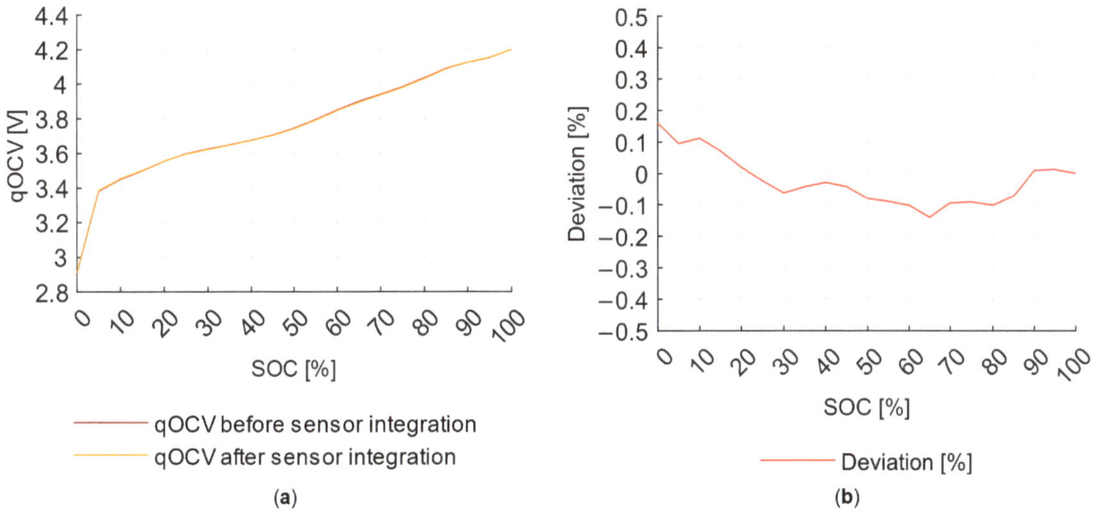

Figure 4. Comparison of (**a**) the qOCV curve for $Cell_{INT}$ before and after sensor integration and (**b**) the relative deviation.

Table 1. Comparison of the discharge capacity and 1 kHz impedance for $Cell_{INT}$ before and after sensor integration.

Reference Values	Before Sensor Integration	After Sensor Integration	Deviation
Discharge capacity [1] at 22 °C 0.15 C	96.09 Ah	97.28 Ah	+1.22%
1 kHz Impedance at 22 °C and 50% SOC	0.921 mOhm	0.893 mOhm	−1.03%

[1] Nominal capacity of 95 Ah according to the cell manufacturer's data sheet. No information is provided on C rate and ambient temperature.

After the initial validation tests, further full charge/discharge cycles were performed to quantify the temperature gradient between the cell center and the cell housing surface at 0.15 C. For this purpose, the course of the internal and external temperature of the cell was analyzed for a reference cycle with 100% DOD as shown in Figure 5. In Figure 5c, the temperature difference between the two thermocouples inside the cell is approximately constant at a value of 0.4 °C during the reference cycle. The distance between the two measuring points is 130 mm. The temperature difference between the center of the cell and the point on the surface of the cell housing with the smallest distance is 1.8 °C at maximum. The absolute distance between the two measuring points is 17.85 mm, whereby 1.2 mm is accounted for by the thickness of the housing wall and 16.65 mm by the total thickness of the jelly roll. This results in an average temperature gradient of 0.1 °C/mm. The comparatively greater temperature gradient can be attributed to the comparatively poorer thermal conductivity perpendicular to the electrode surface compared to the thermal conductivity in the plane of the electrode and the more efficient heat dissipation at the cell surface [26].

3.2. Thermal Runaway with/without Integrated Thermocouples

The evaluation of the thermal runaway behavior on the basis of different characteristic temperatures (T1, T2 and T3) before and during the thermal runaway is carried out according to Feng et al. [27]. T1 is the temperature at which the decomposition of the solid electrolyte interface (SEI) begins. T2 is the temperature at which the thermal runaway is triggered. The time of reaching temperature T2 is also indicated by a temperature rise rate at around 1 °C·s^{-1}. At this point, the side reactions, such as the decomposition of the SEI film, are highly exothermic reactions between Li and organic solvents, melting of the separator and decomposition of the cathode materials. T3 is the maximum cell temperature during the thermal runaway of the battery cell. T2 is the most important evaluation parameter for thermal runaway behavior, as a higher T2 generally correlates with the thermal stability of a battery cell. T3 is a good indicator of the risk of thermal propagation at system level. With increasing temperature T3, the risk of thermal propagation increases [27].

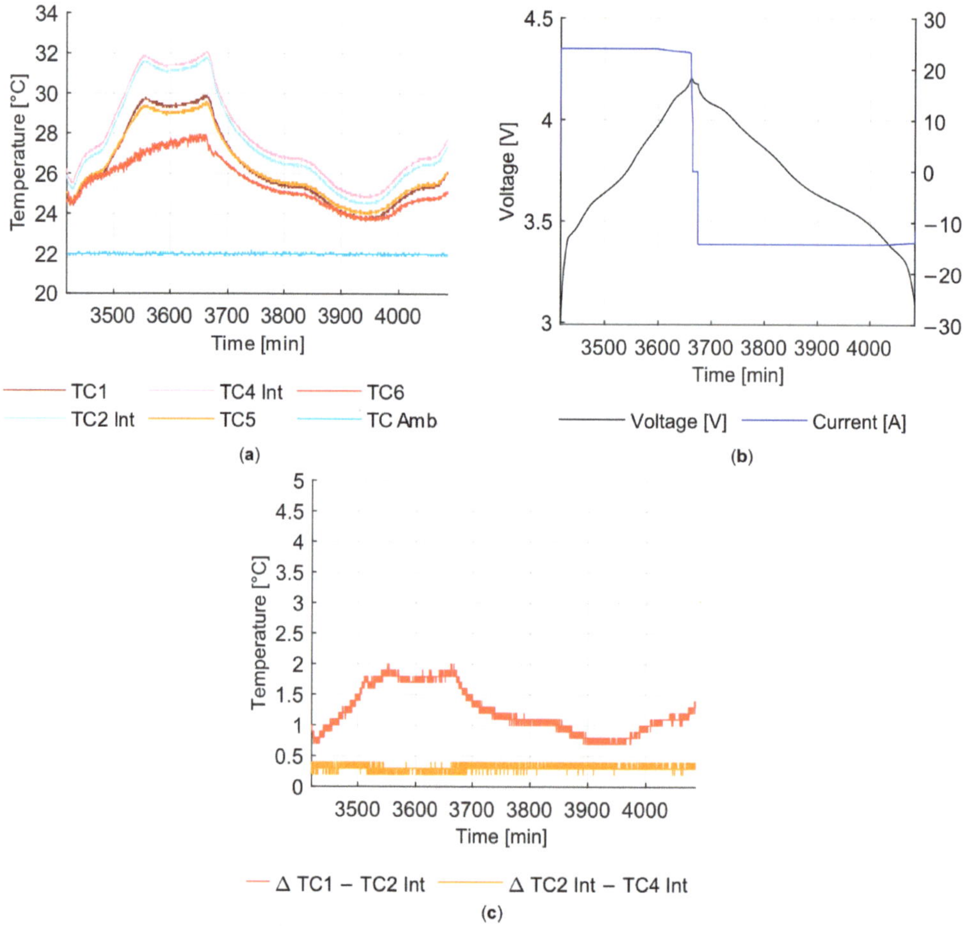

Figure 5. Reference cycle at 100% DOD and detected internal and external cell temperature values over the charge/discharge cycle: (**a**) course of the temperature measured on the cell surface at measuring points 1, 5 and 6 and the temperature measured inside $Cell_{INT}$ at measuring points 2 and 4; (**b**) progression of the cell voltage and charge/discharge current in the reference cycle; (**c**) absolute temperature difference between the internal cell temperature at measuring point 2 and the external cell temperature at measuring point 1 as well as the temperature difference between the two integrated thermocouples at measuring points 2 and 4.

For $Cell_{INT}$, the characteristic temperatures are analyzed based on the temperatures detected inside the cell according to Figure 6d. In this case, T1 is 95.7 °C and is reached 4289 s after the end-of-charge voltage is exceeded. Up to this point, the temperatures at measuring points 2 and 4 do not deviate significantly from each other. This is due to the small charging current of 0.25 C. The cell is in a thermally stable state, whereby the decomposition of the SEI layer has already begun. T2 is reached first at measuring point 4 after 4600 s on the selected time axis. The temperature here is 157.0 °C. For measuring point 2, T2 is reached after 4602 s at 154.7 °C. This supports the findings of Feng et al. that LIBs with high-energy-density materials will release their stored electrochemical and chemical energy at temperatures significantly below 300 °C as thermal energy [27]. For measuring point 2, T3 is reached after 4614 s at 925.5 °C. T3 is also reached at measuring point 4 after 4615 s. The maximum temperature measured inside the cell here is 723.5 °C. This significant difference in the maximum expression of T3 at two measurement points 130 mm apart on the identical electrode position is interesting for a possible consideration of a thermal propagation scenario on battery system level. The results suggest that the positioning of the sensors in the cell has a decisive influence on the detection of the actual maximum temperatures in the cell and that this must be considered when planning experiments to characterize possible propagation scenarios. This is also evident from the data for the temperature difference between different measuring points for $Cell_{INT}$.

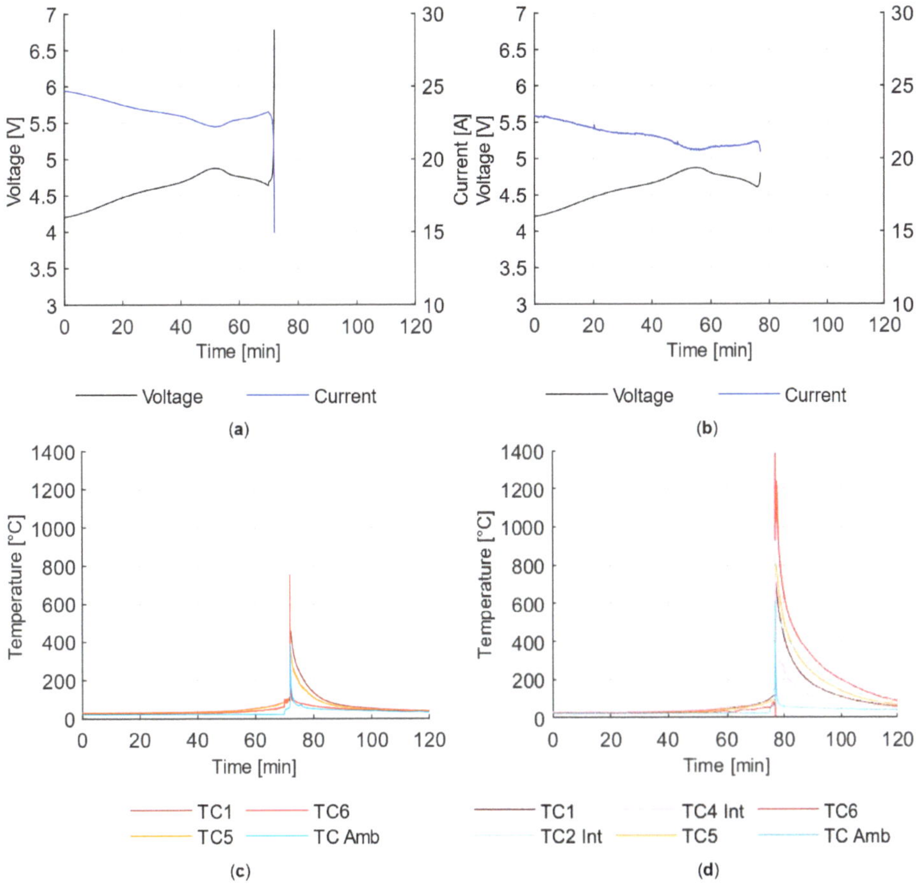

Figure 6. Test data from overcharging $Cell_{INT}$ and $Cell_{REF}$ without integrated sensors: (**a**) voltage and current data from the point of reaching the end-of-charge voltage to the thermal runaway for $Cell_{REF}$; (**b**) voltage and current data from the point of reaching the end-of-charge voltage to the thermal runaway for $Cell_{INT}$; (**c**) course of the surface temperature and the ambient temperature above the vents of $Cell_{REF}$ during overcharging and thermal runaway; (**d**) course of the surface temperature and the ambient temperature above the vents of $Cell_{INT}$ during overcharging and thermal runaway.

The temperature difference between the two integrated thermocouples during the thermal runaway according to Figure 7a is up to approx. 260 °C at the maximum and 200 °C after the first peak has subsided at a spatial distance of 130 mm. The differential temperature between the cell interior and the cell surface stabilizes at only 100 °C after the first peak (see Figure 7b) before cooldown. This is due to the significantly poorer heat conduction behavior perpendicular to the electrode compared to heat conduction along the electrode surface, as already described in Section 3.1.

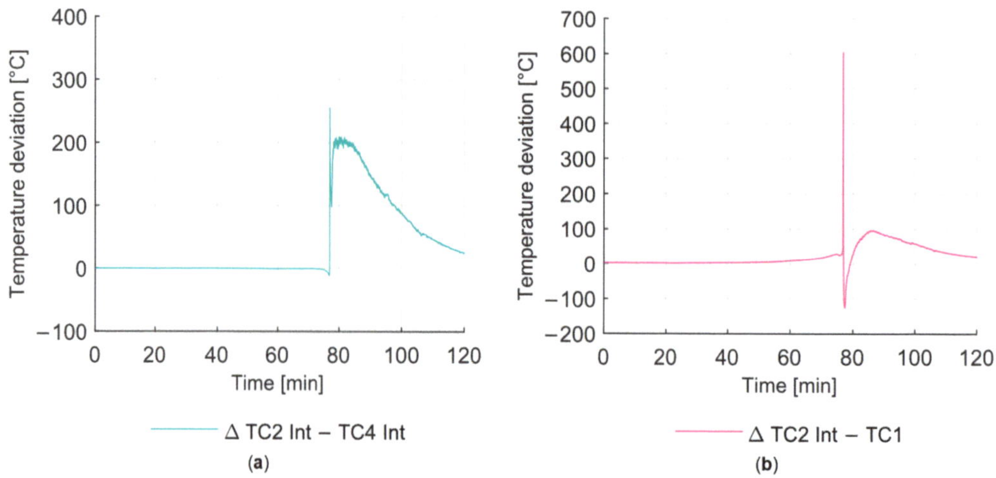

Figure 7. Temperature difference between the thermocouple pairs of $Cell_{INT}$: (**a**) temperature difference between internal measurement points 2 and 4; (**b**) temperature difference between internal measurement point 2 and surface measurement point 1.

At temperatures higher than 1200 °C, the temperature measured at measuring point 6 for $Cell_{INT}$ is considerably higher than the temperatures for T3 (see Figure 6d). Although this temperature was not measured inside the cell, it is particularly significant in the context of a possible thermal propagation scenario. The temperature of the burning gas stream escaping from the vent of the prismatic cell can additionally heat neighboring cells if deflected by, e.g., busbars in the battery system.

To compare the thermal runaway behavior of $Cell_{REF}$ and $Cell_{INT}$, the temperature values measured on the cell surface are analyzed. $Cell_{REF}$ reaches a temperature rise rate of 1 °C*s^{-1} on the cell surface 4.315 s after reaching the cut-off charge voltage according to Figure 8a. The temperature T2 at measuring point 1 is 107.5 °C for $Cell_{REF}$. $Cell_{INT}$ reaches the critical point on the cell surface measuring point 1 at 4.623 s after the end-of-charge voltage has been exceeded (see Figure 8b). The detected temperature T2 for $Cell_{INT}$ is 119.5 °C and, therefore, significantly higher.

At 813.0 °C, the surface temperature T3 at position 1 for $Cell_{INT}$ is also significantly higher than for $Cell_{REF}$, whose maximum temperature T3 is 460.4 °C at the same position (see Figure 6c,d). However, it cannot necessarily be assumed that these temperature differences are due to the change in the cell caused by the implementation of sensors. It is conceivable that the temperature hotspots on $Cell_{REF}$ have formed at positions other than the measurement positions. Studies on the thermal runaway behavior of identical cells under identical boundary conditions and without integrated sensors also show a divergence in thermal runaway behavior [28].

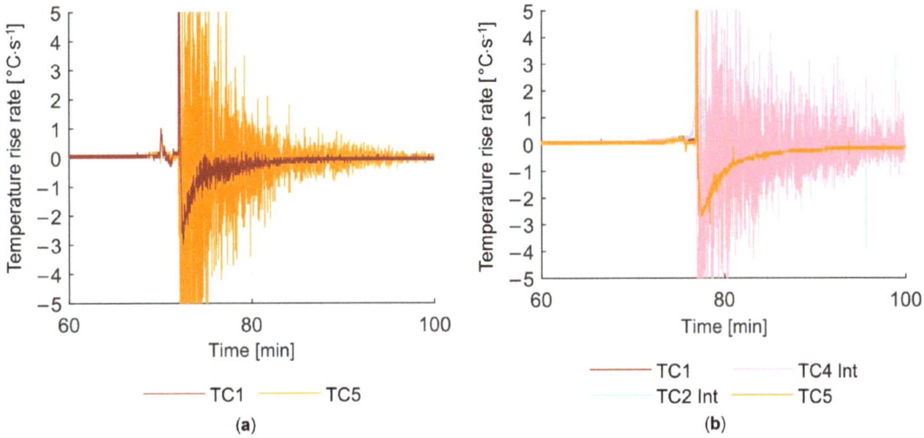

Figure 8. Temperature change rate at various measuring points before, during and after the thermal runaway limited to an observation range of +/−5 °C: (**a**) surface temperature at measurement points 1 and 5 of $Cell_{REF}$; (**b**) surface temperature at measurement points 1 and 5 as well as internal temperature at measurement points 2 and 4 of $Cell_{INT}$.

However, the comparison of the voltage data during the process of overcharging according to Figure 6a,b shows significant similarities between the behavior of $Cell_{REF}$ and $Cell_{INT}$. According to Wang et al., the significant increase in cell voltage shortly before thermal runaway can be attributed to the consumption of Li, which is caused by Li-related side reactions at the interface between electrolyte and electrode and the structural change in the cathode active material [29]. The local voltage maximum for $Cell_{REF}$ is 4.878 V. The local voltage maximum for $Cell_{INT}$ is 4.872 V. A comparison of the temperature data from measuring point 6 of both cells shows that the voltage maximum coincides with the time directly before the vents of both cells open.

For future studies, it is necessary to determine the proportion of the deviation possibly attributable to cell modification by means of further tests to be able to evaluate the comparability.

4. Conclusions

In this work, two thermocouples were inserted into a commercial prismatic high-energy lithium-ion battery cell (NMC811/C, 95 Ah) to characterize the thermal runaway behavior. A cell prepared with thermocouples and an identical reference cell without integrated thermocouples were each brought into thermal runaway by overcharging. Based on the voltage, charge current, surface temperature and internal temperature data of the reference cell and the modified cell, the thermal runaway behavior was investigated. The experimental results provide a data-based insight into the real-time behavior of a prismatic automotive LIB with a Ni-rich cathode and high energy content. The following main conclusions can be drawn from the results:

(1) The methods presented in the existing literature for integrating temperature sensors into small-format battery cells with low energy density and low energy content can be transferred with minor adaptations to large-format prismatic battery cells with high energy density and high energy content without major impact on the cell properties. The loss of solvent from the electrolyte due to evaporation at room temperature during sensor integration can be compensated for by subsequently adding an electrolyte with an identical composition without loss of cell capacity. In future work, there is a need to determine the amount of solvent evaporated as precisely as possible to prevent overcompensation.

(2) In large-format prismatic high-energy battery cells, a temperature difference of up to 1.8 °C between the internal temperature between the two jelly rolls and the external cell surface temperature occurs during cycling under the condition of natural convection, even at low charging currents of 0.25 C, which are likely to be much higher at higher C rates. This confirms the assumption that cell-integrated temperature sensors offer considerable added value for understanding the internal processes in the cell, especially in large-format battery cells with a high energy density.
(3) Using integrated thermocouples, the point of no return can be detected 21 s earlier in the event of a thermal runaway induced by overcharging with a constant charging current of 0.25 C in direct comparison to surface temperature measurement. This confirms the potential of cell-integrated temperature sensors for the early detection of potentially critical conditions for cell chemistries with low thermal stability.

In future research work, the aim is to further standardize the approach of cell-integrated sensor technology, especially for large-format prismatic cells, to fully capture the influence of the sensor technology and its integration. This would enable high-precision comparability of cell behavior with and without integrated sensors and thus better transferability of the results to cell development.

Author Contributions: Conceptualization, N.K. and F.H.; data curation, N.K. and F.H.; funding acquisition, H.H. and A.K.; methodology, N.K.; project administration, H.H. and A.K.; supervision, C.O.; visualization, N.K. and F.H.; writing—original draft, N.K. and F.H.; writing—review and editing, N.K., C.O. and M.F. All authors have read and agreed to the published version of the manuscript.

Funding: This research received no external funding.

Data Availability Statement: The data presented in this study are available on request from the corresponding author. The data are not publicly available due to privacy.

Conflicts of Interest: The authors declare no conflicts of interest.

References

1. Rangarajan, S.S.; Sunddararaj, S.P.; Sudhakar, A.V.; Shiva, C.K.; Subramaniam, U.; Collins, E.R.; Senjyu, T. Lithium-Ion Batteries-The Crux of Electric Vehicles with Opportunities and Challenges. *Clean Technol.* **2022**, *4*, 908–930. [CrossRef]
2. BloombergNEF. *BNEF Electric Vehicle Outlook 2023*; BloombergNEF: New York, NY, USA, 2023.
3. Masias, A.; Marcicki, J.; Paxton, W.A. Opportunities and Challenges of Lithium Ion Batteries in Automotive Applications. *ACS Energy Lett.* **2021**, *6*, 621–630. [CrossRef]
4. Chen, Y.; Kang, Y.; Zhao, Y.; Wang, L.; Liu, J.; Li, Y.; Liang, Z.; He, X.; Li, X.; Tavajohi, N.; et al. A review of lithium-ion battery safety concerns: The issues, strategies, and testing standards. *J. Energy Chem.* **2021**, *59*, 83–99. [CrossRef]
5. Feng, X.; Ouyang, M.; Liu, X.; Lu, L.; Xia, Y.; He, X. Thermal runaway mechanism of lithium ion battery for electric vehicles: A review. *Energy Storage Mater.* **2018**, *10*, 246–267. [CrossRef]
6. Tran, M.-K.; Mevawalla, A.; Aziz, A.; Panchal, S.; Xie, Y.; Fowler, M. A Review of Lithium-Ion Battery Thermal Runaway Modeling and Diagnosis Approaches. *Processes* **2022**, *10*, 1192. [CrossRef]
7. Wei, Z.; Zhao, J.; He, H.; Ding, G.; Cui, H.; Liu, L. Future smart battery and management: Advanced sensing from external to embedded multi-dimensional measurement. *J. Power Sources* **2021**, *489*, 229462. [CrossRef]
8. Jinasena, A.; Spitthoff, L.; Wahl, M.S.; Lamb, J.J.; Shearing, P.R.; Strømman, A.H.; Burheim, O.S. Online Internal Temperature Sensors in Lithium-Ion Batteries: State-of-the-Art and Future Trends. *Front. Chem. Eng.* **2022**, *4*, 804704. [CrossRef]
9. Novais, S.; Nascimento, M.; Grande, L.; Domingues, M.F.; Antunes, P.; Alberto, N.; Leitão, C.; Oliveira, R.; Koch, S.; Kim, G.T.; et al. Internal and External Temperature Monitoring of a Li-Ion Battery with Fiber Bragg Grating Sensors. *Sensors* **2016**, *16*, 1394. [CrossRef]
10. Nascimento, M.; Novais, S.; Ding, M.S.; Ferreira, M.S.; Koch, S.; Passerini, S.; Pinto, J.L. Internal strain and temperature discrimination with optical fiber hybrid sensors in Li-ion batteries. *J. Power Sources* **2019**, *410–411*, 1–9. [CrossRef]
11. Raghavan, A.; Kiesel, P.; Sommer, L.W.; Schwartz, J.; Lochbaum, A.; Hegyi, A.; Schuh, A.; Arakaki, K.; Saha, B.; Ganguli, A.; et al. Embedded fiber-optic sensing for accurate internal monitoring of cell state in advanced battery management systems part 1: Cell embedding method and performance. *J. Power Sources* **2017**, *341*, 466–473. [CrossRef]
12. Lee, C.-Y.; Lee, S.-J.; Hung, Y.-M.; Hsieh, C.-T.; Chang, Y.-M.; Huang, Y.-T.; Lin, J.-T. Integrated microsensor for real-time microscopic monitoring of local temperature, voltage and current inside lithium ion battery. *Sens. Actuators A Phys.* **2017**, *253*, 59–68. [CrossRef]

13. Zhu, S.; Han, J.; An, H.-Y.; Pan, T.-S.; Wei, Y.-M.; Song, W.-L.; Chen, H.-S.; Fang, D. A novel embedded method for in-situ measuring internal multi-point temperatures of lithium ion batteries. *J. Power Sources* **2020**, *456*, 227981. [CrossRef]
14. Fleming, J.; Amietszajew, T.; Charmet, J.; Roberts, A.J.; Greenwood, D.; Bhagat, R. The design and impact of in-situ and operando thermal sensing for smart energy storage. *J. Energy Storage* **2019**, *22*, 36–43. [CrossRef]
15. Mutyala, M.S.K.; Zhao, J.; Li, J.; Pan, H.; Yuan, C.; Li, X. In-situ temperature measurement in lithium ion battery by transferable flexible thin film thermocouples. *J. Power Sources* **2014**, *260*, 43–49. [CrossRef]
16. Zhang, G.; Cao, L.; Ge, S.; Wang, C.-Y.; Shaffer, C.E.; Rahn, C.D. In Situ Measurement of Radial Temperature Distributions in Cylindrical Li-Ion Cells. *J. Electrochem. Soc.* **2014**, *161*, A1499–A1507. [CrossRef]
17. Martiny, N.; Rheinfeld, A.; Geder, J.; Wang, Y.; Kraus, W.; Jossen, A. Development of an All Kapton-Based Thin-Film Thermocouple Matrix for In Situ Temperature Measurement in a Lithium Ion Pouch Cell. *IEEE Sens. J.* **2014**, *14*, 3377–3384. [CrossRef]
18. Osswald, P.J.; Erhard, S.V.; Wilhelm, J.; Hoster, H.E.; Jossen, A. Simulation and Measurement of Local Potentials of Modified Commercial Cylindrical Cells. *J. Electrochem. Soc.* **2015**, *162*, A2099–A2105. [CrossRef]
19. Spinner, N.S.; Hinnant, K.M.; Mazurick, R.; Brandon, A.; Rose-Pehrsson, S.L.; Tuttle, S.G. Novel 18650 lithium-ion battery surrogate cell design with anisotropic thermophysical properties for studying failure events. *J. Power Sources* **2016**, *312*, 1–11. [CrossRef]
20. Waldmann, T.; Wohlfahrt-Mehrens, M. In-Operando Measurement of Temperature Gradients in Cylindrical Lithium-Ion Cells during High-Current Discharge. *ECS Electrochem. Lett.* **2014**, *4*, A1–A3. [CrossRef]
21. Xu, C.; Feng, X.; Huang, W.; Duan, Y.; Chen, T.; Gao, S.; Lu, L.; Jiang, F.; Ouyang, M. Internal temperature detection of thermal runaway in lithium-ion cells tested by extended-volume accelerating rate calorimetry. *J. Energy Storage* **2020**, *31*, 101670. [CrossRef]
22. Parekh, M.H.; Li, B.; Palanisamy, M.; Adams, T.E.; Tomar, V.; Pol, V.G. In Situ Thermal Runaway Detection in Lithium-Ion Batteries with an Integrated Internal Sensor. *ACS Appl. Energy Mater.* **2020**, *3*, 7997–8008. [CrossRef]
23. Mei, W.; Liu, Z.; Wang, C.; Wu, C.; Liu, Y.; Liu, P.; Xia, X.; Xue, X.; Han, X.; Sun, J.; et al. Operando monitoring of thermal runaway in commercial lithium-ion cells via advanced lab-on-fiber technologies. *Nat. Commun.* **2023**, *14*, 5251. [CrossRef]
24. Schmalstieg, J.; Rahe, C.; Ecker, M.; Sauer, D.U. Full Cell Parameterization of a High-Power Lithium-Ion Battery for a Physico-Chemical Model: Part I. Physical and Electrochemical Parameters. *J. Electrochem. Soc.* **2018**, *165*, A3799–A3810. [CrossRef]
25. Günter, F.J.; Burgstaller, C.; Konwitschny, F.; Reinhart, G. Influence of the Electrolyte Quantity on Lithium-Ion Cells. *J. Electrochem. Soc.* **2019**, *166*, A1709–A1714. [CrossRef]
26. Chen, S.C.; Wan, C.C.; Wang, Y.Y. Thermal analysis of lithium-ion batteries. *J. Power Sources* **2005**, *140*, 111–124. [CrossRef]
27. Feng, X.; Zheng, S.; Ren, D.; He, X.; Wang, L.; Cui, H.; Liu, X.; Jin, C.; Zhang, F.; Xu, C.; et al. Investigating the thermal runaway mechanisms of lithium-ion batteries based on thermal analysis database. *Appl. Energy* **2019**, *246*, 53–64. [CrossRef]
28. Willstrand, O.; Pushp, M.; Andersson, P.; Brandell, D. Impact of different Li-ion cell test conditions on thermal runaway characteristics and gas release measurements. *J. Energy Storage* **2023**, *68*, 107785. [CrossRef]
29. Wang, Z.; Yuan, J.; Zhu, X.; Wang, H.; Huang, L.; Wang, Y.; Xu, S. Overcharge-to-thermal-runaway behavior and safety assessment of commercial lithium-ion cells with different cathode materials: A comparison study. *J. Energy Chem.* **2021**, *55*, 484–498. [CrossRef]

batteries

MDPI

Article

The Impact of a Combined Battery Thermal Management and Safety System Utilizing Polymer Mini-Channel Cold Plates on the Thermal Runaway and Its Propagation

Henrik-Christian Graichen [1,*], Gunar Boye [1], Jörg Sauerhering [2], Florian Köhler [3] and Frank Beyrau [1]

[1] Institute of Fluid Dynamics and Thermodynamics, Otto von Guericke University Magdeburg, 39106 Magdeburg, Germany
[2] Department of Applied Biosciences and Process Engineering-Thermal Process and Energy Engineering, Anhalt University of Applied Sciences, 06366 Köthen, Germany
[3] Institute of Apparatus and Environmental Technology, Otto von Guericke University Magdeburg, 39106 Magdeburg, Germany
* Correspondence: henrik.graichen@ovgu.de

Abstract: Lithium-ion batteries are widely used in mobile applications because they offer a suitable package of characteristics in terms of specific energy, cost, and life span. Nevertheless, they have the potential to experience thermal runaway (TR), the prevention and containment of which require safety measures and intensive thermal management. This study introduces a novel combined thermal management and safety application designed for large aspect-ratio battery cells such as pouches and thin prismatics. It comprises polymer-based mini-channel cold plates that can indirectly thermally condition the batteries' faces with liquid. They are lightweight and space-saving, making them suitable for mobile systems. Furthermore, this study experimentally clarifies to which extent the application of polymer mini-channel cold plates between battery cells is suitable to delay TR by heat dissipation and to prevent thermal runaway propagation (TRP) to adjacent cells by simultaneously acting as a thermal barrier. NMC pouch cells of 12.5 Ah capacity were overcharged at 1 C to induce TR. Without cold plates, TR and TRP occurred within one hour. Utilizing the polymer mini-channel cold plates for face cooling, the overcharge did not produce a condition leading to cell fire in the same time frame. When the fluid inlet temperature was varied between 5 and 40 °C, the overcharged cell's surface temperature peaked between 50 and 60 °C. Indications were found that thermal conditioning with the polymer cold plates significantly slowed down parts of the process chain before cell firing. Their peak performance was measured to be just under 2.2 kW/m^2. In addition, thermal management system malfunction was tested, and evidence was found that the polymer cold plates prevented TRP to adjacent cells. In conclusion, a combined thermal management and safety system made of polymer mini-channel cold plates provides necessary TR-related safety aspects in lithium battery systems and should be further investigated.

Keywords: lithium-ion battery; pouch cell; battery safety; thermal runaway; overcharge; propagation prevention; thermal management; face cooling; non-metallic polymer-based cold plate; mini-channel heat sink

Citation: Graichen, H.-C.; Boye, G.; Sauerhering, J.; Köhler, F.; Beyrau, F. The Impact of a Combined Battery Thermal Management and Safety System Utilizing Polymer Mini-Channel Cold Plates on the Thermal Runaway and Its Propagation. *Batteries* **2024**, *10*, 1. https://doi.org/10.3390/batteries10010001

Academic Editor: Mingyi Chen

Received: 15 November 2023
Revised: 7 December 2023
Accepted: 12 December 2023
Published: 20 December 2023

1. Introduction

Lithium-ion batteries heat up under load due to electrochemical reactions and various multi-physical processes, which interact closely. The underlying thermal effects have distinct origins and can be further differentiated. The operation temperature exerts a significant influence on these processes and, thus, on the performance, aging, and, ultimately, the hazard potential of the batteries. Ideally, a cell temperature of approximately 25 °C is maintained during operation to prevent the slow down of reaction and transport processes, increasing overvoltages. In contrast, at elevated temperatures, the battery performance increases, but simultaneously, parasitic reaction mechanisms intensify, which leads to accelerated cell degradation and, in the worst case, to thermal runaway (TR).

Different research groups [1,2] urgently recommend a closer evaluation of integrated safety measures for fire and thermal runaway propagation (TRP) in the battery thermal management system (BTMS). In particular, this applies to mobile applications because of weight and space availability limitations. However, every battery electric vehicle (BEV) must fulfill strict regulatory safety requirements before road approval. Thermal barriers enhance safety during abnormal operation conditions by containing the damage of cell fires and avoiding propagation to adjacent cells. However, they disrupt the regular heat transport path during normal operation. Thermal conditioning, which involves heating and cooling, maintains battery cells within their optimal temperature range and slows down critical processes during their early stages by intensive heat dissipation. Therefore, a trade-off between the rapid dissipation of heat and heat blockage is required in the design of a combined thermal management and safety system. In this article, a novel approach to an application addressing this trade-off is introduced, consisting of polymer-based mini-channel cold plates suitable for the face cooling of large surface-to-volume battery cells. Additionally, the effectiveness of this approach in achieving the trade-off is experimentally investigated.

In a regular use case of lithium-ion battery cells, three fundamental shares of irreversible heat production can be pointed out [3,4] that result from the following effects: (I) ohmic losses based on Joule heating related to charge transport and polarization, as well as overvoltages originating from (II) activation losses of the kinetics at the catalytic surface, and (III) concentration losses due to imperfections in the species transport during diffusion. Additionally, a reversible share related to the entropic heat production of the primary reaction system is noticeable under load [5]. In Equation (1), a simplified expression [3,6] of the released battery heat during cycling, initially proposed by [7], can be found.

$$\dot{Q} = \underbrace{I^2 r_i}_{\dot{Q}_{irrev}} + \underbrace{I\left(T\frac{\partial U_{OCV}}{\partial T}\right)}_{\dot{Q}_{rev}} \tag{1}$$

The irreversible heat increases as the current I does and depends on the cell's internal resistance, r_i. The reversible heat is proportional to the current. Since it is related to the electrochemical reactions' entropy alterations, determinable by the derivative of the open circuit voltage U_{OCV} with respect to the temperature T, it is contingent on the direction of the reactions [5]. Despite both shares of heat production depending on the state of charge (SOC), at charging and discharging rates around 1 C, they are of similar magnitude and, as such, pertinent to heat dissipation [8,9].

Outside the regular operation range in voltage and temperature, rapid battery degradation occurs, accompanied by accelerated aging [10]. The cycle stability of such abused and irreversibly damaged cells is not guaranteed anymore [11], and there is an increased risk of TR and cell fire. A TR's cell fire results from a long and complex process chain, which commences with the intensification of parasitic side reactions at cell temperatures as low as 40 °C [12]. The deposition and oxidation of different species hinder the charge and mass transports inside the cell, and the polarization overvoltage and internal resistance rise, as does the temperature level because of an increased side reaction heat production. The runaway's nature can be depicted well as a sequence of alternating causes and effects. Species degenerate, and a mixture of highly flammable gases is formed and then ignited from internal short circuits upon reaching the onset temperature. The remaining chemically bound energy is released quickly through more heat, and highly toxic substances are emitted while the cell burns out explosively [13,14]. If adjacent cells are exposed to severe heating, with a delay in time, they also undergo the process chain toward TR, referred to as TRP. In the experimental part of this study, battery cells are overcharged to induce TR, which is further explained in Section 2.1.

To comply with TR and cell fire safety regulations, numerous technical approaches have been integrated and tested in battery systems as safety mechanisms at cell, module, and pack levels in recent years, incorporating resistant cell materials [15], sensor-based hazard detection [16], or safety valves [17,18], and thermal barriers [19,20] to reduce criticality and hinder TRP and its impact. BTMS-related safety approaches aim towards a slowdown of the TR's underlying process chain in its early stages by compensating for the temperature rise through intensive heat dissipation since it is subject to Arrhenius law [21]. As a general rule, a temperature rise of 10 K doubles the reaction rates. However, BTMS should regulate battery packs during normal operation conditions to maintain a temperature range of 15–35 °C while ensuring a high degree of uniformity with gradients below 5 K within each cell and between all cells in the pack [3,22]. On the downside, implementing a BTMS increases the battery pack's complexity, weight, and volume. An enormous variety of BTMS options has been developed in the past. Figure 1 offers a potential classification of prevalent thermal management concepts, highlighting the type of system introduced in this article. Further classification can be conducted in terms of flow configuration (serial, parallel), exceptional media cooling (nano-fluids, heat pipes), or a unique location of heat dissipation, such as from the cell tabs [23]. The state-of-the-art applications of each category are discussed in detail in [1,24–29].

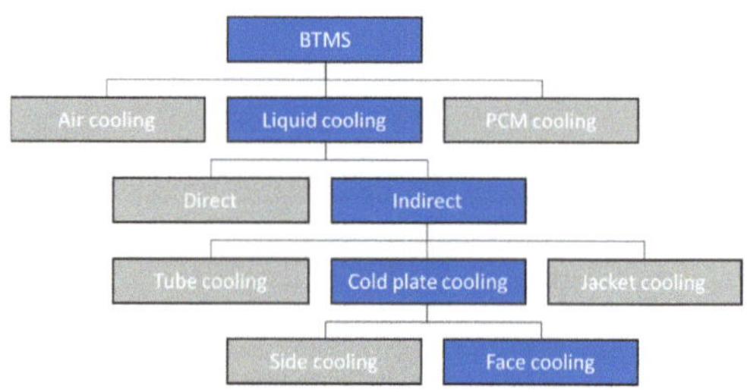

Figure 1. Potential classification of BTMS.

In general, battery cells can be cooled by air, liquids, or Phase Change Materials (PCMs). Air cooling shows poor heat dissipation rates because of its inherent thermophysical limitations. Nevertheless, these systems are simple, low-cost, and lightweight [26]. If high heat dissipation requirements have to be met, such as those experienced during fast charging, liquid cooling remains the preferred method, as recently affirmed by [25]. However, the need for leakage prevention increases the price and complexity [26]. PCM cooling uses the latent heat during the material's melting and solidification. These systems can prevent TRP as PCM media serve as thermal buffers and insulators [30]. However, heat cannot be dissipated actively, so PCMs are mostly coupled with another method into hybrid approaches.

Cold plates are usually flat metal plates with internal fluid channels [2,29] that exhibit a significant cooling performance in combination with rectangular cells. The placement of the cold plates to the cell, either attached to its faces or sides, is of the essence for the quality of the heat dissipation in terms of cell temperature uniformity and, even more importantly, for enhancing the safety potential as a thermal barrier. In this experimental study, face cooling with cold plates is utilized; see Section 2.2 for more details. The method is paired with pouch cells, whose response to thermal conditioning is significant [24].

Integration of small ducts into cold plates is considered where space and weight limitations apply. Mini-channel cold plates of only 5 mm thickness are reported to be capable of demonstrator usage [31,32]. Even approaches with plates as thin as 2 mm, used on a laboratory scale, have been found [33] to cool individual cells up to 4 C. However, its sealing and applicability to battery modules remain unclear. The demonstrators mentioned above are composed of aluminum, an excellent thermal conductor, whose capability to avoid TRP as an intercellular application is weak. However, the implementation of insulation layers between large-capacity prismatic cells of at least 2 mm thickness and 0.1–0.2 W/(mK) thermal conductivity was reported to prevent TRP [34,35].

The concept of non-metallic, polymer mini-channel cold plates is introduced to combine the cold plates as a BTMS with a safety aspect. Channel geometries on a mini-scale can be conveniently machined, and sealing is reached with thermal bonding [36], which results in very thin structures containing internal fluid channels.

The combination of passive heat blockage and active heat dissipation into one component is deemed necessary to comply with forthcoming battery system safety regulations [27]. However, it is seldom mentioned in literature. In Table 1, the most relevant studies addressing this trade-off are outlined. As a further restriction, the studies' applications exclusively utilized a polymer component or thin aluminum structures with mini-channels coupled with layers of thermally insulating materials.

Table 1. Collection of combined thermal management and safety approaches.

Source/ Year	Type of Examination	Type of Cell/ Configuration	Type of BTMS	Findings
[37]/ 2022	Num./exp., Cycling/cooling	Pouch, 10 Ah/ module	Direct cooling by immersion, ABS micro-channel plates as flow separator/ turbulator between cells	Micro-channel plate improves cooling performance in narrow space by reducing the temperature gradient in the system
[38]/ 2021	Num./exp., TR ind./cooling	Prismatic, 37 Ah/ module	Indirect bottom cooling by aluminum mini-channel cold plate and insulation by aerogel layers between cells	Singular liquid cooling or insulation fails to mitigate TRP; coupling of cooling and insulating with 1 mm layers avoids TRP
[39]/ 2021	Num./exp., Cycling and TR ind./cooling	Prismatic, 25 Ah/ array of two cells	Indirect cooling aluminum plate-fin and fluid cooling channels, embedded in PCM layer between cells	Cycling: coupled cooling results in more uniform temperatures than singular techniques; TRP: induction by overheating only prevented by coupled cooling, not by singular techniques; PCM of high thermal conductivity lowers system temperature but increases TRP probability
[40]/ 2021	Num./exp., Cycling/cooling	18650 Cylindrical, Unknown cell capacity/ 50 Ah pack	Indirect cooling by polymer macro-channel tubes, connected at one side with the cells	Heat dissipation requirements met at low discharge rates; <5 °C temperature gradient in the system up to 2 C
[41]/ 2019	Num., TR ind./cooling	18650 Cylindrical, 1.5 Ah/ module	Indirect cooling by aluminum micro-channel cold plates embedded in a PCM matrix between battery rows	TRP induced by nail penetration is prevented by coupled cooling, not by singular techniques
[42]/ 2016	Num., Cycling/cooling	Prismatic, unknown capacity/ Module	Indirect cooling by aluminum mini-channel tubes coupled into PCM layer between cells	Coupled cooling lowers maximum temperatures; decreasing temperatures with rising thermal conductivity of the PCM

The closest resemblance between this study's application and literature comes from hybrid thermal management approaches, whose liquid cooling inlay serves heat dissipation, while the PCM matrix performs as a thermal buffer and insulator during abnormal heat generation but at a much higher weight and volume compared to the polymer mini-channel cold plates.

Consequently, to the authors' best knowledge, this article represents the first experimental study utilizing a polymer mini-channel cold plate as a combined thermal management and safety system for batteries integrated into a single component. This study examines whether non-metallic cold plates are viable for the trade-off between dissipative heat removal from batteries during the early stages of TR while acting as a thermal barrier under critical circumstances in the later stages to prevent propagation to adjacent cells.

In the experiments, NMC pouch cells of 12.5 Ah are overcharged at 1 C while receiving thermal conditioning from the polymer mini-channel cold plates. The following questions regarding the application's functionality will be investigated:

- Is there a difference in battery cell behavior between thermally conditioned and non-conditioned cells when subjected to overcharge-induced TR up to the cell fire? Is there a significant temporal impact on the process chain recognizable, especially before the TR and cell fire?
- Which heat dissipation rate is achievable using the polymer mini-channel cold plates?

- If the TR and cell fire are reached, is there a detectable preventive influence of the polymer cold plates on TRP to the adjacent cells?

2. Theory

Safety testing for different battery abuse scenarios and battery level scales has been developed [43]. However, establishing standardized TR test procedures is challenging due to the large variety of cell formats and chemistries. The methods used must be field-relevant [44], and an essential distinction is made in terms of the thermal, mechanical, or electrical triggering of the TR. The organic electrolyte is flammable, and cells may be exposed to overtemperature due to a fire scenario. The potential for crush or penetration is crash-related, which may result in an internal short circuit occurring at the intrusion site. Excessive overcharging of lithium batteries is reported to be one of the most prevalent safety issues for large-scale applications [45]. An overcharged battery releases more heat and combustible gases during cell fire events than those triggered by overheating or penetration [46] because of an increased energy content [47]. The TR in this study's experiments is overcharge-induced, posing a challenging scenario for the combined thermal management and safety system.

2.1. Overcharge-Induced Thermal Runaway

In the overcharging process of a battery cell, five distinct stages occur before TR and the cell fire, which is widely applicable to cells of varying capacities across different cell chemistries [13,14,48,49]. Regarding NMC chemistries, charging leads to oxidation at the NMC cathode, lithium-ion charge transfer, as well as intercalation and lithium reduction at the graphite anode; see the reactions in Equations (2) and (3).

$$\text{Cathode}: \ Li_aNi_xMn_yCo_zO_2 \rightarrow Li_{a-k}Ni_xMn_yCo_zO_2 + kLi^+ + ke^- \ \text{ with } k < 1 \tag{2}$$

$$\text{Anode}: \ nC + kLi^+ + ke^- \rightarrow Li_kC_n \ \text{ with } n \geq 6 \tag{3}$$

During the first stage of overcharging, the regular charging process continues beyond the upper cutoff voltage because not all available spots in the anode active material are occupied by lithium-ions at 100 % SOC. Stage I comprises the most substantial time until the TR but is strongly connected to cell chemistry. Nickel-rich cathodes, responsible for increased capacities, exhibit a decreasing thermal stability [50]. The external appearance of the cell remains unchanged during stage I, but there is a gradual increase in surface temperatures and voltage. This relatively stable stage ends, and lithium plating occurs if any successive steps involved in the lithium transport and intercalation become sluggish. Initially, diffusion of solvated lithium-ions in the electrolyte takes place, followed by the charge transfer at the electrode/electrolyte interface, and finally, lithium diffusion into the solid electrode material [51].

In stage II, the cell begins to expand, and significant voltage and surface temperature increases are noticeable. Metallic lithium deposits at the anode as mosses or dendrites [52,53]. The SOC, from which this overcharging phenomenon occurs, is a vital function of anode oversizing in terms of capacity and size compared to the cathode, making it cell-specific [54,55]. Experiments are documented, which show no lithium plating up to 120% SOC [56]. With nearly complete lithiation of the graphite anode, there is an increase in the concentration of lithium-ions at the anode/electrolyte interface because, per time step, fewer ions can deposit and intercalate compared to the number moved to the anode [57–59]. The widely accepted criteria conception for lithium plating occurrence necessitates a lithium-ion concentration that exceeds the saturation at the interface and an interface potential equally or below that of Li/Li^+ at 0 V [60,61]. The negative potential shift raises the cell voltage [62]. Electrolytes become unstable at excessively high voltage levels in contact with lithium metal. Spontaneous reactions lead to species consumption [63] and the formation of additional passivation layers at the electrode interface [51,58], similar to the primary Solid Electrolyte Interface (SEI) film formed during the first battery cycle [64,65].

With elevated temperatures, the SEI grows more porous and subsequently less stable and performing [4]. Consequently, the charge and mass transports inside the cell are hindered, coupled with an increase in the polarization overvoltage and the cell's internal resistance, making lithium plating a self-accelerating process [66,67]. Additionally, electrolyte decomposition at high voltages of 4.9–5 V leads to gas generation and cell swelling [68–70], and its consumption decreases the ionic conductivity, causing a further increase in overpotentials, which results in an additional heat generation and significant temperature rise. In stage I and early stage II, Joule and reaction heating contributed the most to the overall heat production. However, as the temperature and voltage levels increase, parasitic side reactions intensify, introducing more side reaction heat into the system and marking the transition to the later stages of overcharging [71,72]. Stages III and IV are characterized by species decomposition, cell deformation due to substantial swelling, and further increases in surface temperature.

The transition to stage III is characterized by a brief voltage plateau, ascribed to a disproportionation reaction and intensive species conversion, as referred to in [73,74]. Electrode surfaces show significant defects like pits and cracks, leading to a substantially non-uniform lithium deposition beneath the SEI layer [75–77]. The resultant mechanical stress causes dendrites to break through the SEI layer, leading to its partial destruction [78]. Destruction and healing processes of the SEI take place simultaneously, supporting the voltage stagnation. As the internal cell temperature surpasses a critical threshold of approximately 60 °C, the thermal electrolyte decomposition is further promoted [79], and a mixture of highly flammable gases is formed [69]. The cell expansion increases the electrode spacing, leading to higher internal resistance, overpotentials, and additional heating [14,48]. Following the voltage plateau, the voltage elevates to a local maximum or crest value.

At the transition to stage IV, the voltage decreases due to various phenomena. For example, excessive delithiation of the cathode led to an irreversible structural change and a collapse of the crystal [80,81]. The cell envelopes or casings reach the tensile limit of the seam because of the internal pressure build-up. They open up and release warm gas during the first venting [16,49]. The core temperature of the cell reaches a point at which the SEI decomposes, and dendrite growth from the deposition of solid materials on the electrodes' surface leads to separator penetration and the emergence of micro short circuits [81]. These result in local hot spots because of high current densities and corresponding Joule heating. The cell voltage increases to a global maximum due to a rising ohmic resistance and then sharply drops to zero as the separator melting and shrinkage leads to an immense short circuit.

Finally, at the beginning of stage V of overcharging, the onset temperature for cell fire is reached. In this context, we define it as a detectable steep surge in surface temperature, the start of the second venting, and rapid thermal runaway. Sparks at the short circuit ignite the highly flammable gas mixture, and the cell undergoes TR while burning out explosively [82–84]. During this fifth stage, there is a short peak in Joule heating, followed by a dominant heat release due to the release of chemically bound energy [48].

2.2. Cold Plate Thermal Management

Designing a performant thermal conditioning system must consider the cell's internal structure; see Figure A1 in the Appendix A.2. The pouch contains a stack of numerous thin layers, which in the smallest repeatable structure, the unit cell, comprises the cathode and anode current collectors with electrodes' coatings and the separator. The porous layers are filled with electrolytes. All other components except the metallic current collectors show inferior thermal conductivity [85]. When it comes to heat dissipation out of the battery volume, acc. to Equation (4), the effective thermal conductivity k_{eff} of the materials being involved in the transport process and hence the path available l for heat transfer is pivotal [86]. Nevertheless, the available heat transfer surface A also plays a crucial role.

$$\dot{Q} = -\frac{k_{eff}}{l} A(T_2 - T_1) \quad (4)$$

The heterogeneous cell stack exhibits a greatly anisotropic thermophysical behavior [85]. The effective thermal conductivity for the direction perpendicular to the stacking direction has to be described as a series connection of thermal resistances. In contrast, the direction parallel to the stack is a parallel connection; see Equations (A1) and (A2) in the Appendix A.1. Consequently, battery cells show significantly higher thermal bulk conductivities in-plane than perpendicular to the stacked layers, typically differing by more than one order of magnitude; see Section 3.1. However, these are still relatively poor and one magnitude short compared to the metallic materials commonly used for heat sinks (Al ~240 W/(mK) and Cu ~400 W/(mK)). In this context, it is concluded that heat dissipation from lithium-ion batteries is not an issue of high heat flux, as seen in microelectronic devices. However, it is influenced by low effective conductivities and high heat flow rates [3]. Heat dissipation from the center of a battery presents a greater difficulty than from the outer layers near the surface. The close arrangement of cells in battery packs, the trend towards large formative cells of high thickness, and fast charging, in general, aggravate the challenge of thermal conditioning [4].

If manufacturers of mobile applications use cold plate thermal management, they typically opt for side cooling, particularly from the bottom. These approaches show easy technical feasibility and are cost-effective, feature a solid safety aspect with the cooling fluid staying outside the high-voltage compartment, and exploit the cells' higher in-plane thermal bulk conductivity. A notable instance is the VW ID.3, whose battery modules are situated atop an aluminum cooling plate with internal macro fluid channels [87]. Furthermore, although marginally more complex and costly, face cooling through cold plates is similarly industrialized. In the high-voltage battery of the Chevrolet Volt, slender fin-like aluminum cold plates hosting multiple mini-channels are placed between the cells [88,89].

To realize a combined thermal management and safety system embodied by a single component of the polymer mini-channel cold plates, these must be positioned between the cells and not at their sides. Only under this condition can the polymer plates fulfill their purpose of a thermal barrier to prevent TRP. Furthermore, trend-setting studies of several research groups have shown far-reaching advantages of a thermal management concept with cold plates applied to the batteries' faces over side-cooling approaches in terms of quantity and quality of heat dissipation. During the cycling of large-format batteries [32,90] and commercially available electric vehicle battery packs [91] of up to 3 C, face cooling resulted in a lower mean cell temperature and a significantly smaller temperature gradient across the cells' central section than bottom cooling. The outcome was attributed to a much larger heat transfer interface of the face cooling system with the batteries and the cell's poor thermal conductivity, which comes into calculation if the complete heat has to be transported internally toward the bottom cold plate, where it is dissipated through a small area. Other advantageous aspects of face cooling concern a short reaction time of the battery system temperatures [92] and more homogeneous aging at the electrode level [93,94].

The term cold plate is widely used but can be misleading, as the thermal management of battery cells is not exclusively about cooling but thermal conditioning. Consequently, the investigated polymer structure is also referred to as a mini-channel heat sink or heat source (MCHS).

3. Materials and Methods

This chapter contains information about the battery cells and polymer mini-channel cold plates used in the overcharging experiments. Additionally, the experimental design, including the setup, procedure, and measurement plan performed, is explained.

3.1. Battery Cell

Batteries manufactured by CTS Technology Co. (Hunan, China) were used; see Figure 2. This pouch cell of 12.5 Ah capacity has an NMC cathode, while the anode is graphite-based. The manufacturer-provided information and properties are shown in Table 2. The cell does not contain a safety vent at the sealing. The pouch cells are thin and have the highest surface-to-volume ratio of any commercially available cell type [24], which makes them an attractive fit for face cooling. The cell components are not surrounded by a hard casing but a resistive composite aluminum envelope. Therefore, the cells require a mechanical pre-stress under load. The pouch is slightly larger than the effective contact surface (84 mm × 182 mm), which is in contact with the adjacent cell or the cold plates in the assembled stack; see Section 3.3.1.

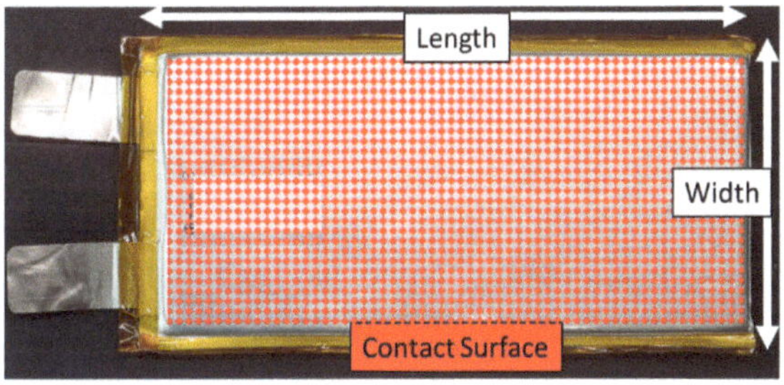

Figure 2. NMC pouch cell of 12.5 Ah capacity, CTS-S7688190.

Table 2. Pouch cell properties.

Parameter	Value
Voltage limits (V)	3–4.2
Nominal capacity (Ah)	12.5
Weight (g)	260 ± 15
Energy density (Wh/kg)	173
Max. charge/discharge current	1 C
Operating temperatures (°C)	0~50
Width/Length/Thickness (mm)	88/192/7.6
Internal resistance (mΩ) @ 1 kHz and 50% SOC	3.5
Cycle stability (-)	2000

The cells' performance was pre-tested for quality control [95], and they were approved for experimental use if the results matched the manufacturer's datasheet. A Delta Elektronika SM 15–200 D power supply and a Höcherl & Hackl PLI6406 electrical load were used for cycling. The cell capacity was determined by charge counting; see Equation (5).

$$q_{el} = \int_0^t I(t)dt \tag{5}$$

The cells were fully CCCV-charged to 4.2 V at 1 C, then cycled and measured during a full 1 C CC-discharge to 3 V at a room temperature of 25 ± 2 °C. The resulting capacities were measured between 12.6 and 12.8 Ah, with no cell showing significant deviations. Additionally, a randomized sample set of cells was characterized with respect to their internal resistance, according to the IEC 62620 standard [96], and their voltage response to CC-discharging at rates of 1 C, 0.5 C, and 0.1 C. As can be taken from Equation (6), the internal resistance is determined as the direct current resistance r_{DC} with currents of I_2 = 0.5 C and I_1 = 0.2 C.

$$r_{DC} = \frac{V_2 - V_1}{I_2 - I_1} \tag{6}$$

Instead, the manufacturer-provided data refers to an Electrochemical Impedance Spectroscopy (EIS) at 1 kHz and 50% SOC and amounts to 3.5 mΩ. However, multiple research groups have shown that this value tends to underestimate internal resistance [97,98]. For a representative cell, in Figure 3a,b, the voltage response and the internal resistance according to DIN EN 62620 can be found. As expected, the voltage response decreases with higher discharge rates. The measured resistance values are between 5 and 7 mΩ, which is up to twice as high as the manufacturer's specification for the abovementioned reason. Below 15% SOC, the resistance increases up to 10 mΩ. All of the tested cells showed this behavior with no significant deviations.

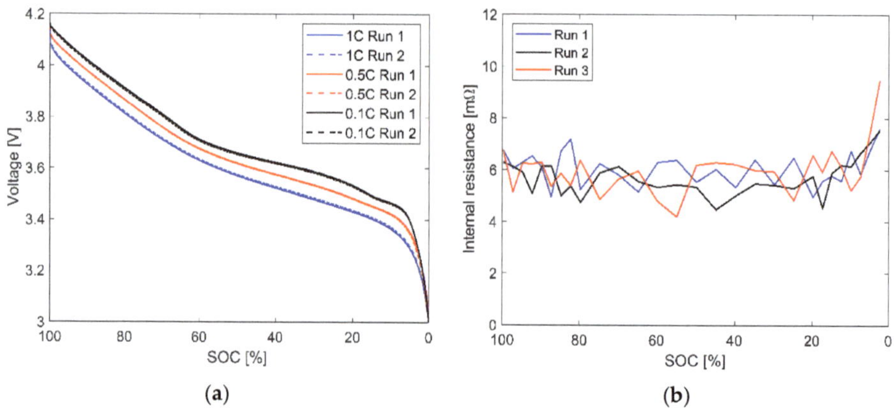

Figure 3. (**a**) Discharging voltage response at different C-rates; (**b**) internal resistance according to DIN 62620.

Transient Plane Source (TPS) method measurements of the effective thermal bulk conductivity were made for cell characterization. Using a 4922 Mica-insulated sensor and 20 averaged single shots at a power of 1 W and a time of 10 s, values of $k_{eff,\perp}$ = 0.72 ± 0.02 W/(m K) perpendicularly $k_{eff,\parallel}$ = 30.3 ± 0.5 W/(m K) in-plane were obtained.

3.2. Polymer Mini-Channel Cold Plates

In recent years, non-metallic mini-channel cold plates, or MCHSs, have been developed at the OVGU Magdeburg. They consist of robust polymer materials to promote lightweight construction and own an internal channel system sealed tightly for leakage-free fluid transport. They are manufactured by micro-milling and thermal bonding at the glass transition temperature [36]. To prevent the fine channels from clogging, adhesives of any kind are averted. The channel geometry is adaptable to the applications' needs. Aluminum adapters with circumferential holes were used for the connection to the macroscopic fluid supply system. Figure 4a,b show the demonstrator used in the experiments.

The mini-channel cold plates are made of standard polycarbonate (Makrolon®), an amorphous thermoplastic polymer. It has a high fatigue strength that does not lead to material flow and deformation of the mini-channels during pre-stressing of the pouch cells. According to the manufacturer, the material has a thermal conductivity of 0.2 W/(mK), a heat capacity of 1170 J/(kg K), a density of 1200 kg/m^3, and the glass transition temperature is reached at about 148 °C.

The cooling channel geometry is divided into two sections; see Figure 4a. The inlet and outlet are designed to distribute and collect the fluid evenly to and from each channel. The fluid flow field dissipates heat from the cells' faces. The dissipated heat rate can be calculated from the fluid mass flow rate $\dot{m}$, the fluid heat capacity c, and the fluid temperature difference between the inlet and outlet; see Equation (7). Deionized water was used as the fluid in the experiments.

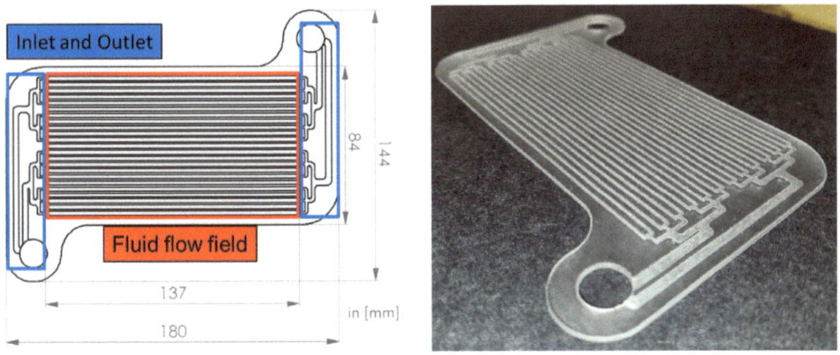

(**a**) (**b**)

Figure 4. (**a**) CAD sketch of the mini-channel cold plate (**b**) experimental demonstrator.

$$\dot{Q} = \dot{m}c(T_{Fluid,out} - T_{Fluid,in}) \tag{7}$$

Regarding the flow field geometry, a multi-channel system results in a more uniform temperature field during thermal conditioning than a single channel, while an increased channel width and height reduce the pressure drop and may result in a lower maximum temperature [29]. At the cost of a slightly higher pressure drop, the inlet and outlet were designed as a bifurcation, leading to better flow uniformity and a more homogeneous heat transfer than a consecutive manifold [99,100], borrowed from fuel cell technology [101]. The width and height of the flow field channels are 1 and 0.5 mm, respectively. The hydraulic diameter of 667 µm classifies the geometry as a mini-channel system according to [102]. The walls between the channels are 2.25 mm wide. The flow field contains 24 channels with a length of 137 mm.

With a total thickness of 2 mm and a relatively low thermal conductivity of the polycarbonate, the manufactured mini-channel cold plates match the properties of the aforementioned insulation material layers tested in [34,35], which prevented TRP.

3.3. Experimental Design

3.3.1. Structure of the Battery Assembly

The batteries and the mini-channel cold plates were placed into a steel clamping device during overcharging; see Figure 5. The setup contains two fully charged pouch cells, tabs to the top, three cold plates in the case of thermal conditioning, and multiple sheath thermocouples (TCs). The cell to be overcharged is consecutively designated as the "OC cell" and its neighbor as "cell 2". Insulation layers reduce the heat transfer from the cell and cold plate stack into the clamping device [103].

Pre-stress applied by the clamping device's bolting serves proper operation during the experiment. Firstly, the swelling behavior of the pouch cell is homogenized, which prevents early non-reproducible short circuit failures, and secondly, thermal contact resistances between batteries and cold plates are reduced. Industrial manufacturers typically use Thermal Interface Materials (TIMs) in the form of conductive pastes and adhesives for contact resistance reduction. A previous paper showed that TIMs usually embody one of the largest thermal resistances in a cooling system [104]. Furthermore, they complicate the recycling of battery systems. Therefore, with large heat transfer areas and moderate heat flux in battery modules [3,26], the TIMs are entirely dispensed with.

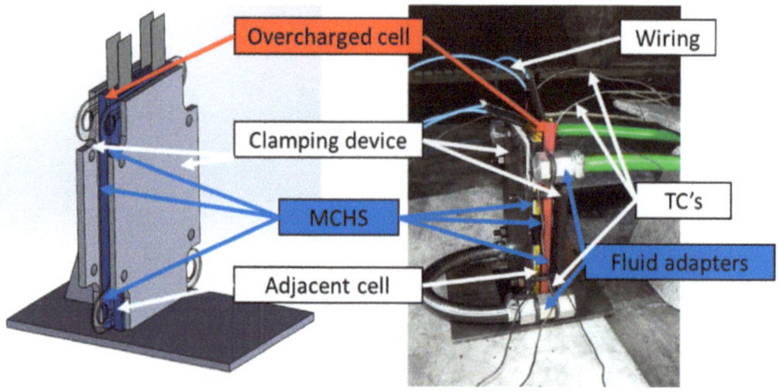

Figure 5. CAD schematic and realization of the experimental setup.

In Figure 6a,b, the cell and cold plate assembly is presented in detail. Three MCHSs are placed next to the cells in the thermally conditioned setup. To achieve a homogeneous conditioning performance, the mini-channel plates were connected in parallel to the fluid supply, and the fluid ran in countercurrent through adjacent plates [26,29].

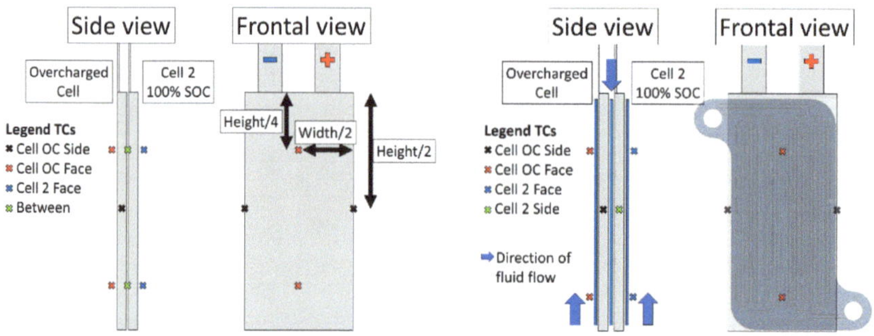

Figure 6. Distribution of TCs (**a**) setup without thermal conditioning; (**b**) setup with MCHS conditioning.

Eight calibrated type-K sheath TCs (1 mm) were installed to observe the temperature changes in the assembly during the overcharging procedure, the TR, and its propagation; cf. Figure 6a,b. In the setup without thermal conditioning, they are located in the upper and lower thirds of the battery faces and laterally in the fold of the aluminum envelope of the overcharged cell. In contrast, good contact between the batteries and the MCHSs is required in the conditioned setup. Consequently, the TCs are located in the upper- and lower thirds between the outer cold plates and the thermal insulation (not displayed) and laterally in the aluminum folds of both batteries. During the measurement campaign, no significant differences were found between the respective TCs at the assembly's head and foot. For clarity, the averaged corresponding temperatures are presented in the results in Section 4. A pre-test for evaluation of the TC positioning showed good agreement between the temperatures measured laterally at the sides of the OC cell and its face between both cells before TR; cf. Figure A2 in the Appendix A.2. During the cell fire, the TCs between both cells show the highest temperatures. Nevertheless, the laterally captured values reasonably indicate the OC cell's surface temperature, which is consequently assumed for the thermally conditioned experiments.

3.3.2. Experimental Setup and Procedure

The experimental setup must be split for reasons of protection and is partially located inside the fire and the control rooms; see Figure 7. The fire room is equipped with an exhaust gas purification system, and its robustness allows experiments up to EUCAR hazard level 7—the highest classification—and includes the explosion and complete disintegration of a battery cell. External monitoring from the control room becomes necessary. The clamping device, including batteries, cold plates, and distributed TCs, is installed in the fire room. A high-speed color camera allows for the visual detection of venting and cell fire events. The battery is charged by a DC power supply, and the voltage is measured at its tabs. A refrigeration thermostat supplies the system with tempered deionized water. The fluid's mass flow and temperature at the inlet and outlet of the cold plate assembly are measured. A difference-pressure sensor signals potential leakage across the MCHSs. Data are recorded at 1 Hz and a camera frame rate of 250 fps. More detailed information about the equipment is provided in Table 3.

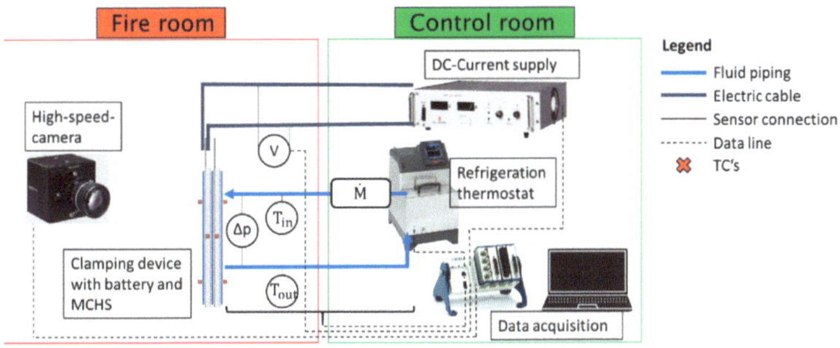

Figure 7. Schematic representation of the experimental setup in the fire and control rooms.

Table 3. Technical details of the experimental equipment.

Device		**Model**		**Comment**
Power Supply		Delta Elektronika SM15–200 D		Range: 0–15 V, 0–200 A
Refrigeration thermostat		Huber Ministat 240		Range: −45–200 °C; Cooling power 550 W @ 0 °C
High-speed camera		Photron Fastcam Mini UX100		1.280 × 1024 pixels, 250 fps frame rate
Sensors	**Model**		**Range**	**Max. uncertainty**
Temperature	Sheath TC Type K, 1 m × 1 mm		Up to 1300 °C short-term	After calibration: ±0.2 K
Mass flow	Krohne Optimass 6400 C		0–450 kg/h	±0.05% of the value
Pressure	Yokogawa EJX110A		5–1000 mbar	±0.04% of the value
Voltage	Voltcraft VC 950 Datalogger		0.001 mV–1000 V	±0.03% of the value
Data acquisition system				
Rack: NI cDAQ-9174; Modules: NI-9203 Current Input, NI-9213 Thermocouple				

Before the experiment, each cell was fully CCCV-charged at a rate of 1 C and a cutoff current of 0.05 C. Thermal pre-conditioning begins, and overcharging at a rate of 1 C or 12.5 Ah is started when the desired fluid inlet temperature is reached and the fluid outlet temperature changes by less than 1 K within 10 min, which takes between 60 and 120 min, depending on the fluid temperature level. Without thermal conditioning, the overcharging procedure can start immediately. The camera starts recording after the detection of the cell opening and first venting, which embodies a critical state and an approaching TR. The experiments end with the potential TR and TRP.

3.3.3. Measurement Plan

Table 4 provides an overview of the experimental overcharging scenarios performed in this study. In the first scenario (#1–9), no thermal conditioning or safety system was installed. This creates a base reference of the temporal process sequence of the overcharged cell towards the TR and the TRP to cell 2 without any safety measures. The second scenario (#1*–4*) captures whether the utilization of the polymer cold plates for thermal conditioning leads to significant temporal changes in the behavior of the OC cell or an overall different outcome from the first scenario. In the case of TR, evidence of protection regarding TRP to the adjacent cell is investigated. A variation in the fluid inlet temperature significantly affects heat transfer [104,105]. At 5 °C, it might impact the cell's internal process chain the most because a significant temperature difference between the cold plate and the battery is created. Fluid inlet temperatures of 20 and 30 °C arise from applicational reasons and the desire to achieve ideal temperatures inside the battery system under load. In the third scenario (#5*^{c} and 6*^{c}), a malfunctioning thermal management system is emulated by raising the fluid inlet temperature to a critically high value of 40 °C, representing a heat dissipation limitation due to a damaged cooling circuit. The OC cells are brought to a high SOC after the pouch opening and first venting, and then, in two steps of escalation, either the current supply (#5*^{c}) or the fluid supply (#6*^{c}) is stopped. The outcome is examined to determine if the compromised combined thermal management and safety system maintains a protective effect, particularly respecting TRP towards the adjacent cell.

Table 4. Experimentally tested overcharging scenarios.

Test Scenario	Experiment Number	Fluid Inlet Temperature (°C)	Comment
Overcharging, no thermal conditioning	#1–9	-	Base reference
Overcharging, thermal conditioning with MCHS	#1*–4*	5, 20, 30	Regular cooling
Overcharging, malfunctioning thermal conditioning with MCHS at critical conditions	#5*^{c}	40	Interruption of overcharging
	#6*^{c}	40	Interruption of conditioning

4. Results and Discussion

In the following, for each scenario, the description of the temperature and voltage characteristics, the visualization of the cell fire, and the evaluation of the results with literature are carried out.

4.1. Overcharging without Thermal Conditioning

In the first scenario without polymer cold plate utilization, the overcharged cell undergoes TR within less than an hour in every experiment. TRP is then followed to the adjacent cell in less than one minute.

4.1.1. Temperature and Voltage Characteristics

Figure 8a presents the temperature and voltage changes during overcharging without thermal conditioning above the SOC based on representative experiment #1. At a C-rate of 1, a cell is charged from 100 to 200% SOC in 60 min. Figure 8b shows the time interval before the TR in more detail. As can be taken from it, certain characteristics of the changes were assigned to specific externally detectable phenomena, the backgrounds of which are discussed in Section 4.1.3.

During stage I of overcharging, the voltage and the measured temperatures increased steadily in a linear manner. At 150% SOC, the voltage started growing superlinearly from 4.84 V to a local maximum of 5.59 V at 162% SOC, which marked the beginnings of stages II and III, respectively. All temperatures, especially at the overcharged cell's outer face, showed a noticeable increase. The phenomena can be associated with initiated lithium plating and cell expansion due to internal gas production. The voltage demonstrated a period of stagnation and then increased to a second local maximum at 178% SOC and 7.22 V, where it abruptly dropped by almost 1.2 V, showing the transition to stage IV. All temperatures spontaneously increased, which affected the TCs next to the OC cell more than those next to cell 2 and suggested the opening process of the pouch and first venting of warm gases from the inside had taken place. However, the camera did not capture any observable visual effect of this first gas ejection.

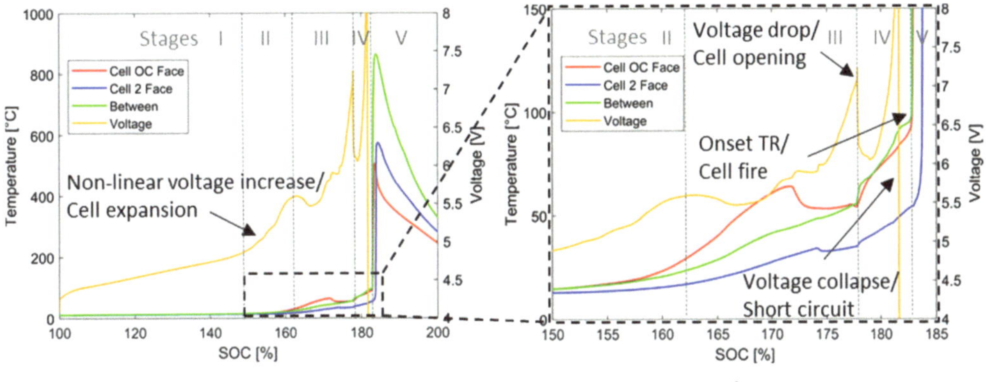

Figure 8. (**a**) Characteristic changes in temperature and voltage in experiment #1; (**b**) magnification before cell fire.

Shortly after, the voltage took a turn, rose progressively, spiked to 11.36 V at 182% SOC, and then collapsed, which embodied an internal short circuit. Around the voltage collapse, white vapor from the top of the overcharged cell was visible in the camera footage. After the first cell opening, the temperatures of the overcharged cell increased steadily to about 90 °C. At an onset surface temperature of 95 °C, the temperatures rose sharply and led to stage V with TR and the actual cell fire due to gas mixture ignition. Internal onset temperatures are known to be significantly higher than surface temperatures [72]. At 183% SOC, the overcharged cell's outer face temperature peaked at 506 °C. An even higher temperature of up to 865 °C was reached between the cells. Less than one minute after the TR, the adjacent cell had heated up significantly, and TRP occurred, which is proven by a peak of the adjacent cell's face temperature of 574 °C. Table 5 shows the averaged temporal characteristics of the selected events and stages regarding the overcharging procedure of experiments #1–9 with voltage noise and cell fire. In the Appendix A.2, Table A1 contains information on the individual experiments. At an average SOC of 145 %, stage II with a non-linear voltage increase, defined as 5 % deviation from linear increase, and cell expansion, at a SOC of 172% stage IV with voltage drop and cell opening including first venting, and at a SOC of 179% stage V with TR and cell fire began. Therefore, the non-linear voltage increase with gas development started after 27 min of overcharging. Another 16.2 min elapsed between the non-linear voltage increase and voltage drop with cell opening and first venting. Then, 4.5 min after the cell opening, at the earliest after 70 s, and the latest after 12 min, TR and cell fire commenced, cf. Table A1.

Table 5. Average temporal characteristics of the first scenario without thermal conditioning.

Exp. #1–9	Non-Linear Increase of Voltage/ Begin Stage II			Voltage Drop/ Begin Stage IV					TR/ Begin Stage V		Duration from Voltage Increase to Drop/ Stages II + III	Duration from Voltage Drop to TR/ Stage IV
	(V)	Time (min)	SOC (%)	Upper (V)	Lower (V)	Diff. (V)	Time (min)	SOC (%)	Time (min)	SOC (%)	Time (min)	Time (min)
Mean	4.87	27.0	145	6.94	5.87	1.07	43.2	172	47.7	179	16.2	4.5
Std. Dev.	0.06	2.0	3	0.89	0.50	0.55	2.2	4	4.4	7	2.3	3.6

4.1.2. Visualization of the Cell Fire

In Figure 9, high-speed camera recordings of the overcharging procedure without thermal conditioning are shown shortly before and during the cell fire. During the first venting, the pouches tended to open at the top, near the cell's tabs. In (a), the beginning of the second venting and the impending cell fire were indicated by white vapor. As the temperature difference between the battery cell and the ambience gradually increased due to further overcharging, the escaping gas mixture's condensation became visible, marking the rising pouch internal temperatures. With an increasing gas production rate, the tensile strength of the seam was reached, and the envelope opened at multiple locations, evidenced in (b). As the internal temperature continued to rise, the vapor changed from white to gray, then to thick black smoke in (c), gradually containing higher proportions of active materials due to pouch internal degradation and melting [106]. In (d), the ignition of the combustible gas composition was triggered by local hot spots due to short circuits. Subsequently, the internal structures dissolved entirely, and a burst of fiery particles, as shown in (e), was forcefully discharged from the cell. The cell burned out completely, see (f), and posed a considerable risk of TRP because of its heat generation.

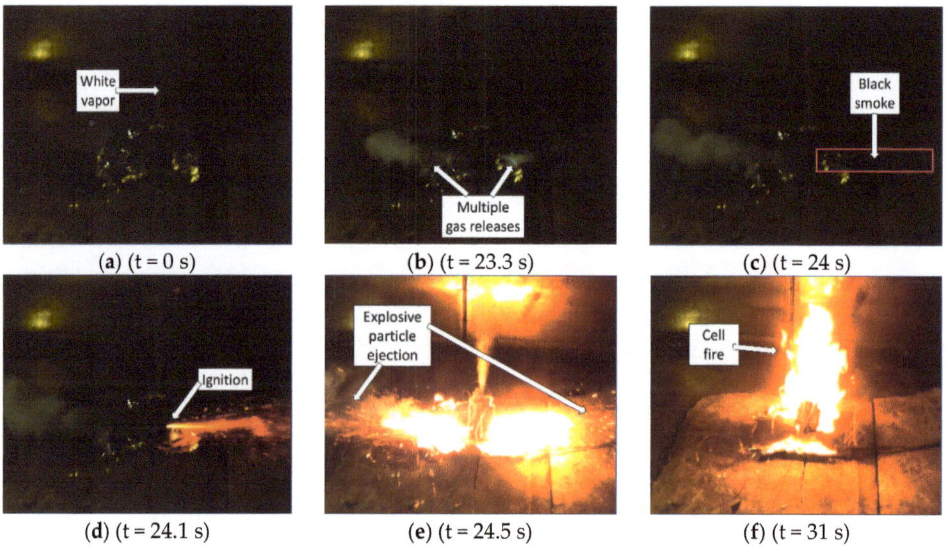

Figure 9. Characteristic processes shortly before and during cell fire: (**a**) Second venting with white vapor emerging from the tabs; (**b**) intensified gas release at various openings; (**c**) active material release with the appearance of black smoke; (**d**) ignition of flammable components; (**e**) explosive particle discharge; and (**f**) cell fire.

4.1.3. Evaluation of the Results

Specific characteristics of the voltage noise were assigned to externally detectable phenomena during overcharging in Section 4.1.1, namely the cell expansion and pouch opening before the cell fire. However, the relation between the mechanical processes and the voltage behavior under overcharging conditions is not considered much in the literature. In [107], pouch cells (NMC 622, 75 Ah) were overcharged at rates ranging from 1–3 C without thermal conditioning, and their swelling behavior up to TR was investigated. The cell's expansion force to the clamping device and the proportional cell displacement were measured next to the voltage and temperature. Cell opening was detected offline with a Fourier Transform-Infrared Spectrometer (FT-IR). The group was able to predict upcoming TR events early by detecting the swelling behavior. However, the qualitative similarity between the change in voltage and cell expansion, of which in Figure 10 an excerpt [107] is shown, had been less emphasized.

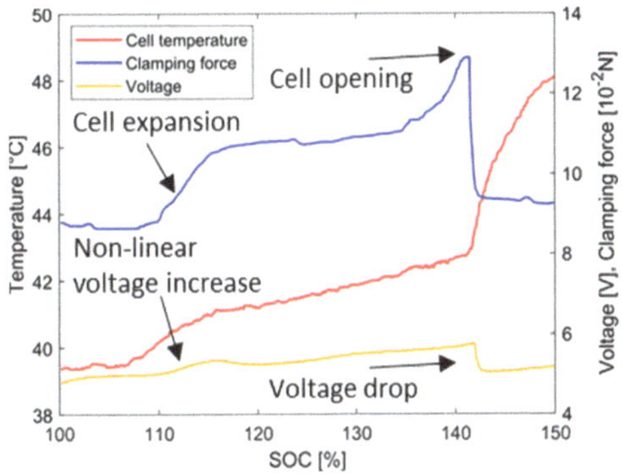

Figure 10. Characteristic changes in cell temperature, voltage, and clamping force during overcharging. Data revisualized with permission from [107].

The voltage and temperature characteristics closely resemble the results of the first experimental scenario without thermal conditioning, displayed in Figure 8. A first cell expansion due to gas formation was detected during the non-linear voltage increase. At the voltage drop, an increase in temperature was measured, and the FT-IR detected the first gaseous electrolyte components, denoting the cell opening and first venting. The clamping force remained with a positive offset compared to the experiment's beginning, which can be attributed to an increased cell thickness because of excessive lithium plating [107]. The voltage correlates with the cell's resistance and the electrode spacing, which varies depending on the cell expansion [16,49] and is one explanation for the related change in the voltage and cell expansion.

According to Arrhenius's law, the gas production rate increases with temperature. Therefore, a quicker cell expansion and a higher gas production rate can be concluded from a steeper voltage increase. Since the voltage behavior directly reflects the cell expansion and pouch opening and allows an estimation of the internal processes during overcharging, it will consequently serve as the comparison between the experimental scenarios of this study.

Nevertheless, to illustrate the impact of the polymer cold plates on overcharge-induced TR, the temporal characteristics of the first experimental scenario without thermal conditioning must be validated with similar studies from the literature.

In the swelling behavior study of [107], it was observed that at 1 C overcharging, the expansion of the cell initiated at 112% SOC on average, the cell opened with a voltage drop at 147%, and a cell fire was experienced at 214%. The first two events were reached at lower SOCs, whereas the cell fire events were reached at a higher SOC than in the first experimental scenario of this study without thermal conditioning. Another group [14] conducted overcharging experiments on 40 Ah pouch cells of distinct chemistries (NMC 111, 622, and 811) at a rate of 1 C. The NMC 622 cell showed a non-linear voltage increase at a 129% SOC, and TR began at a 141% SOC with an onset surface temperature of 92 °C. According to their nickel contents, the NMC 111 and 811 cells showed higher and lower thermal stabilities with TRs at 155% and 134% SOCs, respectively. Consequently, all corresponding processes were reached at lower SOCs than in the first experimental scenario. Instead, in [13], cells of 1200 mAh capacity (NMC 523) were overcharged at 0.4 C and showed TRs ranging from 160 to 205% SOCs, aligning with the range of the first experimental scenario of this study. In both studies [13,14], no abrupt voltage drop and cell opening were detected before TR. Instead, a more gradual voltage drop was found.

The experimental results of the first scenario are generally in line with those findings of similar studies in the literature. The deviations in SOCs concerning certain events in the overcharging process could be significantly related to the cell-specific oversizing of the anode capacity.

4.2. Overcharging with Thermal Conditioning

In the second scenario, the polymer mini-channel cold plates were implemented between the pouch cells to provide thermal conditioning and prevent TR propagation.

4.2.1. Temperature and Voltage Characteristics and Comparison

Figure 11a–c displays the characteristic cell temperature and voltage changes as a result of overcharging and thermal conditioning at different fluid inlet temperatures. The experiments were conducted with different fluid temperature levels, referred to as "low" (5 °C), "medium" (20 °C), and "high" (30 °C). For clarity, the temperatures have been reduced to the laterally measured ones, reflecting the cell's state most accurately.

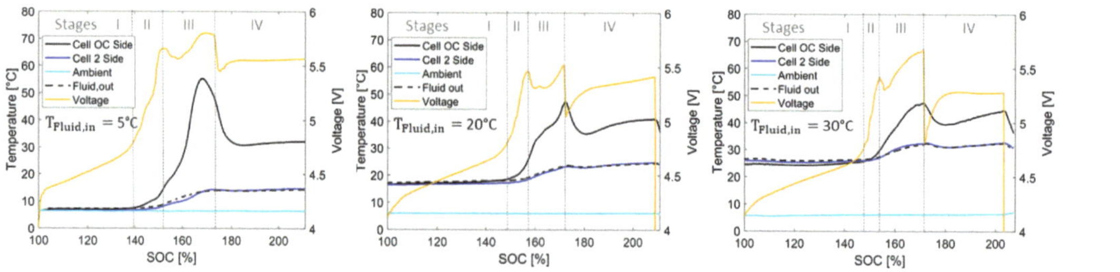

(a) (b) (c)

Figure 11. Characteristic changes in temperature and voltage during overcharging with face conditioning at fluid inlet temperatures in experiments #2*, #3*, and #4*: (**a**) Low at 5 °C; (**b**) medium at 20 °C; and (**c**) high at 30 °C.

As in the first scenario without thermal conditioning (cf. Figure 8), in every experiment, a voltage noise containing a non-linear voltage increase, a local maximum, a voltage plateau, and an abrupt voltage drop, along with a peak in temperature of the OC cell was detected, which can be attributed to the internal gas production, and therefore to the cell expansion and pouch opening with the first venting. Table 6 shows the temporal characteristics of all thermally conditioned experiments. These were conducted in an ambient temperature range of 6 to 7 °C and at 11 °C in experiment #1*. Table 6 indicates that the process behavior was analogous for both experiments #1* and #2* under low fluid temperature conditions.

The temperature of the overcharged cell rose significantly from the start of the non-linear voltage increase. It peaked shortly before the voltage drop if conditioned under low fluid temperature conditions and at its voltage drop at medium and high fluid temperatures, as shown in Figure 11. The peak temperatures of the OC cell reached 59 and 55 °C under low fluid temperature conditions but stayed at lower temperatures of 47 and 48 °C when conditioned by medium- and high-tempered fluid. The reduction in the OC cell's temperature after the voltage drop is more significant with a lower fluid temperature.

Table 6. Temporal characteristics of the second scenario with thermal conditioning.

Exp.	Fluid Inlet Temp.	Starting Temp. of Cell	Non-Linear Increase of Voltage/ Begin Stage II			Voltage Drop/ Begin Stage IV					Duration from Voltage Increase to Drop/ Stage II + III	Max. Temp. OC Cell
	(°C)	(°C)	(V)	Time (min)	SOC (%)	Upper (V)	Lower (V)	Diff. (V)	Time (min)	SOC (%)	Time (min)	(°C)
#1*	5	9	4.91	23.7	139	5.83	5.30	0.53	42.9	172	19.2	59
#2*	5	6	4.75	22.7	138	5.79	5.45	0.34	43.7	173	20.9	55
#3*	20	17	4.76	28.6	148	5.52	5.05	0.47	43.0	172	14.5	47
#4*	30	25	4.73	28.2	148	5.67	4.82	0.85	42.8	171	14.7	48

At a low fluid temperature, after the voltage drop at a 172% SOC, the voltage recovered in a degressive manner. Then, it remained constant, slightly above 5.5 V, for more than 20 min beyond a SOC of 200%, while the cell temperature dropped to 32 °C during the cell opening and then stagnated; see Figure 11a. The experimental time in the fire room is a valuable resource. Since the stage's IV duration of overcharging in experiments #1* and #2*, from the cell opening to the potential TR, of more than 20 min exceeded the respective times of the unconditioned experiments with an average duration of 4.5 min and a maximum of 12 min by far, the overcharging procedure was terminated here.

Before doing so, the camera footage was checked for optical effects of white vapor; cf. Figure 9a. Its presence indicated elevated internal temperatures and the beginning of the second venting in the first scenario without thermal conditioning. Although every overcharged cell in the second thermally conditioned scenario opened at the top near its tabs during the first venting, vapor was not detectable in any experiments with low, medium, and high fluid inlet temperatures, which suggests a certain system stability. Ceasing the overcharging resulted in a cool down and stabilization of the OC cell in each experiment, with no TR and cell fire occurrence.

At a medium fluid temperature, the voltage recovered more significantly after its drop than at a low fluid temperature, yet in a degressive way. Then, it showed a slightly more linear increase, staying below 5.5 V. The temperature showed a degressive increase, reaching a limit of 41 °C after 20 min from the voltage drop, as depicted in Figure 11b. The voltage recovery was even more significant at high fluid temperature yet still degressive. It remained stable below 5.5 V throughout the experiment, while the cell temperature showed a degressive increase up to 44 °C about 20 min from the voltage drop; see Figure 11c. As before, the systems stabilized after the overcharging was terminated.

Comparing the temporal characteristics of the experiments conducted in the first two scenarios without and with thermal conditioning reveals significant findings, as indicated in Tables 5 and 6. Stage II of overcharging, characterized by the initiation of a non-linear voltage increase, commenced at lower SOCs of 138 to 139% when conditioned at a low fluid temperature than at the medium and high levels (148% SOC). It began at an average of a 145% SOC without conditioning. The voltage drop and stage IV were reached at approximately the same SOC of 172% for both sets of experiments, leading to significant differences between the scenarios in the duration of stages II and III between the non-linear voltage increase and voltage drop. At a low fluid temperature, a length of 19 to 21 min was measured, whereas it took a good 16 min on average without thermal conditioning. It should be noted here that overcharging in the unconditioned experiments #1–9 started at ambient and cell temperatures of 7–14 °C; see Table A1. This temperature range falls between the OC cell's starting temperatures in the conditioned experiments at low and medium fluid temperature levels. Finally, stages II and III were decreased to barely 15 min at medium and high fluid temperatures.

4.2.2. Evaluation of Results and Comparison

The disparities in the temporal characteristics of the first two experimental scenarios with and without thermal conditioning, as well as the prolongation of stage IV of overcharging after the cell opening in the thermally conditioned scenario, can be attributed to the cooling effect of the polymer cold plates.

Lithium plating on the anode, besides due to overcharging, is intensified at low-temperature working conditions as well as high-rate charging [108]. The batteries were not exposed to invalidly high currents, only to their maximum charging rate. At low temperatures, the poor battery performance of lithium batteries is related to the increased polarization of the anode, leading to the occurrence of early lithium plating before the anode intercalation sites are even fully utilized [109,110] because of the electrolyte's reduced ionic conductivity and slow lithium diffusion within the graphite [111,112]. Early lithium plating, due to the low-temperature criterion, is promoted the most by using fluid for conditioning at a low temperature of 5 °C in experiments #1* and #2*, to a lesser extent in the thermally unconditioned first scenario and the least at higher fluid temperatures in experiments #3* und #4*. If lithium plating is already more pronounced at lower SOCs, the associated parasitic side reactions and gas production are also promoted earlier. Consequently, the non-linear voltage increase and pouch expansion occurred from lower SOCs of 138 to 139% during low fluid-temperature conditioning, on average at a 145% SOC in the experiments without thermal conditioning and at a 148% SOC with medium and high fluid temperatures.

According to Arrhenius law, the kinetics of the parasitic side reactions and the gas production rate are correlated to the OC cell temperature. At elevated temperatures, more gas volume per timestep is produced. Instead, the cell's pouch opens up as the tensile limit of the seam is reached. Consequently, the time interval between the initial gas production and pouch opening is reduced if the cell is conditioned at elevated fluid temperatures.

Figure 12 presents the voltage and laterally measured temperature changes in the overcharged cell from meaningful experiments of the thermally conditioned and unconditioned scenarios in (a) and (b), respectively.

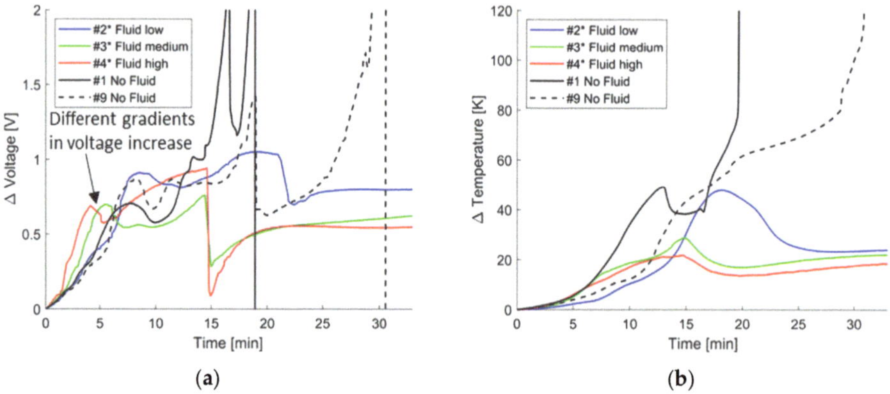

Figure 12. Comparison of thermally conditioned experiments #1*, #3*, and #4* at different fluid temperatures, along with unconditioned experiments #1 and #9: (**a**) Changes in voltage and (**b**) laterally measured OC cell temperature.

Two unconditioned experiments, #1 and #9, are included next to the thermally conditioned experiments, #2*, #3*, and #4*, at low, medium, and high fluid temperatures. Experiment #1 introduced the first experimental scenario in Section 4.1.1. With a duration of 3 min, this cell exhibited a relatively short stage IV of overcharging between the pouch opening and cell fire, while it lasted the longest in experiment #9 with 12 min. To enable comparability between all experiments, every curve's starting point refers to the beginning of stage II of overcharging, marked by the non-linear voltage increase, after deducting the corresponding voltage and temperature offsets.

Firstly, the voltage increased more rapidly if fluid at medium and high temperatures was used, compared to low-tempered fluid or no conditioning, due to the increased gas production rate and faster cell expansion at higher temperatures. Despite the two unconditioned cells and the cells conditioned at a low fluid temperature showing a more significant temperature increase over the subsequent minutes than those conditioned at medium and high fluid temperatures, their pouches opened later.

Secondly, the impact of face conditioning by the polymer MCHS was noticeable after the voltage drop. Without thermal conditioning in experiments #1 and #9, the voltage immediately took a turn and continued to rise progressively until a cell fire event occurred. The cell temperature rose steadily and reached the onset of TR at approximately 100 °C, the latest 12 min after the pouch opening. In contrast, after each cell opening during the thermally conditioned experiments, a degressive voltage increase and a reduction in cell temperature, followed by a limited growth and stagnation of both metrics, were observable for at least 20 min until the overcharging process was terminated. Consequently, the thermally conditioned cells outlasted the unconditioned ones regarding the system's stability. However, although there was no discernible superlinear and progressive voltage increase after the cell opening, the possibility of TR and cell fire cannot be ruled out entirely after an additional overcharging time.

The utilization of the polymer MCHS for thermal conditioning showed a recognizable temporal impact on the cell behavior and the process chain during the attempted overcharge-induction of the TR. Despite their low thermal conductivity material, the cold plates facilitated a significant heat dissipation from the pouch cells. The cell temperatures were lowered in the later stages of overcharging. Stage IV, starting from the pouch opening, was exceptionally prolonged, without reaching a cell fire for the next 20 min. The absence of white vapor at the end of the overcharging procedure indicated a significant reduction in the OC cell's internal temperatures using the polymer cold plates for face cooling.

4.3. *Overcharging with Malfunctioning Thermal Management*

In the third experimental scenario, a malfunctioning thermal management system is emulated. If the cooling system is damaged, the heat dissipation of the mini-channel cold plates can be reduced, and reaching the onset of the TR and cell fire is more likely. The polymer cold plate's function as a thermal barrier in the system is investigated regarding the differences in TRP compared to the first scenario without any safety measures. Two different escalation levels were tested. After setting a "critical" fluid inlet temperature of 40 °C, either the power supply is switched off in experiment #5^{*c} at a specific SOC 12 min after the voltage drop and pouch opening, or the cooling unit is switched off in #6^{*c} at the same SOC.

4.3.1. Temperature and Voltage Characteristics

Figure 13 presents the voltage and temperature characteristics during overcharging with a malfunctioning thermal management system, with (a) and (b) showing the outcomes of experiments #5^{*c} and #6^{*c}, respectively. Table 7 shows the temporal characteristics of the corresponding experiments. As in the first two experimental scenarios, a voltage noise containing a non-linear voltage increase and voltage drop was identified. With 10.6 and 10.1 min, the shortest times between cell expansion and cell opening were found, correlating with the critically high fluid temperature and elevated gas production rates.

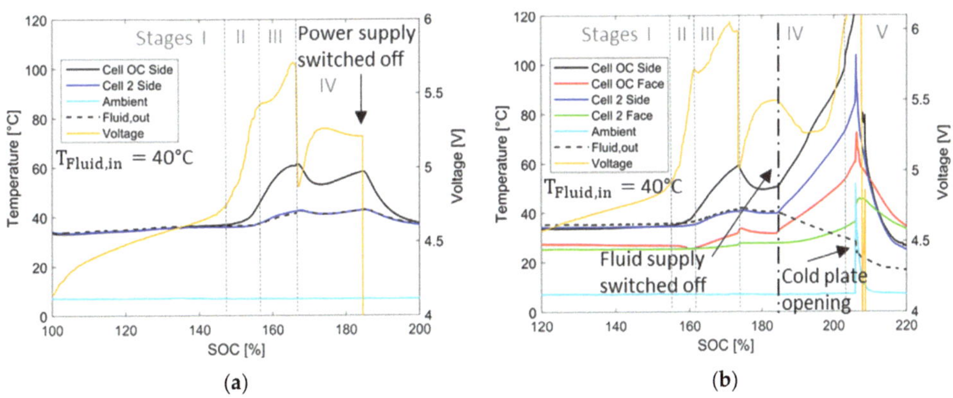

Figure 13. Characteristic changes in temperature and voltage during overcharging with malfunctioning thermal management: (**a**) Power supply switched off at 185% SOC; (**b**) fluid supply switched off at 185% SOC.

Table 7. Temporal characteristics of the third scenario with malfunctioning thermal conditioning.

Exp.	Fluid Inlet Temp.	Starting Temp. of Cell	Non-Linear Increase of Voltage/ Begin Stage II			Voltage Drop/ Begin Stage IV					Duration from Voltage Increase to Drop/ Stage II + III	Max. Temp. OC Cell
	(°C)	(°C)	(V)	Time (min)	SOC (%)	Upper (V)	Lower (V)	Diff. (V)	Time (min)	SOC (%)	Time (min)	(°C)
#5*c	40	33	4.77	28.6	148	5.69	4.87	0.82	39.2	165	10.6	62
#6*c	40	34	5.01	33.9	156	5.99	5.02	0.97	44.0	173	10.1	59

In experiment #5*c, the cell opening occurred at the lowest SOC of 165% among all thermally conditioned experiments; see Figure 13a. The change in voltage then showed strong similarities to the previous experiments #1*–4*. After the drop, the voltage recovered quickly, but degressively, to a local maximum of 5.26 V and then stagnated at that level. Meanwhile, the OC cell's lateral temperature, after peaking to 62 °C during the pouch opening and reducing by 9 K afterward, changed its orientation and linearly increased to 58 °C after 12 min. Here, at a 185% SOC, the power supply was switched off, while no white vapor from the cell and a beginning second venting had been visible before; cf. Figure 9a. All measured temperatures immediately decreased, and the overcharged cell stabilized at 37 °C 9 min later.

In experiment #6*c, the events and their outcomes diverged. In the early overcharging stages, an analogous appearance to experiment #5*c was observed; see Figure 13b. The voltage noise occurred slightly later. Consequently, at a SOC of 185%, the voltage drop was experienced just a good 6 min prior. However, the voltage and OC cell's lateral temperature had already leveled to 5.5 V and 51 °C shortly after. Upon switching off the thermostat's pump and halting the fluid flow, the temperature at every point in the clamping device increased significantly in a linear manner, with the most rapid ascent occurring at the OC cells' sides. The voltage decreased by 0.2 V and subsequently increased significantly. After approximately 10 min, white vapor was observed in the camera footage. About 11 min after switching off the fluid supply, at an onset temperature of 98 °C, the TR and cell fire occurred at a 203% SOC. While the lateral temperatures at the overcharged and now burning-out cell were measured to a maximum of 769 °C, the corresponding TCs of cell 2 only captured a temperature of 104 °C for a concise amount of time. The TCs, located at the face positions of the OC and the adjacent cell between the outer MCHS and the clamping device, recorded peaks of 72 and 46 °C, respectively. Although the temperature on the adjacent cell shortly exceeded a critical value of 60 °C at its sides, no evidence of a TRP was found. A decrease in the fluid outlet temperature was detected upon cell ignition, which revealed the opening of one of the cold plates next to the overcharged burning cell. The remaining cooling fluid flew out of the piping towards the opening, and the TC in the fitting adjusted accordingly to the cooler downstream fluid.

4.3.2. Visualization of the Cell Fire with Thermal Management and Safety System

Figure 14 displays the characteristics of the cell fire in experiment #6*c, including the water-supplied combined thermal management and safety system. Here it must be noted that inside an actual battery pack, alternative non-reactive heat transfer media are needed to supply the cold plates, since water which is in contact with lithium reacts and produces hydrogen gas, which serves the cell fire as additional fuel.

Shortly before the cell caught fire, white vapor was released from the opened cell in (a). The cell ignited spontaneously without the occurrence of black smoke beforehand in (b), and the active material was ejected explosively from the cell in (c). When the polymer cold plates next to the cell melted and opened in (d), a reaction of the water droplets with the hot lithium was likely, which fueled the cell fire with further combustible gases. The cell fire persisted in (e) until an extinguishing of the flame next to the opened cold plates was observed in (f). After the TR of the OC cell, no propagation and second TR were observed.

(**a**) (t = 0 s) (**b**) (t = 33.3 s) (**c**) (t = 33.5 s)
(**d**) (t = 35 s) (**e**) (t = 39.5 s) (**f**) (t = 41.5 s)

Figure 14. Characteristics of the cell fire with implemented MCHS: (**a**) Second venting with white vapor begins at the tabs; (**b**) intensified release and ignition of gaseous electrolyte; (**c**) explosive ejection of active material; (**d**) reaction between lithium and cooling water; (**e**) cell fire; and (**f**) extinguishing of flame by cooling water.

4.3.3. Impact of Polymer-Based Cold Plates as Thermal Barriers

Figure 15 shows the clamping device after the cell fire in experiment #6*[c], the disassembled cell array with the central polymer mini-channel cold plate being attached to the overcharged and now burned-out cell, and the condition of the outer polymer mini-channel cold plate next to the overcharged cell. The overcharged cell had burned out violently, revealing a see-through of its core. The adjacent polymer MCHSs were partly molten and opened at several positions, so cooling water flooded the setup. No evidence of propagation of the TR towards the adjacent cell could be detected.

After the experiment, the vacuum of the adjacent pouch cell remained intact, and the cell maintained a steady voltage of 4.2 V during a 14-day observation period. In summary, implementing a mini-channel cold plate of 2 mm thickness as a thermal barrier next to a pouch cell, which is subject to an overcharge-induced TR, can have a significant effect on the prevention of TRP to the adjacent cell.

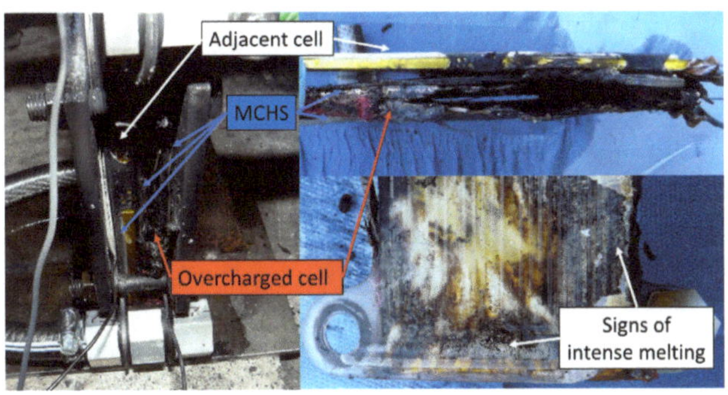

Figure 15. Experimental setup (#6*[c]) with polymer mini-channel cold plates after TR and cell fire of the overcharged cell.

4.4. Performance of the MCHS

Implementing polymer mini-channel cold plates significantly impacted the overcharge-induced TR and TRP of the tested pouch cells, which warrants further investigations. For an evaluation of subsequent development stages of the cold plates and ongoing optimizations, the characteristics of the dissipated rate of heat flow are shown in Figure 16a,b regarding the low and medium fluid temperatures in experiments #2 and #3, respectively. These results are representative and applicable to the other experiments.

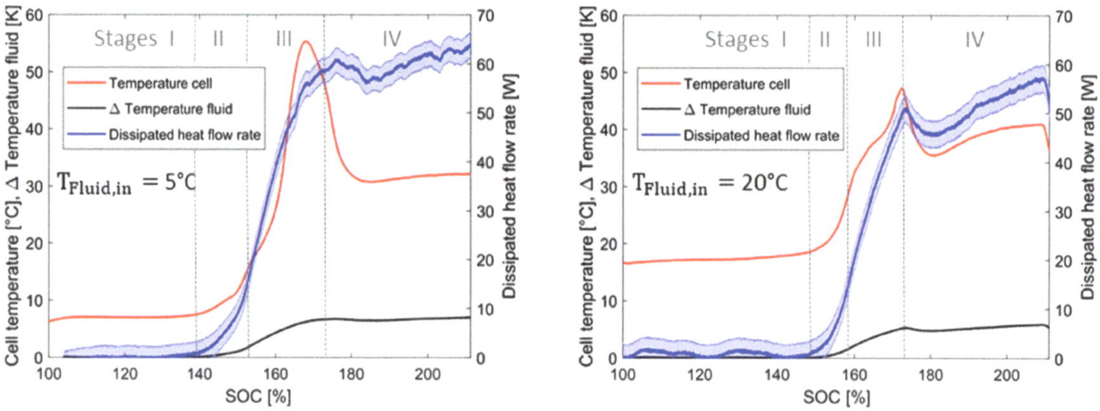

(**a**) (**b**)

Figure 16. Dissipated heat flow during overcharging at different fluid inlet temperatures: (**a**) Low at 5 °C; (**b**) medium at 20 °C.

The mass flow was set to 2 g/s at the lowest fluid inlet temperature level and increased slightly at higher temperatures. Significant heat dissipation began analogously to the cell's temperature increase during cell expansion from stage II of overcharging. A temporal peak of 60.9 ± 2.54 W and 51 ± 2.71 W was achieved at low and medium fluid temperatures, correspondingly. After the cell opening and first venting, the cell's surface cooled down, and the dissipated heat flow reached a local minimum, which was more pronounced at higher fluid temperatures. As soon as the cell temperature rose again during stage IV of overcharging, the dissipated heat flow rate increased until the cell temperature leveled out. Here, dissipated rates of heat flow of 65 ± 2.58 W and 57 ± 2.74 W at low and medium fluid temperatures were obtained. Based on the uncertainties from Table 3 pertaining to the mass flow and temperature sensors, the heat flow is estimated to have uncertainties of 4–5%. Even with a calibration of the TCs, around 90% of the uncertainty originates from the fluid temperature measurement.

Incorporating not only the flow field but the entire contact surface between cells and cold plates for heat transfer (see Section 3.2), the maximum dissipated heat flux was calculated to be 2150 and 1900 W/m^2 at low and medium fluid temperatures, respectively. At this point, a determination of the exact heat transfer coefficients at the wall–fluid interface does not seem appropriate based on the experimental results since it has been found that none of the generally applied ideal thermal boundary conditions are valid, and the characteristic length of the flow cross-section changes in the inlet and outlet regions.

Nevertheless, this first development stage of the polymer cold plates showed an adequate and satisfactory dissipation performance if heat fluxes in lithium-ion battery modules between 200 and 3000 W/m^2 are assumed according to [3,26].

5. Conclusions

This study presents a novel combined thermal management and safety system for thin prismatic and pouch battery cells and tests its impact on delaying and preventing TR and TRP through experimentation. Such a system must solve an inevitable trade-off. On the one hand, it must successfully dissipate a sufficient heat flow rate during regular operation and the early stages of an impending TR to slow and delay its progress. On the other hand, in the event of a cell fire, it must serve as a thermal barrier to avoid irregular heating of adjacent cells to prevent TRP. The introduced combined thermal management and safety system consists of polymer mini-channel cold plates for application between battery cells.

The experimental setup in this study contained a cell array of two fully charged pouch cells with NMC chemistry, of which one cell was overcharged at 1 C to induce TR and TRP. In detail, three scenarios were distinguished: (1) no application of cold plates, (2) cold plate application between the cells at regular fluid inlet temperatures ranging from 5 to 30 °C, and (3) a cold plate application as before, but with a malfunction in the cooling system, resulting in a critically high fluid inlet temperature of 40 °C and an interruption of the fluid supply during overcharging.

In the first scenario, the TR and TRP occurred within one hour of overcharging in each experiment. Prior to the cell fire, a voltage noise indicated a cell expansion from an average SOC of 145%, and a pouch opening with first venting due to internal gas development was found at an average SOC of 172%. The time interval between both events averaged 16.2 min across nine experiments. After another 4.5 min, on average, at a SOC of 179%, but at a maximum after 12 min, the onset of TR and cell fire occurred. The sensed temperatures between the two cells exceeded 800 °C, and within less than one minute, the TR propagated to the adjacent cell.

In the second scenario, utilizing the cold plates as a combined thermal management and safety application at different fluid inlet temperatures of 5, 20, and 30 °C, overcharging the cell did not produce a TR and a subsequent TRP in the same time scale. Instead, a temporal shift in the characteristic process chain was observed. At low fluid temperatures, gas production and cell expansion commenced at a lower SOC of 138% than without thermal conditioning. Conversely, they began later at 148% with medium and high fluid temperatures. The gas production rate is directly correlated with the fluid inlet temperature level, subsequently affecting the time from the first expansion to the cell opening. On average, this process took 20 min at a low fluid temperature and approximately 14.5 min at medium and high levels.

Altogether, no significant impact of thermal conditioning was observed on the cell opening itself, which consistently occurred at SOCs ranging between 171% and 173% with and without thermal conditioning. Instead, with thermal conditioning, a significant temperature decrease at the cell's surface was detectable after the pouch opening and first venting, which was not observable without thermal conditioning. Stage IV of overcharging between the pouch opening and a potential TR with cell fire was most affected by thermal conditioning. In the subsequent experimental runtime, the polymer cold plates dissipated an increasing rate of heat flow. This enabled a significant stabilization in the overcharged cell's voltage and temperature, which was not detectable without thermal conditioning. Even 20 min after the cell opening, no thermal runaway had occurred. Additionally, there was no visual evidence indicating the initiation of the second venting, such as white vapor being released from the overcharged cell at this point of the overcharging experiment if the cells were thermally conditioned by the polymer cold plates between a 5–30 °C fluid temperature. After the overcharging procedure was terminated, the system managed to stabilize.

In the third scenario, a defective thermal management system was simulated with a critical fluid inlet temperature of 40 °C. A similar outcome to the second scenario was observed in the early overcharging stages. The period from initial cell expansion to pouch opening was reduced to 10.8 min, according to an increased gas production rate at elevated temperatures. Despite the critically high fluid temperature, the cell stabilized utterly after the overcharging procedure was terminated 12 min after the cell opening at a SOC of 185%. At this point, no visual indications of a second venting initiation were detected. If the fluid supply was halted at the same SOC, the overcharged cell heated up promptly, and then second venting and TR occurred approximately 10 and 11 min later at a SOC of 203%. Since the polymer cold plates served as an excellent thermal barrier, despite a violent cell fire that started from the overcharged cell in the assembly, the lateral temperatures of the adjacent cell just peaked momentarily at 100 °C, which did not result in a TRP. While the overcharged cell burned out entirely, the adjacent cell maintained a stable voltage and an intact vacuum within a fortnightly observation period.

In summary, the polymer mini-channel cold plates serve the requirements of a combined thermal management and safety application and effectively address the trade-off mentioned before. Despite the low thermal conductivity of the polymer, making use of extensive heat transfer surfaces, thermal conditioning and heat dissipation of the mini-channel cold plates are so intense that a significant delay of the overcharge-induced TR, especially in the form of prolongation of stage IV of overcharging, between the cell opening with first venting and TR with a cell fire, is observed. A heat flux of up to 2150 W/m^2 is dissipated by the polymer cold plates, and the overcharged cells are stabilized so that no cell fire occurs within a time, leading to safe TR without thermal conditioning. In the event of TR and a cell fire, the polymer's low thermal conductivity, typically considered a drawback, becomes beneficial since the cold plates serve as thermal barriers that protect the adjacent cells from excessive short-term heat generation during TR and prevent TRP.

The polymer-based mini-channel cold plates, next to their viability as a combined thermal management and safety application, are among the thinnest and lightest of their kind. They are robust and less expensive than comparable aluminum components, making them an appealing option for mobile applications. In the future, they should be further investigated as an alternative to conventional applications.

6. Outlook

Using the polymer cold plates within a battery system could extend the evacuation time during a hazardous situation inside a BEV. Numerous entry points regarding further investigations arise from the experimental study performed.

The experiments under thermal conditioning should be repeated to investigate the cell's long-term behavior and stability during stage IV of overcharging. Online electrochemical mass spectrometry [113] is advised to acquire crucial insights into the effect of thermal conditioning on internal cell processes. According to Figure A3 in the Appendix A.2, a first long-term overcharging test at a low fluid temperature showed no TR for almost 300 min of overcharging. The pouch opened during the experiment with a first venting, but no white vapor and second venting were visually detected. A post-mortem examination revealed a completely dry cell without liquid electrolyte. A maximum heat flow of 162 ± 4.16 W was dissipated at a fluid temperature difference between the inlet and outlet of 11.2 K.

Further investigations should consider a higher quantity of cells in the experimental setup with a larger capacity situated in a more encapsulated environment to represent a mobile application-oriented scenario. Additionally, thermal and mechanical TR induction should be considered. The overcharging behavior of thicker cells should be studied comparatively. The focus should be on the internal temperature gradients and the TR risk potential at different cell thicknesses using the face-conditioning approach.

The investigation of the heat dissipation potential of the polymer cold plates during regular cycling or super-fast charging is advisable. In this context, a variation in the polymer material and the cooling channel geometry should be considered. If fluid mixing and boundary layer interruptions within the channel geometry can be intensified without disproportionately increasing the turbulence and pressure drop, a performance increase is expectable, compared to a flow field with straight channels [29]. Hybrid cooling approaches, such as the combination of face and tab cooling, should be evaluated, aiming for a high-temperature uniformity inside the cell volume at sufficiently high heat dissipation rates under a heavy load of the cells [23].

Nevertheless, it is imperative not to overlook the contradictory heat dissipation and protection requirements. Future work should explore material modifications, and the cold plate geometry should be redesigned to maintain the external temperature of the adjacent cells below 60 °C.

A month after the overcharging experiments, the thermally conditioned cells from experiments #1*–#4*^{c} still posed a safety hazard since their voltages remained above the nominal upper limit of 4.2 V, and liquid electrolyte was discovered near the pouch opening, as shown in Figure A4 in the Appendix A.2. Their resistance was found to have raised by magnitudes, from 6–10 mΩ to 0.54–1.04 Ω, embodying substantial cell damage. Regular discharging failed, and alternative strategies must be considered in the context of recycling damaged cells. CCCV-discharging down to meager rates of 0.05 C allowed for greater discharge capacities but at the expense of longer discharging times.

Author Contributions: Conceptualization, H.-C.G., J.S., G.B. and F.B.; methodology, H.-C.G. and G.B.; software, H.-C.G. and G.B.; validation, H.-C.G.; formal analysis, H.-C.G. and G.B.; investigation, H.-C.G., G.B. and J.S.; resources, J.S., F.K. and F.B.; data curation, H.-C.G.; writing—original draft preparation, H.-C.G.; writing—review and editing, G.B., J.S., F.K. and F.B.; visualization, H.-C.G.; supervision, G.B., J.S. and F.B.; project administration, G.B. and F.B.; funding acquisition, G.B. and F.B. All authors have read and agreed to the published version of the manuscript.

Funding: Ministry of Science, Energy, Climate Protection, and Environment, Saxony-Anhalt, Germany.

Data Availability Statement: All data that justify and support the statements made in this article are included in the article.

Acknowledgments: Funding of parts of this work by the federal state of Saxony-Anhalt (Germany) is gratefully acknowledged. The authors thank the staff of IBK Heyrothsberge (Saxony-Anhalt, Germany) and C. Vogel particularly for their hospitality and support during the TR experiments. The authors would also like to thank P. Vorwerk, L. Bläubaum, L. Schmidt, A. Dlugosch, A. Weiß, and L. Block for discussions of diverse nature and ideas that have added value to this article. Further thanks to K. Amano, E. Gimadieva, and U. Krause for experiment-related support and to S. Thiele for support in preparing the experimental demonstrators. We acknowledge the support of Magdeburg University's Open Access Publication Fund with the article processing charge.

Conflicts of Interest: The authors declare no conflict of interest.

Nomenclature

Symbol	Name	Unit
A	Area	m^2
c	Heat capacity	J/kg K
I	Current	A
k	Thermal conductivity	W/m K
l	length	m
$\dot{m}$	Mass flow rate	kg/m^3
t	Time	s
T	Temperature	K
q_{el}	Electric charge	Ah
Q	Heat	J
$\dot{Q}$	Heat flow rate	W
r	Electric resistance	Ω
U_{OCV}	Open circuit voltage	V
V	Electric voltage	V

Subscripts

DC	Direct current
eff	effective
el	electric
Fluid,in	Fluid inlet
Fluid,out	Fluid outlet
i	internal
irrev	irreversible
layer,j	layer number j
OCV	Open Circuit Voltage
rev	reversible
$\parallel$	parallel
$\perp$	perpendicular

Abbreviation

Exp.	Experimental
Ind.	Induction
Max.	Maximum
Num.	Numerical
OC	Overcharged
Std. dev.	Standard deviation
Temp.	Temperature

Acronyms

ABS	Acrylonitrile butadiene styrene
CC	Constant Current
CCCV	Constant Current Constant Voltage
BEV	Battery electric vehicle
BTMS	Battery thermal management system
EIS	Electrochemical Impedance Spectroscopy
EUCAR	European Council for Automotive Research & Development
FT-IR	Fourier Transform-Infrared Spectrometer
MCHS	Mini-channel heat sink or heat source
NMC	Nickel-Manganese-Cobalt
OVGU	Otto-von-Guericke University
PCM	Phase Change Material
SEI	Solid Electrolyte Interphase
SOC	State of charge
TCs	Thermocouples
TIM	Thermal Interface Material
TPS	Transient Plane Source
TR	Thermal runaway
TRP	Thermal runaway propagation

Appendix A.

Appendix A.1. Equations

$$k_{eff,\perp} = \frac{\sum_{j=1}^{n} l_{layer,j} k_{layer,j}}{l_{total}} \tag{A1}$$

$$k_{eff,\parallel} = \frac{l_{total}}{\sum_{j=1}^{n} \frac{l_{layer,j}}{k_{layer,j}}} \tag{A2}$$

Appendix A.2. Figures and Tables

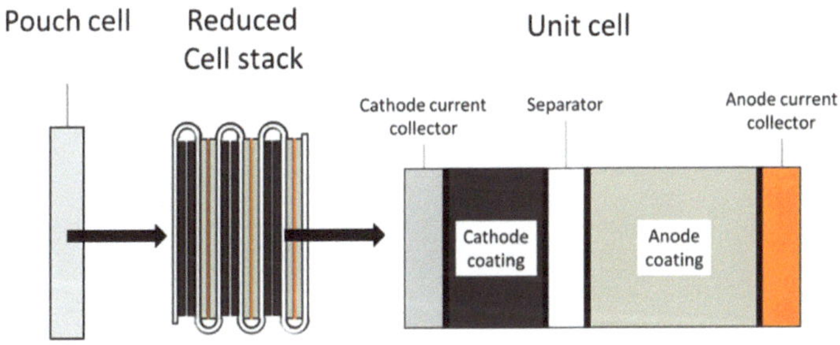

Figure A1. Simplified structure of a battery cell, the reduced stack, and the unit cell.

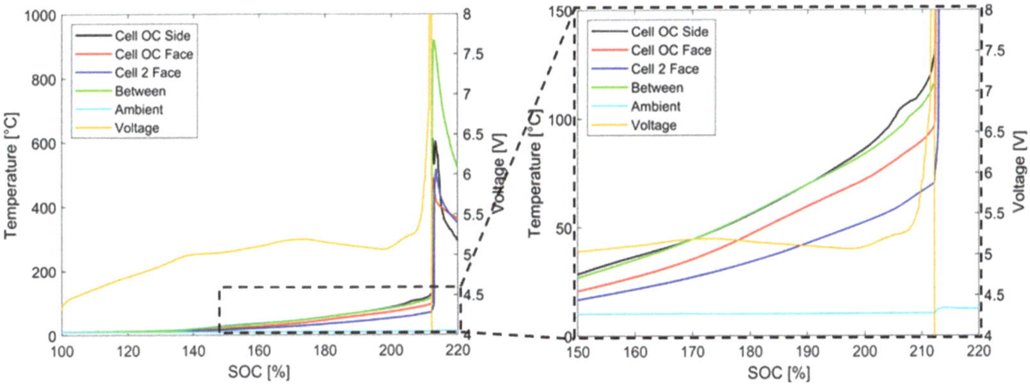

(a) (b)

Figure A2. (**a**) Temperature and voltage characteristics during TC positioning test; (**b**) magnification before TR.

Table A1. Individual temporal characteristics during overcharging without thermal conditioning.

Exp.	Ambient Temp.	Non-Linear Increase of Voltage/ Begin Stage II			Voltage Drop/ Begin Stage IV					TR/ Begin Stage V		Duration from Voltage Increase to Drop/ Stages II + III	Duration from Voltage Drop to TR/ Stage IV
	(°C)	(V)	Time (min)	SOC (%)	Upper (V)	Lower (V)	Diff. (V)	Time (min)	SOC (%)	Time (min)	SOC (%)	Time (min)	Time (min)
#1	8	4.89	30.1	150	7.22	6.05	1.17	46.7	178	49.7	183	16.6	3.0
#2	10	4.89	27.8	146	6.25	5.25	1.00	43.7	173	50.0	183	15.9	6.3
#3	8	4.85	26.1	143	6.84	5.93	0.91	43.8	173	45.0	175	17.7	1.2
#4	10	4.96	24.8	141	8.33	6.91	1.42	39.0	165	40.5	168	14.2	1.5
#5	12	4.91	26.4	144	5.77	5.28	0.49	42.5	171	50.0	183	16.1	7.5
#6	11	4.80	29.1	149	6.83	6.23	0.60	40.6	168	41.7	170	11.5	1.1
#7	14	4.86	29.2	149	8.52	6.08	2.44	45.2	175	46.5	178	16.0	1.3
#8	10	4.88	25.3	142	6.51	5.69	0.82	44.0	173	50.3	184	18.7	6.3
#9	7	4.76	24.3	140	6.17	5.43	0.74	43.2	172	55.2	192	18.9	12.0

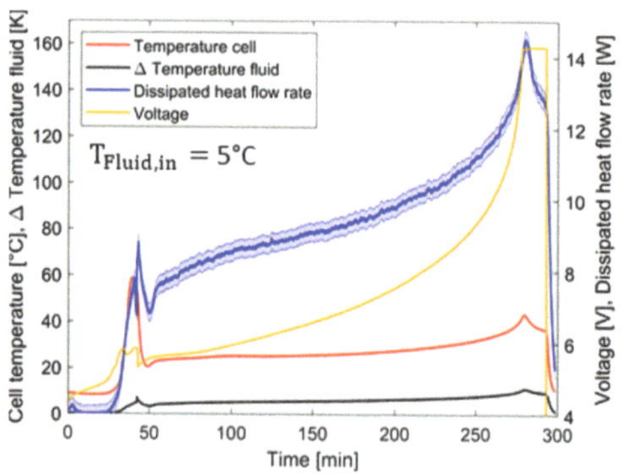

Figure A3. Long-term overcharging at 1 C at 5 °C fluid temperature and a mass flow of 3 g/s.

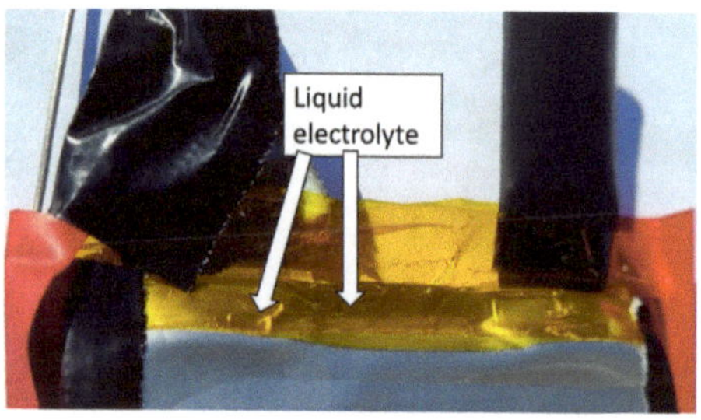

Figure A4. Liquid electrolyte under the Kapton® tape near the pouch opening at the tabs.

References

1. Mallick, S.; Gayen, D. Thermal behaviour and thermal runaway propagation in lithium-ion battery systems—A critical review. *J. Energy Storage* **2023**, *62*, 106894. [CrossRef]
2. Li, W.; Zhou, Y.; Zhang, H.; Tang, X. A Review on Battery Thermal Management for New Energy Vehicles. *Energies* **2023**, *16*, 4845. [CrossRef]
3. Bandhauer, T.M.; Garimella, S.; Fuller, T.F. A Critical Review of Thermal Issues in Lithium-Ion Batteries. *J. Electrochem. Soc.* **2011**, *158*, R1. [CrossRef]
4. Tomaszewska, A.; Chu, Z.; Feng, X.; O'Kane, S.; Liu, X.; Chen, J.; Ji, C.; Endler, E.; Li, R.; Liu, L.; et al. Lithium-ion battery fast charging: A review. *eTransportation* **2019**, *1*, 100011. [CrossRef]
5. Newman, J.; Balsara, N.P. *Electrochemical Systems*, 4th ed.; Wiley: Hoboken, NJ, USA, 2021; ISBN 9781119514596.
6. Gu, W.B.; Wang, C.Y. Thermal-Electrochemical Modeling of Battery Systems. *J. Electrochem. Soc.* **2000**, *147*, 2910. [CrossRef]
7. Bernardi, D.; Pawlikowski, E.; Newman, J. A General Energy Balance for Battery Systems. *J. Electrochem. Soc.* **1985**, *132*, 5–12. [CrossRef]
8. Heubner, C.; Schneider, M.; Lämmel, C.; Michaelis, A. Local Heat Generation in a Single Stack Lithium Ion Battery Cell. *Electrochim. Acta* **2015**, *186*, 404–412. [CrossRef]
9. Hu, Y.; Choe, S.-Y.; Garrick, T.R. Measurement of heat generation rate and heat sources of pouch type Li-ion cells. *Appl. Therm. Eng.* **2021**, *189*, 116709. [CrossRef]
10. Duan, J.; Tang, X.; Dai, H.; Yang, Y.; Wu, W.; Wei, X.; Huang, Y. Building Safe Lithium-Ion Batteries for Electric Vehicles: A Review. *Electrochem. Energy Rev.* **2020**, *3*, 1–42. [CrossRef]
11. Kleiner, K.; Jakes, P.; Scharner, S.; Liebau, V.; Ehrenberg, H. Changes of the balancing between anode and cathode due to fatigue in commercial lithium-ion cells. *J. Power Sources* **2016**, *317*, 25–34. [CrossRef]
12. Korthauer, R. *Lithium-Ion Batteries: Basics and Applications*; Springer: Berlin/Heidelberg, Germany, 2019; ISBN 9783662530719.
13. Liu, J.; Huang, Z.; Sun, J.; Wang, Q. Heat generation and thermal runaway of lithium-ion battery induced by slight overcharging cycling. *J. Power Sources* **2022**, *526*, 231136. [CrossRef]
14. Wang, Z.; Yuan, J.; Zhu, X.; Wang, H.; Huang, L.; Wang, Y.; Xu, S. Overcharge-to-thermal-runaway behavior and safety assessment of commercial lithium-ion cells with different cathode materials: A comparison study. *J. Energy Chem.* **2021**, *55*, 484–498. [CrossRef]
15. Liu, B.; Jia, Y.; Yuan, C.; Wang, L.; Gao, X.; Yin, S.; Xu, J. Safety issues and mechanisms of lithium-ion battery cell upon mechanical abusive loading: A review. *Energy Storage Mater.* **2020**, *24*, 85–112. [CrossRef]

16. Cai, T.; Valecha, P.; Tran, V.; Engle, B.; Stefanopoulou, A.; Siegel, J. Detection of Li-ion battery failure and venting with Carbon Dioxide sensors. *eTransportation* **2021**, *7*, 100100. [CrossRef]
17. Wang, Q.; Ping, P.; Zhao, X.; Chu, G.; Sun, J.; Chen, C. Thermal runaway caused fire and explosion of lithium ion battery. *J. Power Sources* **2012**, *208*, 210–224. [CrossRef]
18. Battery Pack Directed Venting System. Available online: https://patents.google.com/patent/US8557416B2/en (accessed on 12 November 2023).
19. Vehicle Battery Pack Thermal Barrier. Available online: https://patents.google.com/patent/US8875828B2/en (accessed on 12 November 2023).
20. Bai, Q.; Li, K.; Zan, J.; Liu, J.; Ou, J.; Liu, J. Influence of Insulation Material Thickness on Spread of Thermal Runaway in Battery Packs. *Processes* **2023**, *11*, 1321. [CrossRef]
21. Arrhenius, S. XXXI. On the influence of carbonic acid in the air upon the temperature of the ground. *Lond. Edinb. Dublin Philos. Mag. J. Sci.* **1896**, *41*, 237–276. [CrossRef]
22. Jalkanen, K.; Karppinen, J.; Skogström, L.; Laurila, T.; Nisula, M.; Vuorilehto, K. Cycle aging of commercial NMC/graphite pouch cells at different temperatures. *Appl. Energy* **2015**, *154*, 160–172. [CrossRef]
23. Hunt, I.A.; Zhao, Y.; Patel, Y.; Offer, J. Surface Cooling Causes Accelerated Degradation Compared to Tab Cooling for Lithium-Ion Pouch Cells. *J. Electrochem. Soc.* **2016**, *163*, A1846–A1852. [CrossRef]
24. Gungor, S.; Gocmen, S.; Cetkin, E. A review on battery thermal management strategies in lithium-ion and post-lithium batteries for electric vehicles. *J. Therm. Eng.* **2023**, *9*, 1078–1099. [CrossRef]
25. Liu, J.; Chen, H.; Huang, S.; Jiao, Y.; Chen, M. Recent Progress and Prospects in Liquid Cooling Thermal Management System for Lithium-Ion Batteries. *Batteries* **2023**, *9*, 400. [CrossRef]
26. Xia, G.; Cao, L.; Bi, G. A review on battery thermal management in electric vehicle application. *J. Power Sources* **2017**, *367*, 90–105. [CrossRef]
27. Shahid, S.; Agelin-Chaab, M. A review of thermal runaway prevention and mitigation strategies for lithium-ion batteries. *Energy Convers. Manag. X* **2022**, *16*, 100310. [CrossRef]
28. Kumar, A.; Kumar, A. Identification and Mitigation of Shortcomings in Direct and Indirect Liquid Cooling-Based Battery Thermal Management System. *Energies* **2023**, *16*, 3857. [CrossRef]
29. Sarvar-Ardeh, S.; Rashidi, S.; Rafee, R.; Karimi, N. A review on the applications of micro-/mini-channels for battery thermal management. *J. Therm. Anal. Calorim.* **2023**, *148*, 7959–7979. [CrossRef]
30. Wilke, S.; Schweitzer, B.; Khateeb, S.; Al-Hallaj, S. Preventing thermal runaway propagation in lithium ion battery packs using a phase change composite material: An experimental study. *J. Power Sources* **2017**, *340*, 51–59. [CrossRef]
31. Jiaqiang, E.; Han, D.; Qiu, A.; Zhu, H.; Deng, Y.; Chen, J.; Zhao, X.; Zuo, W.; Wang, H.; Chen, J.; et al. Orthogonal experimental design of liquid-cooling structure on the cooling effect of a liquid-cooled battery thermal management system. *Appl. Therm. Eng.* **2018**, *132*, 508–520. [CrossRef]
32. Tang, A.; Li, J.; Lou, L.; Shan, C.; Yuan, X. Optimization design and numerical study on water cooling structure for power lithium battery pack. *Appl. Therm. Eng.* **2019**, *159*, 113760. [CrossRef]
33. Li, Q.; Shi, H.; Xie, G.; Xie, Z.; Liu, H. Parametric study and optimization on novel fork-type mini-channel network cooling plates for a Li-ion battery module under high discharge current rates. *Int. J. Energy Res.* **2021**, *45*, 17784–17804. [CrossRef]
34. Yu, Y.; Huang, Z.; Mei, W.; Jia, Z.; Song, L.; Wang, Q. Preventing effect of different interstitial materials on thermal runaway propagation of large-format lithium iron phosphate battery module. *J. Energy Storage* **2023**, *63*, 107082. [CrossRef]
35. Feng, X.; He, X.; Ouyang, M.; Lu, L.; Wu, P.; Kulp, C.; Prasser, S. Thermal runaway propagation model for designing a safer battery pack with 25 Ah $LiNi_xCo_yMn_zO_2$ large format lithium ion battery. *Appl. Energy* **2015**, *154*, 74–91. [CrossRef]
36. Tsao, C.-W.; DeVoe, D.L. Bonding of thermoplastic polymer microfluidics. *Microfluid Nanofluid* **2009**, *6*, 1–16. [CrossRef]
37. Xu, Z.; Xu, J.; Guo, Z.; Wang, H.; Sun, Z.; Mei, X. Design and Optimization of a Novel Microchannel Battery Thermal Management System Based on Digital Twin. *Energies* **2022**, *15*, 1421. [CrossRef]
38. Rui, X.; Feng, X.; Wang, H.; Yang, H.; Zhang, Y.; Wan, M.; Wei, Y.; Ouyang, M. Synergistic effect of insulation and liquid cooling on mitigating the thermal runaway propagation in lithium-ion battery module. *Appl. Therm. Eng.* **2021**, *199*, 117521. [CrossRef]
39. Zhang, W.; Liang, Z.; Yin, X.; Ling, G. Avoiding thermal runaway propagation of lithium-ion battery modules by using hybrid phase change material and liquid cooling. *Appl. Therm. Eng.* **2021**, *184*, 116380. [CrossRef]
40. Zhang, T.; Gao, Q.; Gu, Y.; Li, Y. Studies on thermal management of lithium-ion battery using non-metallic heat exchanger. *Appl. Therm. Eng.* **2021**, *182*, 116095. [CrossRef]
41. Kshetrimayum, K.S.; Yoon, Y.-G.; Gye, H.-R.; Lee, C.-J. Preventing heat propagation and thermal runaway in electric vehicle battery modules using integrated PCM and micro-channel plate cooling system. *Appl. Therm. Eng.* **2019**, *159*, 113797. [CrossRef]
42. Rao, Z.; Wang, Q.; Huang, C. Investigation of the thermal performance of phase change material/mini-channel coupled battery thermal management system. *Appl. Energy* **2016**, *164*, 659–669. [CrossRef]
43. Ruiz, V.; Pfrang, A.; Kriston, A.; Omar, N.; van den Bossche, P.; Boon-Brett, L. A review of international abuse testing standards and regulations for lithium ion batteries in electric and hybrid electric vehicles. *Renew. Sustain. Energy Rev.* **2018**, *81*, 1427–1452. [CrossRef]
44. Feng, X.; Zheng, S.; Ren, D.; He, X.; Wang, L.; Liu, X.; Li, M.; Ouyang, M. Key Characteristics for Thermal Runaway of Li-ion Batteries. *Energy Procedia* **2019**, *158*, 4684–4689. [CrossRef]

45. Ren, D.; Feng, X.; Lu, L.; He, X.; Ouyang, M. Overcharge behaviors and failure mechanism of lithium-ion batteries under different test conditions. *Appl. Energy* **2019**, *250*, 323–332. [CrossRef]
46. Essl, C.; Golubkov, A.W.; Fuchs, A. Comparing Different Thermal Runaway Triggers for Two Automotive Lithium-Ion Battery Cell Types. *J. Electrochem. Soc.* **2020**, *167*, 130542. [CrossRef]
47. Larsson, F.; Mellander, B.-E. Abuse by External Heating, Overcharge and Short Circuiting of Commercial Lithium-Ion Battery Cells. *J. Electrochem. Soc.* **2014**, *161*, A1611–A1617. [CrossRef]
48. Chen, Y. Recent advances of overcharge investigation of lithium-ion batteries. *Ionics* **2022**, *28*, 495–514. [CrossRef]
49. Huang, L.; Zhang, Z.; Wang, Z.; Zhang, L.; Zhu, X.; Dorrell, D.D. Thermal runaway behavior during overcharge for large-format Lithium-ion batteries with different packaging patterns. *J. Energy Storage* **2019**, *25*, 100811. [CrossRef]
50. Noh, H.-J.; Youn, S.; Yoon, C.S.; Sun, Y.-K. Comparison of the structural and electrochemical properties of layered $Li[Ni_xCo_yMn_z]O_2$ (x = 1/3, 0.5, 0.6, 0.7, 0.8 and 0.85) cathode material for lithium-ion batteries. *J. Power Sources* **2013**, *233*, 121–130. [CrossRef]
51. Liu, Q.; Du, C.; Shen, B.; Zuo, P.; Cheng, X.; Ma, Y.; Yin, G.; Gao, Y. Understanding undesirable anode lithium plating issues in lithium-ion batteries. *RSC Adv.* **2016**, *6*, 88683–88700. [CrossRef]
52. Broussely, M.; Biensan, P.; Bonhomme, F.; Blanchard, P.; Herreyre, S.; Nechev, K.; Staniewicz, R.J. Main aging mechanisms in Li ion batteries. *J. Power Sources* **2005**, *146*, 90–96. [CrossRef]
53. Vetter, J.; Novák, P.; Wagner, M.R.; Veit, C.; Möller, K.-C.; Besenhard, J.O.; Winter, M.; Wohlfahrt-Mehrens, M.; Vogler, C.; Hammouche, A. Ageing mechanisms in lithium-ion batteries. *J. Power Sources* **2005**, *147*, 269–281. [CrossRef]
54. Bugga, R.V.; Smart, M.C. Lithium Plating Behavior in Lithium-Ion Cells. *ECS Trans.* **2010**, *25*, 241–252. [CrossRef]
55. Kim, C.-S.; Jeong, K.M.; Kim, K.; Yi, C.-W. Effects of Capacity Ratios between Anode and Cathode on Electrochemical Properties for Lithium Polymer Batteries. *Electrochim. Acta* **2015**, *155*, 431–436. [CrossRef]
56. Arai, J.; Nakahigashi, R.; Sugiyama, T. Study on Cycle Life of Lithium-Ion Batteries Using in Situ^7 li Solid-State Nuclear Magnetic Resonance. *ECS Trans.* **2015**, *MA2015-03*, 449. [CrossRef]
57. Grimsmann, F.; Brauchle, F.; Gerbert, T.; Gruhle, A.; Parisi, J.; Knipper, M. Impact of different aging mechanisms on the thickness change and the quick-charge capability of lithium-ion cells. *J. Energy Storage* **2017**, *14*, 158–162. [CrossRef]
58. Petzl, M.; Kasper, M.; Danzer, M.A. Lithium plating in a commercial lithium-ion battery—A low-temperature aging study. *J. Power Sources* **2015**, *275*, 799–807. [CrossRef]
59. Hossain, S.; Kim, Y.-K.; Saleh, Y.; Loutfy, R. Overcharge studies of carbon fiber composite-based lithium-ion cells. *J. Power Sources* **2006**, *161*, 640–647. [CrossRef]
60. Purushothaman, B.K.; Landau, U. Rapid Charging of Lithium-Ion Batteries Using Pulsed Currents A Theoretical Analysis. *J. Electrochem. Soc.* **2006**, *153*, A533–A542. [CrossRef]
61. Arora, P.; Doyle, M.; White, R.E. Mathematical Modeling of the Lithium Deposition Overcharge Reaction in Lithium-Ion Batteries Using Carbon-Based Negative Electrodes. *J. Electrochem. Soc.* **1999**, *146*, 3543–3553. [CrossRef]
62. Burns, J.C.; Stevens, D.A.; Dahn, J.R. In-Situ Detection of Lithium Plating Using High Precision Coulometry. *J. Electrochem. Soc.* **2015**, *162*, A959–A964. [CrossRef]
63. Aurbach, D. A short review of failure mechanisms of lithium metal and lithiated graphite anodes in liquid electrolyte solutions. *Solid State Ion.* **2002**, *148*, 405–416. [CrossRef]
64. An, S.J.; Li, J.; Daniel, C.; Mohanty, D.; Nagpure, S.; Wood, D.L. The state of understanding of the lithium-ion-battery graphite solid electrolyte interphase (SEI) and its relationship to formation cycling. *Carbon* **2016**, *105*, 52–76. [CrossRef]
65. Palacín, M.R. Understanding ageing in Li-ion batteries: A chemical issue. *Chem. Soc. Rev.* **2018**, *47*, 4924–4933. [CrossRef] [PubMed]
66. Bhattacharyya, R.; Key, B.; Chen, H.; Best, A.S.; Hollenkamp, A.F.; Grey, C.P. In situ NMR observation of the formation of metallic lithium microstructures in lithium batteries. *Nat. Mater.* **2010**, *9*, 504–510. [CrossRef] [PubMed]
67. Aurbach, D.; Zinigrad, E.; Teller, H.; Dan, P. Factors Which Limit the Cycle Life of Rechargeable Lithium (Metal) Batteries. *J. Electrochem. Soc.* **2000**, *147*, 1274. [CrossRef]
68. Streich, D.; Guéguen, A.; Mendez, M.; Chesneau, F.; Novák, P.; Berg, E.J. Online Electrochemical Mass Spectrometry of High Energy Lithium Nickel Cobalt Manganese Oxide/Graphite Half- and Full-Cells with Ethylene Carbonate and Fluoroethylene Carbonate Based Electrolytes. *J. Electrochem. Soc.* **2016**, *163*, A964–A970. [CrossRef]
69. Liu, Q.Q.; Xiong, D.J.; Petibon, R.; Du, C.Y.; Dahn, J.R. Gas Evolution during Unwanted Lithium Plating in Li-Ion Cells with EC-Based or EC-Free Electrolytes. *J. Electrochem. Soc.* **2016**, *163*, A3010–A3015. [CrossRef]
70. Feng, X.; Fang, M.; He, X.; Ouyang, M.; Lu, L.; Wang, H.; Zhang, M. Thermal runaway features of large format prismatic lithium ion battery using extended volume accelerating rate calorimetry. *J. Power Sources* **2014**, *255*, 294–301. [CrossRef]
71. Fleischhammer, M.; Waldmann, T.; Bisle, G.; Hogg, B.-I.; Wohlfahrt-Mehrens, M. Interaction of cyclic ageing at high-rate and low temperatures and safety in lithium-ion batteries. *J. Power Sources* **2015**, *274*, 432–439. [CrossRef]
72. Yuan, Q.; Zhao, F.; Wang, W.; Zhao, Y.; Liang, Z.; Yan, D. Overcharge failure investigation of lithium-ion batteries. *Electrochim. Acta* **2015**, *178*, 682–688. [CrossRef]
73. Zhu, X.; Wang, Z.; Wang, Y.; Wang, H.; Wang, C.; Tong, L.; Yi, M. Overcharge investigation of large format lithium-ion pouch cells with $Li(Ni_{0.6}Co_{0.2}Mn_{0.2})O_2$ cathode for electric vehicles: Thermal runaway features and safety management method. *Energy* **2019**, *169*, 868–880. [CrossRef]

74. Hestenes, J.C.; Sadowski, J.T.; May, R.; Marbella, L.E. Transition Metal Dissolution Mechanisms and Impacts on Electronic Conductivity in Composite $LiNi_{0.5}Mn_{1.5}O_4$ Cathode Films. *ACS Mater. Au* **2023**, *3*, 88–101. [CrossRef]
75. Yamaki, J.-I.; Tobishima, S.-I. Rechargeable lithium anodes. In *Handbook of Battery Materials*; Daniel, C., Besenhard, J.O., Besenhard, J.O., Eds.; Wiley: Weinheim, Germany, 2012; pp. 377–404, ISBN 9783527326952.
76. Gireaud, L.; Grugeon, S.; Laruelle, S.; Yrieix, B.; Tarascon, J.-M. Lithium metal stripping/plating mechanisms studies: A metallurgical approach. *Electrochem. Commun.* **2006**, *8*, 1639–1649. [CrossRef]
77. Aurbach, D.; Cohen, Y. Morphological Studies of Li Deposition Processes in LiAsF6/PC Solutions by In Situ Atomic Force Microscopy. *J. Electrochem. Soc.* **1997**, *144*, 3355–3360. [CrossRef]
78. Ohsaki, T.; Kishi, T.; Kuboki, T.; Takami, N.; Shimura, N.; Sato, Y.; Sekino, M.; Satoh, A. Overcharge reaction of lithium-ion batteries. *J. Power Sources* **2005**, *146*, 97–100. [CrossRef]
79. Nowak, S.; Winter, M. Review—Chemical Analysis for a Better Understanding of Aging and Degradation Mechanisms of Non-Aqueous Electrolytes for Lithium Ion Batteries: Method Development, Application and Lessons Learned. *J. Electrochem. Soc.* **2015**, *162*, A2500–A2508. [CrossRef]
80. Chen, S.; Wang, Z.; Yan, W. Identification and characteristic analysis of powder ejected from a lithium ion battery during thermal runaway at elevated temperatures. *J. Hazard. Mater.* **2020**, *400*, 123169. [CrossRef] [PubMed]
81. Wang, S.; Rafiz, K.; Liu, J.; Jin, Y.; Lin, J.Y.S. Effects of lithium dendrites on thermal runaway and gassing of $LiFePO_4$ batteries. *Sustain. Energy Fuels* **2020**, *4*, 2342–2351. [CrossRef]
82. Feng, X.; Ouyang, M.; Liu, X.; Lu, L.; Xia, Y.; He, X. Thermal runaway mechanism of lithium ion battery for electric vehicles: A review. *Energy Storage Mater.* **2018**, *10*, 246–267. [CrossRef]
83. Jiang, L.; Luo, Z.; Wu, T.; Shao, L.; Sun, J.; Liu, C.; Li, G.; Cao, K.; Wang, Q. Overcharge Behavior and Early Warning Analysis of $LiNi_{0.5}Co_{0.2}Mn_{0.3}O_2$/C Lithium-Ion Battery with High Capacity. *J. Electrochem. Soc.* **2019**, *166*, A1055–A1062. [CrossRef]
84. Feng, X.; He, X.; Ouyang, M.; Wang, L.; Lu, L.; Ren, D.; Santhanagopalan, S. A Coupled Electrochemical-Thermal Failure Model for Predicting the Thermal Runaway Behavior of Lithium-Ion Batteries. *J. Electrochem. Soc.* **2018**, *165*, A3748–A3765. [CrossRef]
85. Richter, F.; Kjelstrup, S.; Vie, P.J.; Burheim, O.S. Thermal conductivity and internal temperature profiles of Li-ion secondary batteries. *J. Power Sources* **2017**, *359*, 592–600. [CrossRef]
86. Oehler, D.; Bender, J.; Seegert, P.; Wetzel, T. Investigation of the Effective Thermal Conductivity of Cell Stacks of Li-Ion Batteries. *Energy Technol.* **2021**, *9*, 2000722. [CrossRef]
87. Wassiliadis, N.; Steinsträter, M.; Schreiber, M.; Rosner, P.; Nicoletti, L.; Schmid, F.; Ank, M.; Teichert, O.; Wildfeuer, L.; Schneider, J.; et al. Quantifying the state of the art of electric powertrains in battery electric vehicles: Range, efficiency, and lifetime from component to system level of the Volkswagen ID.3. *eTransportation* **2022**, *12*, 100167. [CrossRef]
88. Kurmaev, R.K.; Struchkov, V.S.; Novak, V.V. Experience in the development of an effective thermal management system for the high-voltage battery of the vehicle. *IOP Conf. Ser. Mater. Sci. Eng.* **2020**, *819*, 12020. [CrossRef]
89. Parrish, R.; Elankumaran, K.; Gandhi, M.; Nance, B.; Meehan, P.; Milburn, D.; Siddiqui, S.; Brenz, A. Voltec Battery Design and Manufacturing. In Proceedings of the SAE 2011 World Congress & Exhibition, Detroit, MI, USA, 12–14 April 2011; SAE Technical Paper. SAE International: Warrendale, PA, USA, 2011. [CrossRef]
90. Schmidt, A.; Oehler, D.; Weber, A.; Wetzel, T.; Ivers-Tiffée, E. A multi scale multi domain model for large format lithium-ion batteries. *Electrochim. Acta* **2021**, *393*, 139046. [CrossRef]
91. Darcovich, K.; MacNeil, D.D.; Recoskie, S.; Cadic, Q.; Ilinca, F. Comparison of cooling plate configurations for automotive battery pack thermal management. *Appl. Therm. Eng.* **2019**, *155*, 185–195. [CrossRef]
92. Loges, A.; Herberger, S.; Seegert, P.; Wetzel, T. A study on specific heat capacities of Li-ion cell components and their influence on thermal management. *J. Power Sources* **2016**, *336*, 341–350. [CrossRef]
93. Paarmann, S.; Schuld, K.; Wetzel, T. Inhomogeneous Aging in Lithium-Ion Batteries Caused by Temperature Effects. *Energy Technol.* **2022**, *10*, 2200384. [CrossRef]
94. Werner, D.; Paarmann, S.; Wiebelt, A.; Wetzel, T. Inhomogeneous Temperature Distribution Affecting the Cyclic Aging of Li-Ion Cells. Part II: Analysis and Correlation. *Batteries* **2020**, *6*, 12. [CrossRef]
95. Kong, L.; Aalund, R.; Alipour, M.; Stoliarov, S.I.; Pecht, M. Evaluating the Manufacturing Quality of Lithium Ion Pouch Batteries. *J. Electrochem. Soc.* **2022**, *169*, 40541. [CrossRef]
96. *IEC 62620*; Secondary Cells and Batteries Containing Alkaline or Other Non-Acid Electrolytes—Secondary Lithium Cells and Batteries for Use in Industrial Applications. International Electrotechnical Commission (IEC): Geneva, Switzerland, 2014.
97. Onda, K.; Kameyama, H.; Hanamoto, T.; Ito, K. Experimental Study on Heat Generation Behavior of Small Lithium-Ion Secondary Batteries. *J. Electrochem. Soc.* **2003**, *150*, A285. [CrossRef]
98. Abdul-Quadir, Y.; Laurila, T.; Karppinen, J.; Jalkanen, K.; Vuorilehto, K.; Skogström, L.; Paulasto-Kröckel, M. Heat generation in high power prismatic Li-ion battery cell with $LiMnNiCoO_2$ cathode material. *Int. J. Energy Res.* **2014**, *38*, 1424–1437. [CrossRef]
99. Alihosseini, Y.; Zabetian Targhi, M.; Heyhat, M.M.; Ghorbani, N. Effect of a micro heat sink geometric design on thermo-hydraulic performance: A review. *Appl. Therm. Eng.* **2020**, *170*, 114974. [CrossRef]
100. Liu, H.; Li, P. Even distribution/dividing of single-phase fluids by symmetric bifurcation of flow channels. *Int. J. Heat Fluid Flow* **2013**, *40*, 165–179. [CrossRef]

101. Li, P.; Coopamah, D.; Ki, J.-P. Uniform distribution of species in fuel cells using a multiple flow bifurcation design. In Proceedings of the ASME 2008 6th International Conference on Fuel Cell Science, Engineering and Technology, Denver, CO, USA, 16–18 June 2008; pp. 897–902. [CrossRef]
102. Kandlikar, S.G.; Grande, W.J. Evolution of Microchannel Flow Passages—Thermohydraulic Performance and Fabrication Technology. *Heat Transf. Eng.* **2003**, *24*, 3–17. [CrossRef]
103. López Bonilla, J.; Fond, B.; Graichen, H.; Hamann, J.; Beyrau, F.; Boye, G. Thermal characterization of high-performance battery cells during charging and discharging using optical temperature measurement methods. In Proceedings of the FISITA World Congress 2021—Technical Programme, Prague, Czech Republic, 13–17 September 2021; ISBN 9781916025929.
104. Graichen, H.-C.; Sauerhering, J.; Stamann, O.; Beyrau, F.; Boye, G. Influence of Adhesive Tapes as Thermal Interface Materials on the Thermal Load of a Compact Electrical Machine. *WEVJ* **2022**, *13*, 42. [CrossRef]
105. Graichen, H.-C.; Sauerhering, J.; Reuber, F.; Dlugosch, A.; Beyrau, F.; Chinni, M.; Boye, G. Advanced thermal management optimization of a liquid cooled compact electrical engine with air gap winding. In Proceedings of the FISITA World Congress 2021—Technical Programme, Prague, Czech Republic, 13–17 September 2021; ISBN 9781916025929.
106. Feng, X.; Ren, D.; He, X.; Ouyang, M. Mitigating Thermal Runaway of Lithium-Ion Batteries. *Joule* **2020**, *4*, 743–770. [CrossRef]
107. Vorwerk, P.; Hahn, S.-K.; Daniel, C.; Krause, U.; Keutel, K. Detection of Critical Conditions in Pouch Cells Based on Their Expansion Behavior. *Batteries* **2022**, *8*, 42. [CrossRef]
108. Liu, P.; Yang, L.; Xiao, B.; Wang, H.; Li, L.; Ye, S.; Li, Y.; Ren, X.; Ouyang, X.; Hu, J.; et al. Revealing Lithium Battery Gas Generation for Safer Practical Applications. *Adv. Funct. Mater.* **2022**, *32*, 2208586. [CrossRef]
109. Lin, H.; Chua, D.; Salomon, M.; Shiao, H.-C.; Hendrickson, M.; Plichta, E.; Slane, S. Low-Temperature Behavior of Li-Ion Cells. *Electrochem. Solid-State Lett.* **2001**, *4*, A71. [CrossRef]
110. Huang, C.-K.; Sakamoto, J.S.; Wolfenstine, J.; Surampudi, S. The Limits of Low-Temperature Performance of Li-Ion Cells. *J. Electrochem. Soc.* **2000**, *147*, 2893. [CrossRef]
111. Yaqub, A.; Lee, Y.-J.; Hwang, M.J.; Pervez, S.A.; Farooq, U.; Choi, J.-H.; Kim, D.; Choi, H.-Y.; Cho, S.-B.; Doh, C.-H. Low temperature performance of graphite and $LiNi_{0.6}Co_{0.2}Mn_{0.2}O_2$ electrodes in Li-ion batteries. *J. Mater. Sci.* **2014**, *49*, 7707–7714. [CrossRef]
112. Li, J.; Yuan, C.F.; Guo, Z.H.; Zhang, Z.A.; Lai, Y.Q.; Liu, J. Limiting factors for low-temperature performance of electrolytes in $LiFePO_4$/Li and graphite/Li half cells. *Electrochim. Acta* **2012**, *59*, 69–74. [CrossRef]
113. Misiewicz, C.; Lundström, R.; Ahmed, I.; Lacey, M.J.; Brant, W.R.; Berg, E.J. Online electrochemical mass spectrometry on large-format Li-ion cells. *J. Power Sources* **2023**, *554*, 232318. [CrossRef]

batteries

Article

Organic and Inorganic Hybrid Composite Phase Change Material for Inhibiting the Thermal Runaway of Lithium-Ion Batteries

Jie Mei [1,2], Guoqing Shi [1,2,*], He Liu [1,2] and Zhi Wang [1,2]

1 College of Safety Engineering, China University of Mining and Technology, Xuzhou 221116, China; jmei@cumt.edu.cn (J.M.); liuhe@cumt.edu.cn (H.L.); zhiwang@cumt.edu.cn (Z.W.)
2 State Key Laboratory of Coal Resources and Safe Mining, China University of Mining and Technology, Xuzhou 221116, China
* Correspondence: sgq@cumt.edu.cn

Abstract: To deal with the flammability of PA (paraffin), this paper proposes a CPCM (composite phase change material) with a high heat-absorbing capacity for mitigating the thermal runaway of lithium-ion batteries. Two heating power levels were used to trigger thermal runaway in order to investigate the influence of heating power on thermal runaway characteristics and the mitigation effect of the PCM (phase change material). Thermal runaway processes and temperature changes were recorded. The results showed that heating results in a violent reaction of the battery, generating a high temperature and a bright flame, and the burning of PA increases the duration of a steady flame, indicating an increased threat. SA (sodium acetate trihydrate) effectively inhibited PA combustion, and the combustion time was reduced by 40.5%. PA/SA effectively retarded the rise in temperature of the battery, and the temperature rise rate was reduced by 87.3%. Increased heating power caused faster thermal runaway, and the thermal runaway mitigation effect of the CPCM was dramatically reduced. This study may provide a reference for the safe design and improvement of thermal management systems.

Keywords: lithium-ion battery; paraffin; thermal runaway; phase change material; heat absorption

Citation: Mei, J.; Shi, G.; Liu, H.; Wang, Z. Organic and Inorganic Hybrid Composite Phase Change Material for Inhibiting the Thermal Runaway of Lithium-Ion Batteries. *Batteries* **2023**, *9*, 513. https://doi.org/10.3390/batteries9100513

Academic Editors: Carlos Ziebert and Thomas Wetzel

Received: 11 September 2023
Revised: 10 October 2023
Accepted: 16 October 2023
Published: 17 October 2023

1. Introduction

Because of their high energy and power density, lithium-ion batteries have been used in a variety of applications in recent decades, including electric vehicles, large-scale energy storage, and power grids [1–3]. However, the long-term operation of batteries at a high temperature reduces their lifetime and may induce thermal runaway, leading to great danger [4–6]. As a result, the construction of a thermal management system for lithium-ion batteries has been proposed in order to assure the safety of the batteries [7–9].

PA (paraffin) is considered an ideal thermal management material for lithium-ion batteries due to its suitable phase change temperature, large latent heat and corrosion resistance. Abbas and An et al. respectively used PA as a PCM (phase change material) for the thermal management system of lithium-ion batteries and employed different liquid-cooled plates in combination with PA to form an efficient BTMS (battery thermal management system). The results showed that the heat-absorbing effect of PA suppressed the temperature rise of the batteries and gained a more homogeneous temperature distribution among the module, and the liquid-cooled plates further improved the thermal management performance of PA [10,11]. In order to improve the thermal management performance of PA, PA is blended with other materials to obtain improved properties, such as thermal conductivity, and this blend is called a composite phase change material. To address the low thermal conductivity of PA, Hussain et al. introduced PA into graphene-coated nickel and found that the thermal conductivity of the CPCM (composite phase change material) was significantly improved and the battery temperature was further reduced [12]. According to research conducted by Wang et al., a graphite powder/paraffin/nickel foam ternary CPCM can not only regulate the temperature increase of the battery surface, but can also decrease the heat dissipation of the battery at low temperatures to assure normal usage [13]. Zhao et al. designed a BTMS with a combination of a copper foam/paraffin CPCM and liquid cooling channels, showing that the introduction of liquid cooling reduces the maximum temperature of the battery while also leading to a bigger temperature difference; thus, a balance between the maximum temperature and temperature uniformity is required [14]. Greco et al. employed compressed expanded natural graphite to increase the thermal conductivity of paraffin and demonstrated that the composite performed significantly better than forced air cooling [15]. Kang et al. introduced a new high thermal conductivity and insulating CPCM made of PA and silicon carbide for the thermal management of lithium-ion batteries and found that the temperature of the battery pack was significantly reduced [16]. Wang et al. investigated the effects of PA with different melting points and different mass fractions of EG (expandable graphite) on the thermal management performance of a CPCM using numerical simulation and the results showed that 48 °C was a suitable melting point, whereas the larger the mass fraction of EG, the higher the battery temperature [17]. Chen et al. used EG and silicon carbide as thermally conductive materials, blending them with PA to form a CPCM, and found that the thermal conductivity could reach up to 4.086 $W/(m \cdot K)$, exhibiting an extremely high cooling efficiency [18]. The battery temperature was lowest with 9 fins, according to Chen et al.'s investigation of the influence of the number of fins on the thermal management performance of paraffin [19]. Fins and expanded graphite can significantly improve the thermal management performance of paraffin, according to research by Mei et al. [20].

However, batteries may still experience thermal runaway due to aging, design flaws, and the presence of some extreme conditions. Liu et al. showed that lithium plating is generated during the charging and discharging of lithium-ion batteries, which is highly likely to lead to thermal runaway [21]. According to research by Liu et al., even slight overcharging can cause the battery's internal resistance to rise, which in turn causes the battery to heat up more, eventually leading to thermal runaway [22]. Further research on the thermal runaway caused by overcharging was conducted by Mao et al. The findings indicated that after overcharging, the deposited lithium reacts with the electrolyte to produce a significant amount of heat, and the stability of the cathode material also diminishes [23]. Hu et al. investigated the thermal runaway characteristics of electrical abuse and showed that with an increasing charge rate, the battery presents a higher thermal risk, with a lower onset temperature and higher maximum temperature [24]. Liu et al. investigated the thermal runaway of lithium-ion batteries under overcharging and showed that high-current overcharging results in two violent combustion–explosion reactions [25]. External heat sources are another frequent cause of thermal runaway in lithium-ion batteries, which not only produce a great deal of heat, but also result in the production of hazardous combustible and toxic gases [26–29]. Huang et al. investigated the effect of heating power on the thermal runaway of batteries, and the results showed that with the increase in heating power, the onset time advances and the heat release rate increases [30]. When Md Said et al. simulated the effects of the mechanical impact on lithium-ion batteries, they found that the battery was rapidly damaged and produced an extremely high temperature [31]. Therefore, even though the BTMS regulates the rise in battery temperature, thermal runaway may still occur, and it is necessary to investigate the impact of a PCM during thermal runaway. In the case of a battery's thermal runaway, PA, a flammable PCM, can increase the threat. Zhang et al. found that a composite phase change material consisting of paraffin and expanded graphite helps dissipate heat, but thermal runaway propagates more widely once it has occurred [32]. Dai et al. conducted thermal runaway experiments on a 18650 Li-ion battery containing PA and found that the heat-absorbing effect of PA prolongs the onset of thermal runaway, but its flammability greatly increases the heat release, and the use of a mixture of flame retardants can reduce the heat release rate [33]. Weng et al. showed that the addition of EG to PA significantly suppresses the combustion flame, but accelerates the propagation of thermal runaway [34]. Huang et al. used a composite flame retardant in combination with paraffin wax and other materials to form a new type of flame-retardant flexible CPCM and discovered that the addition of 15 wt% is able to inhibit heat diffusion and resist flames, achieving a better flame retardant effect [35]. Graphene-enhanced hybrids are effective at preventing the spread of thermal runaway, according to Wang et al. [36]. However, there are few detailed studies on the impact of PA on the thermal runaway of lithium-ion battery processes, and the use of a CPCM for inhibiting the flammability of PA and the thermal runaway threat is rarely investigated. In addition, the effect of different heating powers on thermal runaway and validation of the effectiveness of a CPCM have rarely been reported.

Therefore, in this paper, the inorganic material SA (sodium acetate trihydrate) was mixed with PA (paraffin) to form a CPCM, and its inhibitory effect on PA's flammability, as well as the effect on the thermal runaway of batteries, was investigated. Due to the high latent heat and non-flammability of SA, the evaporation of its water of crystallization absorbs heat, which theoretically inhibits the combustion of PA and reduces the threat of a battery's thermal runaway. Additionally, the effectiveness of the CPCM was evaluated at various heat intensities, and data on the thermal runaway process, flame, and battery temperature were gathered and analyzed. Some characteristic parameters such as the temperature rise rate, temperature rise time after thermal runaway, and peak battery temperature were compared and analyzed. This study serves as a guide for creating a secure PCM lithium-ion BTMS (battery temperature management system).

2. Experimental Section

2.1. Materials

The battery used in this study, a Sony US18650VTC5, had a 2600 mAh capacity, weighed approximately 45.4 g, and had discharge and charge cutoff voltages of 2.0 and 4.2 V, respectively. The positive electrode of the battery was nickel–manganese–cobalt ternary material and the negative electrode was graphite. Before testing, the battery was cycled three times, and was then fully charged to 100% SOC to show the greatest hazard of thermal runaway. The two PCMs used in this paper were PA and SA, and their detailed parameters are shown in Table 1. The PA was supplied by Henan Baihuali Chemical Product Co., Ltd., Zhengzhou, China, and its phase transition temperature was 48–50 °C, with a latent heat of 136 J/g. The SA was supplied by Sinopharm Chemical Reagent Co., Ltd., Shanghai, China, and its phase transition temperature was slightly higher than that of the PA at 58 °C, with a latent heat of 260 J/g, which was nearly twice as much as that of the PA. In addition, the SA contained crystal water, which can absorb a large amount of heat in the case of thermal runaway, thus lowering the battery temperature.

Table 1. Physical and thermochemical properties of the materials.

Name	Molecular Formula	ρ/(g/cm^3)	Melting Point/°C	Boiling Point/°C	Specific Heat Capacity (J/(kg·°C))
Paraffin	C_nH_{2n+2}	0.88	45–48	322	2140
SA	$CH_3COONa{\cdot}3H_2O$	1.45	58	400	1970

2.2. Experimental Setup

Figure 1 depicts a schematic of the experimental setup used to study thermal runaway. The heater was the same size as the battery and provided 200 and 500 W of heating power, respectively. The heater stopped heating after the battery suffered thermal runaway. The battery was placed in a battery tube made of an Mg–Al (magnesium–aluminum) alloy (type 5052) with a thermal conductivity of 138 W/m·°C, a diameter of 36 mm, a height of 65 mm, and a thickness of 2 mm. There were three screws (8 mm in diameter) on the top and bottom of the tube to hold the battery in place. To measure the temperature change during thermal runaway, thermocouples were placed in the middle of the battery surface, as well as 5 and 15 cm above the battery. The flame temperatures at 5 and 15 cm above the battery were termed T_{f1} and T_{f2}, respectively, while the temperature at the battery surface was assigned the designation T_b. The SA and PA were placed up-and-down to maximize the heat absorption effect of the SA, with the SA occupying a third of the volume. The temperature of the material was denoted T_m. K-type thermocouples were utilized to measure the temperature, and C-7018 was used to record data. Digital cameras were used to record the thermal runaway process.

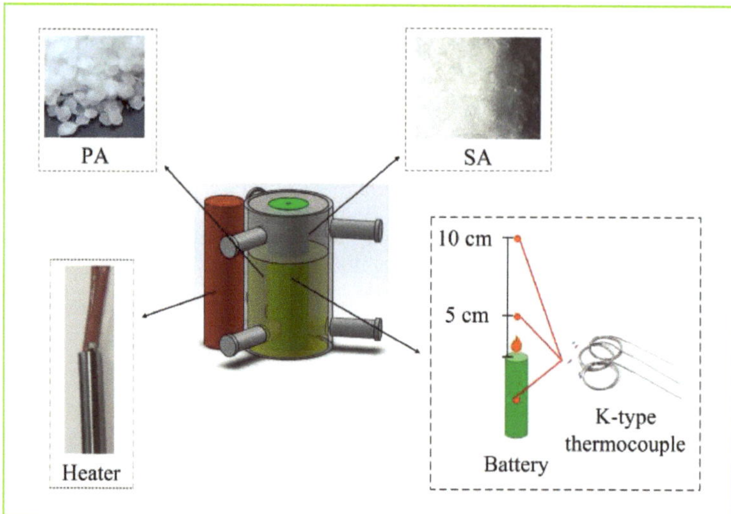

Figure 1. Schematic diagram of the experimental platform.

2.3. Experimental Conditions

The tests were categorized into three groups, i.e., Group I (single battery), Group II (PA), and Group III (SA/PA). Excluding the volume occupied by the fixation screws, the amount of PA added alone was 14.55 g, whereas the amount of PA and SA added to the CPCM was 9.7 and 7.99 g, respectively. The impact of heating power on the thermal runaway was examined using two heating powers, 200 and 500 W. Each test was performed at least three times and the relative standard deviations of characteristic parameters were calculated for all groups. A smaller relative standard deviation means more accurate experimental data. The standard deviations of the characteristic times and temperature during thermal runaway were less than 15%, so the experimental results are highly consistent and reliable.

3. Results and Discussions

3.1. Thermal Runaway Behaviors of a Single Battery

The thermal runaway process of a Group I with a single battery is depicted in Figure 2a, which can be divided into four stages: Heating, jet spark, steady flame, and extinguishing. It can be seen that the thermal runaway of the battery generated a great danger—jet sparks and flame combustion. The t_{com} (the duration of a stable combustion stage) of a single battery was 10 s, and the main burning substances were internal electrolyte and separator, alongside other combustible substances, with the combustion generating a great threat to the surrounding environment. According to Figure 2b, the battery suffered more severe damage following thermal runaway, as the top safety valve ruptured, and the battery case contained more carbon black as a result of combustion.

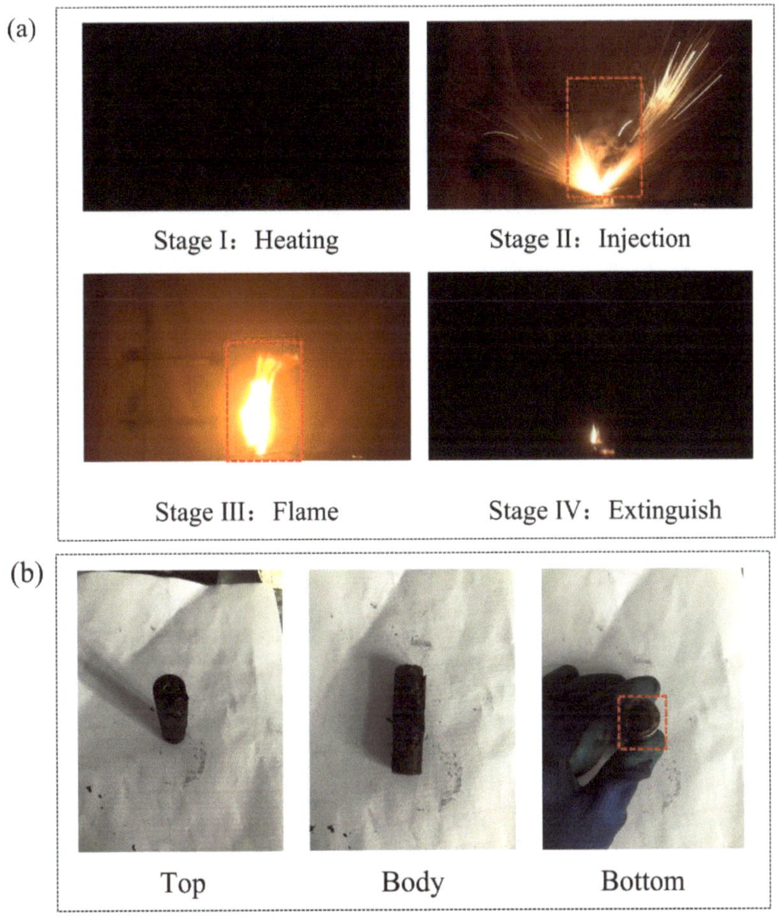

Figure 2. Single battery thermal runaway process and residues: (**a**) The thermal runaway process; (**b**) the residues.

As shown in Figure 3a, the battery heated up to 133.9 °C after 393 s, at which point it entered the thermal runaway stage, causing a sudden rise in T_b, jet sparks, and a continuous flame. Around 432 s, T_b reached a peak value of 681.9 °C and then decreased, indicating the ending of the thermal runaway. Due to the heater rod's constant heating, the battery experienced thermal runaway, which was characterized by an extraordinarily high temperature and a potentially dangerous flame. The $T_{f2\text{-}max}$ (maximum value of T_{f2}) was 1031.9 °C, which was much higher than T_{f1}, indicating that the battery flame height was above 15 cm, which would cause serious damage to the surroundings and even a secondary disaster.

3.2. Effect of the PCMs on Thermal Runaway

The thermal runaway process and residues of the battery for Groups II (PA) and III (PA/SA) are shown in Figure 4. It can be seen that the addition of PA provided a more violent injection process and a higher flame than the single battery, while the process of PA/SA was almost the same as that of a single battery. A probable reason for violent jetting is that the heat absorption led to a reduction in the internal reaction rate of the battery, active combustibles accumulated inside the battery, and when the temperature reached the onset temperature of thermal runaway, the internal combustibles were instantly released, leading to a more violent jetting. The t_{com} for Groups II (PA) and III (PA/SA) was 42 and 25 s, respectively, while the t_{com} for a single battery was 10 s. The thermal runaway was thus made more harmful by the addition of PA, whereas the t_{com} of PA/SA reduced by 40.5% compared to that of PA, demonstrating SA's efficiency in inhibiting PA combustion. Both PCM additions provided relatively well-preserved battery shells, as shown by the residue plots. Two potential explanations for this include the fact that PA has a relatively low boiling point and would not cause significant damage to the battery case and that the heat absorption ability of PA prevents the breakdown of the battery case.

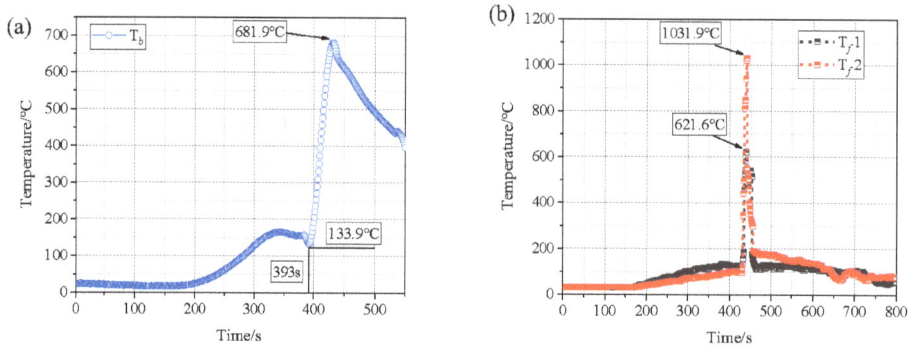

Figure 3. Temperature of Group I: (**a**) The battery surface temperature; (**b**) the flame temperature (redrawn from [37]).

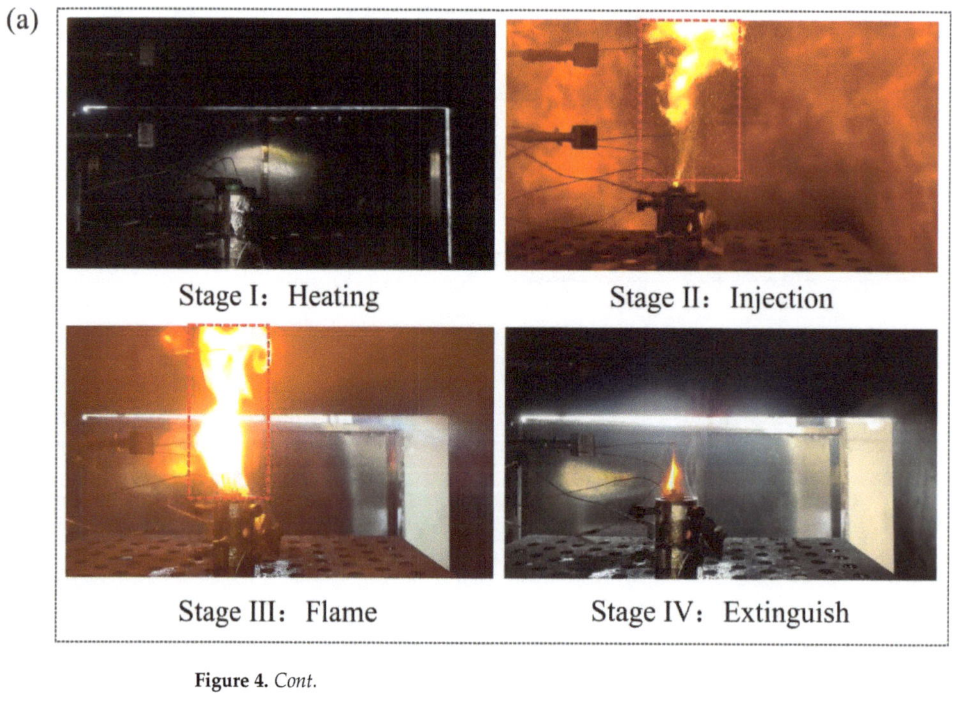

Figure 4. *Cont.*

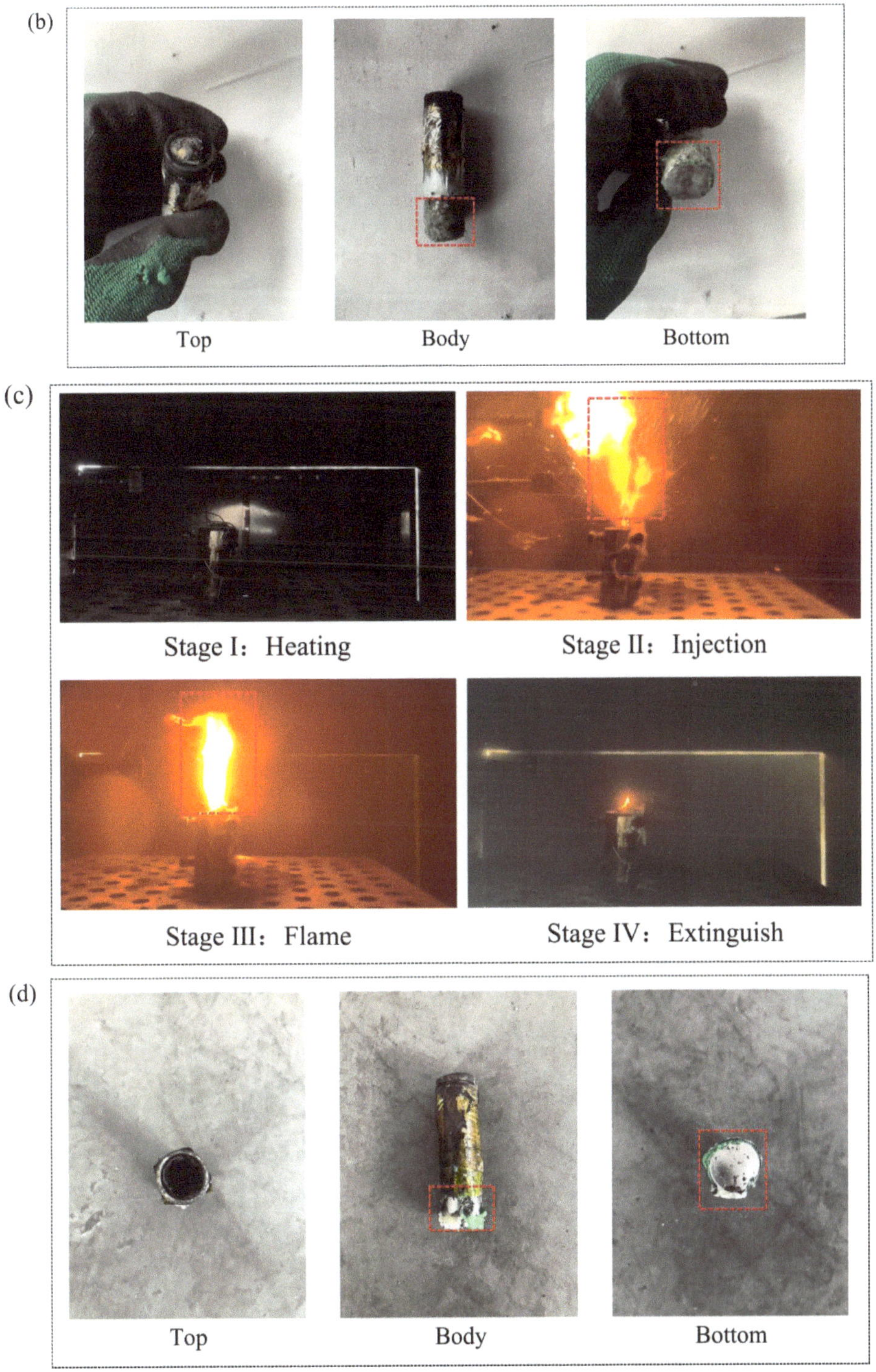

Figure 4. Thermal runaway process and residues of Groups II and III: (**a**,**c**) The thermal runaway process of Groups II and III; (**b**,**d**) the battery residues of Groups II and III.

Figure 5 shows the battery and flame temperature curves during thermal runaway for both PCMs. According to Figure 5a, the addition of PA increased the thermal runaway onset temperature by 33.5%, while also prolonging the onset time by 58.6%. The $T_{b\text{-}max}$ (maximum value of T_b) of Group II (PA) was 644.9 °C, which shows no significant change from the single battery. In comparison to pure PA, the addition of SA substantially delayed the thermal runaway onset time and temperature, with an increase of 6.5% in onset temperature and 47.1% in thermal runaway onset time; the $T_{b\text{-}max}$ dropped by 23.4% as well. The latent heat was calculated based on the properties of the two PCMs in the Materials section, with an ΔH (maximum latent heat absorption) of 1.98 kJ for PA and 3.4 kJ for SA/PA. As a result, the huge latent heat of PA was able to absorb the heat transferred by the heating rod, thus delaying the thermal runaway. The ΔH of SA/PA was increased by 41.7% more than that of PA, which resulted in a substantial increase in the thermal runaway onset time. Figure 5b demonstrates that, despite the prolonged steady combustion duration, PA had little impact on the flame threat, as seen by the tiny difference in the peak battery flame temperature of Groups II (PA) and I (single battery). It is noteworthy that the $T_{f2\text{-}max}$ of PA/SA was significantly increased, related to the fluctuation of the flame during the combustion process, resulting in both a higher T_{f1} and a higher T_{f2}.

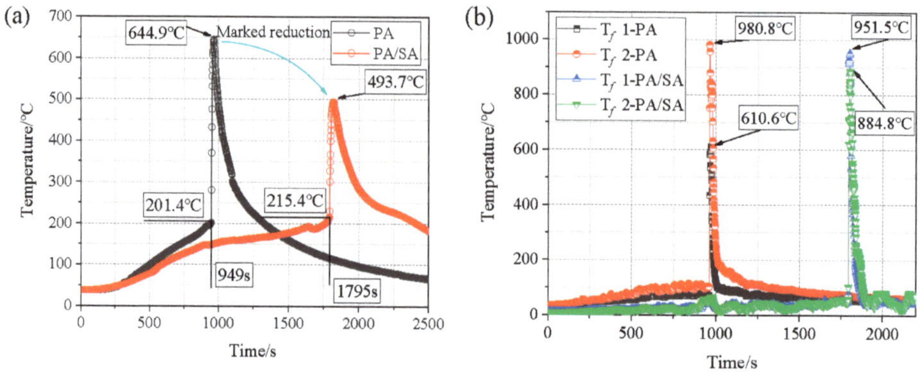

Figure 5. Temperature of Groups II and III: (**a**) The battery surface temperature; (**b**) the flame temperature.

The $t_{one\text{-}max}$ (time interval between thermal runaway onset and the maximum temperature) of the three groups is shown in Figure 6. The $t_{one\text{-}max}$ of Group I (single battery) was 39 s, which was the longest among the three groups, while the $t_{one\text{-}max}$ of Group II (PA) was only 16 s, which was reduced by 59%. Therefore, the PA was ignited after the battery thermal runaway, which accelerated the inside chemical reaction and caused the battery temperature to peak more quickly. The $t_{one\text{-}max}$ of Group III (PA/SA) was 37 s, which was 56.8% longer than that of Group II (PA), although slightly shorter than that of Group I (single battery). A significant quantity of heat was absorbed by SA due to its high latent heat, which, on the one hand prevented a temperature rise in the battery and, on the other, hindered the combustion of PA, slowing the battery temperature rise rate. Therefore, the introduction of SA for inhibiting PA combustion was effective and helpful in reducing the rise in battery temperature.

The material temperature of the two PCMs is shown in Figure 7a, indicating that the T_m of PA/SA was much lower than that of PA, which shows that the heat absorption effect of SA greatly lowered the material temperature of the CPCM, with a maximum reduction of 45.8%. The inhibition mechanism of SA is shown in Equations (1) and (2) [38,39]. Because of its enormous latent heat capacity and the contribution of its crystal water to absorb heat, the battery temperature decreased substantially and the thermal runaway onset time was greatly prolonged. Additionally, the concentration of combustibles could be reduced due to the evaporation of crystal water, which also partially prevented the thermal runaway reaction of the battery. Figure 7b shows the battery $T_{rate\text{-}ave}$ (average temperature rise rate) before thermal runaway of the three groups of experiments, and the $T_{rate\text{-}ave}$ of Group II (PA) was reduced by 38.0% more than that of Group I (single battery), while the $T_{rate\text{-}ave}$ of Group III (PA/SA) was only 0.1 °C/s, which was 87.3% and 44.4% lower than that of the single battery and PA, respectively. Therefore, the excellent heat absorption effect of SA greatly prevented the battery temperature rise, thus prolonging the thermal runaway onset, and SA also had an inhibitory effect on the combustion of PA, which decreased the material's temperature.

$$CH_3COONa{\cdot}3H_2O \underset{cooling(T>T_{mel})}{\overset{heating(T>T_{mel})}{\rightleftarrows}} CH_3COONa{\cdot}yH_2O + (3-y)H_2O \quad (1)$$

$$CH_3COONa{\cdot}3H_2O \underset{cooling(T>T_{mel})}{\overset{heating(T>T_{mel})}{\rightleftarrows}} CH_3COONa + 3H_2O \quad (2)$$

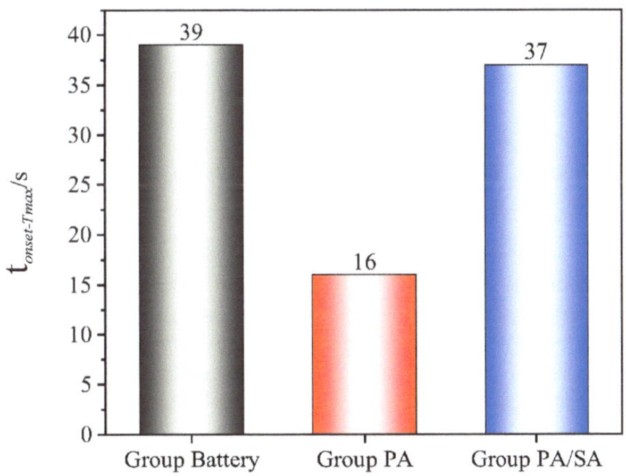

Figure 6. Time of the different groups from thermal runaway onset to maximum temperature of the battery.

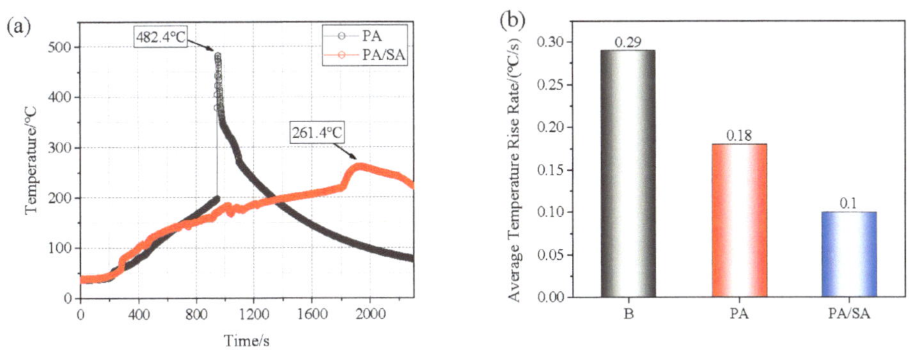

Figure 7. Temperature of the material and battery temperature rise rate of Groups II and III: (**a**) The material temperature; (**b**) the battery temperature rise rate before thermal runaway.

3.3. *Influence of Heating Power*

In actuality, overheating triggers thermal runaway accidents, and the heating power acting on the battery is usually random and variable. Different heating powers lead to variations in the characteristic time and temperature of the battery's thermal runaway, which may affect the thermal runaway suppression effect of the CPCM. To verify the suitability of the new CPCM at higher heating powers, thermal runaway experiments at a heating power of 500 W are presented in this section. Figure 8 shows the temperature curves of a single battery at 500 W, for which the onset of thermal runaway occurred 21.9% earlier than that at 200 W, while the $T_{b\text{-}max}$ and $T_{f\text{-}max}$ (maximum value of T_{f1} or T_{f2}) did not change significantly. Therefore, the increased heating power merely accelerated the onset time of thermal runaway, with less effect on the battery and flame temperature.

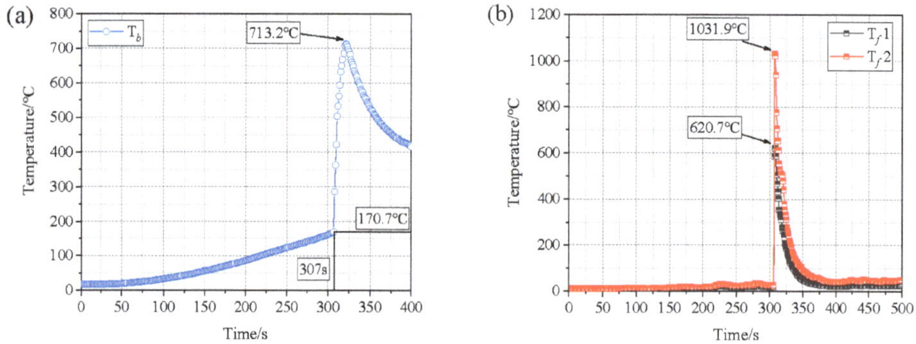

Figure 8. Temperature of Group I at 500 W: (**a**) The battery surface temperature; (**b**) the flame temperature.

Figure 9a shows that the thermal runaway onset time of Group II (PA) was 635 s, and the temperature was 215.3 °C. Comparing the results with those in Section 3.2, the thermal runaway onset temperature was less varied, while the thermal runaway onset time was reduced by 33.1%. The thermal runaway onset time was decreased by 54.7% in Group III (PA/SA), which showed a more noticeable difference. Increasing the heating power had little impact on the temperature of the battery or the flame. In addition, it is worth noting that the T_{f2} of the PA/SA at 500 W was significantly lower than that at 200 W. The main reason is that the greater heating power resulted in a more intense internal reaction, and thus the occurrence of a lower flame was greatly reduced.

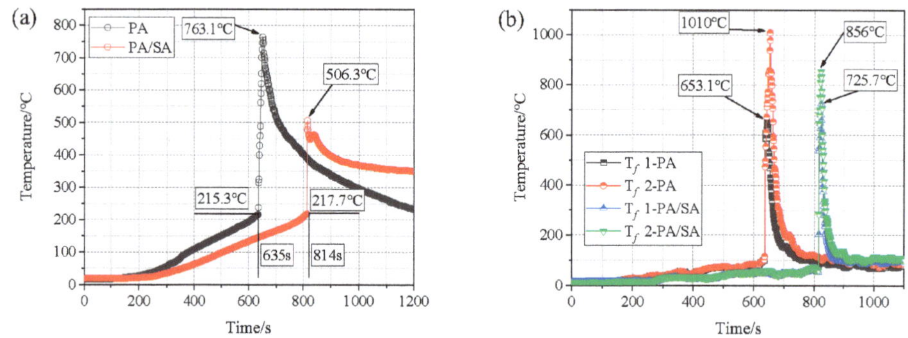

Figure 9. Temperature of Groups II and III at 500 W: (**a**) The battery surface temperature; (**b**) the flame temperature.

Figure 10 shows a comparison of the combustion duration and temperature rise rate for two heating powers, highlighting that an increased heating power is a greater threat. As seen in Figure 10a, the t_{com} of the single battery nearly remained unchanged, while that of Group II (PA) increased by 16%, indicating that a higher heating power causes more PA to ignite, leading to a bigger t_{com}. Group III (PA/SA) changed most significantly, with an increase of 37.5%, indicating that an increased heating power significantly weakens the inhibition of PA combustion by SA. Figure 10b shows that increasing the heating power leads to a rapid increase in the battery temperature, with $T_{rate\text{-}ave}$ increasing by 42% for a single battery, and 43.8% and 60% for PA and PA/SA, respectively. An increased heating power thus results in more severe thermal runaway consequences and also diminishes the heat-absorbing capabilities of both PCMs, decreasing the efficiency of thermal runaway retardation.

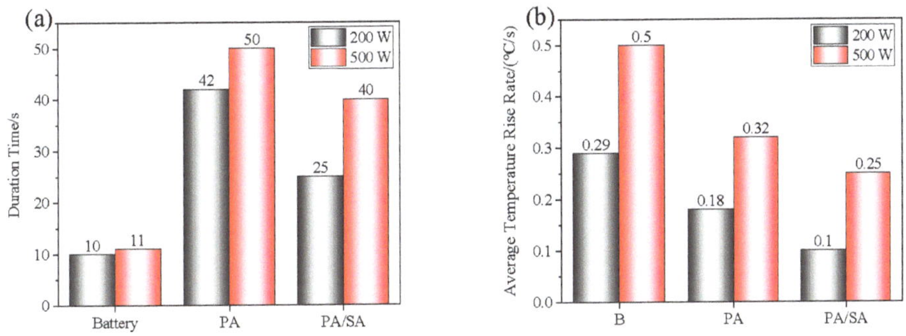

Figure 10. Comparison of two heating powers: (**a**) Burning duration; (**b**) battery temperature rise rate before thermal runaway.

Figure 11 shows the comparison of two heating powers after thermal runaway. The temperature of the single battery was significantly affected by the heating power; $t_{one\text{-}max}$ dropped by 64.1%, demonstrating that increasing the heating power made the internal reaction rate during thermal runaway faster and caused a slight increase in $T_{b\text{-}max}$, which was unfavorable for suppressing thermal runaway. While the 15.5% increase in $T_{b\text{-}max}$ demonstrated that a larger heating power led to a fuller combustion of the PA and the battery, which produced a higher peak temperature of the battery, the unchanged $t_{one\text{-}max}$ of Group II (PA) at both heating powers indicates that the accelerating effect of PA combustion on the rise in battery temperature was almost independent of the heating power. It is interesting to note that the $t_{one\text{-}max}$ for Group III (PA/SA) was only 2 s, indicating that the increased heating power significantly reduced the inhibition effect of PA/SA on the rise in the battery temperature after thermal runaway, which is also related to the increased heating power and the lower peak temperature of the battery. The T_b of Group III (PA/SA) was the smallest among the three groups, suggesting the efficiency of PA/SA in reducing the battery's peak temperatures, and it was nearly unaffected by the heating power, despite the rapid temperature rise of the battery following thermal runaway under 500 W heating.

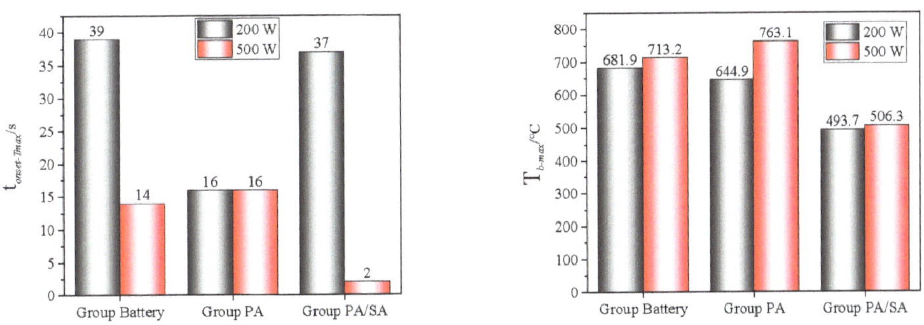

(**a**) (**b**)

Figure 11. Comparison of two heating powers: (**a**) $t_{one-max}$; (**b**) T_{b-max}.

4. Conclusions

To counter the thermal runaway of lithium-ion batteries and to reduce the flammable risk of PA, a new CPCM was proposed with an enhanced heat absorption capability to mitigate thermal runaway and to reduce the threat of PA. Heating rods of the same size as the battery were used to trigger thermal runaway. The combustion process and flame and battery surface temperatures during thermal runaway were recorded, and the thermal runaway mitigation mechanism of the CPCM was revealed. Additionally, comparisons and analyses of the characteristics of thermal runaway and the impact of the PCMs at various heating powers were conducted. The main conclusions of this study are summarized as follows:

(1) The thermal runaway of the battery was accompanied by violent combustion behaviors and a high temperature. The heat absorption of PA delayed the thermal runaway by 33.5%, but its flammable characteristic led to more violent combustion and a longer combustion duration. The t_{com} of Group II (PA) was 42 s, while that of Group I (single battery) was only 10 s.

(2) The heat absorption of SA reduced the combustion of PA, and Group III (PA/SA) had a 40.5% reduction in t_{com} than that of Group II (PA). The rise in battery temperature was significantly slowed down by PA/SA, and its $T_{rate\text{-}ave}$ was decreased by 87.3% and 44.4% in comparison to PA and a single battery, respectively. The temperature of the battery flame was essentially unaffected by either PCM.

(3) Although the battery and flame temperatures of the two PCM groups were essentially unaffected by the increasing heating power, the thermal runaway mitigation effect was significantly reduced, while the thermal runaway onset time was advanced by at least 33.1%. The heating power had a pronounced impact on SA/PA, with t_{com} and $T_{rate\text{-}ave}$ increasing by 37.5% and 60%, respectively. The fact that the $t_{one\text{-}max}$ for Group III (PA/SA) changed from 37 to 2 s further illustrates the tremendous impact of the heating power on the suppression effect of PA/SA on a battery' temperature rise.

Herein, an experimental study was conducted on the effect of PA on the thermal runaway of lithium-ion batteries and a CPCM was proposed for inhibiting the combustion of PA and mitigating the thermal runaway. However, this paper did not conduct a detailed study of different ratios of the CPCM, nor did it conduct experiments to verify its thermal management performance. Therefore, the future work requires an in-depth study of different material ratios and thermal management performance.

Author Contributions: Conceptualization, J.M. and G.S.; methodology, J.M. and H.L.; writing—original draft preparation, J.M.; writing—review and editing, J.M., G.S., H.L. and Z.W. All authors have read and agreed to the published version of the manuscript.

Funding: This research was funded by the National Natural Science Foundation of China (No. 52074279), the National Natural Science Foundation of China (No. 52204253), the Jiangsu Provincial Double-Innovation Doctor Program (No. 140923027), the Opening Foundation of Civil Aircraft Fire Science and Safety Engineering Key Laboratory of Sichuan Province (No. MZ2023KF06), and the Fundamental Research Funds for the Central Universities (No. 2022QN1009).

Data Availability Statement: All data are contained within the article.

Conflicts of Interest: The authors declare no conflict of interest.

References

1. Adnan, M. The Future of Energy Storage: Advancements and Roadmaps for Lithium-Ion Batteries. *Int. J. Mol. Sci.* **2023**, *24*, 7457. [CrossRef]
2. Li, M.; Lu, J.; Chen, Z.; Amine, K. 30 Years of Lithium-Ion Batteries. *Adv. Mater.* **2018**, *30*, 1800561. [CrossRef] [PubMed]
3. Kim, T.; Song, W.; Son, D.-Y.; Ono, L.K.; Qi, Y. Lithium-ion batteries: Outlook on present, future, and hybridized technologies. *J. Mater. Chem. A* **2019**, *7*, 2942–2964. [CrossRef]
4. Ouyang, D.; Weng, J.; Chen, M.; Wang, J. Impact of high-temperature environment on the optimal cycle rate of lithium-ion battery. *J. Energy Storage* **2020**, *28*, 101242. [CrossRef]

5. Leng, F.; Tan, C.M.; Pecht, M. Effect of Temperature on the Aging rate of Li Ion Battery Operating above Room Temperature. *Sci. Rep.* **2015**, *5*, 12967. [CrossRef]
6. Feng, X.; Sun, J.; Ouyang, M.; He, X.; Lu, L.; Han, X.; Fang, M.; Peng, H. Characterization of large format lithium ion battery exposed to extremely high temperature. *J. Power Sources* **2014**, *272*, 457–467. [CrossRef]
7. Lu, L.; Han, X.; Li, J.; Hua, J.; Ouyang, M. A review on the key issues for lithium-ion battery management in electric vehicles. *J. Power Sources* **2013**, *226*, 272–288. [CrossRef]
8. Ping, P.; Peng, R.; Kong, D.; Chen, G.; Wen, J. Investigation on thermal management performance of PCM-fin structure for Li-ion battery module in high-temperature environment. *Energy Convers. Manag.* **2018**, *176*, 131–146. [CrossRef]
9. Liu, J.; Chen, H.; Huang, S.; Jiao, Y.; Chen, M. Recent Progress and Prospects in Liquid Cooling Thermal Management System for Lithium-Ion Batteries. *Batteries* **2023**, *9*, 400. [CrossRef]
10. Abbas, S.; Ramadan, Z.; Park, C.W. Thermal performance analysis of compact-type simulative battery module with paraffin as phase-change material and flat plate heat pipe. *Int. J. Heat Mass Transf.* **2021**, *173*, 121269. [CrossRef]
11. An, Z.; Luo, Y.; Zhang, C.; Li, D. Performance of chocolate bar-shaped modular thermal management system combined metal lattice liquid-cooling plate with paraffin in high-rate discharge. *J. Energy Storage* **2022**, *56*, 106017. [CrossRef]
12. Hussain, A.; Abidi, I.H.; Tso, C.Y.; Chan, K.C.; Luo, Z.; Chao, C.Y.H. Thermal management of lithium ion batteries using graphene coated nickel foam saturated with phase change materials. *Int. J. Therm. Sci.* **2018**, *124*, 23–35. [CrossRef]
13. Wang, Z.; Du, C.; Qi, R.; Wang, Y. Experimental study on thermal management of lithium-ion battery with graphite powder based composite phase change materials covering the whole climatic range. *Appl. Therm. Eng.* **2022**, *216*, 119072. [CrossRef]
14. Zhao, Y.; Zou, B.; Ding, J.; Ding, Y. Experimental and numerical investigation of a hybrid battery thermal management system based on copper foam-paraffin composite phase change material and liquid cooling. *Appl. Therm. Eng.* **2023**, *218*, 119312. [CrossRef]
15. Greco, A.; Jiang, X.; Cao, D. An investigation of lithium-ion battery thermal management using paraffin/porous-graphite-matrix composite. *J. Power Sources* **2015**, *278*, 50–68. [CrossRef]
16. Kang, W.; Zhao, Y.; Jia, X.; Hao, L.; Dang, L.; Wei, H. Paraffin/SiC as a Novel Composite Phase-Change Material for a Lithium-Ion Battery Thermal Management System. *Trans. Tianjin Univ.* **2021**, *27*, 55–63. [CrossRef]
17. Wang, S.; Zhang, D.; Li, C.; Wang, J.; Zhang, J.; Cheng, Y.; Mei, W.; Cheng, S.; Qin, P.; Duan, Q.; et al. Numerical optimization for a phase change material based lithium-ion battery thermal management system. *Appl. Therm. Eng.* **2023**, *222*, 119839. [CrossRef]
18. Chen, M.; Zhang, S.; Zhao, L.; Weng, J.; Ouyang, D.; Chen, Q.; Kong, Q.; Wang, J. Preparation of thermally conductive composite phase change materials and its application in lithium-ion batteries thermal management. *J. Energy Storage* **2022**, *52*, 104857. [CrossRef]
19. Chen, H.; Abidi, A.; Hussein, A.K.; Younis, O.; Degani, M.; Heidarshenas, B. Investigation of the use of extended surfaces in paraffin wax phase change material in thermal management of a cylindrical lithium-ion battery: Applicable in the aerospace industry. *J. Energy Storage* **2022**, *45*, 103685. [CrossRef]
20. Mei, J.; Shi, G.; Liu, H.; Wang, Z.; Chen, M. Investigation on the optimization strategy of phase change material thermal management system for lithium-ion battery. *J. Energy Storage* **2022**, *55*, 105365. [CrossRef]
21. Liu, J.; Wang, Z.; Bai, J.; Gao, T.; Mao, N. Heat generation and thermal runaway mechanisms induced by overcharging of aged lithium-ion battery. *Appl. Therm. Eng.* **2022**, *212*, 118565. [CrossRef]
22. Liu, J.; Huang, Z.; Sun, J.; Wang, Q. Heat generation and thermal runaway of lithium-ion battery induced by slight overcharging cycling. *J. Power Sources* **2022**, *526*, 231136. [CrossRef]
23. Mao, N.; Zhang, T.; Wang, Z.; Gadkari, S.; Wang, J.; He, T.; Gao, T.; Cai, Q. Revealing the thermal stability and component heat contribution ratio of overcharged lithium-ion batteries during thermal runaway. *Energy* **2023**, *263*, 125786. [CrossRef]
24. Hu, J.; Liu, T.; Wang, X.; Wang, Z.; Wu, L. Investigation on thermal runaway of 18,650 lithium ion battery under thermal abuse coupled with charging. *J. Energy Storage* **2022**, *51*, 104482. [CrossRef]
25. Liu, Z.; Guo, X.; Meng, N.; Yu, Z.; Yang, H. Study of thermal runaway and the combustion behavior of lithium-ion batteries overcharged with high current rates. *Thermochim. Acta* **2022**, *715*, 179276. [CrossRef]
26. García, A.; Zhao, P.; Monsalve-Serrano, J.; Villalta, D.; Martinez-Boggio, S. Optical diagnostics of the venting spray and combustion behaviour during Li-ion battery thermal runaway induced by ramp heating. *Appl. Therm. Eng.* **2023**, *218*, 119308. [CrossRef]
27. Golubkov, A.W.; Planteu, R.; Krohn, P.; Rasch, B.; Brunnsteiner, B.; Thaler, A.; Hacker, V. Thermal runaway of large automotive Li-ion batteries. *RSC Adv.* **2018**, *8*, 40172–40186. [CrossRef]
28. He, C.X.; Yue, Q.L.; Chen, Q.; Zhao, T.S. Modeling thermal runaway of lithium-ion batteries with a venting process. *Appl. Energy* **2022**, *327*, 120110. [CrossRef]
29. Li, W.; Wang, H.; Zhang, Y.; Ouyang, M. Flammability characteristics of the battery vent gas: A case of NCA and LFP lithium-ion batteries during external heating abuse. *J. Energy Storage* **2019**, *24*, 100775. [CrossRef]
30. Huang, Z.; Shen, T.; Jin, K.; Sun, J.; Wang, Q. Heating power effect on the thermal runaway characteristics of large-format lithium ion battery with $Li(Ni_{1/3}Co_{1/3}Mn_{1/3})O_2$ as cathode. *Energy* **2022**, *239*, 121885. [CrossRef]
31. Md Said, M.S.; Mohd Tohir, M.Z. Visual and thermal imaging of lithium-ion battery thermal runaway induced by mechanical impact. *J. Loss Prev. Process Ind.* **2022**, *79*, 104854. [CrossRef]
32. Zhang, W.; Huang, L.; Zhang, Z.; Li, X.; Ma, R.; Ren, Y.; Wu, W. Non-uniform phase change material strategy for directional mitigation of battery thermal runaway propagation. *Renew. Energy* **2022**, *200*, 1338–1351. [CrossRef]

33. Dai, X.; Kong, D.; Du, J.; Zhang, Y.; Ping, P. Investigation on effect of phase change material on the thermal runaway of lithium-ion battery and exploration of flame retardancy improvement. *Process Saf. Environ. Prot.* **2022**, *159*, 232–242. [CrossRef]
34. Weng, J.; Ouyang, D.; Yang, X.; Chen, M.; Zhang, G.; Wang, J. Alleviation of thermal runaway propagation in thermal management modules using aerogel felt coupled with flame-retarded phase change material. *Energy Convers. Manag.* **2019**, *200*, 112071. [CrossRef]
35. Huang, Q.; Li, X.; Zhang, G.; Weng, J.; Wang, Y.; Deng, J. Innovative thermal management and thermal runaway suppression for battery module with flame retardant flexible composite phase change material. *J. Clean. Prod.* **2022**, *330*, 129718. [CrossRef]
36. Wang, Z.; Wang, J. Investigation of external heating-induced failure propagation behaviors in large-size cell modules with different phase change materials. *Energy* **2020**, *204*, 117946. [CrossRef]
37. Mei, J.; Shi, G.; Liu, H.; Wang, Z.; Chen, M. Experimental study on the effect of passive retardation method for thermal runaway mitigation of lithium-ion battery. *Appl. Therm. Eng.* **2023**, *230*, 120861. [CrossRef]
38. Wang, Y.; Yu, K.; Peng, H.; Ling, X. Preparation and thermal properties of sodium acetate trihydrate as a novel phase change material for energy storage. *Energy* **2019**, *167*, 269–274. [CrossRef]
39. Hu, P.; Lu, D.-J.; Fan, X.-Y.; Zhou, X.; Chen, Z.-S. Phase change performance of sodium acetate trihydrate with AlN nanoparticles and CMC. *Sol. Energy Mater. Sol. Cells* **2011**, *95*, 2645–2649. [CrossRef]

batteries

MDPI

Article

Evolution of Safety Behavior of High-Power and High-Energy Commercial Li-Ion Cells after Electric Vehicle Aging

Pierre Kuntz [1,2,*], Loïc Lonardoni [2], Sylvie Genies [2], Olivier Raccurt [2] and Philippe Azaïs [2]

1 Entroview, 38400 Saint-Martin-d'Hères, France
2 CEA—Commissariat à l'Energie Atomique et aux Energies Alternatives, LITEN—Laboratoire d'Innovation pour les Technologies des Energies Nouvelles et les Nanomatériaux, DEHT—Département de l'Electricité et de l'Hydrogène pour les Transports, Grenoble Alpes University, 38000 Grenoble, France; loic.lonardoni@cea.fr (L.L.); sylvie.genies@cea.fr (S.G.); olivier.raccurt@cea.fr (O.R.); philippe.azais@cea.fr (P.A.)
* Correspondence: pierre.kuntz@entroview.com

Abstract: The Li-ion battery is one of the key components in electric car development due to its performance in terms of energy density, power density and cyclability. However, this technology is likely to present safety problems with the appearance of cell thermal runaway, which can cause a car fire in the case of propagation in the battery pack. Today, standards describing safety compliance tests, which are a prerequisite for marketing Li-ion cells, are carried out on fresh cells only. It is therefore important to carry out research into the impact of cell aging on battery safety behavior in order to ensure security throughout the life of the battery, from manufacturing to recycling. In this article, the impact of Li-ion cell aging on safety is studied. Three commercial 18,650 cells with high-power and high-energy designs were aged using a Battery Electric Vehicle (BEV) aging profile in accordance with the International Electrotechnical Commission standard IEC 62-660. Several thermal (Accelerating Rate Calorimetry—ARC) and standardized safety (short-circuit, overcharge) tests were performed on fresh and aged cells. This study highlights the impact of aging on safety by comparing the safety behavior of fresh and aged cells with their aging conditions and the degradation mechanisms involved.

Keywords: Li-ion; battery; aging; degradation mechanisms; abuse test; safety behavior

Citation: Kuntz, P.; Lonardoni, L.; Genies, S.; Raccurt, O.; Azaïs, P. Evolution of Safety Behavior of High-Power and High-Energy Commercial Li-Ion Cells after Electric Vehicle Aging. *Batteries* **2023**, *9*, 427. https://doi.org/10.3390/batteries9080427

Academic Editor: Mingyi Chen

Received: 7 June 2023
Revised: 21 July 2023
Accepted: 9 August 2023
Published: 16 August 2023

1. Introduction

Li-ion batteries with graphite-based negative electrodes are now widespread in electric mobility applications thanks to their higher energy density and durability compared to other storage systems. Presently, the main developments in Li-ion batteries concern increasing their energy and power density by developing new positive electrode materials or blending silicon with graphite in the negative electrode. Despite the fact that Li-ion technology has numerous advantages, it has been proven that the Li-ion battery poses a safety risk [1] and is the source of many car fires [2]. Therefore, battery safety assessment is a key issue that must be dealt with in order to continue developing more efficient and durable vehicles, as well as ensuring the user's safety.

The lifetime of Li-ion cells is being continuously improved by innovations, but it is well known that their use can cause degradation mechanisms inside the cell [3,4]. Depending on the storage and cycling conditions, several aging mechanisms can be triggered and induce physical and chemical modifications of the internal components. They can provoke physico-chemical changes, like active material damage [5–11], lithium consumption by SEI growth [5,12–14] and lithium plating [15–19]. These physical and chemical modifications induced by aging first impact cell performance, but can also influence cell safety. Therefore, cell aging conditions have a strong influence on a cell's safety behavior.

One research group has already studied the safety of aged Li-ion cells coupled with *post*-mortem analysis, especially in terms of thermal stability, by performing Accelerating Rate Calorimetry (ARC) tests after cycle aging at 0.5 C and at 0 °C, 5 °C, 25 °C and 45 °C [20]. The tests showed that low-temperature cycle aging reduces cell thermal stability [20,21]. The Li plating phenomenon that takes place on the negative electrode at low temperature ranges is principally responsible for the degradation of safety behavior, because metallic Li reacts exothermally with the electrolyte solvent. On the contrary, high-temperature aging, which induces the degradation mechanism of SEI growth on the negative electrode, does not have a negative impact on the thermal stability of the cell [20,21].

Other studies have also treated the subject of thermal stability but were not always supplemented by *post*-mortem studies. Some studies observed the decrease of thermal stability of cycled cells at low temperatures [22,23], and compared the thermal stability of several cell chemistries, the State Of Health (SOH) and the State of Charge (SOC) [24].

In an accident caused by a Li-ion cell, a temperature increase is always responsible for triggering thermal runaway [25], possibly leading to fire and/or explosion. Electrical solicitations like overcharging and short-circuiting can be responsible for this temperature increase, which is why international safety standards have standardized electrical safety tests (overcharge and short-circuit) to ensure the safety of fresh cells before their sale on the market. To our knowledge, the safety behavior of aged cells during overcharging and short-circuiting has not been discussed in the literature. Studies appear to have focused only on overcharge mechanisms [26] or compared the safety behavior of fresh cells for different chemistries [27].

Our research work has two different parts. The first one concerns the *ante*-mortem analysis of the Li-ion cells studied, the aging of these cells and their *post*-mortem analysis through the identification of the degradation mechanisms involved. A paper on the first part of our work has already been published and bears the title "Identification of Degradation Mechanisms by Post-Mortem Analysis for High Power and High Energy Commercial Li-Ion Cells after Electric Vehicle Aging" [28]. In this first part, the aging process, *ante*-mortem and *post*-mortem analyses (half coin cell at the electrode level, Scanning Electron Microscope (SEM), Energy-Dispersive X-ray (EDX), Glow Discharge-Optical Emission spectrometer (GD-OES), X-Ray Diffraction (XRD), Nuclear Magnetic Resonance (NMR), etc.) have already been performed [28]. This work allows us to understand which main degradation mechanisms occur according to aging conditions, cell chemistry, and design.

The second part of our research work, which is the subject of this article, has the objective of understanding the effects of aging on cell safety. The use of real experimental data from cell aging and aged cell safety tests, coupled with *post*-mortem analysis, highlight the impact on safety of cell internal degradation mechanisms due to aging.

The novelty of this article consists in the understanding on the impact of aging on Li-ion cell safety, through the realization of a large panel of experimental tests: *ante*-mortem study, several aging conditions according to the standards, *post*-mortem study to identify the aging mechanisms involved, and finally abuse and safety tests on fresh and aged cells to highlight the impact of aging on Li-ion cell safety. In addition, this study was carried out on a large amount of cells, divided into three commercial cell references, with 100 units for each reference.

To assess the impact of aging on safety behavior, and not only of thermal behavior, several abuse tests were performed on each fresh and aged cell. Thermal stability was characterized by the ARC test and electrical safety by short-circuiting and overcharging according to the international standard IEC 62-660, part 2 [29]. Data cross-referencing between aging degradation mechanisms [28] and the safety behavior observed enables understanding of the influence of each aging mechanism on the safety of Li-ion cells.

2. Material and Methods

2.1. Samples

The three commercial cells chosen were from SAMSUNG SDI: INR18650-35E (written "35E" afterward), INR18650-32E (written "32E" afterward) and INR18650-30Q (written "30Q" afterward). Each cell of the same reference used in this study comes from the same manufacturer production batch. To ensure that the internal mechanical design, separator and the electrolyte composition of the three samples were similar, the three cells were provided by the same manufacturer.

All the internal components of the fresh and aged cells were analyzed in the first part of our work [28]. The detailed description of the fresh cell chemistry and design, and of the degradation mechanisms occurring inside the aged cells, is described in this paper.

2.2. Aging Tests

Three cell references were investigated using BEV (Battery Electric Vehicle) representative aging at various temperatures (−20 °C, 0 °C, 25 °C, 45 °C), according to the international standard IEC 62-660 part 1 [30], including cycle aging and calendar aging.

Cycle aging was performed at various temperatures (−20 °C, 0 °C, 25 °C, 45 °C). The cycling described in the international standard IEC 62-660 is representative of a typical driving cycle, including a CC-CV charge and a discharge profile with current pulses that simulate braking (charge) and acceleration (discharge) event. SOH assessments using electrical performance measurements were performed every 28 days at 25 °C. The cycling test was stopped once the cell had lost at least 20% of its initial performance in terms of capacity, energy or power density, or after 6 months of cycling.

Calendar aging was achieved at 45 °C, either at constant voltage (CV) at 4.2 V or at 100% SOC in Open Circuit Voltage (OCV, meaning that the voltage was not maintained constantly). The voltage difference between OCV and CV calendar aging was between 0.02 V and 0.1 V maximum after 6 weeks aging. According to the standard, SOH assessments using electrical performance measurements were performed at 25 °C every 6 weeks, and after 18 weeks of storage, the calendar test was stopped.

These electrical measurements were used to trace the evolution of appropriate cell characteristics: capacity, internal resistance and nominal voltage. Each aging procedure was performed on a batch of 15 cells. The periodical electrical performance measurements and cycling test were achieved on a PEC® SBT 05250 test bench (6 V, 50 A).

2.3. Abuse Tests

Thermal stability tests were performed with an ARC EV with Thermal Hazard Technology® equipment. The Heat–Wait–Seek [31] protocol was applied with 5 °C temperature steps, 35 min rest and the detection of the onset temperature (T_{onset}) for a temperature increase of more than 0.02 °C/min. T_{onset} is the temperature at which exothermal reactions start in the cell and the cell starts to heat up by itself.

Overcharging and short-circuiting were carried out on a abuse test bench made specially for this study (shown in Figure 1) at an ambient temperature between 20 °C and 30 °C. According to the standard IEC 62-660 part 2 [29], the short-circuit set-up had a maximum resistance of 5 mΩ and had to be applied for 10 min or until cell failure. Overcharging consisted of applying a current of C (considering the cell capacity at the moment of the test) up to 8.4 V or cell failure. This new abuse test bench allows one to perform up to 6 tests in a row and is fully remote-controllable. The bench was placed in a reinforced box and in a closed room under air suction to ensure the user's safety. The release cylinder works with compressed air and is controlled by an Arduino card. Only one cell can be connected at the same time; this system allows one to connect the chosen cell and change it when the test is finished without entering the room.

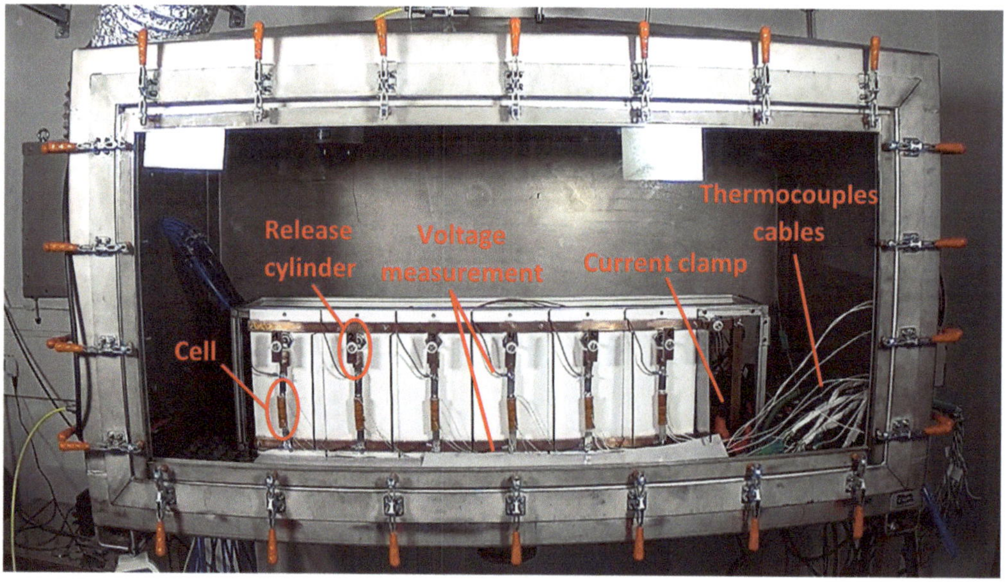

Figure 1. Picture of the abuse test bench specifically developed for this study.

A GM10-2E0 data acquisition unit (Yokogama®) was used for the acquisition of cell voltage, current and temperature. Voltage was measured directly by the unit with a measurement uncertainty of $\leq\pm2.5$ mV. K-type thermocouples were fixed in the middle of each cell using a Kapton tape to ensure the measurement of the surface cell temperature with a measurement uncertainty of $\leq\pm0.3$ °C. The current was measured using an aerometric clamp with measurement uncertainties of $\leq\pm95$ mA for overcharging and $\leq\pm4$ A for short-circuiting.

All the thermal and safety tests were carried out at 100% state of charge (SOC), because it is the worst possible case in terms of safety [24]. Before each test, the cells were charged by a constant current at C/3 until 4.2 V, followed by a constant voltage until C/20. A battery's C-rate is defined by the rate of time in which it takes to charge or discharge; for example, at C/3 it takes 3 h to charge or discharge.

3. Cell Aging and Degradation Mechanism

Before describing a picture of the aging state of sample for each protocol, an overview of the cell materials and design at the initial state is required. One cell of each reference was opened to perform physico-chemical analysis. All the results of the *ante*-mortem analysis are given in the first part of this study, published in a previous paper [28]. Regarding the electrochemical characterization, the three fresh cell batches (100 units) had a very low standard deviation regarding their characteristics in terms of weight, capacity, nominal voltage, internal resistance and energy density.

The chemistry of the three cells showed similarities, but also some differences. 30Q is composed of a blend of NCA ($Li_XNi_yCo_zAl_aO_2$ with $y + z + a = 1$ and $0 < x \leq 1$) + NC ($Li_XNi_wCo_bO_2$ with $w + b = 1$ and $0 < x \leq 1$) on the positive electrode and of a blend of graphite + silicon on the negative electrode. 32E contains only NCA on the positive electrode and only graphite on the negative electrode. 35E also contains only NCA on the positive electrode but a blend of graphite + silicon on the negative electrode. All the detailed chemical compositions of the three cells' electrodes are detailed in our previous paper [28].

The design of the cells was different for the three references: cell 30Q has the thinnest and largest electrodes compared to 35E and 32E. The 30Q cell has a "power" design (i.e., "power cells"); indeed, the larger and thinner electrode allows for delivering higher current. On the contrary, the 35E and 32E cells have an "energy" design (i.e., "energy cells"), with thicker and smaller electrodes, which allows optimizing the energy density.

Electrolyte compositions are different but contain similar solvents: ethylene carbonate (EC), dimethyl carbonate (DMC) and fluoroethylene carbonate (FEC). The separator is based on polyethylene (PE) with AlOOH coating on the negative electrode side and is similar for all references.

The *post*-mortem analysis is detailed in the first part of this study published in [28], dedicated to understanding degradation mechanisms after aging. The main conclusions of this previous study are summarized in Table 1.

Table 1. Main degradation mechanisms detected by post-mortem analysis after aging.

Aging Condition		30Q		32E		35E	
		SOH	Degradation Mechanisms	SOH	Degradation Mechanisms	SOH	Degradation Mechanisms
Cycling	−20 °C	85	Li plating ++ Si cracking and disaggregation ++	59	Li plating +++	83	Li plating ++ Si cracking and disaggregation ++
	0 °C	91	Si cracking and disaggregation +	N/A	Li plating ++	70	Si cracking and disaggregation +
	25 °C	76	SEI growth ++	81	SEI growth +	92	SEI growth +
	45 °C	78	SEI growth +++	80	SEI growth ++	86	SEI growth ++
Calendar	45 °C, CV	87	SEI growth +++	80	SEI growth ++	87	SEI growth ++
	45 °C, OCV	91	SEI growth ++	88	SEI growth +	91	SEI growth +

For tests at higher temperatures (25 °C and above), the Solid Electrolyte Interphase (SEI) growth at the negative electrode was a predominant degradation mechanism observed for both types of test (calendar and cycling). This phenomenon is caused by the solvent degradation for calendar aging and salt degradation for cycle aging [28]. During low-temperature cycling (lower than 25 °C), the main aging mechanism identified was the deposition of metallic lithium leading to the formation of a secondary SEI. In addition, silicon can significantly degrade during aging, especially during cycle aging at low temperatures through particle cracking and disaggregation.

The internal design of the cell seemed to have an influence on the aging mechanisms. Energy cells are more sensitive to low-temperature aging, which induces more lithium metal deposition partly due to their higher internal resistance [28]. Power cells, which work well at lower temperatures, are more affected by high temperature with increased growth of SEI due to their larger electrode surface [28].

The nature of the internal components of the cell also has an important influence on the aging mechanisms, since the nature of the electrolyte influences the growth of SEI through solvent degradation and the silicon sometimes present in the negative electrode seems to be a privileged site of degradation.

All the *ante*- and *post*-mortem work carried out enabled the complete analysis of new and aged cells. Thus, the cause-and-effect relationships between the aging conditions and the aging mechanisms were highlighted by this work. For more details about the *ante*- and *post*-mortem work, please refer to our previous paper [28].

Abusive tests were carried out on the three fresh cell references. These results were then compared with those obtained on cells aged according to different aging modes (different temperatures, calendar, cycling) in order to identify the impact of aging on the abusive test behavior.

The 32E cells cycled at 0 °C did not survive the aging subjected by triggering the opening of the current interrupting device—CID, so they could not be tested for the safety study.

4. Results

4.1. Thermal Stability-ARC

Since separator melting (~140 °C) signals the point of no return of thermal runaway, we decided to stay under this point and stop the ARC test at around 100 °C (90 °C for the 30Q, 100 °C for the 35E and 110 °C for the 32E). We focused on the reactions that take place under the separator melting, because they are responsible for the temperature increase leading to the separator melting and therefore to thermal runaway. In the present study, the onset of self-heating was considered in particular because we were interested in the earliest exothermal reactions responsible for cell heating. It is known that the two reactions that occur at the beginning of self-heating are the reaction of inserted and plated lithium if present with solvent [32] after the thermal degradation of the SEI [2,32,33]. Both of those reactions should take place under 90 °C and can be affected by the chemical changes due to aging. Another objective was to not extensively damage the cell for possible further investigations.

Two cells per aging condition were tested in ARC and the results are shown in Figure 2 for T_{onset} and in Figure 3 for the temperature rate. The results are compared to the fresh cell.

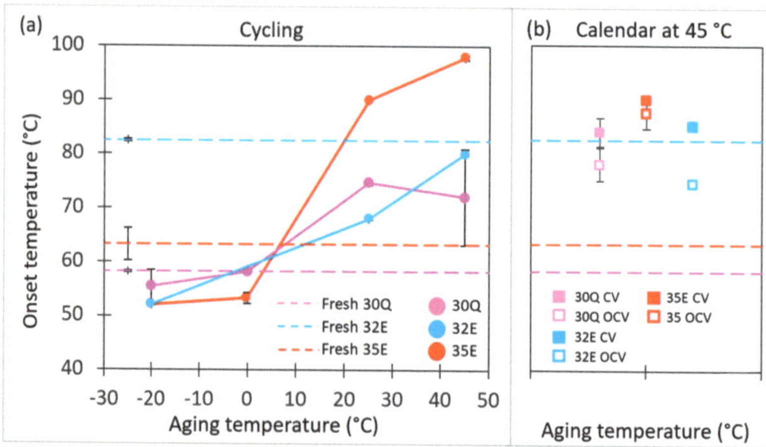

Figure 2. Onset temperature of fresh and aged cells in terms of aging temperature for references 35E, 32E and 30Q, after (**a**) cycle aging and (**b**) calendar aging.

The onset temperature is the beginning of self-heating of the cell due to exothermal reactions during the ARC test. It can be seen that for cells aged at −20 °C and 0 °C, T_{onset} is between 50 °C and 60 °C, and for cells aged at 25 °C and 45 °C T_{onset} is between 65 °C and 100 °C. The change of T_{onset} during aging for the three references T_{onset} is lower after low-temperature cycling than after high-temperature aging (cycling and calendar).

Compared to the fresh cell, two trends can be distinguished. First, the cells containing Si within the negative electrode (30Q and 35E), which has a T_{onset} around 60 °C for the fresh cell, show a decrease of T_{onset} after low-temperature cycling (−20 °C and 0 °C) and, conversely, an increase after high-temperature (45 °C and 25 °C) cycling or calendar storage. Secondly, the cells containing only graphite within the negative electrode (32E), which have a T_{onset} around 80 °C for the fresh cell, show a decrease of T_{onset} after all aging conditions, except after calendar CV storage.

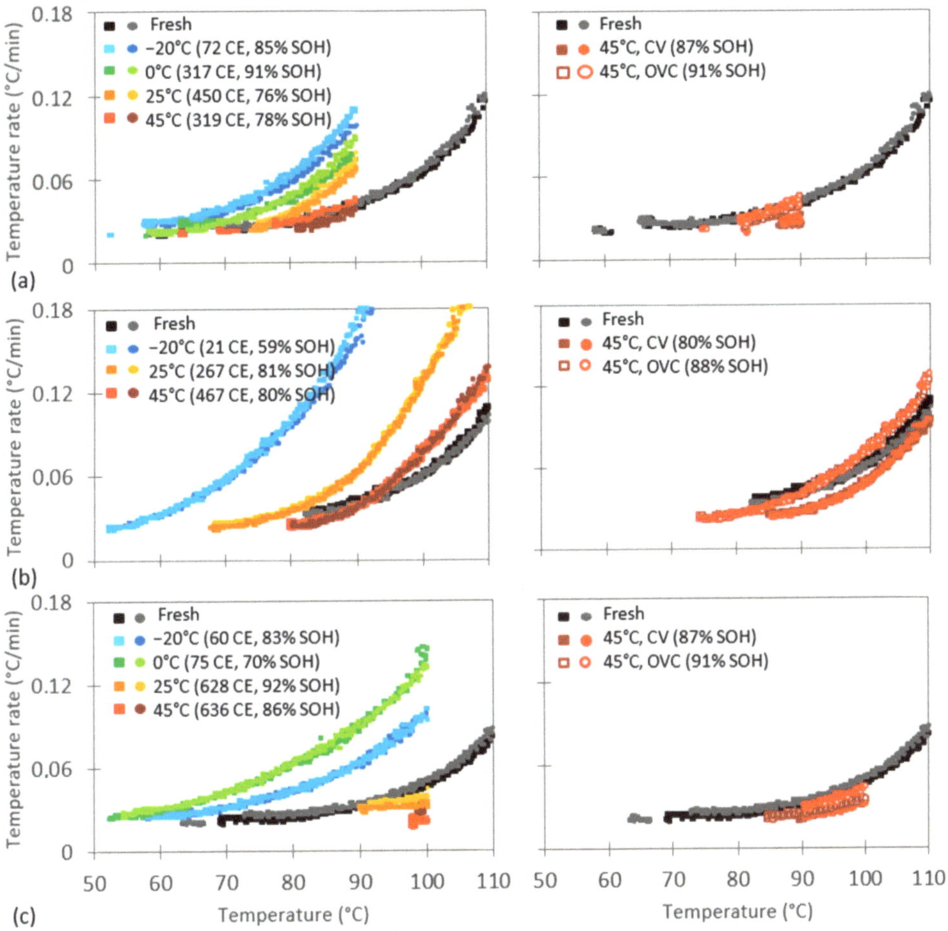

Figure 3. Temperature increase rate during the ARC test of aged and fresh cells for (**a**) 30Q, (**b**) 32E and (**c**) 35E, for 2 cells (square and circle) per aging condition.

The rate of temperature change during the test was also significantly altered by aging. The more the aging was carried out at low temperatures, the faster the temperature rate increased, signifying that the cell was more reactive to self-heating. For example, at 90 °C, the temperature rate increased about 3 times faster for cell 30Q and 6 times faster for cell 32E for a cell aged at −20 °C compared to a cell aged in calendar conditions.

The only exception was for the 35E cell: the temperature of the cells aged at 0 °C increased faster compared to cells aged at −20 °C. However, the behavior at 0 °C can be explained by their lower SOH linked to a more severe damage state (lithium metal deposition and silicon particle degradation). This highlights that the method and the conditions in which the cell is aged had a strong impact on their thermal stability. Systematically, aging at low temperature degraded the thermal stability of the cell.

It is noticeable that for all the cells aged at 45 °C, the temperature rate of the cells was either equivalent to that of the fresh cell or lower for the calendar aging conditions (especially for that set in CV at 4.2 V). This observation reveals that aging at high temperature (especially calendar aging) increased the thermal stability.

Taking into account the *post*-mortem results and the aging effect on thermal stability (T_{onset} and temperature rate), we can assume that the degradation mechanisms of low-temperature cycling (Li plating [15–19,28] and Si cracking [28,34–36]) degraded the thermal stability of the cell. On the contrary, the degradation mechanisms of high-temperature cycling and storage (SEI growth by salt and solvent degradation) improved the thermal stability of the cell.

4.2. Safety Tests

To evaluate the impact of aging on Li-ion cell safety, short-circuit and overcharge tests were performed on fresh and aged cells. Voltage, current and temperature were tracked during the tests. Particular attention was paid to the temperature because it could show if an accident would occur or not. The results of all the safety tests are described here.

4.2.1. Assessment of the Risk of Short-Circuit

Figure 4 shows the results of short-circuit tests. Data were compared with four parameters: CID opening duration, maximum current, maximum temperature and maximum temperature rate increase. The CID opening duration of a fresh cell is between 10 and 20 s, but after aging it increases to 50 s. CID opening is always longer for aged cells than for new ones. We observed a significant increase of the CID opening duration in particular for energy-designed cells after cycling at low temperature.

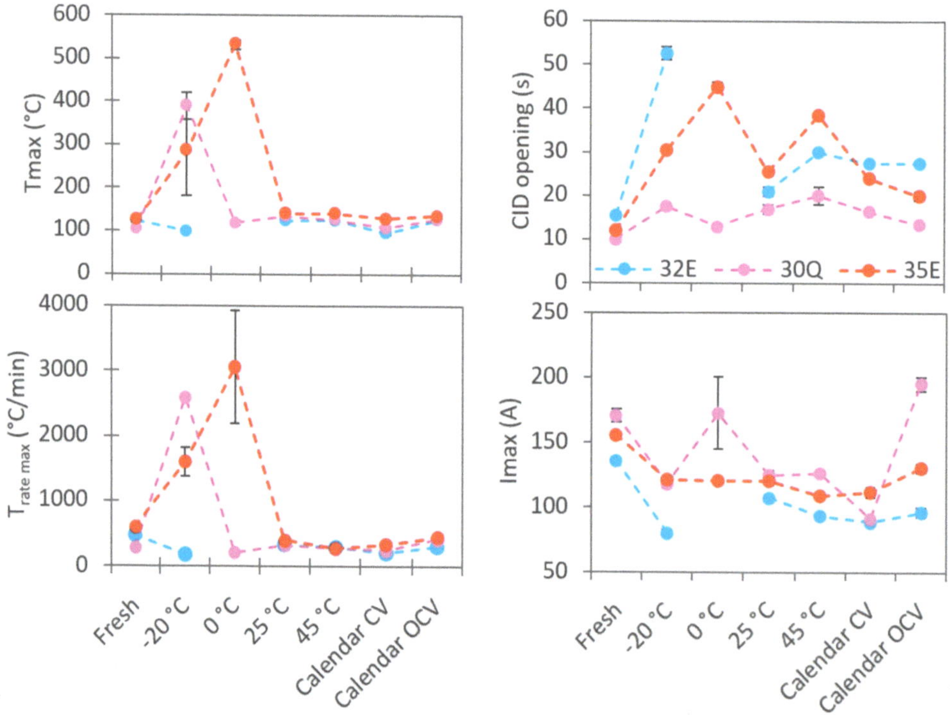

Figure 4. Characteristics of the short-circuit test in terms of CID opening duration, I_{max}, T_{max} and $T_{speed\ max}$ for 30Q, 32E and 35E fresh and aged cells.

The maximum short-circuit current is generally between 100 A and 200 A. Most of the time, the maximum short-circuit current will decrease after aging in accordance with the increase of cell internal resistance. The short-circuit current also depends on the short-circuit resistance, here between 0 mΩ and 5 mΩ, and its variation can explain non-expected values like for 30Q cells after −20 °C and calendar OCV aging.

The maximum temperature reached during the test was between 100 °C and 140 °C for fresh cells. The temperature reached during the short-circuit test was sufficiently high to initiate exothermal reactions inside the cell [2,32,33,37–40], as shown during the ARC test (Figure 3). After aging at 45 °C and 25 °C, the maximum temperature did not change significantly. We could even observe a small decrease of the maximal temperature for the CV calendar-aged cells at 45 °C, confirming that high-temperature aging can be beneficial for cell safety. After aging at low temperature, some cells went into thermal runaway (35E aged at −20 °C and 0 °C and 30Q aged at −20 °C). The maximum temperature reached during short-circuiting was more than 500 °C for 35E cells aged at 0 °C.

The maximum temperature increase rate was between 300 °C/min and 600 °C/min for fresh cells and decreased slightly between 200 °C/min and 500 °C/min for aged cells which did not go into thermal runaway. For the cells that went into thermal runaway, the maximum temperature increase rate could increase up to 3000 °C/min at the cell surface.

Table 2 shows a comparison of the aging degradation mechanisms identified for each aged cell during the *post*-mortem studies and the observation during the short-circuit tests. First, all the fresh cells passed the short-circuit safety test successfully because they did not vent and remained intact. After aging, the CID was no longer efficient enough because venting and explosion of the cells was observed.

Table 2. Short-circuit test results, compared to the SOH and the aging mechanisms identified for each aging condition for cells 30Q, 35E and 32E.

Ref	Fresh	−20 °C	0 °C	25 °C	45 °C	CV	OCV
30Q	✓	85% SOH Li plating Si cracking 💥	91% SOH Si cracking ☁	76% SOH SEI ☁	78% SOH SEI ☁ ✓	87% SOH SEI ✓	91% SOH SEI ☁
35E	✓	83% SOH Li plating Si cracking 💥	70% SOH Si cracking 💥	92% SOH SEI ☁	86% SOH SEI ☁	87% SOH SEI ☁	91% SOH SEI ☁
32E	✓	59% SOH Li plating ✓	Li plating	81% SOH SEI ✓	80% SOH SEI ✓	80% SOH SEI ✓	88% SOH SEI ✓

Meaning of symbols: 💥 explosion; ☁ gas and smoke; ✓ NTR.

The 32E cell, unlike the other cells tested, does not contain silicon in the electrode and is not the cause of any negative events. However, we have shown during the post-mortem study of other cell references that silicon was particularly degraded on the surface of the negative electrode. Its considerable volume variations during cycling induced its fragmentation, thus favoring the formation of lithium silicates [28]. The silicon-containing cells exploded after cold aging and often released gases and fumes. It is therefore likely that silicon plays an important role in the safe behavior of the cells, as all the negative events took place in silicon-containing cells, whether or not lithium plating was present.

The energy-type design also appears to be a detrimental factor in terms of safety, as the 35E cell presented more explosions and outgassing than the 30Q cell during the short-circuit tests.

4.2.2. Assessment of the Risk of Overcharge

Figure 5 shows the results of the overcharge tests. Data were compared through three parameters: the final SOC reached at the end of the test, the temperature at the CID opening and the maximum temperature. The final SOC of the overcharge test was between 110% and 130%. The final overcharge SOC can evolve during aging; a significant increase of the final SOC was observed for the energy-designed cell after aging at low temperatures.

This parameter was lower for all the cells after calendar aging, especially for the calendar condition at OCV.

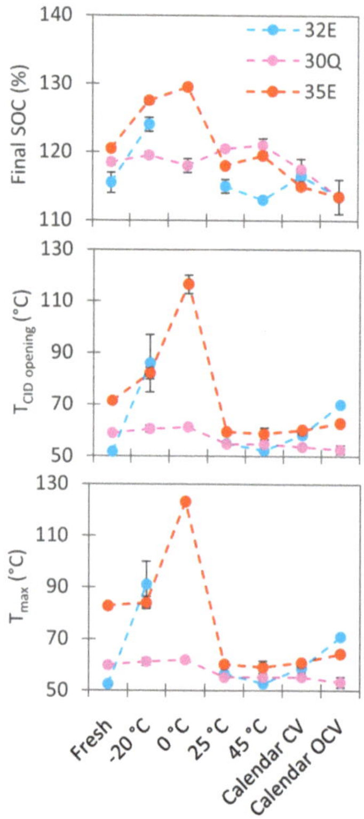

Figure 5. Characteristics of the overcharge test in terms of final SOC, $T_{CID\ opening}$ and T_{max} for 30Q, 32E and 35E fresh and aged cells.

The temperature at the CID opening and the maximum temperature reached were between 50 °C and 120 °C, most often under 70 °C. The temperature at the CID opening was very close to the maximum temperature of the test. It revealed that CID triggers early enough to avoid irrevocable exothermal decomposition reactions. During overcharging, no accident was observed and all cells stayed intact (no venting). However, the cells aged at low temperature had the biggest temperature rise during the test, especially for the energy-designed cells (up to 120 °C), confirming that aging at low temperature decreases cell safety.

5. Discussion

The set of experimental results detailed above highlighted the thermal behavior of fresh and aged cells following standardized protocols and a wide temperature range. An assessment of the risk of the technology throughout its use was thus made possible.

At fresh state, the first two exothermic reactions that cause the cell temperature rise and initiate the thermal runaway are known. They imply SEI decomposition and the direct reaction of inserted lithium with electrolytes, since lithiated graphite is no longer protected by this passivation layer.

The decomposition of the SEI starts at a temperature between 90 °C and 120 °C [2,32,33,37]. The enthalpy of reaction was between 180 and 350 J/g [41]. A representative decomposition reaction considering LiEDC ($(CH_2OCO_2Li)_2$) as the main compound of the SEI is the following:

$$(CH_2OCO_2Li)_2 \rightarrow Li_2CO_3 + C_2H_4 + CO_2 + \frac{1}{2}O_2 \quad (1)$$

The enthalpy of the direct reaction between organic solvents and lithium inserted within graphite (free of SEI, decomposed beforehand) [32,38] is between 350 and 1714 J/g [41]. The reaction proposed considering only one of the solvents present in the electrolyte (Ethylene carbonate, EC) takes the following form:

$$2Li_xC_6 + C_3H_4O_3(EC) \rightarrow 2Li_{x-1}C_6 + Li_2CO_3 + C_2H_4 \quad (2)$$

Both the reactions mentioned here can take place simultaneously and generate other reactions when the temperature rises.

After aging, the degradation phenomena generate new exchange surfaces that impact the exothermicity of the cells. One of these phenomena is lithium metal deposition, favored by cycling at low temperatures. Cells aged at very low temperatures (−20 °C), for which we identified the lithium metal deposit on the negative electrode as the main aging mechanism, effectively showed the least thermal stability. When the cell heats up, the lithium metal present on the surface of the negative electrode is directly available to react with the organic solvents in the electrolyte as early as 68 °C [32]. These reactions are exothermic and contribute to the temperature increase of the cell. The presence of electrodeposited lithium metal in the cell is therefore responsible for its greater "reactivity", as the cell heats up earlier and faster. These findings are consistent with the results obtained in the ARC tests with T_{onset} lowering for all cells containing lithium metal. Indeed, whatever the reference studied (30Q, 32E or 35E), the cells in which a lithium metal deposit was observed had a T_{onset} between 50 °C and 60 °C. Exothermic reactions involving lithium metal thus appeared to start as early as 50–60 °C. The following reactions can take place as early as 68 °C according to the literature [32] and have enthalpies between 350 and 1714 J/g [41].

$$2Li + C_3H_4O_3(EC) \rightarrow Li_2CO_3 + C_2H_4 \quad (3)$$

$$2Li + C_4H_6O_3(PC) \rightarrow Li_2CO_3 + C_3H_6 \quad (4)$$

$$2Li + C_3H_6O_3(DMC) \rightarrow Li_2CO_3 + C_2H_6 \quad (5)$$

In some cases, as a function of the amount of plated lithium, when the cell is brought to higher temperatures, the electro-deposited lithium in contact with the electrolyte may oxidize partially or entirely and form a "secondary" SEI [17]. It will no longer be able to react with solvents according to the above-mentioned reactions, or it may be covered with a passivation layer, creating lithium metal islands isolated from the electrolyte and consequently become electrochemically inactive.

SEI growth is the major degradation phenomenon at high temperatures. Cells aged at 45 °C in cycling and calendar conditions are the most thermally stable because when the cell heats up, the first exothermic reaction encountered will be the decomposition of SEI at a temperature between 90 °C and 120 °C [2,32,33,37]. Cells aged at high temperatures are more thermally stable due to a thicker, homogeneous layer of SEI, which explains the start of thermal runaway at higher temperatures than for a fresh cell or for a low-temperature-aged cell.

Regarding the impact of silicon in thermal runaway, it can be noted that the only explosions observed in short-circuit concerned 30Q cells aged at −20 °C and 35E cells aged at −20 °C and 0 °C, i.e., low-temperature-aged cells containing silicon in their negative electrodes. On the other hand, for the 32E cell which did not contain silicon, there was never any fire, even in the case of aging at −20 °C. Silicon therefore has a significant effect on the thermal stability of the cells. To interpret this result, the mechanical behavior of silicon must be considered. Silicon cycling causes a volume expansion during the lithiation phases and a shrinkage during the delithiation phases [35]. As we observed in post-mortem analysis, these successive volume variations cause cracks and fractionate the particles into smaller and smaller pieces [36,42]. However, it has been shown that the particle size of lithiated silicon has a significant impact on the exothermicity of its degradation between 100 °C and 150 °C [39,40,43]. The disaggregation of lithiated particles of silicon causes a considerable increase in the exchange surface with the electrolyte. This increases the kinetics of the reactions involved and thus the acceleration of cell heating. These reactions start between 100 °C and 150 °C and can therefore lead to additional heating up to reach the separator shrinking temperature, then the triggering of an internal short-circuit, and finally the cell runaway.

Figure 6 shows an illustration of the exothermic reactions present in a low-temperature-aged cell (with lithium metal deposition) containing silicon in the negative electrode. The coupling of lithium metal deposition and silicon degradation is therefore the worst-case scenario in terms of safety. It was in this configuration that we observed cell explosions during the short-circuit tests, the worst stability during the ARC tests and significant heating during overcharging (30Q and 35E cells aged at −20 °C).

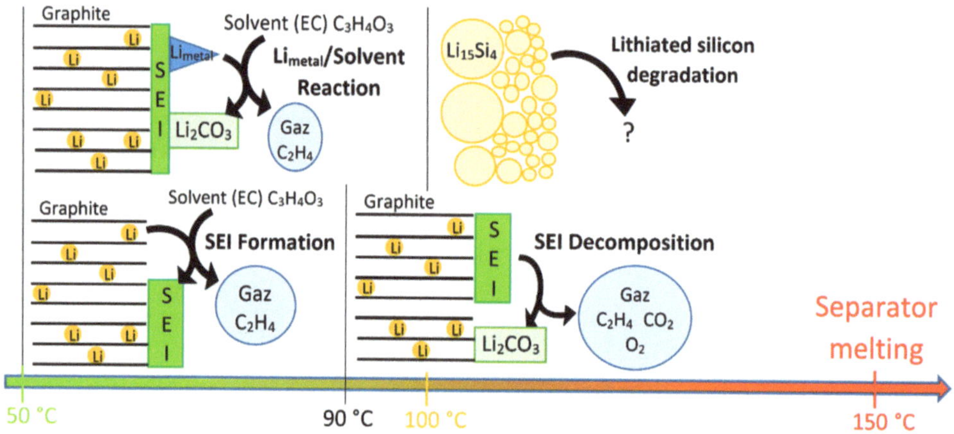

Figure 6. Illustration of the exothermic reactions in a cell containing silicon in the negative electrode and aged at low temperature, with lithium metal deposition.

Finally, this study carried out on energy- and power-designed cells allowed us to assess the impact of cell design on safety. It has been shown that fresh power cells seem to be slightly less stable than fresh energy cells considering their T_{onset} and runaway speed being higher in ARC tests. This observation is still valid after aging at high temperatures (25 °C and above) in calendar and cycling life. But this must be qualified after ageing at low temperatures. We indeed saw that safety hazards occur more after aging at low temperatures (−20 °C and 0 °C). Energy-type cells age poorly at low temperatures compared to power-type cells. As we saw in our previous study [28], they are more exposed to lithium metal deposition, so their thermal stability degrades more rapidly as the cycling temperature decreases. In addition, energy cells have two major safety disadvantages: a higher internal resistance, which gives them the capacity to heat more easily than a power cell, and a higher energy density (amount of combustible active material), which potentially represents a greater risk in case of accident. The operating temperature range of the battery must therefore be considered when selecting a cell and not only its initial performances to ensure that safety is maintained throughout the aging of the cell. A minimum operating temperature must be defined to avoid the deposition of metallic lithium for each cell. This minimum temperature is lower for power cells.

6. Conclusions

The influence of aging was correlated with the thermal behavior of cells in the preliminary phase of thermal runaway through three safety tests, i.e., thermal stability, short-circuiting and overcharging. This study showed that the thermal behavior in the runaway phase depends on aging conditions that favor and intensify certain aging mechanisms (SEI growth, Li-plating). Power and energy cell designs must be considered because they do not have the same ability to accept high current, especially at low temperature ranges. Specific experimental devices were set up to perform a large number of tests, which certified that the observed behaviors were clearly repeatable. The exothermic phenomenon involved above 150 °C is the consequence of the separator melting, after which thermal runaway can no longer be controlled. The whole issue of maintaining the safety of this accumulator technology is played out before the separator melts. After aging at room and elevated temperatures, where the main degradation mechanism is the growth of SEI on the surface of the negative electrode, the thermal stability of the cell was improved. Thicker SEI served as a better protective layer by delaying the onset of further exothermic reactions until it decomposed itself. Thermal degradation of SEI is an exothermic phenomenon that can begin as early as 90 °C, but the associated energy is relatively low. After low-temperature aging, in which the main degradation mechanism was the deposition of metallic lithium on the surface of the negative electrode, the thermal stability of the cell was degraded. The electrodeposited lithium in metallic form reacted with solvents as early as about 60 °C. This was the first exothermic reaction that took place with a relatively high reaction enthalpy. It can be delayed if lithium metal is surrounded by a secondary "SEI". The chemistry of the negative active materials also had an influence on the development of its safety behavior. Indeed, we showed that silicon breaks down into smaller particles and that this phenomenon is exacerbated during cold aging. The reactivity of lithiated silicon particles increased significantly as their size decreased. Thus, an additional exothermic reaction took place between 100 and 150 °C in cells containing silicon, and this reaction increased after low-temperature cycling due to the degradation of silicon by fragmentation.

Energy-type cells are more thermally stable in a fresh state, but the significant degradations they undergo during cycling at low temperatures drastically increased the risk of thermal runaway. Power cells, on the other hand, were less stable when fresh, but their better performance in low-temperature cycling made them slightly less exposed to thermal runaway after aging.

The abusive behavior qualification of the Li-ion cells and in particular the corresponding standards must also take aging into account to be able to qualify the safety of the batteries. Finally, these results could contribute towards improving safety standards.

One way of improving the results of this work would be to monitor cell parameters more closely during testing, and achieve greater reproducibility by testing a larger number of cells. However, this is a time-consuming and costly process.

Author Contributions: P.K.: Conceptualization, methodology, experimentation, data curation, writing and supervision. L.L.: experimentation. S.G.: Conceptualization, methodology, experimentation and supervision. O.R. and P.A.: Conceptualization, methodology and supervision. All authors have read and agreed to the published version of the manuscript.

Funding: This research was funded by the research center CEA LITEN.

Data Availability Statement: The data are not publicly available.

Conflicts of Interest: The authors declare no conflict of interest.

References

1. Garche, J.; Brandt, K. Safety Considerations with Lithium-Ion Batteries. In *Encyclopedia of Electrochemistry*; Bard, A.J., Ed.; Wiley: Hoboken, NJ, USA, 2020; pp. 1–25. ISBN 978-3-52730-250-5.
2. Lisbona, D.; Snee, T. A Review of Hazards Associated with Primary Lithium and Lithium-Ion Batteries. *Process Saf. Environ. Prot.* **2011**, *89*, 434–442. [CrossRef]
3. Schlasza, C.; Ostertag, P.; Chrenko, D.; Kriesten, R.; Bouquain, D. Review on the Aging Mechanisms in Li-Ion Batteries for Electric Vehicles Based on the FMEA Method. In Proceedings of the 2014 IEEE Transportation Electrification Conference and Expo (ITEC), Dearborn, MI, USA, 15–18 June 2014; pp. 1–6.
4. Waldmann, T.; Iturrondobeitia, A.; Kasper, M.; Ghanbari, N.; Aguesse, F.; Bekaert, E.; Daniel, L.; Genies, S.; Gordon, I.J.; Löble, M.W.; et al. Review—Post-Mortem Analysis of Aged Lithium-Ion Batteries: Disassembly Methodology and Physico-Chemical Analysis Techniques. *J. Electrochem. Soc.* **2016**, *163*, A2149–A2164. [CrossRef]
5. Klett, M.; Eriksson, R.; Groot, J.; Svens, P.; Ciosek Högström, K.; Lindström, R.W.; Berg, H.; Gustafson, T.; Lindbergh, G.; Edström, K. Non-Uniform Aging of Cycled Commercial LiFePO4//Graphite Cylindrical Cells Revealed by Post-Mortem Analysis. *J. Power Sources* **2014**, *257*, 126–137. [CrossRef]
6. Buchberger, I.; Seidlmayer, S.; Pokharel, A.; Piana, M.; Hattendorff, J.; Kudejova, P.; Gilles, R.; Gasteiger, H.A. Aging Analysis of Graphite/$LiNi_{1/3}Mn_{1/3}Co_{1/3}O_2$ Cells Using XRD, PGAA, and AC Impedance. *J. Electrochem. Soc.* **2015**, *162*, A2737–A2746. [CrossRef]
7. Liu, L.; Li, M.; Chu, L.; Jiang, B.; Lin, R.; Zhu, X.; Cao, G. Layered Ternary Metal Oxides: Performance Degradation Mechanisms as Cathodes, and Design Strategies for High-Performance Batteries. *Prog. Mater. Sci.* **2020**, *111*, 100655. [CrossRef]
8. Wohlfahrt-Mehrens, M.; Vogler, C.; Garche, J. Aging Mechanisms of Lithium Cathode Materials. *J. Power Sources* **2004**, *127*, 58–64. [CrossRef]
9. Mukhopadhyay, A.; Sheldon, B.W. Deformation and Stress in Electrode Materials for Li-Ion Batteries. *Prog. Mater. Sci.* **2014**, *63*, 58–116. [CrossRef]
10. Zhou, W.; Hao, F.; Fang, D. The Effects of Elastic Stiffening on the Evolution of the Stress Field within a Spherical Electrode Particle of Lithium-Ion Batteries. *Int. J. Appl. Mech.* **2013**, *05*, 1350040. [CrossRef]
11. Zhou, W. Effects of External Mechanical Loading on Stress Generation during Lithiation in Li-Ion Battery Electrodes. *Electrochim. Acta* **2015**, *185*, 28–33. [CrossRef]
12. Peled, E.; Menkin, S. Review—SEI: Past, Present and Future. *J. Electrochem. Soc.* **2017**, *164*, A1703–A1719. [CrossRef]
13. Grolleau, S.; Delaille, A.; Gualous, H.; Gyan, P.; Revel, R.; Bernard, J.; Redondo-Iglesias, E.; Peter, J. Calendar Aging of Commercial Graphite/$LiFePO_4$ Cell–Predicting Capacity Fade under Time Dependent Storage Conditions. *J. Power Sources* **2014**, *255*, 450–458. [CrossRef]
14. Eddahech, A.; Briat, O.; Vinassa, J.-M. Performance Comparison of Four Lithium–Ion Battery Technologies under Calendar Aging. *Energy* **2015**, *84*, 542–550. [CrossRef]
15. Ecker, M.; Shafiei Sabet, P.; Sauer, D.U. Influence of Operational Condition on Lithium Plating for Commercial Lithium-Ion Batteries–Electrochemical Experiments and Post-Mortem-Analysis. *Appl. Energy* **2017**, *206*, 934–946. [CrossRef]
16. Petzl, M.; Danzer, M.A. Nondestructive Detection, Characterization, and Quantification of Lithium Plating in Commercial Lithium-Ion Batteries. *J. Power Sources* **2014**, *254*, 80–87. [CrossRef]
17. Rangarajan, S.P.; Barsukov, Y.; Mukherjee, P.P. In Operando Signature and Quantification of Lithium Plating. *J. Mater. Chem. A* **2019**, *7*, 20683–20695. [CrossRef]
18. Matadi, B.P.; Geniès, S.; Delaille, A.; Chabrol, C.; de Vito, E.; Bardet, M.; Martin, J.-F.; Daniel, L.; Bultel, Y. Irreversible Capacity Loss of Li-Ion Batteries Cycled at Low Temperature Due to an Untypical Layer Hindering Li Diffusion into Graphite Electrode. *J. Electrochem. Soc.* **2017**, *164*, A2374–A2389. [CrossRef]
19. Senyshyn, A.; Mühlbauer, M.J.; Dolotko, O.; Ehrenberg, H. Low-Temperature Performance of Li-Ion Batteries: The Behavior of Lithiated Graphite. *J. Power Sources* **2015**, *282*, 235–240. [CrossRef]
20. Waldmann, T.; Quinn, J.B.; Richter, K.; Kasper, M.; Tost, A.; Klein, A.; Wohlfahrt-Mehrens, M. Electrochemical, Post-Mortem, and ARC Analysis of Li-Ion Cell Safety in Second-Life Applications. *J. Electrochem. Soc.* **2017**, *164*, A3154–A3162. [CrossRef]
21. Börner, M.; Friesen, A.; Grützke, M.; Stenzel, Y.P.; Brunklaus, G.; Haetge, J.; Nowak, S.; Schappacher, F.M.; Winter, M. Correlation of Aging and Thermal Stability of Commercial 18650-Type Lithium Ion Batteries. *J. Power Sources* **2017**, *342*, 382–392. [CrossRef]
22. Carter, R.; Klein, E.J.; Atkinson, R.W.; Love, C.T. Mechanical Collapse as Primary Degradation Mode in Mandrel-Free 18650 Li-Ion Cells Operated at 0 °C. *J. Power Sources* **2019**, *437*, 226820. [CrossRef]
23. Friesen, A.; Horsthemke, F.; Mönnighoff, X.; Brunklaus, G.; Krafft, R.; Börner, M.; Risthaus, T.; Winter, M.; Schappacher, F.M. Impact of Cycling at Low Temperatures on the Safety Behavior of 18650-Type Lithium Ion Cells: Combined Study of Mechanical and Thermal Abuse Testing Accompanied by Post-Mortem Analysis. *J. Power Sources* **2016**, *334*, 1–11. [CrossRef]
24. Wu, P.; Romberg, J.; Cheng, X.; Hao, W.; Si, H.; Li, H.; Qiu, X. A Study on Thermal Runaway of Commercial Lithium-Ion Cells: Influence of SOC, Cell Chemistry and Ageing Status on Safety Performance. In *Proceedings of the 19th Asia Pacific Automotive Engineering Conference & SAE-China Congress 2017: Selected Papers*; Society of Automotive Engineers (SAE-China), Ed.; Lecture Notes in Electrical Engineering; Springer: Singapore, 2019; Volume 486, pp. 619–627. ISBN 978-981-10-8505-5.
25. Feng, X.; Ouyang, M.; Liu, X.; Lu, L.; Xia, Y.; He, X. Thermal Runaway Mechanism of Lithium Ion Battery for Electric Vehicles: A Review. *Energy Storage Mater.* **2018**, *10*, 246–267. [CrossRef]

26. Belov, D.; Yang, M.-H. Investigation of the Kinetic Mechanism in Overcharge Process for Li-Ion Battery. *Solid State Ion.* **2008**, *179*, 1816–1821. [CrossRef]
27. Brand, M.; Glaser, S.; Geder, J.; Menacher, S.; Obpacher, S.; Jossen, A.; Quinger, D. Electrical Safety of Commercial Li-Ion Cells Based on NMC and NCA Technology Compared to LFP Technology. In Proceedings of the 2013 World Electric Vehicle Symposium and Exhibition (EVS27), Barcelona, Spain, 17–20 November 2013; pp. 1–9.
28. Kuntz, P.; Raccurt, O.; Azaïs, P.; Richter, K.; Waldmann, T.; Wohlfahrt-Mehrens, M.; Bardet, M.; Buzlukov, A.; Genies, S. Identification of Degradation Mechanisms by Post-Mortem Analysis for High Power and High Energy Commercial Li-Ion Cells after Electric Vehicle Aging. *Batteries* **2021**, *7*, 48. [CrossRef]
29. *IEC 62660-2:2018*; Secondary Lithium-Ion Cells for the Propulsion of Electric Road Vehicles-Part 2: Reliability and Abuse Testing. International Electrotechnical Commission: Geneva, Switzerland, 2018.
30. *IEC 62660-1:2018*; Secondary Lithium-Ion Cells for the Propulsion of Electric Road Vehicles-Part 1: Performance Testing. International Electrotechnical Commission: Geneva, Switzerland, 2018.
31. Shelkea, A.V.; Buston, J.E.H.; Gill, J.; Howard, D.; Williams, R.C.E.; Read, E.; Abaza, A.; Cooper, B.; Richards, P.; Wen, J.X. Combined Numerical and Experimental Studies of 21700 Lithium-Ion Battery Thermal Runaway Induced by Different Thermal Abuse. *Int. J. Heat Mass Transf.* **2022**, *194*, 123099. [CrossRef]
32. Wang, Q.; Ping, P.; Zhao, X.; Chu, G.; Sun, J.; Chen, C. Thermal Runaway Caused Fire and Explosion of Lithium Ion Battery. *J. Power Sources* **2012**, *208*, 210–224. [CrossRef]
33. Yang, H.; Bang, H.; Amine, K.; Prakash, J. Investigations of the Exothermic Reactions of Natural Graphite Anode for Li-Ion Batteries during Thermal Runaway. *J. Electrochem. Soc.* **2005**, *152*, A73. [CrossRef]
34. Liu, X.H.; Zhong, L.; Huang, S.; Mao, S.X.; Zhu, T.; Huang, J.Y. Size-Dependent Fracture of Silicon Nanoparticles During Lithiation. *ACS Nano* **2012**, *6*, 1522–1531. [CrossRef]
35. Feng, K.; Li, M.; Liu, W.; Kashkooli, A.G.; Xiao, X.; Cai, M.; Chen, Z. Silicon-Based Anodes for Lithium-Ion Batteries: From Fundamentals to Practical Applications. *Small* **2018**, *14*, 1702737. [CrossRef]
36. Choi, P.; Parimalam, B.S.; Su, L.; Reeja-Jayan, B.; Litster, S. Operando Particle-Scale Characterization of Silicon Anode Degradation during Cycling by Ultrahigh-Resolution X-Ray Microscopy and Computed Tomography. *ACS Appl. Energy Mater.* **2021**, *4*, 1657–1665. [CrossRef]
37. Mikolajczak, C.; Kahn, M.; White, K.; Long, R.T. *Lithium-Ion Batteries Hazard and Use Assessment*; Exponent Failure Analysis Associates, Inc. for Fire Protection Research Foundation: Menlo Park, CA, USA, 2011.
38. von Kolzenberg, L.; Latz, A.; Horstmann, B. Solid–Electrolyte Interphase During Battery Cycling: Theory of Growth Regimes. *ChemSusChem* **2020**, *13*, 3901–3910. [CrossRef] [PubMed]
39. Yoon-Soo, P.; Sung-Man, L. Thermal Stability of Lithiated Silicon Anodes with Electrolyte. *Bull. Korean Chem. Soc.* **2011**, *32*, 145–148. [CrossRef]
40. Profatilova, I.A.; Stock, C.; Schmitz, A.; Passerini, S.; Winter, M. Enhanced Thermal Stability of a Lithiated Nano-Silicon Electrode by Fluoroethylene Carbonate and Vinylene Carbonate. *J. Power Sources* **2013**, *222*, 140–149. [CrossRef]
41. Spotnitz, R.; Franklin, J. Abuse Behavior of High-Power, Lithium-Ion Cells. *J. Power Sources* **2003**, *113*, 81–100. [CrossRef]
42. Kim, M.; Yang, Z.; Bloom, I. Review—The Lithiation/Delithiation Behavior of Si-Based Electrodes: A Connection between Electrochemistry and Mechanics. *J. Electrochem. Soc.* **2021**, *168*, 010523. [CrossRef]
43. Li, C.; Shi, T.; Li, D.; Yoshitake, H.; Wang, H. Dependence of Thermal Stability of Lithiated Si on Particle Size. *J. Power Sources* **2016**, *335*, 38–44. [CrossRef]

batteries

MDPI

Article

Experimental Investigation on the Thermal Management for Lithium-Ion Batteries Based on the Novel Flame Retardant Composite Phase Change Materials

Yue Yu [1,2,†], Jiaxin Zhang [1,2,†], Minghao Zhu [1,2], Luyao Zhao [1,2], Yin Chen [1,2,*] and Mingyi Chen [1,2,*]

[1] School of Emergency Management, Jiangsu University, Zhenjiang 212013, China; yuyue@stmail.ujs.edu.cn (Y.Y.); 3210904015@stmail.ujs.edu.cn (J.Z.); zhuminghao@stmail.ujs.edu.cn (M.Z.); zhaoly@ujs.edu.cn (L.Z.)

[2] School of the Environment and Safety Engineering, Jiangsu University, Zhenjiang 212013, China

* Correspondence: 1000005956@ujs.edu.cn (Y.C.); chenmy@ujs.edu.cn (M.C.)

† These authors contributed equally to this work.

Abstract: Thermal management systems are critical to the maintenance of lithium-ion battery performance in new energy vehicles. While phase change materials are frequently employed in battery thermal management systems, it's important to address the concerns related to their leakage and flammability, as they can pose hazards to the safety performance of batteries. This paper proposes a novel flame retardant composite phase change material (CPCM) consisting of paraffin, high-density polyethylene, expanded graphite, ammonium polyphosphate, red phosphorus, and zinc oxide. The performance of CPCMs containing different ratios of flame retardants is investigated, and their effects when applied to battery thermal management systems are compared. The results demonstrate that the leakage rate of the flame retardant CPCMs is maintained within 1%, indicating excellent flame retardant performance and thermal management efficiency. The combination of ammonium polyphosphate and red phosphorus in the flame retardant exhibits effective synergistic effects, while zinc oxide may help phosphate compounds create their bridging bonds, which would then make it possible to construct a char layer that would separate heat and oxygen. Under a 2C discharge rate, the maximum temperature of the battery pack remains below 50 °C, and the temperature difference can be controlled within 5 °C. Even under a 3C discharge rate, the maximum temperature and temperature difference are reduced by 30.31% and 29.53%, respectively.

Keywords: lithium-ion battery; composite phase change materials; flame retardant properties; thermal properties; thermal management systems

Citation: Yu, Y.; Zhang, J.; Zhu, M.; Zhao, L.; Chen, Y.; Chen, M. Experimental Investigation on the Thermal Management for Lithium-Ion Batteries Based on the Novel Flame Retardant Composite Phase Change Materials. *Batteries* **2023**, 9, 378. https://doi.org/10.3390/batteries9070378

Academic Editor: Carlos Ziebert

Received: 14 June 2023
Revised: 9 July 2023
Accepted: 13 July 2023
Published: 14 July 2023

1. Introduction

In recent years, sustainable development has received widespread attention around the world, and carbon peaking and carbon neutrality have emerged as critical strategic goals for national development [1,2]. There is a large amount of green energy in nature, such as solar, wind, geothermal, and biomass energy, but their use has not yet reached the expected level [3,4]. To reduce environmental pollution, many countries have chosen to switch from conventional to renewable energy sources and develop clean and efficient new energy vehicle technologies. By replacing conventional gasoline cars with electric or hybrid electric ones, less carbon dioxide is released into the atmosphere. Currently, the safety of electric and hybrid vehicles has become one of the hotspots in the research field and should be given high priority and improved in terms of its thermal safety performance [5]. The battery, being the primary source of power in new energy vehicles, has a significant impact on the driving performance and safety performance of new energy vehicles [6]. Lithium-ion batteries (LIBs) are frequently employed as energy storage devices for new energy vehicles due to their long cycle life, high energy density, low self-discharge rate, and high charge/discharge rate. However, LIBs are extremely temperature-sensitive chemical products. The temperature has a direct impact on the chemical processes that take place during the charging and discharging of LIBs. LIBs are subject to short-circuiting or thermal abuse due to the lack of effective heat dissipation at high temperatures, generating large amounts of heat and triggering thermal runaway, seriously endangering the life and health of car drivers [7,8]. In addition, as the development of electric vehicles demands, the LIBs energy density was enhanced, which also means higher power and more potential risks. Therefore, the design of effective thermal management systems capable of maintaining the excellent performance of LIBs is essential in the application.

To effectively address these issues, a large number of cooling systems have been proposed in previous research, including air cooling, liquid cooling, heat pipes, and phase change material cooling [9–12]. Based on energy consumption, cooling techniques can be categorized as active cooling and passive cooling. Active cooling requires additional energy consumption, such as fans, pumps, and other equipment, while passive cooling, such as phase change material (PCM) cooling, does not require additional energy consumption. However, low thermal conductivity and low efficiency are drawbacks of air cooling-based battery pack modules [13]. Liquid cooling is currently the predominant method for commercial new energy vehicles managing battery temperature; however, the large temperature difference of the batteries and the increased heat dissipation from external devices to the battery limit the development of liquid cooling systems. PCM-based battery thermal management is a new type of thermal management technology. PCM offers the benefits of large thermal energy storage capacity and zero energy consumption since it does not produce temperature change during the phase change process, with strong thermal management capability and high efficiency [14]. Liu et al. [15] investigated the effect of different heat generation powers, air cooling airflow rates, melting point, thermal conductivity, and filling thickness of PCM on the thermal management of fast charging modules and recommended the use of PCM with a suitable melting point and a high thermal conductivity. However, the limited thermal conductivity of PCM materials is an inherent disadvantage. The coefficient for the majority of PCM materials is around 0.2 $W \cdot m^{-1} \cdot K^{-1}$, which leads to some problems when using PCM to absorb the battery heat generation. On the other hand, after the PCM has completely melted, the lower thermal conductivity of PCM makes it an insulator, inhibiting heat transmission and leading to a fast rise in temperature [16]. High thermal conductors and high melting point polymers are often added to improve the performance of PCMs. High levels of expanded graphite (EG) increase the thermal conductivity but also reduce the density of thermal energy storage. The addition of EG also affects the leakage of the composite PCM/EG during the melt solidification process [17]. Therefore, the shape stability of the PCM is improved by adding materials such as high-density polyethylene (HDPE) to reduce the amount of leakage during the cycle. Li et al. [18] prepared a thermally conductive insulating composite phase change material (CPCM), and when DM and PCM were combined in a 1:1 ratio, the CPCM thermal conductivity increased significantly. The results of the experiments show that the PCM battery module based on h-BN has a well-controlled maximum temperature and temperature difference. Wu et al. [19] prepared a PCM plate reinforced by a copper mesh sandwiched between two PCM plates, keeping a portion exposed outside the PCM plate. The copper mesh imparted good thermal conductivity and cooling efficiency to the PCM. Zou et al. [20] prepared CPCM with different ratios of graphene and multi-walled carbon nanotubes. The combination of three-dimensional nanostructured multi-walled carbon nanotubes (MWCNT) and two-dimensional nanostructured graphene was able to reduce the thermal boundary resistance. The thermal conductivity increased to 0.87 $W \cdot m^{-1} \cdot K^{-1}$ with an almost constant temperature, which facilitated the battery heat dissipation. Atinafu et al. [21] used paraffin (PA) as the PCM and metal-organic gel (cMOG) as the support material and then used the excellent thermal conductivity of BN to make a PA/BN/cMOG CPCM by impregnation. At a correspondingly high PCM load and with a negligible effect on latent heat loss after 200 thermal cycles, the CPCM had an appreciable thermal conductivity and latent heat storage capacity, with good thermal durability. Xiao et al. [22] combined the thermosetting hydrophobic polymer (THP) backbone with PA/EG via in situ radical polymerization. This new CPCM module had excellent reliability and temperature control. The maximum temperature and temperature difference could be kept below 50.9 °C and 5.0 °C, respectively. Zhang et al. [23] prepared a new flexible PCM by adding BN and silicone rubber (SR) to PA/EG. The thermal conductivity of the sample was as high as 0.95 $W \cdot m^{-1} \cdot K^{-1}$. The thermal management system using CPCM showed that at high discharge, the maximum cell temperature has dropped considerably, lower than under natural cooling conditions and that the temperature difference could be kept below 5 °C. Hu et al. [24] prepared a CPCM based on lauric acid, EG, and graphene and examined the cooling effectiveness at various discharge rates and ambient temperatures, and the battery module using this material was able to reduce the battery temperature under extreme conditions. However, the phase change components and polymer framework in CPCMs are generally combustible organic compounds, and when the battery thermal runaway develops, the CPCM can intensify the combustion and severely damage the battery module. In general, adding flame retardant powders to CPCMs may effectively improve flame retardancy [25].

Halogen-free flame retardants are commonly used because they do not produce corrosive gases and do not pose a secondary threat to human life or pollute the environment. Flame retardants can hasten the dehydration and carbonization of materials under extreme temperatures, and create a liquid and char layer to separate volatiles and heat, ultimately terminating the combustion process [26]. Zhang et al. [27] proposed a flame retardant-shaped PCM consisting of PA, EG, HDPE, and intumescent flame retardant (IFR). The interaction between EG and the complex structure formed by the PA/HDPE/IFR system at high temperatures was a phenomenon that increased the strength and stability of the char layer and formed a denser char layer. Sittisart et al. [28] studied flame retardants for PCMs and tested materials based on aluminum hydroxide, magnesium hydroxide, EG, ammonium polyphosphate (APP), pentaerythritol (PER), montmorillonite clay (MMT), and IFR, and found that PA, HDPE, APP, and EG were the most effective. Zhang et al. [29] created a flame retardant CPCM consisting of PA, ethoxyline resin (ER), APP/red phosphorus (RP), and EG and investigated the effects of different APP/RP flame retardant ratios on the thermophysical properties, thermal stability, and flame retardant properties of PCM and analyzed the morphology and structure of the carbon residue after combustion. They found that APP and RP exhibited an obvious synergistic impact. Li et al. [30] created a flame retardant flexible PCM in which melamine (MA) and triphenyl phosphate (TPP) had a strong synergistic impact and demonstrated an excellent flame retardant effect. The flame retardant effect was also good at high discharge rates, providing an effective way to suppress the thermal runaway of the battery. Xu et al. [31] microencapsulated APP to improve compatibility with PCM. The PCM with 19% microencapsulated APP content could achieve a V-0 rating. They then applied the flame retardant material to battery thermal management, the maximum module temperature dropped by 18.83 °C under 2C cycles, and the maximum temperature difference dropped by 7.5 °C. Based on the above research, it is possible to conclude that adding flame retardants to PCM can effectively delay and inhibit thermal runaway. The flame retardant CPCM has strong thermal stability, which may ensure the security of the utilization process of LIBs, and has a broad application prospect in the domains of energy storage and material thermal management.

The main objective of this research is to synthesize PCMs with an appropriate melting point range, high latent heat, and with fire retardant effects to efficiently absorb the heat generation from the battery. Although there are many types of compounds that can be selected as PCMs, PA-based PCMs were chosen for this study due to their high latent heat storage capacity across a small melting temperature range. However, PA, although having sufficient thermal storage density, has a low thermal conductivity and is highly flammable. Therefore, HDPE was used as a support material, EG was added to enhance thermal conductivity, and flame retardant powder was mixed in to enhance heat transfer and flame retardancy. Furthermore, the potential of flame retardant CPCMs for BTMS applications was investigated by comparing the thermal management effects of batteries at different discharge rates.

2. Materials Preparation and Experiment

2.1. Materials

From Haoyu New Energy Technology Co., Ltd. (Cangzhou, China), PA was acquired. From Macklin Biochemical Co., Ltd. (Shanghai, China), HDPE was acquired. EG with 99% purity was purchased from TengShengDa Carbon Graphite Co., Ltd. (Qingdao, China). The composite flame retardant powder was composed of APP, RP, and ZnO with a mass fraction of 60 wt%, 30 wt%, and 10 wt%, respectively. APP was obtained from Yien Chemical Technology Co., Ltd. (Shanghai, China). RP and ZnO were both acquired from Macklin Biochemical Co., Ltd. (Shanghai, China).

2.2. Preparation of the Hybrids Containing PA, HDPE, EG, and FR

As shown in Figure 1, flame retardant CPCMs were synthesized by the melt-blending method. The specific steps for implementation are as follows: (1) A certain amount of PA pellets were put into a beaker and placed in an oven at 100 °C for 30 min until melted. (2) The beaker was put into a heat-collecting thermostatic magnetic stirrer with a temperature set to 90 °C to melt, HDPE was added, and after continuously stirring for 30 min, the HDPE totally melted and uniformly blended with PA. (3) The EG was progressively added to the beaker and stirred for 30 min. (4) Then, the flame retardant powder was added to the beaker and stirred continuously until it was evenly mixed. (5) PA/HDPE/EG/FR was obtained by cooling at room temperature. Table 1 shows the composition of each component of the flame retardant CPCMs.

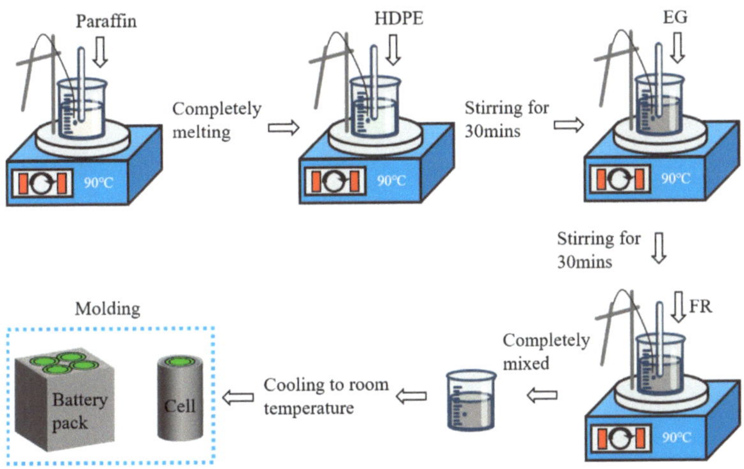

Figure 1. The preparation process of CPCMs and the produced samples.

Table 1. Proportion of the flame retardant composite PCMs.

Sample	Mass Content (wt%)			
	PA	HDPE	EG	FR
PA	100	0	0	0
CPCM15	60	20	5	15
CPCM20	55	20	5	20
CPCM25	50	20	5	25
CPCM30	45	20	5	30

2.3. Characterization of the Flame Retardant CPCMs

2.3.1. Chemical Characterization and Thermophysical Properties

The CPCM was heated in a thermostat at 70 °C for eight hours and analyzed for macroscopic thermal stability, and the mass of CPCM was recorded at two hourly intervals to give a mass loss rate.

The crystalloid phase of CPCMs was tested using the Bruker D8 ADVANCE X-ray diffractometer with Cu Kα radiation. The step size was 0.02°, the scanning angle range was 10–80°, and the scanning speed was 8°/min. The diffraction pattern was compared to observe whether the CPCM was physically mixed and whether a new substance was generated.

Hitachi Regulus-8100 field emission scanning electron microscopy was used to observe the microstructure and composition distribution of CPCMs. The sample was glued to the conductive adhesive and gold-coated for better observation.

The Xiangtan Dra-III multi-functional thermal conductivity tester was chosen to test the thermal conductivity of CPCMs. Two identical samples were obtained after the material was crushed into a cylindrical sample with a 40 mm diameter and 8 mm thickness. The final data was the average of multiple tests.

The phase change temperature and latent heat were tested using a simultaneous thermal analyzer (TA, NETZSCH) in a nitrogen (N_2) atmosphere. The sample weights were 3–5 mg. The experimental temperature range was 10–100 °C, with a heating rate of 10 °C·min^{-1}, an accuracy of ±0.1 °C, and an enthalpy of ±1%. The materials were thermogravimetrically tested using a simultaneous thermal analyzer (TA, NETZSCH) in a nitrogen (N_2) atmosphere. The sample weights were 3–10 mg, the ramp rates were 10 °C·min^{-1}, the temperature ranges were 30–800 °C, the temperature accuracy was ±0.1 °C, and the equilibrium accuracy was ±1%.

2.3.2. Flame Retardant Properties

The UL-94 vertical combustion test was carried out on the CZF-1 vertical combustion tester under the normal air atmosphere. The PCMs were pressed into a mold with dimensions of 130 mm × 11 mm × 3.2 mm. V-0 is the best rank for the UL-94 vertical combustion, while the rank was divided into four levels.

The limiting oxygen index (LOI) was determined by the JF-3 oxygen index instrument in an oxygen/nitrogen atmosphere. In the case of ignition, the oxygen concentration decreased, while with no ignition, the oxygen concentration increased until the LOI was found.

2.4. Experimental Setup

The flame retardant CPCM battery thermal management system was prepared by using the CPCMs. The flame retardant CPCMs' properties to control the temperature were studied by the temperature change of the battery module. Figure 2 depicts the specific settings of the battery module charging and discharging experiment. An Agilent data acquisition module, a computer terminal, and a thermostat made up the whole system. The cell used in this work was Sanyo NCR18650GA. It has a rated voltage of 3.7 V and a typical capacity of 3.2 Ah. The battery module was made up of four 18,650 cells soldered together in a 2 (series) × 2 (parallel) configuration. Before using the new cells, the cells were charged and discharged in cycles and left for 24 h to guarantee stability. In this experiment, the charging process is constant current and constant voltage 1C rate charging. The thermal management system was tested in a thermostat set to 25 °C (±0.5 °C). The NO-PCM battery module was naturally air-cooled only. Flame retardant CPCM served as the cooling medium for the battery module's heat dissipation system. Thermocouples were used to measure the surface temperatures of these cells, and a temperature data collection unit from National Instruments and a 7018 temperature acquisition module were used to collect the data. The thermocouples were distributed as shown in Figure 2, with the thermocouples fixed symmetrically to the cell surfaces. At 1C, 2C, and 3C discharge rates, the temperature changes of the individual cells and the battery pack were monitored, and the average temperature ($T_{battery}$) and the maximum temperature difference (ΔT) of each battery module were calculated separately from the data.

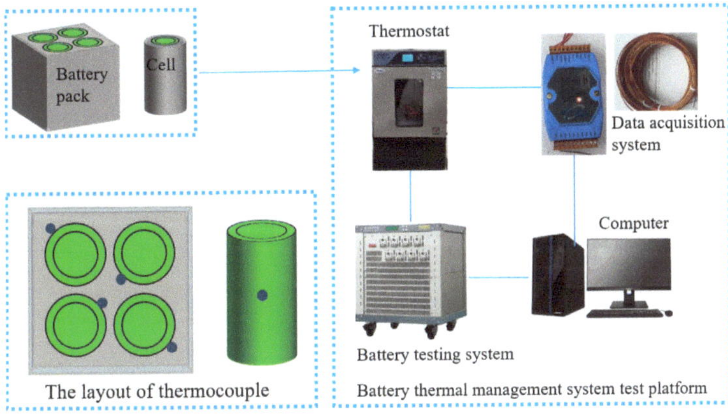

Figure 2. Diagram of the battery experimental test system.

3. Results and Discussions

3.1. The Morphology, Chemical Characterization, and Thermal Properties

As displayed in Figure 3a, at a constant temperature of 70 °C after 2 h, the PA has totally melted, thus, it has a leakage rate of 100% and does not possess any structural stability. CPCM, on the other hand, possesses excellent structural stability and retains its original shape after eight hours, exhibiting only slight leakage at the surface. Figure 3b shows that the leakage rate for all materials except CPCM15 remained within 1% due to the ability of the added EG and HDPE to encapsulate and absorb the PA, while the high level of PA in CPCM15 caused precipitation. From this, it can be deduced that good compatibility of the added materials with the appropriate proportion of PA can, to some extent, reduce the fluidity of the composite.

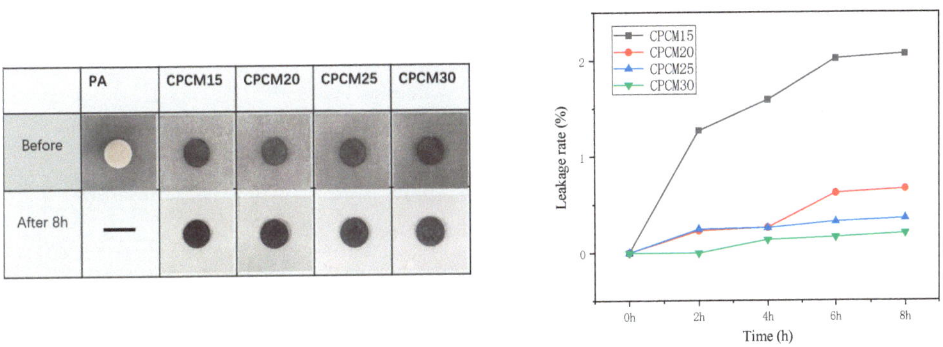

(**a**) (**b**)

Figure 3. (**a**) Images of different samples before and after continuous heating at 70 °C; (**b**) Leakage rate curve diagram of different samples.

Figure 4 depicts the chemical compatibility of the components as determined by XRD analysis. For pure PA, the diffraction peaks appearing at 21.48° and 23.85° represent the crystalline PA, and the diffraction peaks at the corresponding positions of the CPCMs are attributed to the diffraction of crystalline planes such as (110), (200) [32]. HDPE has a similar diffraction peak to PA, appearing at 21.24° and 23.77°, respectively. The diffraction peaks of EG at 26.38° and FR at 14.86°, 15.70°, 29.32°, 31.93°, and 36.42° are in general agreement with those of CPCMs. No other significant diffraction peaks appear in the CPCMs, indicating that the introduction of EG, HDPE, and flame retardants does not affect the original atomic structure of the PA. This shows that the interaction between the PA and the additives is mainly physical, and no chemical reactions that produce new substances occur.

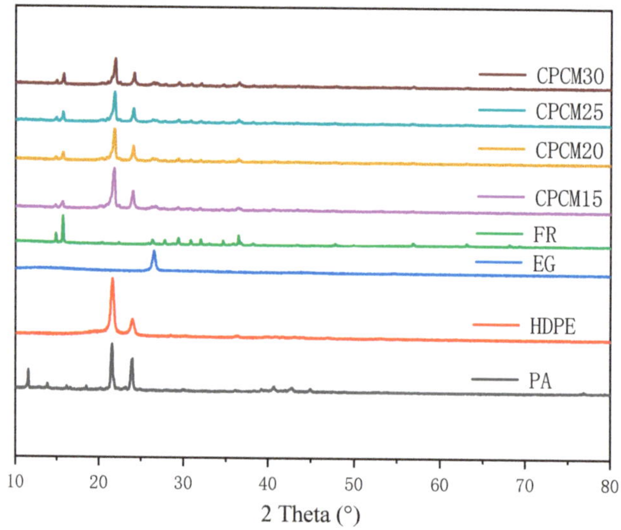

Figure 4. XRD patterns of different samples.

The microscopic morphology of the flame retardant CPCMs can be observed by SEM microscopy. The worm-like microstructure and loose surface structure of EG can be seen in Figure 5a,b, and is able to adsorb PA, HDPE, and flame retardant powder to form an effective thermally conductive network. As shown in Figure 5c–f, PA successfully adsorbs into the microporous structure of EG, HDPE effectively fills the cracks of EG as a supporting material, and the flame retardant powder is dispersed on the voids on the surface of EG, constituting a good thermal conductivity channel. It is therefore concluded that the individual components in these CPCMs are effectively combined.

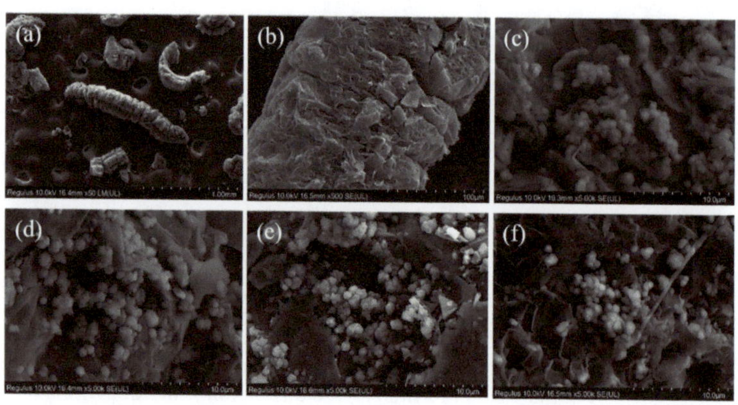

Figure 5. SEM morphology of (**a**,**b**) the worm-like structure of EG; (**c**) CPCM15; (**d**) CPCM20; (**e**) CPCM25; (**f**) CPCM30.

Figure 6 displays the flame retardant CPCMs' thermal conductivity test results. Pure PA has a thermal conductivity of 0.27 W·m^{-1}·K^{-1}. It has been discovered that the EG content is mostly responsible for the increase in thermal conductivity. EG may significantly increase the thermal conductivity of CPCMs, mainly because EG has a stable worm-like structure. This allows CPCMs to establish an efficient heat conduction network and achieve fast heat transfer. The thermal conductivity rose to a maximum of 0.77 W·m^{-1}·K^{-1} with the addition of EG, which is 2.85 times that of pure PA. Flame retardant powders fill the gaps between composite PCMs, forming a continuous heat conduction network. However, the improvement of thermal conductivity mainly depends on the EG content, and the influence of flame retardant powder is minimal, so the difference in thermal conductivity of composite PCMs is very small.

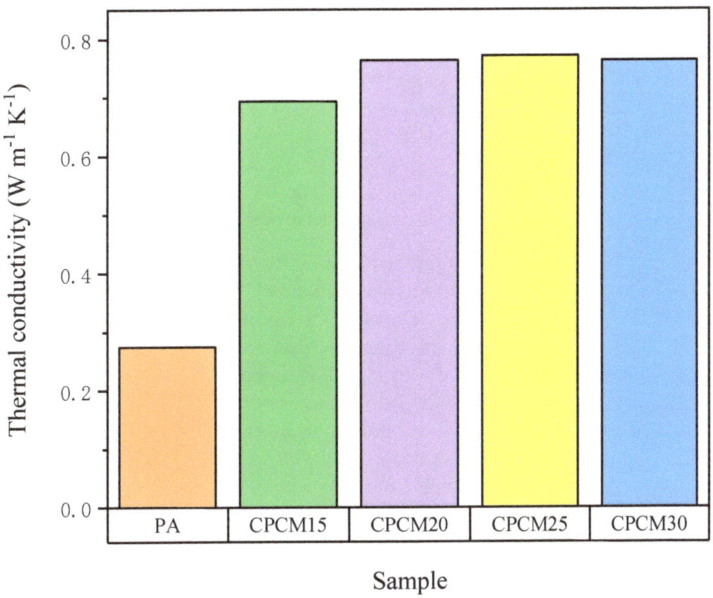

Figure 6. Thermal conductivity of different CPCM samples.

As depicted in Figure 7, there are two phase change peaks on the DSC curve. The chain is often rotated along the long axis to produce the solid-solid phase change peak, which is the initial weak phase change peak. The solid-liquid phase change, which is the cause of the second strong phase change peak, is characterized by a significant latent heat of PA. Since the phase change latent heat is mainly determined by PA, the latent heat value reduced to 132.3 J/g, 119.7 J/g, 87.63 J/g, and 85.78 J/g in the CPCM samples, respectively, presumably because the addition of HDPE and flame retardant particles reduces the PA molecules' ability to migrate thermally during the phase change process, resulting in a decrease in the latent heat of CPCM [33]. The pure PA's melting point is 42.42 °C, and after adding flame retardant powder, the CPCMs' melting point decreases slightly, remaining between 39.79–40.68 °C. This is probably due to the increased thermal conductivity and the fact that the flame retardant powder is filled with PA, resulting in continuous melting and release of latent heat at relatively low temperatures.

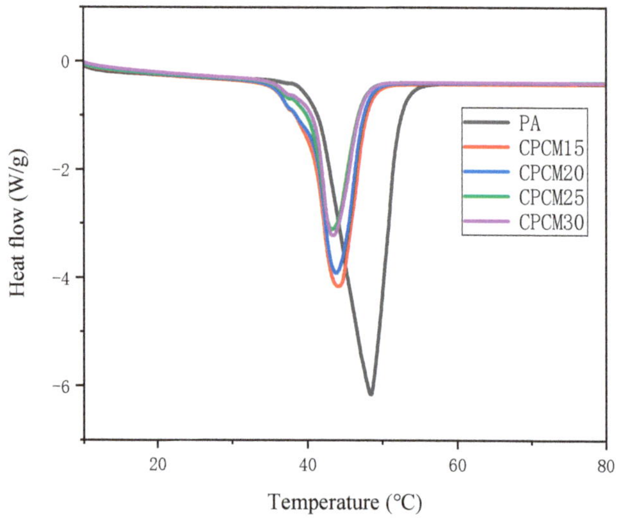

Figure 7. DSC curves of different CPCM samples.

The high thermal stability of PCMs is a requirement for their application in battery thermal management systems [34]. The TG curve in Figure 8 shows a typical two-step degradation. The first step occurs at 200–300 °C and is probably the evaporation of the PA and the thermo-oxidative degradation of the flame retardant system [35]. The decomposition of HDPE is responsible for the second step of degradation occurring at 300–500 °C. CPCMs have greater initial and complete decomposition temperatures than pure PA. Additionally, the thermal stability of CPCMs rises significantly as the flame retardant proportion rises during thermal decomposition. It can be seen that the residue of pure PA is only 1.86%. As a result of the flame retardant content increasing, the residue of CPCMs increased significantly to 14.16%, 18.16%, and 19.09% for CPCM15, CPCM20, and CPCM30, respectively, with the residue of CPCM25 reaching 25.22%. APP, RP, and ZnO as synergistic flame retardant components with EG promoted the formation of a stable char layer as a useful thermal insulation layer, building a stable three-dimensional thermally conductive skeleton and improving the flame retardant properties.

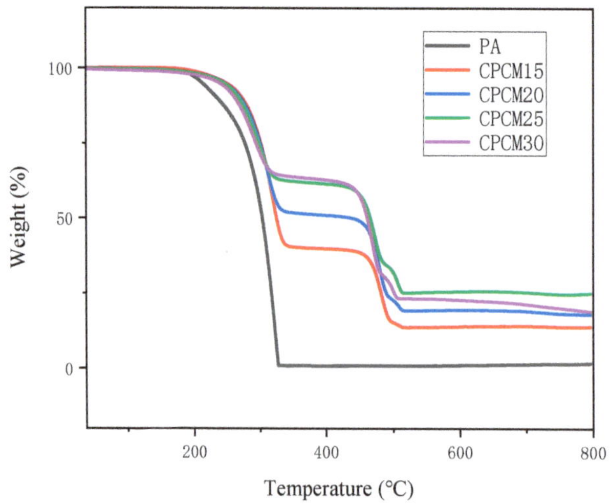

Figure 8. TG curve of different samples.

3.2. The Flammability Test of CPCMs

UL-94 vertical combustion test requires two 10 s combustion tests on the sample and the counting of the flame of the first test flame-out time (T_1) and the second test flame-out time (T_2).When the sum of T_1 and T_2 two periods of time is within 10 s, the sample achieves V-0 rating [36]. In Figure 9, the labeled periods of T_1 and T_2 are indicated with "0 s" markers. The recorded time of sample extinguishment in the two combustion tests reveals that when the flame retardant content reaches 25%, the total combustion time of T_1 and T_2 remains below 7 s. This indicates that the CPCM25 achieves the V-0 grade, demonstrating its excellent flame retardant properties. However, in the second burning test, the remaining samples were unable to be extinguished within a specific time period, leading to a judgment of "no results" (NR) for their grade assessment. Table 2 shows that UL-94 vertical combustion tests displays that CPCMs other than CPCM25 are not rated because PA itself is extremely flammable and cannot effectively suppress flame when the flame retardant content is insufficient. The APP/RP in the flame retardant has an effective synergistic effect [29] and ZnO may help phosphate compounds create the bridging bonds, which would then make it possible to construct a char layer that would separate heat and oxygen [37,38]. However, when the flame retardant content reaches 30%, the first ignition is quickly extinguished, and the second ignition continues to burn. The reason may be that too much flame retardant causes a large amount of gas to be produced, destroys the char layer, and reduces the ability to separate oxygen and dissipate heat, so the flame retardant performance is also correspondingly deteriorated. The LOI value for PA, 17.7%, would indicate that PA is flammable as a core component of CPCM. If there is not enough flame retardant present, it will not form a char layer with complete density and morphology, and therefore CPCM15 and CPCM20 have limited flame retardant effects. As the flame retardant content increased from CPCM15 to CPCM25, the LOI value of CPCM gradually increased. However, for CPCM30, the LOI value decreased to 26.1%. This decrease can be attributed to the excessive flame retardant content in CPCM30, which resulted in increased gas output and the destruction of the char layer. Consequently, the heat-oxygen blocking ability was compromised, leading to a reduction in LOI value.

3.3. Thermal Management Characteristics

3.3.1. Cell

To evaluate the impact of CPCM with different flame retardant contents on the thermal management performance of individual cells, in this experiment, the cooling effectiveness of four flame retardant CPCMs and natural convection (NO-PCM) were investigated separately. Figure 10a–c shows the charging and discharging temperature curves of a single cell under different operating conditions (1C, 2C, 3C), respectively. As shown, the battery temperature dramatically rises during the constant current discharge phase, which is due to the exponential growth of ohmic heat inside the battery and poor heat dissipation on the external surface. The heat production is larger than the heat dissipation, and the temperature reaches the peak at the end of discharge [39]. As observed in Figure 10a–c, the maximum temperature (T_{max}) of NO-PCM cell reaches 34.19 °C, 46.43 °C, and 63.71 °C at 1C, 2C, 3C, respectively. It can be inferred that the NO-PCM cell is unable to quickly transmit the heat produced by the battery, which causes heat to build up inside the battery. The CPCM acts as a buffer against temperature changes during the charging and discharging process, absorbing heat as the $T_{battery}$ rises and releasing heat as the $T_{battery}$ falls. However, the thermal management of the CPCM system at 1C and 2C rates is only slightly better than natural convection. This is because, at low currents, the T_{max} does not exceed the phase change temperature, the latent heat is not utilized, and only the thermal conductivity affects the battery's thermal management. Compared to natural convection, the cells with CPCM did not experience significant heat accumulation during 2C discharge, with the CPCM25-Cell and CPCM30-Cell keeping the T_{max} at 43.12 °C and 40.5 °C, respectively, a reduction of 3.31 °C and 5.93 °C. A comparison of the temperature data during 3C discharge showed that the maximum temperature of the cells using CPCM decreased by 9.1 °C (CPCM15), 11.27 °C (CPCM20), 13.45 °C (CPCM25), and 13.15 °C (CPCM30), respectively. This is because, at high current rates, the latent heat is completely used, as the T_{max} is higher than the phase change temperature. The results show that this CPCM not only has excellent flame retardant properties but also good temperature control.

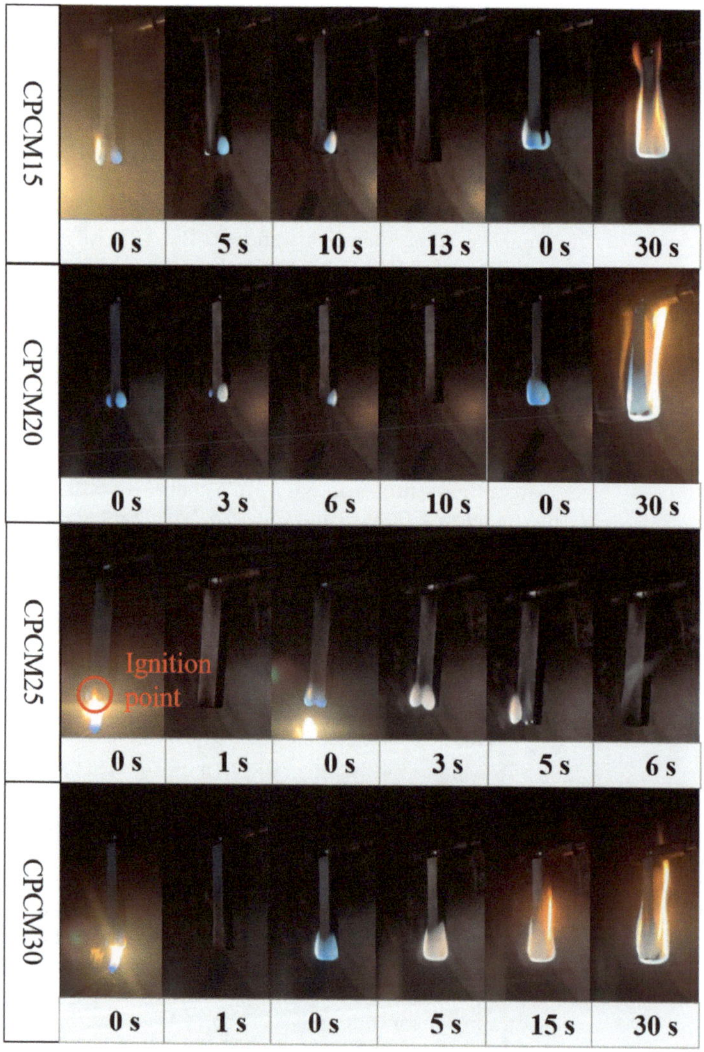

Figure 9. Vertical combustion photos of different CPCMs.

Table 2. Flammability of different samples.

	PA	CPCM15	CPCM20	CPCM25	CPCM30
UL-94	NR	NR	NR	V-0	NR
LOI	17.7%	22.4%	23.8%	28.5%	26.1%

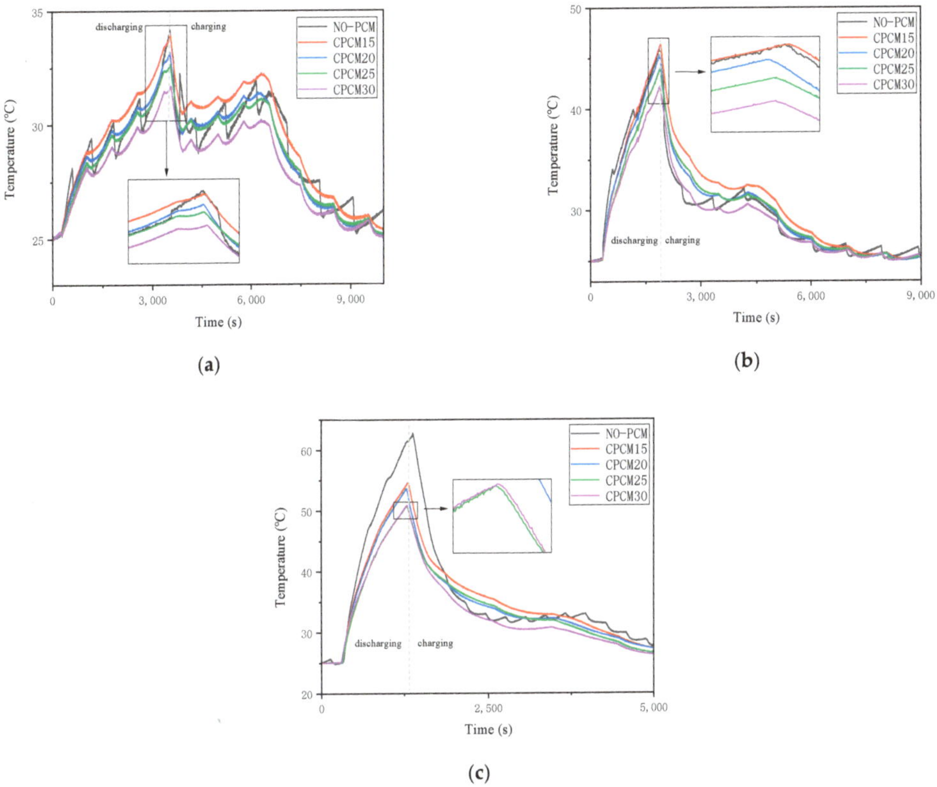

Figure 10. Cell temperature curve (**a**) 1C; (**b**) 2C; (**c**) 3C.

3.3.2. Module

To further investigate the effectiveness of flame retardant CPCMs in battery thermal management, CPCMs with flame retardant properties were applied to LIB modules to study the cooling effect. Figure 11a–c depicts the temperature fluctuation of battery packs at different discharge rates at 25 °C (±0.5 °C), respectively. Irregular temperatures in one cell of the cell module may adversely affect the entire cycling process of the cell [40]. To examine the temperature uniformity, the temperature difference curves are provided in Figure 11d–f. The battery module discharges at 1C, 2C, and 3C, and the T_{max} of the NO-PCM module reaches 41.31 °C, 61.93 °C, and 76.11 °C, respectively. Worse still, the battery module shows severe temperature inhomogeneity, with ΔT_{max} reaching 3 °C, 6.87 °C, and 8.23 °C, respectively, which leads to severe degradation of the battery module's electrochemical performance and tends to trigger thermal runaway. It is clear that by employing the CPCM, the module's T_{max} were all effectively reduced. Particularly, the T_{max} of the battery module, which uses the flame retardant CPCM25 decreased to 37.71 °C, 47.66 °C, and 53.04 °C at different discharge rates, respectively, showing a significant downward trend compared to the natural convection cooling method. The decrease was 3.6 °C and 14.27 °C at 1C and 2C discharge rates, respectively. In particular, the T_{max} dropped significantly by 23.07 °C at the high 3C discharge rate. In addition, CPCM significantly reduces the ΔT between cells at various discharge rates, which helps the battery modules maintain a uniform temperature. The small rate caused little temperature inhomogeneity at a 1C discharge rate. The ΔT_{max} reaches 6.87 °C for NO-PCM as a 2C discharge rate, while CPCM25 reduces the ΔT_{max} to 4.56 °C. In particular, for the 3C discharge rate, the ΔT_{max} increases to 8.23 °C, while the ΔT_{max} for the modules using the CPCM is effectively reduced for both, in particular, the ΔT_{max} for the modules using the CPCM25 is reduced by 29.53%, remaining within

5.8 °C. As can be observed, the ΔT of the CPCM modules steadily slows down at 2C and 3C discharge rates. This is mainly due to the fact that the battery module temperature exceeds the phase change range of the CPCM, and the heat generation is absorbed by CPCM, thus achieving the purpose of controlling the $T_{battery}$. It is important to note that although the flame retardant CPCM has a good cooling effect, its heat dissipation performance decreases when the battery heat generation is much greater than the heat dissipation efficiency of the CPCM due to its limited heat transmit efficiency.

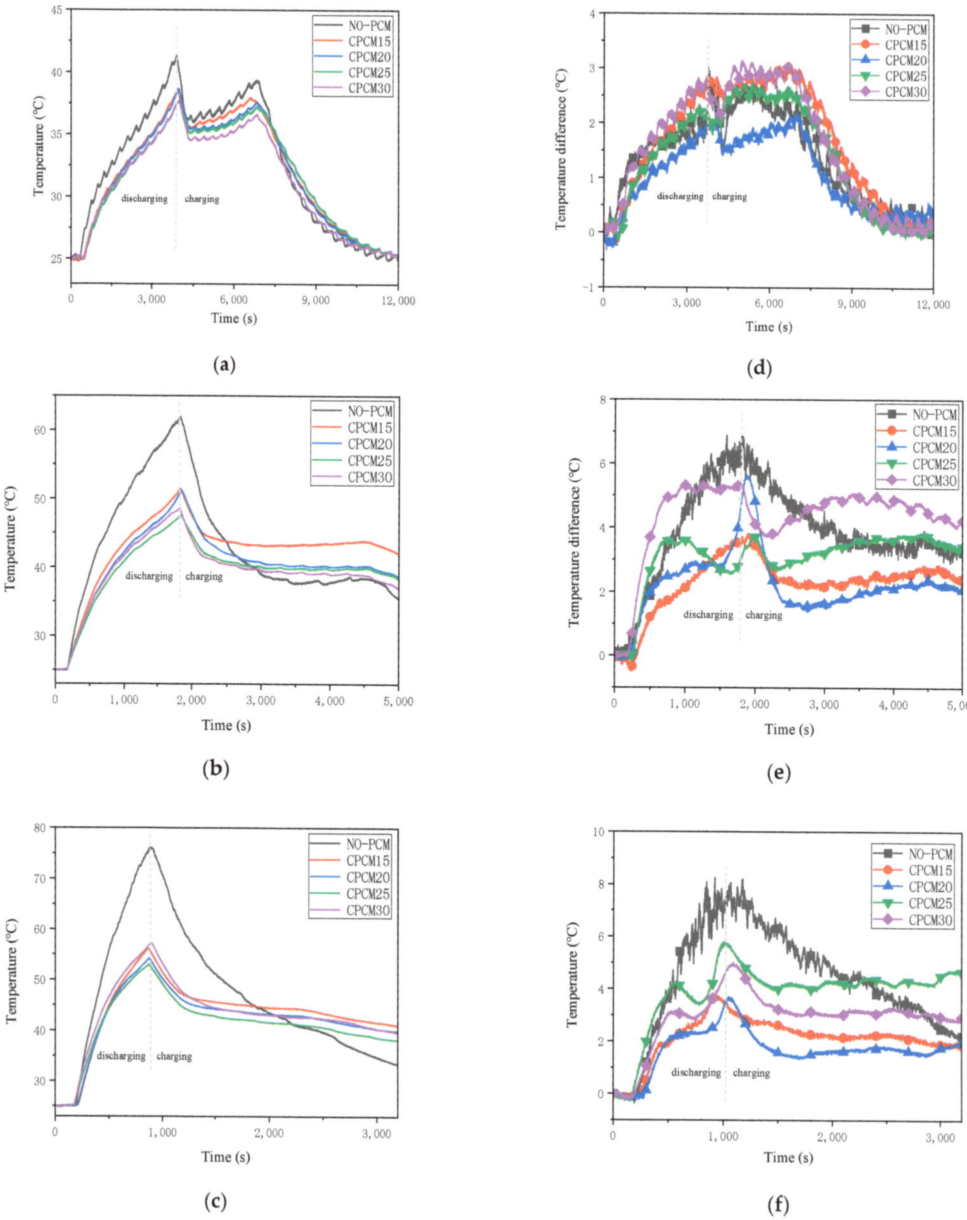

Figure 11. Temperature curve of the battery module (**a**) 1C; (**b**) 2C; (**c**) 3C. Temperature difference curve of the battery module (**d**) 1C; (**e**) 2C; (**f**) 3C.

4. Conclusions

To increase the thermal safety of LIBs, a novel flame retardant CPCM consisting of PA/HDPE/EG/APP/RP/ZnO was developed by the melt blending method, and how varying flame retardant ratios affected the thermophysical and thermal management characteristics of the CPCM were investigated. Specific conclusions are as follows:

(1) XRD results show that the mixture between the PCM and the additives in the flame retardant CPCM made in this experiment is a physical mixture, and no chemical reaction occurs. The flame retardant CPCM has excellent structural stability, and after eight hours of continuous heat, the leakage rate is kept within 1%.

(2) CPCM25 with 25 wt% flame retardant achieved a V-0 rating in the UL-94 test, a thermal conductivity of 0.77 $W \cdot m^{-1} \cdot K^{-1}$, and a latent heat value of 87.63 J/g. In the thermogravimetric test, the residue at 800 °C reached 25.22%.

(3) Battery modules with flame retardant CPCMs offer excellent thermal management. Under a 2C discharge rate, the T_{max} of the battery pack remains below 50 °C, and the ΔT_{max} can be controlled within 5 °C. Even under a 3C discharge rate, the T_{max} and ΔT_{max} are reduced by 30.31% and 29.53%, respectively.

This work has the significance of enhancing safety performance and extending battery life.

Author Contributions: Conceptualization, M.C. and Y.Y.; Methodology, M.C. and Y.C.; Investigation, J.Z. and Y.Y.; Writing—Review & Editing, Y.Y., M.C., M.Z., J.Z. and L.Z. All authors have read and agreed to the published version of the manuscript.

Funding: This research was funded by the National Natural Science Foundation of China (52204213), the Special Scientific Research Project of the School of Emergency Management of Jiangsu University (KY-B-09, KY-D-03), and the Project of Research on Educational Reform and Talent Development of School of Emergency Management of Jiangsu University (JG-03-03, JG-03-05, JG-04-08, JG-04-10).

Data Availability Statement: Data are contained within the article.

Conflicts of Interest: The authors declare no conflict of interest.

References

1. Caineng, Z.O.; Xiong, B.; Huaqing, X.U.; Zheng, D.; Zhixin, G.E.; Ying, W.A.; Jiang, L.; Songqi, P.; Songtao, W. The role of new energy in carbon neutral. *Pet. Explor. Dev.* **2021**, *48*, 411–420.
2. Wei, Y.M.; Chen, K.; Kang, J.N.; Chen, W.; Wang, X.Y.; Zhang, X. Policy and Management of Carbon Peaking and Carbon Neutrality: A Literature Review. *Engineering* **2022**, *14*, 52–63. [CrossRef]
3. Holechek, J.L.; Geli, H.M.E.; Sawalhah, M.N.; Valdez, R. A Global Assessment: Can Renewable Energy Replace Fossil Fuels by 2050? *Sustainability* **2022**, *14*, 4792. [CrossRef]
4. Chen, J.; Su, F.; Jain, V.; Salman, A.; Tabash, M.I.; Haddad, A.M.; Zabalawi, E.; Abdalla, A.A.; Shabbir, M.S. Does Renewable Energy Matter to Achieve Sustainable Development Goals? The Impact of Renewable Energy Strategies on Sustainable Economic Growth. *Front. Energy Res.* **2022**, *10*, 829252. [CrossRef]
5. Liu, W.; Placke, T.; Chau, K.T. Overview of batteries and battery management for electric vehicles. *Energy Rep.* **2022**, *8*, 4058–4084. [CrossRef]
6. Dai, H.; Jiang, B.; Hu, X.; Lin, X.; Wei, X.; Pecht, M. Advanced battery management strategies for a sustainable energy future: Multilayer design concepts and research trends. *Renew. Sustain. Energy Rev.* **2021**, *138*, 110480. [CrossRef]
7. Feng, X.; Ren, D.; He, X.; Ouyang, M. Mitigating Thermal Runaway of Lithium-Ion Batteries. *Joule* **2020**, *4*, 743–770. [CrossRef]
8. Wang, Q.; Ping, P.; Zhao, X.; Chu, G.; Sun, J.; Chen, C. Thermal runaway caused fire and explosion of lithium ion battery. *J. Power Sources* **2012**, *208*, 210–224. [CrossRef]
9. Qi, X.; Sidi, M.O.; Tlili, I.; Ibrahim, T.K.; Elkotb, M.A.; El-Shorbagy, M.A.; Li, Z. Optimization and sensitivity analysis of extended surfaces during melting and freezing of phase changing materials in cylindrical Lithium-ion battery cooling. *J. Energy Storage* **2022**, *51*, 104545. [CrossRef]
10. Mehrabi-Kermani, M.; Houshfar, E.; Ashjaee, M. A novel hybrid thermal management for Li-ion batteries using phase change materials embedded in copper foams combined with forced-air convection. *Int. J. Therm. Sci.* **2019**, *141*, 47–61. [CrossRef]
11. Kong, D.; Peng, R.; Ping, P.; Du, J.; Chen, G.; Wen, J. A novel battery thermal management system coupling with PCM and optimized controllable liquid cooling for different ambient temperatures. *Energy Convers. Manag.* **2020**, *204*, 112280. [CrossRef]
12. Chen, K.; Hou, J.; Song, M.; Wang, S.; Wu, W.; Zhang, Y. Design of battery thermal management system based on phase change material and heat pipe. *Appl. Therm. Eng.* **2021**, *188*, 116665. [CrossRef]

13. Chen, F.; Huang, R.; Wang, C.; Yu, X.; Liu, H.; Wu, Q.; Qian, K.; Bhagat, R. Air and PCM cooling for battery thermal management considering battery cycle life. *Appl. Therm. Eng.* **2020**, *173*, 115154. [CrossRef]
14. Weng, J.; Huang, Q.; Li, X.; Zhang, G.; Ouyang, D.; Chen, M.; Yuen, A.C.Y.; Li, A.; Lee, E.W.M.; Yang, W.; et al. Safety Issue on PCM-based Battery Thermal Management: Material Thermal Stability and System Hazard Mitigation. *Energy Storage Mater.* **2022**, *53*, 580–612. [CrossRef]
15. Liu, X.; Zhang, Z.; Zhu, C.; Wang, F.; Zhao, D.; Li, Z.; Liu, Y.; Zhang, H.; Jiang, H. Experimental investigation on the effect of phase change materials for thermal management improvement of the fast charging power module. *Case Stud. Therm. Eng.* **2023**, *42*, 102711. [CrossRef]
16. Heyhat, M.M.; Mousavi, S.; Siavashi, M. Battery thermal management with thermal energy storage composites of PCM, metal foam, fin and nanoparticle. *J. Energy Storage* **2020**, *28*, 101235. [CrossRef]
17. Jiang, G.; Huang, J.; Fu, Y.; Cao, M.; Liu, M. Thermal optimization of composite phase change material/expanded graphite for Li-ion battery thermal management. *Appl. Therm. Eng.* **2016**, *108*, 1119–1125. [CrossRef]
18. Li, J.; Tang, A.; Shao, X.; Jin, Y.; Chen, W.; Xia, D. Experimental evaluation of heat conduction enhancement and lithium-ion battery cooling performance based on h-BN-based composite phase change materials. *Int. J. Heat Mass Transf.* **2022**, *186*, 122487. [CrossRef]
19. Wu, W.; Yang, X.; Zhang, G.; Ke, X.; Wang, Z.; Situ, W.; Li, X.; Zhang, J. An experimental study of thermal management system using copper mesh-enhanced composite phase change materials for power battery pack. *Energy* **2016**, *113*, 909–916. [CrossRef]
20. Zou, D.; Ma, X.; Liu, X.; Zheng, P.; Hu, Y. Thermal performance enhancement of composite phase change materials (PCM) using graphene and carbon nanotubes as additives for the potential application in lithium-ion power battery. *Int. J. Heat Mass Transf.* **2018**, *120*, 33–41. [CrossRef]
21. Atinafu, D.G.; Dong, W.; Wang, J.; Huang, X.; Wang, J.; Gao, H.; Wang, G. Synthesis and Characterization of Paraffin/Metal Organic Gel Derived Porous Carbon/Boron Nitride Composite Phase Change Materials for Thermal Energy Storage. *Eur. J. Inorg. Chem.* **2018**, *48*, 5167–5175. [CrossRef]
22. Xiao, C.; Wu, X.; Dong, X.; Ye, G.; Zhang, G.; Yang, X. Ultrareliable Composite Phase Change Material for Battery Thermal Management Derived from a Rationally Designed Phase Changeable and Hydrophobic Polymer Skeleton. *ACS Appl. Energy Mater.* **2021**, *4*, 3832–3841. [CrossRef]
23. Zhang, Y.; Huang, J.; Cao, M.; Liu, Z.; Chen, Q. A novel flexible phase change material with well thermal and mechanical properties for lithium batteries application. *J. Energy Storage* **2021**, *44*, 103433. [CrossRef]
24. Hu, S.; Wang, S.; Zhang, Y.; Ma, C.; Wu, S.; Li, L. Effect of passive thermal management system on the electro-thermal performance of battery module. *Int. J. Therm. Sci.* **2023**, *183*, 107842. [CrossRef]
25. Dai, X.; Kong, D.; Du, J.; Zhang, Y.; Ping, P. Investigation on effect of phase change material on the thermal runaway of lithium-ion battery and exploration of flame retardancy improvement. *Process Saf. Environ. Prot.* **2022**, *159*, 232–242. [CrossRef]
26. Huang, Q.; Li, X.; Zhang, G.; Weng, J.; Wang, Y.; Deng, J. Innovative thermal management and thermal runaway suppression for battery module with flame retardant flexible composite phase change material. *J. Clean. Prod.* **2022**, *330*, 129718. [CrossRef]
27. Zhang, P.; Hu, Y.; Song, L.; Ni, J.; Xing, W.; Wang, J. Effect of expanded graphite on properties of high-density polyethylene/paraffin composite with intumescent flame retardant as a shape-stabilized phase change material. *Sol. Energy Mater. Sol. Cells* **2010**, *94*, 360–365. [CrossRef]
28. Sittisart, P.; Farid, M.M. Fire retardants for phase change materials. *Appl. Energy* **2011**, *88*, 3140–3145. [CrossRef]
29. Zhang, J.; Li, X.; Zhang, G.; Wu, H.; Rao, Z.; Guo, J.; Zhou, D. Experimental investigation of the flame retardant and form-stable composite phase change materials for a power battery thermal management system. *J. Power Sources* **2020**, *480*, 229116. [CrossRef]
30. Li, Y.; Wang, T.; Li, X.; Zhang, G.; Chen, K.; Yang, W. Experimental investigation on thermal management system with flame retardant flexible phase change material for retired battery module. *Appl. Energy* **2022**, *327*, 120109. [CrossRef]
31. Xu, Z.; Chen, W.; Wu, T.; Wang, C.; Liang, Z. Thermal management system study of flame retardant solid–solid phase change material battery. *Surf. Interfaces* **2023**, *36*, 102558. [CrossRef]
32. Lin, X.; Zhang, X.; Ji, J.; Liu, L.; Wu, Y.; Yang, M.; Lu, D.; Zheng, H. Development of flexible form-stable phase change material with enhanced electrical resistance for thermal management. *J. Clean. Prod.* **2021**, *311*, 127517. [CrossRef]
33. Liao, H.; Duan, W.; Liu, Y.; Wang, Q.; Wen, H. Flame retardant and leaking preventable phase change materials for thermal energy storage and thermal regulation. *J. Energy Storage* **2021**, *35*, 102248. [CrossRef]
34. Ma, Y.; Yang, H.; Zuo, H.; Zuo, Q.; He, X.; Chen, W.; Wei, R. EG@Bi-MOF derived porous carbon/lauric acid composite phase change materials for thermal management of batteries. *Energy* **2023**, *272*, 127180. [CrossRef]
35. Song, G.; Ma, S.; Tang, G.; Yin, Z.; Wang, X. Preparation and characterization of flame retardant form-stable phase change materials composed by EPDM, paraffin and nano magnesium hydroxide. *Energy* **2010**, *35*, 2179–2183. [CrossRef]
36. Kempel, F.; Schartel, B.; Marti, J.M.; Butler, K.M.; Rossi, R.; Idelsohn, S.R.; Onate, E.; Hofmann, A. Modelling the vertical UL 94 test: Competition and collaboration between melt dripping, gasification and combustion. *Fire Mater.* **2015**, *39*, 570–584. [CrossRef]
37. Liu, W.; Wang, Z.; Su, S.; Wu, H.; Sun, M.; Tang, L. Synergistic flame retardancy of ZnO and piperazine pyrophosphate/melamine cyanurate in polypropylene. *J. Vinyl Addit. Technol.* **2023**, *29*, 202–219. [CrossRef]
38. Ma, Y.; Wang, J.; Xu, Y.; Wang, C.; Chu, F. Effect of zinc oxide on properties of phenolic foams/halogen-free flame retardant system. *J. Appl. Polym. Sci.* **2015**, *132*, 42730. [CrossRef]

39. Wu, W.; Liu, J.; Liu, M.; Rao, Z.; Deng, H.; Wang, Q.; Qi, X.; Wang, S. An innovative battery thermal management with thermally induced flexible phase change material. *Energy Convers. Manag.* **2020**, *221*, 113145. [CrossRef]
40. Pradeep, G.M.; Sankaramoorthy, T.; Elango, M.; Kumar, T.N.; Girimurugan, R. Structural analysis and mechanical properties of thermal battery by flexible phase change materials [P.C.M.]. *Mater. Today Proc.* **2021**, *56*, 3196–3200. [CrossRef]

batteries

MDPI

Review

Recent Advancements in Battery Thermal Management Systems for Enhanced Performance of Li-Ion Batteries: A Comprehensive Review

Amin Rahmani, Mahdieh Dibaj and Mohammad Akrami *

Department of Engineering, University of Exeter, Exeter EX4 4PJ, UK; ar949@exeter.ac.uk (A.R.); m.dibaj3@exeter.ac.uk (M.D.)
* Correspondence: m.akrami@exeter.ac.uk

Abstract: Li-ion batteries are crucial for sustainable energy, powering electric vehicles, and supporting renewable energy storage systems for solar and wind power integration. Keeping these batteries at temperatures between 285 K and 310 K is crucial for optimal performance. This requires efficient battery thermal management systems (BTMS). Many studies, both numerical and experimental, have focused on improving BTMS efficiency. This paper presents a comprehensive review of the latest BTMS designs developed in 2023 and 2024, with a focus on recent advancements and innovations. The primary objective is to evaluate these new designs to identify key improvements and trends. This review categorizes BTMS designs into four cooling methods: air-cooling, liquid-cooling, phase change material (PCM)-cooling, and thermoelectric cooling. It provides a detailed analysis of each method. It also offers a unique examination of hybrid cooling BTMSs, classifying them based on their impact on the cooling process. A hybrid-cooling BTMS refers to a method that combines at least two of the four types of BTMS (air-cooling, liquid-cooling, PCM-cooling, and thermoelectric-cooling) to enhance thermal management efficiency. Unlike previous reviews, this study emphasizes the novelty of recent designs and the substantial results they achieve, offering significant insights and recommendations for future research and development in BTMS. By highlighting the latest innovations and providing an in-depth analysis, this paper serves as a valuable resource for researchers and engineers aiming to enhance battery performance and sustainability through advanced thermal management solutions.

Keywords: battery thermal management systems; Lithium-ion batteries; air cooling; liquid cooling; PCM-cooling; thermoelectric-cooling

Citation: Rahmani, A.; Dibaj, M.; Akrami, M. Recent Advancements in Battery Thermal Management Systems for Enhanced Performance of Li-Ion Batteries: A Comprehensive Review. *Batteries* **2024**, *10*, 265. https://doi.org/10.3390/batteries10080265

Academic Editor: Mingyi Chen

Received: 24 June 2024
Revised: 22 July 2024
Accepted: 23 July 2024
Published: 26 July 2024

1. Introduction

Due to the global energy crisis and environmental pollution, there is an urgent need to shift to safer, cleaner, and more efficient renewable energy sources, which necessitates effective energy storage solutions. Transportation is a major contributor to the global energy problem since it mainly depends on fossil fuels, resulting in excessive energy consumption and the emission of greenhouse gases [1]. This highlights the importance of moving towards cleaner and more sustainable transportation options. Electric vehicles (EVs) powered by Lithium-ion (Li-ion) batteries present a promising solution to the energy crisis by reducing dependence on fossil fuels and lowering greenhouse gas emissions in the transportation sector. The operating temperature and internal heat generation of Li-ion batteries have a significant impact on their performance, lifespan, and safety [2]. Hence, a battery thermal management system (BTMS) is crucial to protect batteries from the negative impacts of increased temperatures and internal heat generation.

The present review provides the basic concept of experimental and numerical works conducted in 2023 and 2024, including air-cooling, liquid-cooling, PCM-cooling, and thermoelectric-cooling base hybrid BTMSs.

1.1. Importance of BTMS

A battery thermal management system (BTMS) is vital for maintaining the optimal performance and longevity of lithium-ion battery packs, which consist of multiple cells arranged in various configurations. The efficiency of these batteries is highly temperature-dependent, as internal heat generated during charge and discharge cycles can cause uneven temperature distribution, reducing the battery's lifespan and effectiveness [3]. Studies have shown that hotspots often form near the electrodes, leading to temperature non-uniformity [4]. To address these challenges and enhance battery performance in electric and hybrid vehicles, effective BTMS is essential, as highlighted by numerous researchers in the field.

1.2. Recent Advances and Critical Analysis of BTMS

In recent years, significant advancements have been made in the field of battery thermal management systems (BTMS), driven by the need to enhance the performance, safety, and longevity of lithium-ion batteries, particularly in electric vehicles and renewable energy storage systems. This section provides a comprehensive analysis of these advancements, critically evaluating the latest research and technological innovations.

Air-Cooling: Air-cooling methods have evolved with various modifications to enhance performance. Recent designs, such as honeycomb structures and multiple inlet/outlet air cooling systems, have shown substantial improvements in cooling efficiency and temperature uniformity.

Liquid-Cooling: Liquid-cooling systems, particularly those with advanced cold plate and cooling channel designs, offer superior thermal management capabilities. Studies on bionic spiral fins and liquid cooling plates have demonstrated significant enhancements in heat dissipation and temperature control.

PCM-Cooling: The use of PCM in BTMS provides passive thermal management, effectively absorbing heat during phase transitions. Hybrid PCM systems that combine PCM with air or liquid cooling have shown improved thermal performance and energy efficiency.

Thermoelectric Cooling: Recent advancements in thermoelectric cooling (TEC) have focused on integrating TEC with other cooling methods for precise temperature control and enhanced efficiency. Studies have highlighted the potential of TEC to maintain battery temperatures within optimal ranges, even under high thermal loads.

1.3. The Motivation for This Work

The current study presents new, innovative BTMS ideas in various categories. Figure 1 shows the battery pack cooling classification used in this study. The reason for this review is that, during 2023 and 2024, more innovative ideas have been presented by researchers as simulation software and methods have been developed at a rapid pace.

This study provides a pioneering and comprehensive analysis of the most recent advancements in battery thermal management systems (BTMS) for lithium-ion batteries, focusing on the innovations developed in 2023 and 2024. Unlike previous reviews, this study not only categorizes BTMS into traditional methods such as air-cooling, liquid-cooling, PCM-cooling, and thermoelectric cooling, but also emphasizes the groundbreaking hybrid systems that integrate multiple cooling technologies for superior thermal management. This work stands out by critically analyzing the performance improvements achieved through novel design modifications, such as the implementation of biomimetic structures, advanced materials like graphene-enhanced PCM, and optimized geometric configurations. By presenting a detailed comparison of the techno-economic aspects of these innovative cooling methods, this study offers significant insights that can drive future research and development, ultimately enhancing the efficiency, safety, and longevity of lithium-ion batteries in electric vehicles and renewable energy storage systems. The thorough evaluation and unique categorization of the latest BTMS designs underscore the importance of this study as a valuable resource for researchers and engineers dedicated to advancing battery technology.

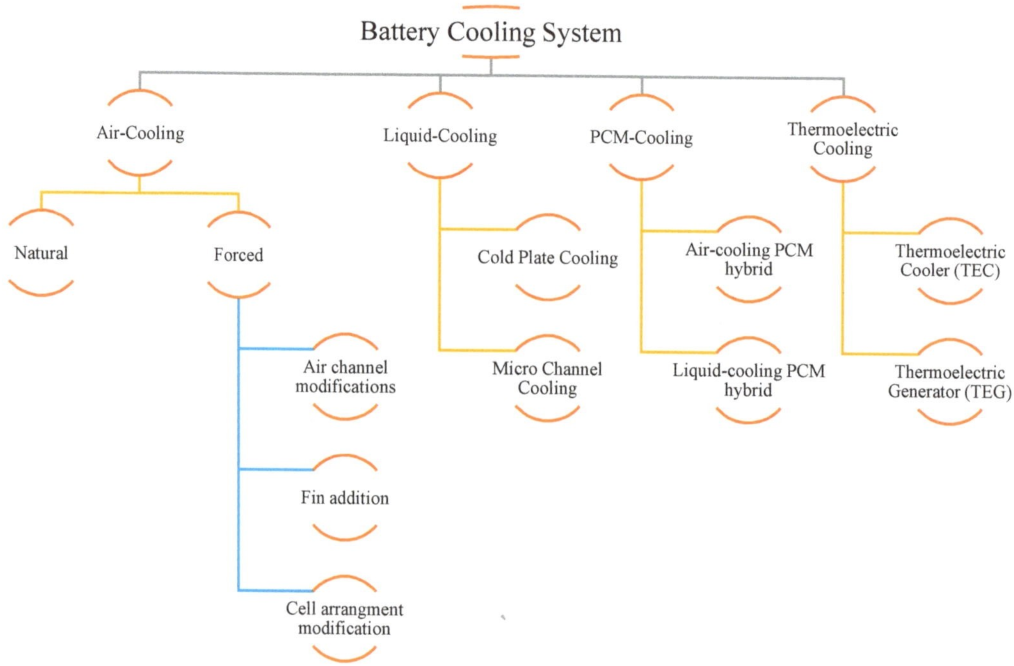

Figure 1. The battery pack cooling classification used in this study is based on the reviewed papers in this study.

2. Recent Progress in BTMS Studies

There have been many documented studies that have extensively explored various forms of battery thermal management systems (BTMS) through experimentation and numerical analysis. Scientists have carried out experiments at many levels, including cell, module, and pack to examine how batteries operate in terms of temperature under various conditions. Additionally, many researchers have utilized Computational Fluid Dynamics (CFD) for the purpose of mathematical modelling. Experimental research is essential for determining the relationships between dependent and independent variables, and for comprehending the effects of parameters. This section emphasizes the efforts of researchers in advancing computational and experimental methods to effectively control the thermal performance of lithium-ion batteries in different categories.

2.1. *Air-Cooled BTMS*

The air-cooling method, which is the natural method of BTMS cooling, can be categorized by free and forced convection. Many advantages are reported for this method, including: its simplicity, low cost, no leakage concern, and easier maintenance. The air-cooling method cannot respond to the demand of battery-pack cooling in high ambient temperatures, especially in natural air cooling. Thus, many modifications have been suggested by researchers, such as modifying air-flow channels, using fin structures, and modifying cell arrangements. Electric vehicles such as the Nissan Leaf, Volkswagen e-Golf, Chevrolet Spark EV, and early models of the BMW i3 utilize air cooling in their battery thermal management systems. Table 1 shows the recent, documented studies on air-cooling BTMS in 2023 and 2024. As detailed in Table 1, experimental and numerical studies have been conducted on air-cooled BTMS with different modifications. A summary of each modification type is presented in this section. The presented data in Table 1 show that, among different methods used to modify the air-cooling method, modifying the air channel can have more impact on the performance of BTMS.

Air-channel modifications: This method of air cooling, which involves altering the patterns of airflow to cool batteries, has been extensively researched and examined by several scientists in the last few years [4]. Continuing their idea, Luo et al. [5] developed an innovative X-type double inlet and outlet, symmetrical, air-cooled battery thermal management system (BTMS) designed to address issues of high temperature, temperature differences, and power dissipation in battery packs. The study showed that the proposed X-type BTMS significantly improves performance by reducing the maximum temperature, temperature difference, and power dissipation by 4.33 K, 74%, and 62.9%, respectively, compared to symmetrical air-cooling BTMS. The optimization and heat transfer correlations provide valuable insights for future BTMS design and improvement. Based on the optimization process, a modified X-type channel was designed, which decreased the maximum temperature by 2 K in comparison with the basic design. Furthermore, Yang et al. [6] investigated the thermal performance of a honeycomb-type, cylindrical lithium-ion battery pack incorporating an air-distribution plate (ADP) and bionic heat sinks. Their study demonstrated that the ADP significantly reduces maximum temperature and temperature differences within the battery pack by 1.7 K and 7.0 K, respectively. The addition of bionic heat sinks further enhances thermal performance and maintains temperature differences within 2 K, thus providing valuable insights for the design of efficient battery thermal management systems (BTMS). Duan et al. [7] developed a novel design method for a multiple inlet/outlet air-cooling frame for pouch lithium-ion batteries using thermal–fluid coupling topology optimization. Their research demonstrated that the optimized cooling frame significantly enhances cooling efficiency and temperature uniformity, reducing the maximum temperature by 4.79% and the temperature difference by 36.40%, compared to traditional frames, making it highly applicable to electric vehicles and hybrid electric vehicles.

Cell-arrangement modification: Cell arrangement is another field of interest among scientists for improving the efficiency of air-cooling BTMS [8]. As an innovative idea, Shen et al. [9] designed a modified Z-shaped, air-cooled battery thermal management system (BTMS) with a non-vertical structure to enhance the thermal behavior of lithium-ion power batteries in electric vehicles. Their study showed that this new system reduced the maximum temperature from 38.15 °C to 34.14 °C, and the temperature difference from 2.59 °C to 1.97 °C. This modified design improves cooling performance and temperature uniformity, offering significant engineering value for the advancement of BTMS in electric vehicles. Furthermore, Kashyap et al. [10] presented an optimized design for a staggered-arranged battery thermal management system (BTMS) using physics-based simulations and evolutionary algorithms. Their research showed that this integrated approach reduced maximum temperature by 0.627%, maximum temperature difference by 49.18%, increased pressure drop by 102.379%, and volume by 6.804%, compared to traditional configurations. This work provides significant improvements in thermal management for lithium-ion battery packs, offering valuable insights for future BTMS designs.

Adding fin structure: Adding metal fins to an air-cooled battery thermal management system (BTMS) significantly enhances cooling efficiency by increasing the surface area for heat dissipation, which improves heat-transfer rates and reduces the maximum temperature and temperature gradients within the battery pack [11]. Chaudhari et al. [12] conducted an experimental and computational analysis of a lithium-ion battery thermal management system (BTMS) using radial fins for air cooling. Their study revealed that forced convection with radial fins significantly enhanced cooling efficiency, reducing the maximum battery temperature by up to 39.23%, compared to natural convection. This research highlights the effectiveness of radial fins in improving the thermal performance and safety of lithium-ion battery packs in electric vehicles. Luo et al. [13] proposed a direct flow cooling battery thermal management system (DFC-BTMS) with bod baffles and a lipid organic liquid coolant to enhance thermal performance in electric vehicles. Their experimental and simulation study showed that the optimized DFC-BTMS achieved a maximum average surface temperature of 24.789 °C and a maximum temperature difference of 2.734 °C, providing an innovative solution for efficient battery thermal management.

Table 1. An outline of the work that has been done on air-cooled BTMS.

No.	Modification/ Novelty	Type of Study	Remarks	Geometry	Ref.	Modification Method
1	X-type double inlet and outlet BP is used.	CFD	T_{max} and ΔT_{max} are deceased 4.33 K and 74% respectively. A heat transfer correlation is obtained.	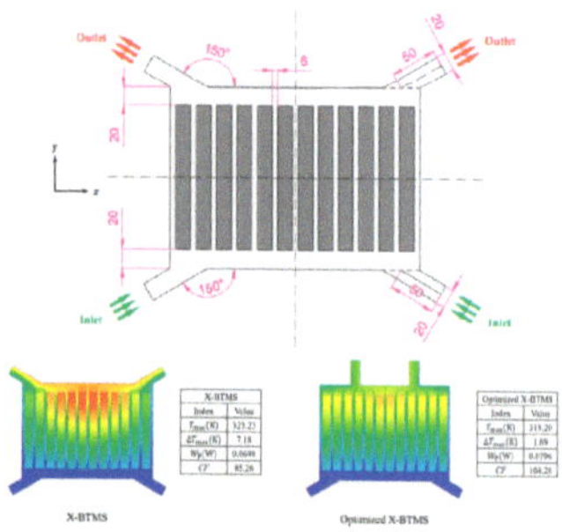	[5,13]	Air channel
2	A new design with two extra inlets is used.	CFD	Maximum temperature and temperature difference are decreased by 17.93% and 12.22%, respectively.	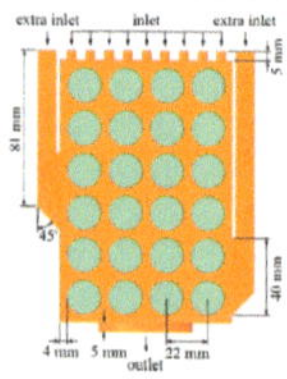	[14]	Air channel
3	A spoiler is used in the air inlet manifold in U and Z-type BPs.	CFD	Maximum temperature and temperature difference are decreased by 2.97 K and 4.98 K, respectively.	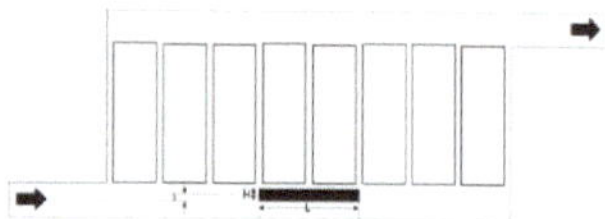	[15]	Air channel
4	A Z-type BP with a new stepwise divergence plenum is suggested.	CFD	Maximum temperature and temperature difference are decreased by 34.65% and 77.51%, respectively. The optimum height for each step and velocity is h = 0.125 mm and v = 3 m/s.	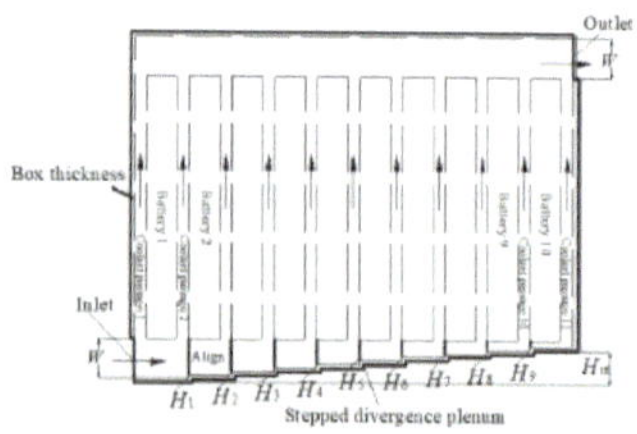	[16]	Air channel
5	A BTMS in dry-out condition is compared with a simple one.	Exp.	Maximum temperature and temperature difference are decreased by 0.31% and 0.48%, respectively.	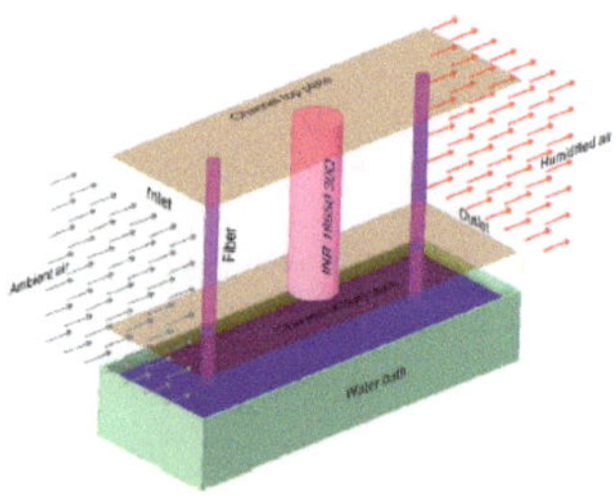	[17]	Air channel
6	A bypass cold-air channel is added to BP.	CFD	Temperature difference in the BP is decreased from 31.2 °C to 3.2 °C, and maximum temperature is decreased from 30.5 °C to 24.7 °C.	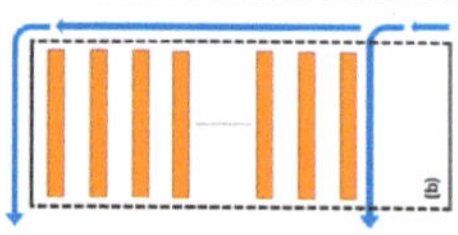	[18]	Air channel

Table 1. *Cont.*

No.	Modification/ Novelty	Type of Study	Remarks	Geometry	Ref.	Modification Method
7	A multiple inlet/outlet air-cooling frame is used for pouch Li-ion BP.	Exp.	This method reduces the module maximum temperature Tmax and temperature differential ΔTmax, which will be beneficial for the cost of small logistical vehicles, thus improving the cooling performance of the LIB module.	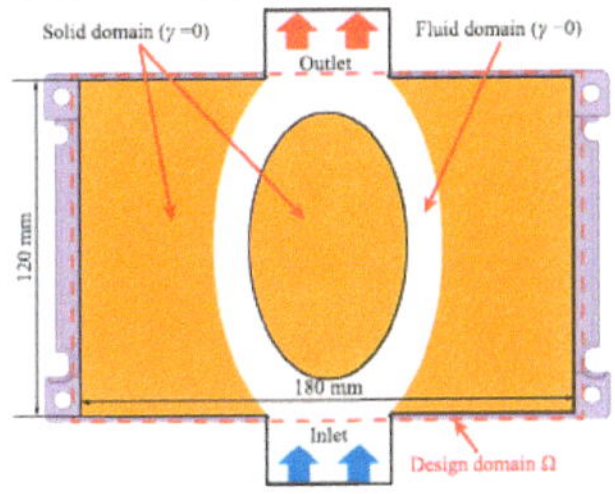	[7]	Air channel
8	A honeycomb-type, cylindrical BP air distribution design.	CFD	Maximum temperature and temperature difference are decreased by 1.7 K and 7.0 K, respectively.	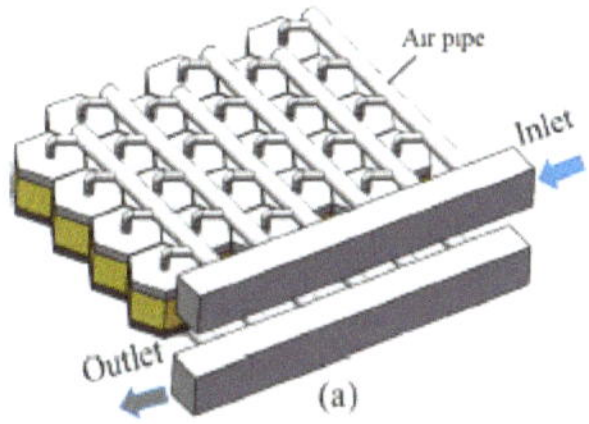	[6]	Air channel
9	Air-flow pattern is changed, based on Z-type and U-type BPs.	CFD	Maximum temperature and temperature difference are decreased by 0.8 K and 2.4 K, respectively.	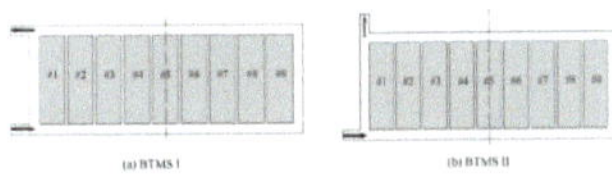	[19]	Air channel
10	A vortex generator is placed in the air channel.	CFD	Maximum temperature and temperature difference are decreased by 0.85 K and 0.77 K, respectively. Pressure drop in air channel is increased by 17.88 Pa.	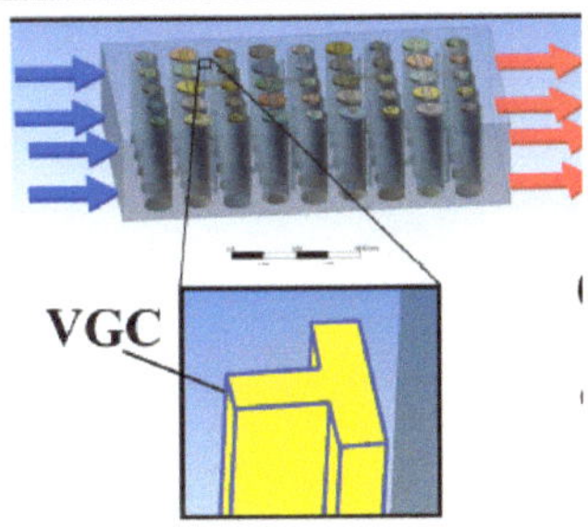	[20]	Air channel
11	A J-type BP is proposed	CFD	Maximum temperature and temperature difference are decreased by 1.57 K and 0.80 K, respectively.	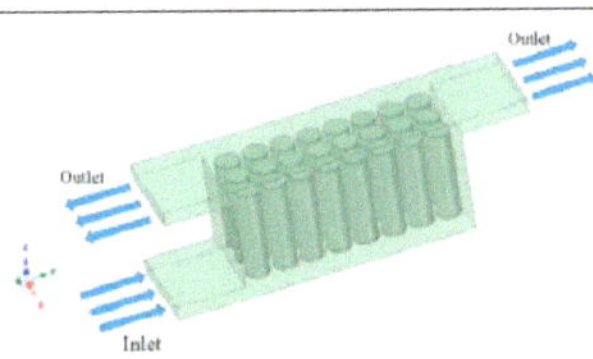	[21]	Air channel
12	A modified Z-type BP is proposed, which tilts the arrangement of battery packs.	CFD	Maximum temperature and temperature difference are decreased by 10.5% and 23.9%, respectively.	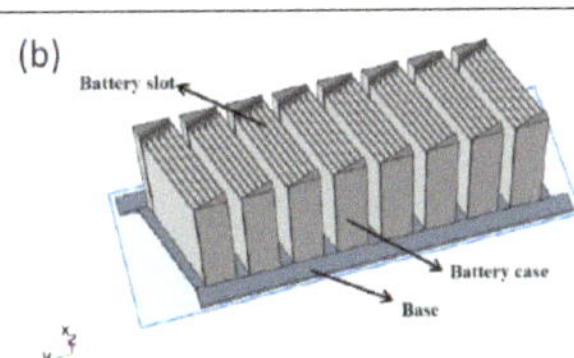	[9]	Air channel

Table 1. *Cont.*

No.	Modification/ Novelty	Type of Study	Remarks	Geometry	Ref.	Modification Method
13	A staggered-arranged BTMS	GA optimization CFD	The optimum architecture is found, which decreases maximum temperature by 49.18% and temperature difference by 102.37%.	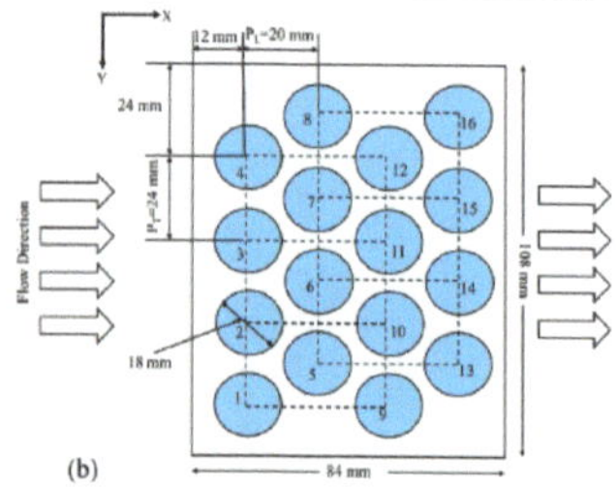	[10]	Cell arrangement
14	An innovative cell arrangement for an EV is proposed.	CFD	The average temperature drop of lithium-ion batteries' cell surface temperature is 5.6% to 7.8% when cold fluid is flowed through them.	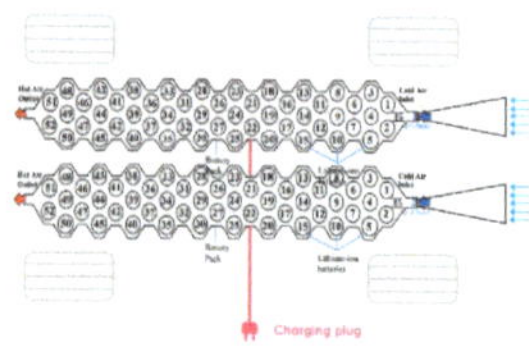	[22]	Cell arrangement
15	The effect of different cell arrangements is evaluated for a 30-cell battery pack.	CFD Exp.	The 5 × 6 BP works better than 15 × 2 BP, and offers better heat dissipation.	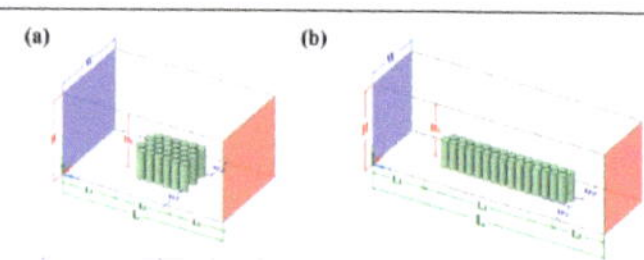	[23]	Cell arrangement
16	Innovative cell holders are used to act as a fin	CFD	Maximum temperatures for the basic model are 301.5, 303.9, and 308.8 K; for the modified model, they are 300.6, 302.1, and 304 K.	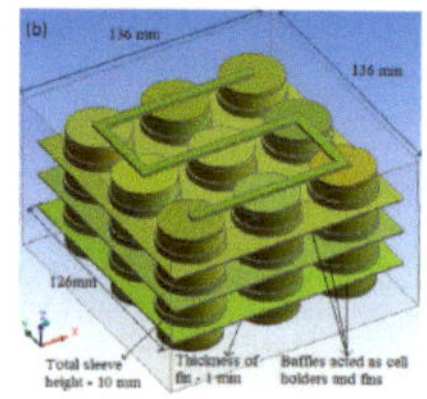	[24]	Adding fin
17	An innovative circular fin is used around cells.	Exp.	Maximum temperature is decreased by 27.26%.	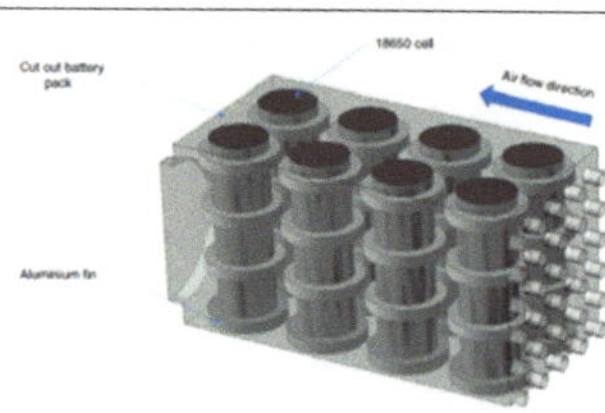	[12]	Adding fin
18	Innovative bod baffles for EV are used.	CFD	For 3C discharge, the new design results in a maximum temperature of 24.7 °C and temperature difference of 2.7 °C.	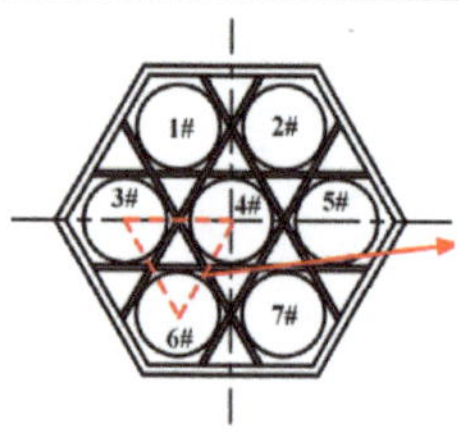	[13]	Adding fin

Table 1. *Cont.*

No.	Modification/ Novelty	Type of Study	Remarks	Geometry	Ref.	Modification Method
19	A separator plate with fish-shaped holes is used between batteries.	CFD	The new design results in a maximum temperature drop of 9.2% and temperature difference drop of 12.2%.	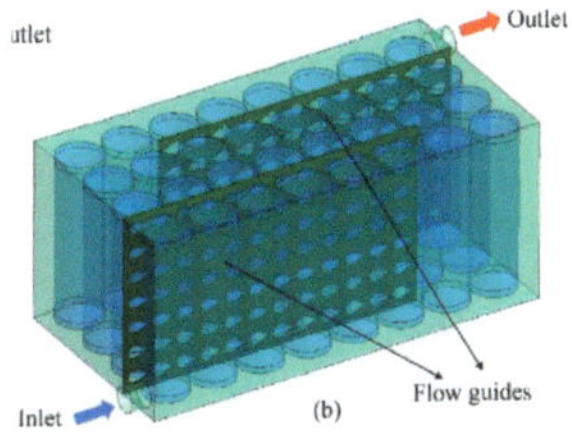	[25]	Adding fin
20	A helical coil is coupled with air jet in LIB.	Exp.	The new design results in a maximum temperature drop of 5.7 °C and temperature difference drop of 4 °C.	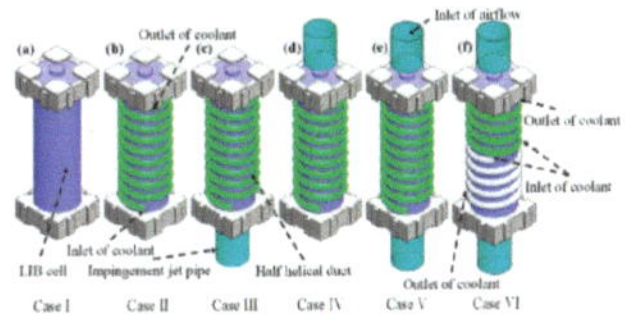	[26]	Adding fin

2.2. Liquid-Cooled BTMS

Liquid cooling battery thermal management systems (LC-BTMS) are a very efficient approach for cooling batteries, especially in demanding applications like electric vehicles. LC-BTMS may be classified into two types: liquid indirect cooling battery thermal management systems (LIDC-BTMS) and liquid direct cooling battery thermal management systems (LDC-BTMS), which are also referred to as immersion cooling systems. LIDC-BTMS employ a liquid-cooling plate that meets the battery module, facilitating the absorption and dissipation of heat produced during charging and discharging cycles. This approach capitalizes on a well-established manufacturing process and has a high capacity for transferring heat efficiently. However, its effectiveness may be limited by the thermal resistance that occurs when the cooling plate meets the battery. On the other hand, LDC-BTMS facilitate direct interaction between the coolant and the battery, resulting in a significant decrease in thermal resistance and an improvement in cooling effectiveness. This technology is very efficient at maintaining ideal battery temperatures, hence extending battery lifespan and reducing thermal runaway. However, the design of LDC-BTMS must be precise to successfully tackle the concerns of coolant containment and system sealing. LC-BTMS have exceptional thermal management capabilities, making them ideal for applications that need reliable and effective cooling solutions.

Electric vehicles such as Tesla's Model S, Model X, Model 3, General Motors' Chevrolet Bolt, and Jaguar's I-PACE use liquid cooling in their battery thermal management systems (BTMS) to effectively regulate heat and improve battery performance and safety. Many innovative designs were reported by researchers in 2023 and 2024, and some of the designs are presented in Table 2. The provided designs in Table 2 can be categorized into modifying cold plate and modifying cooling channels.

Cold plate modification: Optimizing the cold plate in a liquid-cooling system is essential for improving thermal efficiency. To enhance heat transfer, one should choose materials with high thermal conductivity, optimizing surface finishing and increasing the surface area in contact with the heat source. In addition, the heat dissipation can be further improved by designing the internal channels to optimize fluid flow and utilizing high-quality thermal interface materials. Yao et al. [27] introduced an innovative hybrid Battery Thermal Management System (BTMS) that integrates phase change materials (PCM) and a liquid cooling channel inspired by a spider web. This system effectively dissipates heat and keeps the battery module temperature below 40 °C even during high discharge rates. As a result, it greatly improves the efficiency of thermal management. The spider web pattern in the BTMS is employed to maximize heat distribution, enhance surface area for improved heat transfer, and ensure the efficient flow of cooling fluid, thereby enhancing the overall efficiency of thermal management. Li et al. [28] used the same idea and designed a diamond-type cold plate to increase the efficiency of liquid-cooled BTMS.

Cooling channel modification: Modifying cooling channels in battery thermal management systems enhances heat dissipation, ensures uniform temperature distribution, reduces energy consumption, and optimizes overall system performance, thereby improving battery efficiency and longevity. Improving battery thermal management requires implementing changes to the shape of the cooling channels, increasing the amount of exposed surface area, optimizing the paths through which the coolant flows, choosing materials with high conductivity, adjusting the rate at which the coolant flows, and designing effective locations for the coolant to enter and exit. These modifications aim to enhance heat dissipation, ensure consistent temperatures, and improve the system's efficiency. Yates et al. [29] analyzed the performance of liquid cooling designs in cylindrical lithium-ion batteries, focusing on two specific designs: a mini channel cylinder (MCC) and a channel-cooled heat sink (CCHS). Their study investigated the effects of channel number, hole diameter, mass flow rate, and inlet locations on the thermal management performance of these designs. They found that, while the MCC provided superior cooling performance, it also resulted in greater temperature variation and higher manufacturing complexity compared to the CCHS, which offered a more uniform temperature distribution across the battery pack. A novel BTMS that integrates bionic spiral fins inspired by natural vines and spirulina, with an embedded, integrated cold plate, was proposed by Chen et al. [30]. They used a set of numerical experiments to find the optimum structure design and showed that the optimum design can improve cooling and preheating proficiency. They also compared different cold plate designs. These findings provide valuable insights into the design and optimization of BTMS, offering potential improvements in battery performance and lifespan, particularly under high-temperature operating conditions. Furthermore, an innovative parallel sandwich cooling structure improving thermal uniformity and reducing pressure loss in lithium-ion battery packs was designed by Zhao et al. [31]. The study investigated three different cooling water cavities: the series one-way flow corrugated flat tube-cooling structure (Model 1), the series two-way flow corrugated flat tube cooling structure (Model 2), and the parallel sandwich cooling structure (Model 3). Compared to the series cooling systems, Model 3 decreased the average temperature by 26.2% and the maximum temperature by 26.9%. Furthermore, it reduced the temperature difference within the battery pack by 62%.

Table 2. An outline of the work that has been done on liquid-cooled BTMS.

No.	Modification/Novelty	Type of Study	Remarks	Geometry	Ref.	Modification Method
1	A spider web liquid channel is designed. This innovation increases the area surface in contact.	CFD	The novel design maintains the maximum battery temperature below 40 °C, even under high discharge rates.	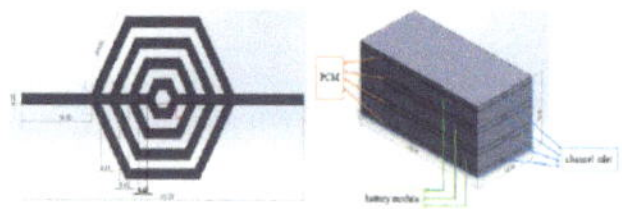	[27]	Cooling channel
2	A novel mini channel cold plate (MCP) with circular slot and zigzag channel is designed.	Exp.	Using this design can decrease maximum temperature by up to 5 °C.	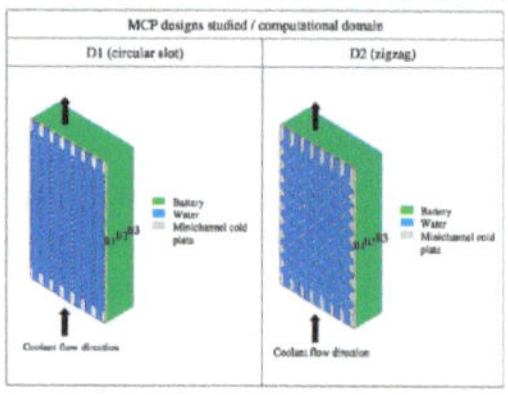	[32]	Cooling channel

Table 2. *Cont.*

No.	Modification/Novelty	Type of Study	Remarks	Geometry	Ref.	Modification Method
3	A new inlet/outlet layout is proposed.	CFD	An optimal BTMS scheme that utilizes aluminium cooling plates with a serpentine flow channel and an inlet flow velocity of 0.5 m/s achieves the best cooling performance, energy efficiency, and material cost-effectiveness.	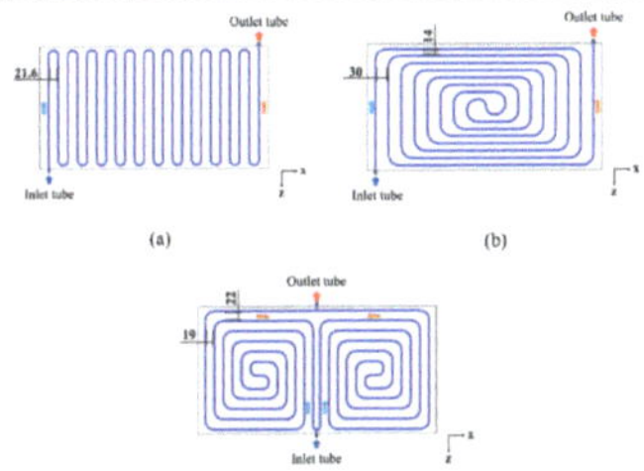	[33]	Cold plate
4	A novel BTMS that integrates bionic spiral fins wrapped with phase change material (PCM) and embedded in a liquid cooling plate is designed.	CFD	This design significantly reduces the maximum battery temperature by 3.1 °C and increases preheating efficiency by 5.6 °C, compared to BTMS without fins	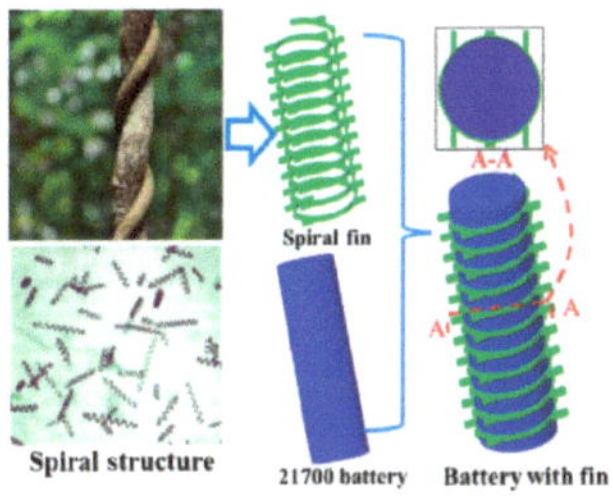	[30]	Cooling channel
5	A diamond flow-type channel cold plate is designed.	CFD	In comparison with usual cold plates, maximum temperature decreases from 313.33 K to 308.98 K, and pressure loss decreases from 1708 Pa to 1180 Pa.	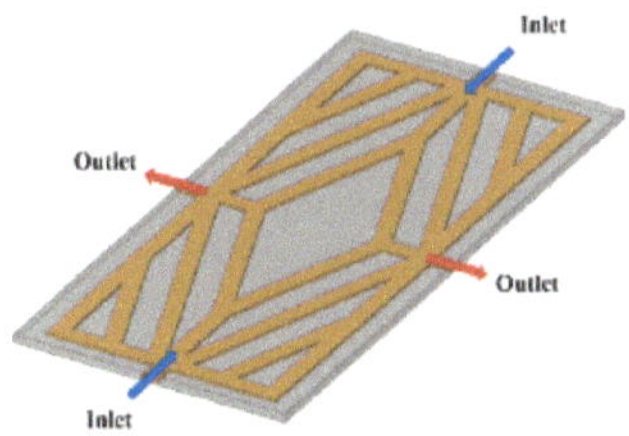	[28]	Cold plate
6	Five different cold plates are compared, and the optimal one is found.	CFD Topology optimization method	The optimum design can decrease maximum temperature and temperature difference by 51% and 42%, respectively.	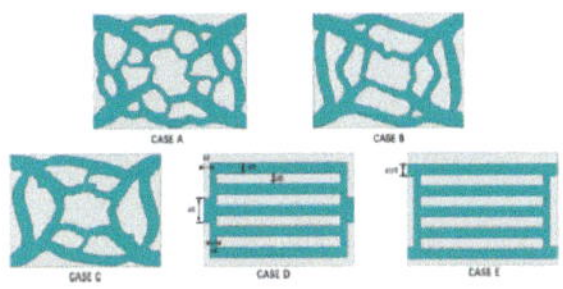	[34]	Cold plate
7	A composite battery thermal management system that integrates both air-cooling and liquid-cooling methods is presented	CFD	The composite thermal management system reduced the highest battery temperature to 317.38 K. It minimized the temperature difference to 3.73 K, and significantly decreased entropy production. This improvement was achieved with increased air and liquid flow rates.	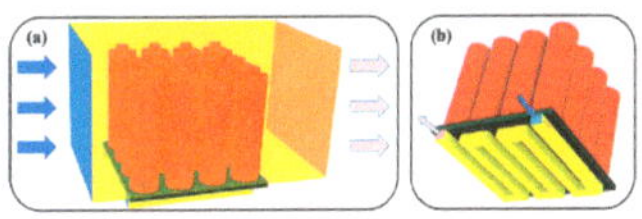	[35]	Cold plate
8	The paper investigates various cooling configurations' impact on LIB temperature behavior and heat transfer, advancing optimal BTMS for EVs.	CFD	The study demonstrates that liquid cooling significantly enhances the thermal performance of the battery pack (BP). Curved channels showed the most promising results, achieving a temperature reduction to 317.38 K and a minimum temperature difference of 3.73 K among the configurations tested.	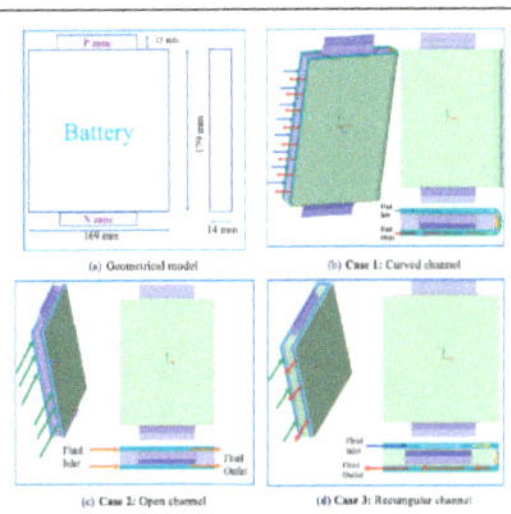	[36]	Cold plate

Table 2. *Cont.*

No.	Modification/Novelty	Type of Study	Remarks	Geometry	Ref.	Modification Method
9	This study investigates a hybrid battery thermal management system (BTMS) that integrates phase change material/copper foam with air jet pipe and liquid channel to enhance the thermal performance of cylindrical lithium-ion batteries (LIBs).	Exp.	The BTMS reduces the maximum temperature and temperature difference of LIBs by 14.6% and 64.7%, respectively,	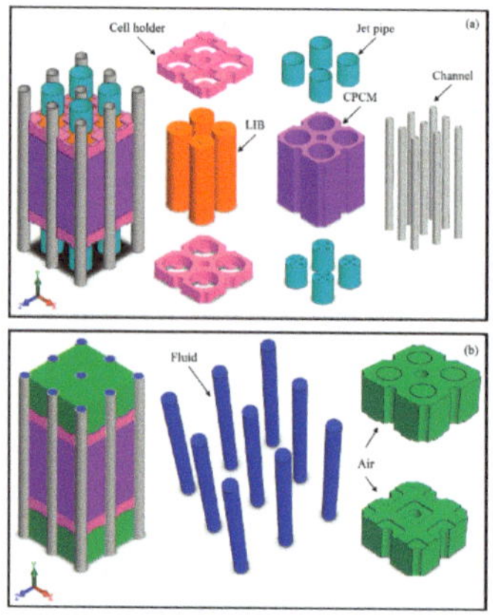	[37]	Cooling channel
10	A hybrid battery thermal management system combining phase change material and copper foam with air jet pipe and liquid channel is designed.	CFD	The indirect liquid cooling microchannel BTMS with nanofluids achieves superior temperature uniformity and rapid cooling, significantly enhancing the thermal performance of the battery module.	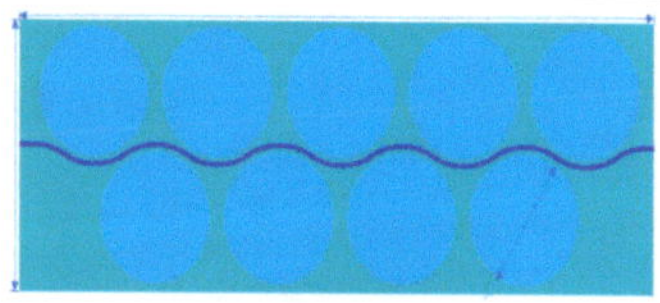	[38]	Cooling channel
11	Many liquid circulation types are compared to minimize energy consumption.	CFD	The reciprocating flow approach reduces the temperature differential and energy consumption by 55.3% and 15.6%, respectively.	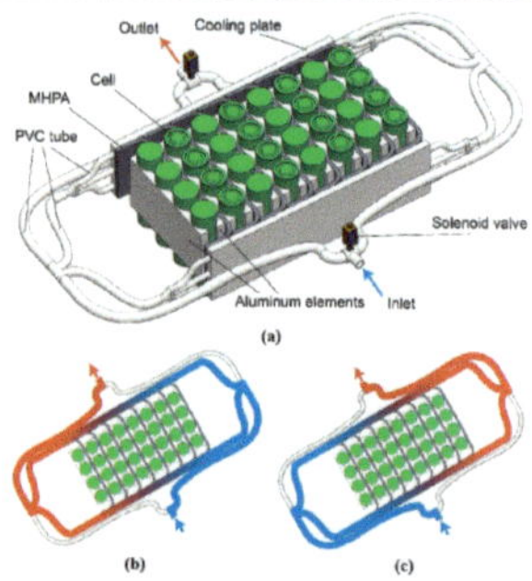	[39]	Cooling channel
12	Innovative Hybrid Nano/Dielectric Fluid Cooling System for Cylindrical Li-Ion Batteries: Enhanced Thermal Management Across Operational Conditions	CFD	This system exhibits excellent thermal efficiency at high discharge rates, particularly when using a 4% Alumina nanofluid and high input velocities. Thermal management was further improved with the addition of curved cooling channels and separator plates. These enhancements effectively lowered both maximum and non-homogeneity temperatures.	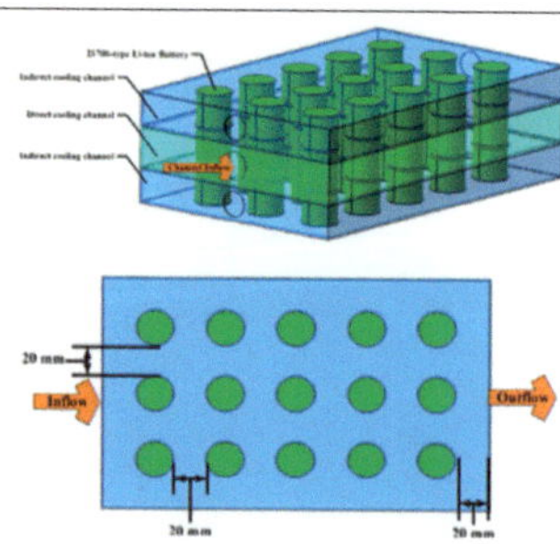	[40]	Cooling channel
13	Hybrid BTMS for Cylindrical Lithium-Ion Batteries: Optimizing Thermal Performance and Ensuring Cell Temperature Uniformity	CFD Exp.	Three heat-conducting blocks (HCBs) and 6 mm cooling channels for excellent cooling efficiency and lightweight construction were used. Periodic air cooling might save energy and preserve battery safety.	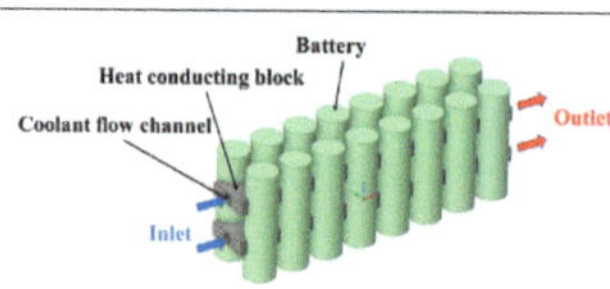	[41]	Cooling channel

Table 2. *Cont.*

No.	Modification/Novelty	Type of Study	Remarks	Geometry	Ref.	Modification Method
14	Battery Pack with 6 × 8 Cell Configuration: Cooling Jacket with Aluminum Plates and Transformer Oil for Direct Liquid Cooling	CFD	A two-pipeline liquid-cooling structure provides the best balance of cooling performance and energy consumption. This structure achieved effective temperature reduction and improved uniformity under high-rate discharge conditions with the hybrid system.	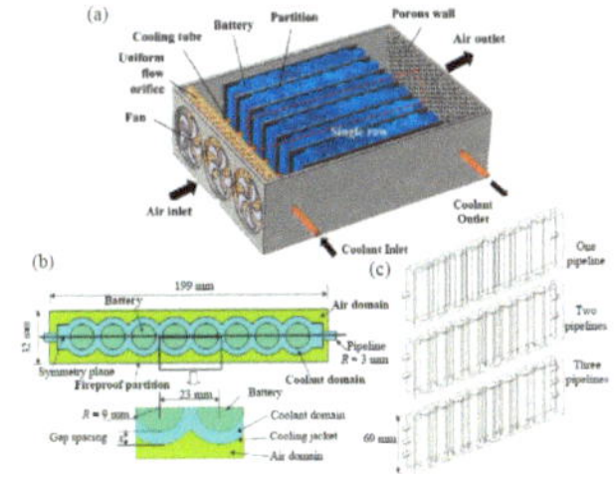	[42]	Cooling channel
15	BTMS with Hollow Copper Sleeves and Bent Liquid Copper Pipes for Cooling: Micro-Controller Optimization for Enhanced Performance	Exp.	the novel Battery Thermal Management System (BTMS), combining CPCM and liquid cooling, effectively controlled battery temperatures. It maintained a maximum temperature below 44.8 °C and a temperature difference under 2 °C. The optimal coolant flow rate was identified as 250 mL/min, balancing cooling efficiency and energy use. The system's design is scalable, making it applicable to various battery types for enhanced thermal management.	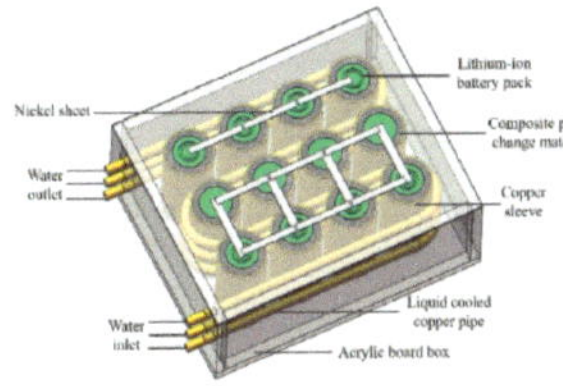	[43]	Cooling channel
16	Hybrid BTMS for Superior Thermal Management: Integrating PCM Modules with Copper and Graphite, Micro Cooling Plates, and Interconnected Cooling Channels.	Exp.	The hybrid Battery Thermal Management System (BTMS) reduced the maximum battery temperature by 13.78%. It also maintained the temperature difference below 3.9 °C and increased energy density by 11.23%. The optimal cooling configuration combined PCM, MHPA, and liquid cooling. This combination outperformed single-method cooling systems in both thermal performance and energy efficiency.	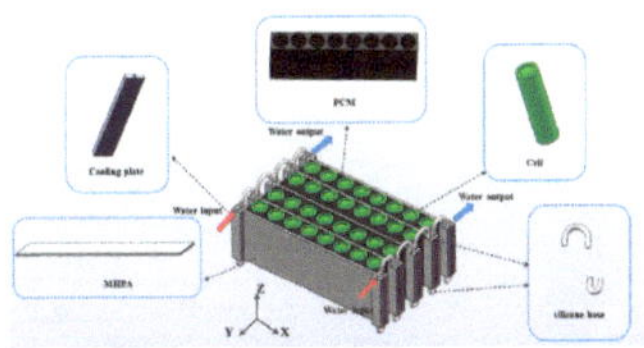	[44]	Cooling channel
17	Comparative Study of Four Designs with Various Inlet/Outlet Positions: Design D Features Circular Slots for Enhanced Heat Transfer and Optimized Channels for Higher Discharge Rates.	CFD	It was found that Design D demonstrated superior comprehensive performance compared to other designs. This design significantly lowered the maximum temperature of the battery pack and improved temperature uniformity. These features make Design D particularly suitable for high discharge-rate conditions.	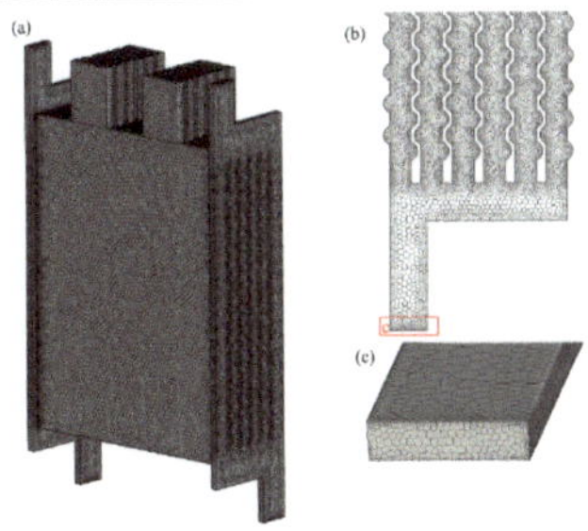	[45]	Cooling channel

Table 2. *Cont.*

No.	Modification/Novelty	Type of Study	Remarks	Geometry	Ref.	Modification Method
18	Optimized Bionic Limulus-Like Fins for Liquid-Cooled Plates: Enhanced Heat Dissipation and Reduced Pressure Loss Compared to Conventional Designs.	CFD	The key finding of this work is that the optimized bionic limulus-like fin design significantly reduced the average temperature by 1.69 °C (4.61%). It also decreased the pressure drop by 6.81 Pa (54.26%). These improvements enhanced overall cooling performance compared to the initial model.	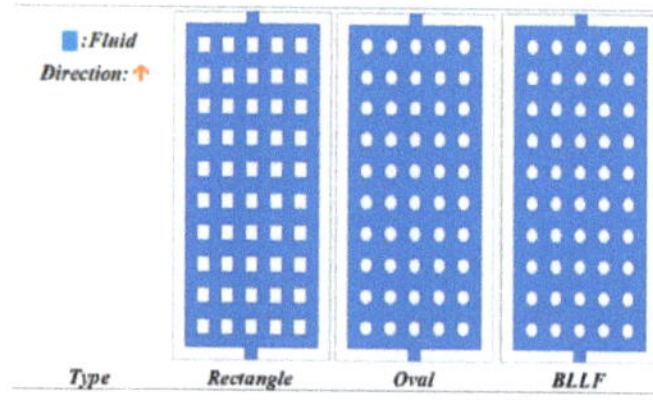	[46]	Cold plate
19	Innovative Battery Pack Cold Plates: Radial Inlet/Outlet Flow Channels and Axial Microchannel Shell with S-Shaped Metal Tubes for Enhanced Cooling and Heat Dissipation.	Exp. Multi objective optimiza-tion	The study optimized a hybrid cooling plate with axial S-shaped microchannels, which enhanced thermal management and energy efficiency in EV batteries. Key findings include improved temperature uniformity, reduced energy consumption, and effective performance in both high-temperature and cold environments. A delayed cooling strategy was also used to save energy. Increasing the delay time before activating liquid cooling significantly decreased energy consumption. A delay of 300 s reduced energy consumption by 46.3% while maintaining a maximum temperature below 40 °C and a temperature difference within 5 °C.	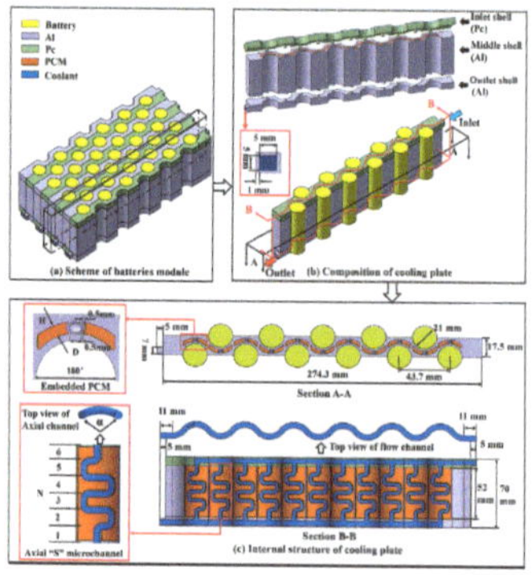	[47]	Cooling channel
20	Novel Parallel Sandwich Cooling Structure: Enhancing Thermal Uniformity and Reducing Pressure Loss in Lithium-Ion Battery Packs	CFD	The parallel sandwich cooling structure significantly enhanced thermal uniformity in battery packs. It reduced the average temperature by 26.2% and the maximum temperature by 26.9%, compared to series cooling systems. The new model also achieved a 62% decrease in temperature difference within the battery pack, demonstrating improved thermal uniformity over series cooling structures.	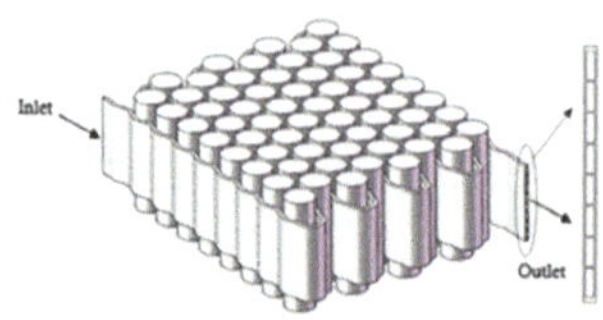	[31]	Cooling channel

2.3. PCM-Cooled BTMS

The concerns related to liquid cooling systems, such as the potential for coolant leakage and the proper disposal of coolant, have been effectively addressed during the commercialization step. Passive thermal management utilizing Phase Change Materials (PCMs) has emerged as an alternative to active cooling systems [48]. Phase Change Materials (PCMs) provide a promising solution as they can absorb and store significant quantities of thermal energy during phase transitions. This allows them to effectively regulate battery temperatures without requiring the use of active cooling components. This approach not only reduces the risks related to coolant management but also simplifies the thermal management system, potentially decreasing maintenance needs and improving overall system reliability. PCMs enhance battery performance, increase battery life, and enhance safety by maintaining consistent and optimal battery temperatures. This makes PCMs a feasible option for contemporary battery thermal management. PCM simulation is considered a complex flow simulation. Mesoscopic methods are used by researchers to model PCM in battery thermal management systems. Rahmani et al. [49] conducted a computational investigation of magnetohydrodynamic flow and the melting process of phase change material in a battery pack.

PCMs provide passive thermal management. They can absorb heat as they change from solid to liquid, but once fully melted they no longer absorb heat effectively. This can result in temperature spikes during long, high-power operation or in high-temperature environments. PCM-based systems often need to be supplemented with active cooling systems, such as liquid or air cooling, to effectively manage the heat generated by the batteries, especially during high discharge rates or rapid charging scenarios. Active cooling systems can ensure continuous thermal regulation, even after the PCM has fully melted. Table 3 shows some innovative PCM-based BTMS which were presented in 2023 and 2024.

Hybrid PCM-air cooling systems: The battery pack can minimize temperature gradients and hot spots by integrating phase change material (PCM) with air-cooling. The PCM serves as a thermal buffer, absorbing and retaining excessive heat during phase change to achieve this. The continuous air supply from the air-cooling system distributes the absorbed heat, which stops any specific areas from becoming overheated. The combined effect guarantees a more uniform distribution of temperature through all cells, which increases battery performance and safety. In addition, air-cooling systems offer advantages in terms of cost, weight, simplicity, and maintenance, compared to liquid cooling systems. When combined with PCM, they effectively handle high thermal loads, enhancing the efficiency and reliability of the hybrid system. Chen et al. [50] presented a new hybrid Battery Thermal Management System (BTMS) that combines phase change material (PCM) with air cooling. It incorporates biomimetic variable-section fins to enhance thermal performance. It is shown that by implementing a delayed air-cooling strategy, power consumption can be reduced by 59% while still maintaining maximum temperature and temperature difference within 40 °C and 3 °C, respectively. A similar study is conducted by Rahmani et al. [51]. They investigated the enhancement of heat storage cooling systems via the implementation of a honeycomb-inspired design, focusing on efficiency and performance. Suo et al. [52] proposed an innovative design in which PCM covers prismatic battery tabs, and air passes batteries through Z-type air channels. Four cases were compared, and it was found that when PCM directly contacts the battery box, the volume of the battery box reduces by 3.48% and thermal performance is enhanced.

Hybrid PCM-liquid cooling systems: Phase change materials (PCM) and liquid cooling together provide substantial benefits for battery thermal management systems (BTMS) in terms of thermal regulation and heat transfer efficiency. Without requiring constant energy input, PCM can efficiently regulate battery temperature during phase transitions by absorbing and storing thermal energy. Nevertheless, the PCM's capacity to absorb heat decreases as it completes its transition. Therefore, by offering both passive and active cooling techniques, integrating PCM with liquid cooling improves the overall thermal management [53]. Hybrid PCM-liquid cooling systems leverage the high thermal conductivity and specific heat capacity of liquid coolants to rapidly remove heat from battery cells. Liquid cooling systems provide superior heat transfer compared to air cooling, making them highly effective for high-power density applications such as electric vehicles (EVs). The integration of PCM into these systems ensures that temperature spikes are mitigated during periods of high thermal load, such as rapid charging or discharging cycles. Wang et al. [54] developed a novel hybrid system combining wavy microchannel cold plates with PCM, which significantly improved both active and passive cooling of cylindrical lithium-ion battery packs. This design not only enhanced the heat-dissipation capabilities but also maintained uniform temperatures across the battery pack, minimizing thermal gradients and hotspots. Li et al. [55] developed a novel passive thermal regulator for lithium-ion batteries, utilizing the volume change during phase transitions of composite phase-change materials (cPCM) to control cooling water flow. This regulator significantly reduced the maximum battery temperature by up to 7.94% at high ambient temperatures, maintaining it below 39 °C, and improved temperature uniformity and fluctuation control during dynamic cycling. This design highlights the potential of passive thermal management systems in enhancing battery performance and longevity under various operating conditions. Hybrid PCM-liquid cooling systems offer significant advantages, including enhanced heat transfer efficiency due to the high thermal conductivity of liquid coolants, leading to faster and more efficient cooling. They improve temperature uniformity across the battery pack, reducing temperature differences and preventing localized overheating. These systems also increase reliability by handling peak thermal loads passively, reducing the strain on the liquid cooling system. Moreover, they are energy efficient, as PCM can absorb peak thermal loads, minimizing the need for continuous liquid cooling. Additionally, these systems are scalable and customizable for various battery sizes and configurations, making them versatile for applications ranging from small electronics to large electric vehicle battery packs [56].

Table 3. An outline of the work that has been done on PCM-cooled BTMS.

No.	Modification/Novelty	Type of Study	Remarks	Geometry	Ref.	Modification Method
1	Lithium-ion (LIB)-PCM Configurations: Square, Circular, Rectangular.	CFD	PCM-RT35 is ideal for temperatures of 20–30 °C. PCM-RT50 is optimal for 40 °C. Copper shells provide superior temperature control. Aluminium shells are more practical. Higher heat transfer coefficients improve temperature stabilization. Rectangular battery packs offer the best thermal management.	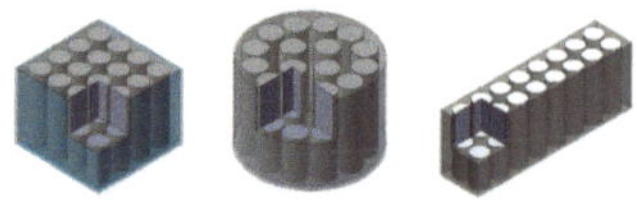	[57]	PCM-air
2	Biomimetic Fin with PCM and Delayed Air Cooling	Exp.	Maximum temperature reduced by 3.4 °C using fins. Power consumption reduced by 33% with beak fins. Optimal fin parameters: 75° angle, 40 mm length. Delayed air-cooling reduces power consumption by 59%.	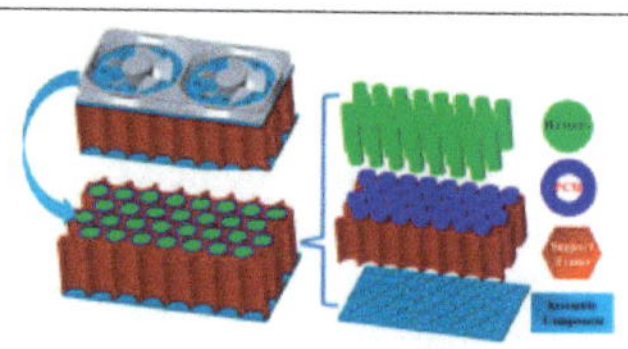	[50]	PCM-air
3	Hybrid PCM with Secondary Air and Liquid Coolants	Exp.	With a discharge rate of 7C, the paraffin with copper foam showed a phase-change percentage of only 6.87%. This suggests effective thermal management without excessive phase change, resulting in reduced pumping power requirements and improved energy efficiency.	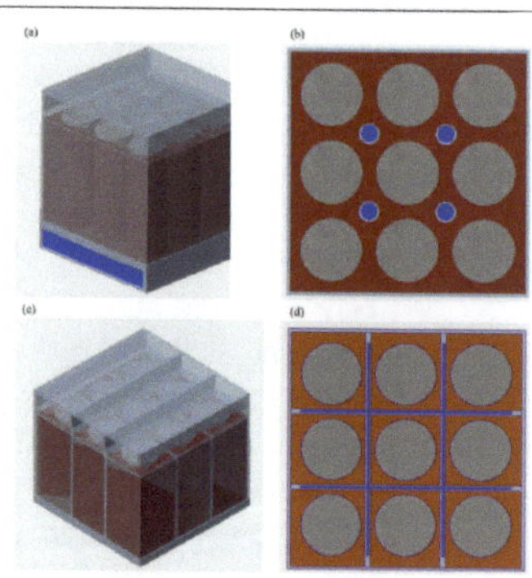	[58]	PCM-liquid
4	Optimized PCM Configurations for Prismatic Batteries	CFD	PCM significantly enhances thermal performance. Case 4 saves PCM while maintaining thermal efficiency. Optimized PCM proportion is 64%.	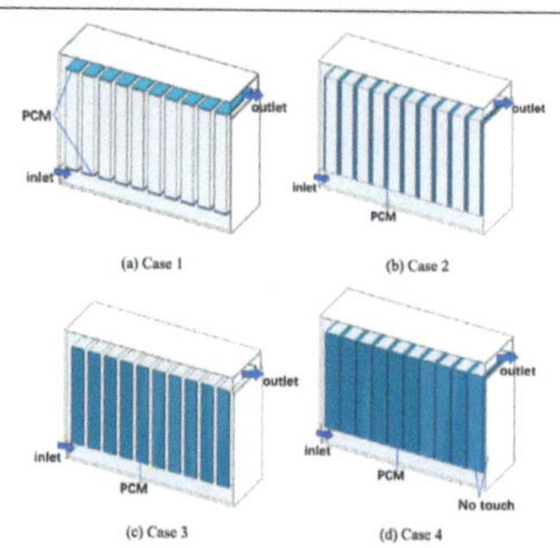	[52]	PCM-air

Table 3. *Cont.*

No.	Modification/ Novelty	Type of Study	Remarks	Geometry	Ref.	Modification Method
5	Hybrid PCM-Air/Fluid Coolants for Enhanced Thermal Performance	CFD	Hybrid strategy with paraffin, air, and fluid coolants significantly improves thermal performance. Eliminates need for pumps. Achieves high temperature uniformity.	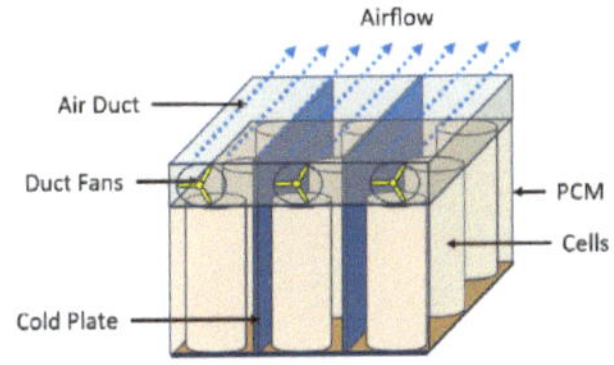	[59]	PCM-air
6	Graphene-Enhanced Paraffin PCM	Exp.	In higher temperature scenarios, researchers have found that a hybrid cooling system that combines phase-change material (PCM) with other cooling methods like forced-air or thermoelectric module cooling may be more effective. The inclusion of graphene-enhanced paraffin as a phase-change material (PCM) has a substantial impact on lowering the temperature of the lithium-ion battery pack. It was shown in the study that as the thickness of the PCM layer increases, the maximum temperature decreases.		[60]	PCM-air
7	Dual PCM System with Petal Design and Optimized Fins	CFD	The optimized solution with an asymmetric fin arrangement significantly decreases the maximum temperature difference (ΔT_max) by 5.53% at 30 °C. The optimized solution decreases the maximum temperature difference (ΔT_max) by 29.19% at 40 °C. It reduces the maximum temperature rise (ΔT_(m-rise)) by 36.15% at 30 °C. It reduces the maximum temperature rise (ΔT_(m-rise)) by 42.76% at 40 °C.	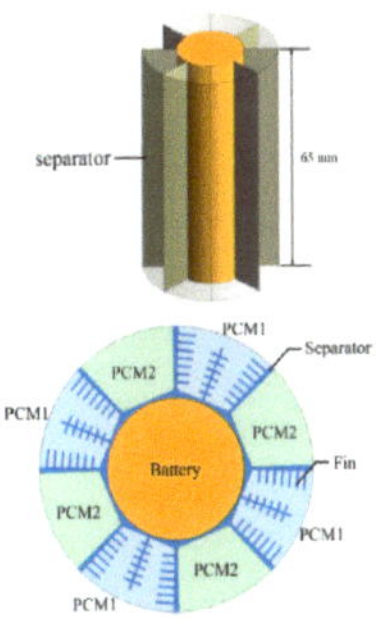	[61]	PCM-air
8	PCM-Based BTMS with Air-Cooling and Cold Plate	CFD	Throughout January, the battery temperature was carefully regulated with the help of a BTMS, staying consistently between 20–28 °C. The maximum temperature difference (ΔT) was only 2.6 °C. This demonstrates the successful implementation of efficient thermal management.	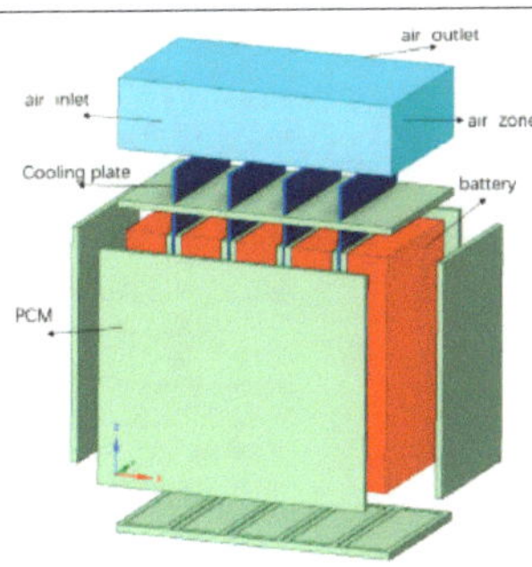	[62]	PCM-air

Table 3. *Cont.*

No.	Modification/ Novelty	Type of Study	Remarks	Geometry	Ref.	Modification Method
9	Hybrid PCM with Wavy Microchannel Cold Plate (HWMCP)	CFD	In cold weather, HWMCP helps prevent battery temperature loss, allowing for longer battery temperature maintenance during short-term parking, compared to WMCP. By reducing the weight of HWMCP by 45%, the system's energy efficiency is significantly improved.	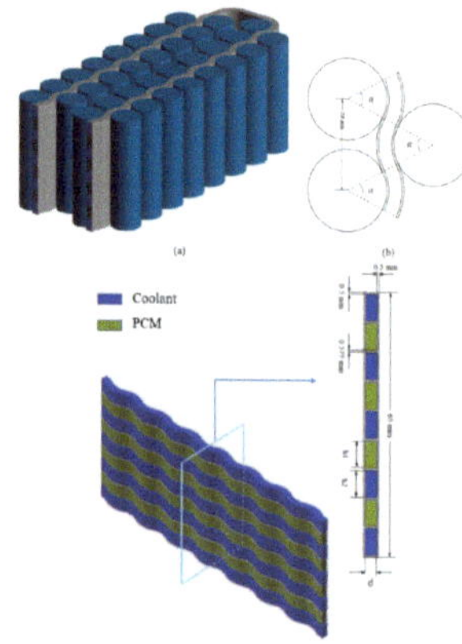	[54]	PCM-liquid
10	Nine Innovative Branch-Fin Designs for Enhanced PCM Thermal Management	CFD	The study presents nine new branch-fin designs for battery thermal management, showing superior thermal performance. These designs increase heat transfer efficiency by 14.98%. They extend operating time by 131.5%. The new designs reduce system weight by 10.28%, compared to conventional designs.	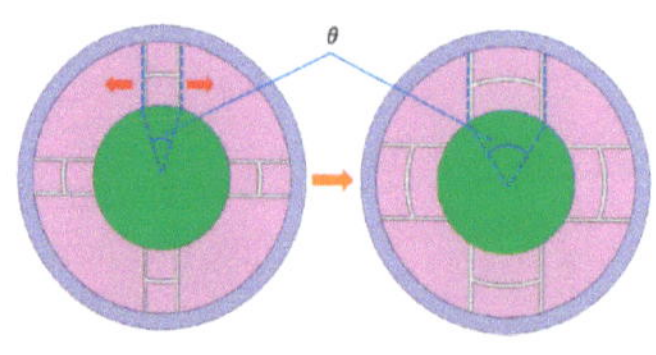	[63]	PCM-air
11	Hybrid PCM with Multi-Stage Tesla Liquid Valve-Cooling	Exp. CFD	The hybrid battery thermal management system with PCM and multistage Tesla valve-cooling significantly decreases energy consumption by 79.9%, compared to traditional systems. It also proves to be more efficient at maintaining battery temperature, especially during cold stops. This results in a doubling of warmth-retention time.	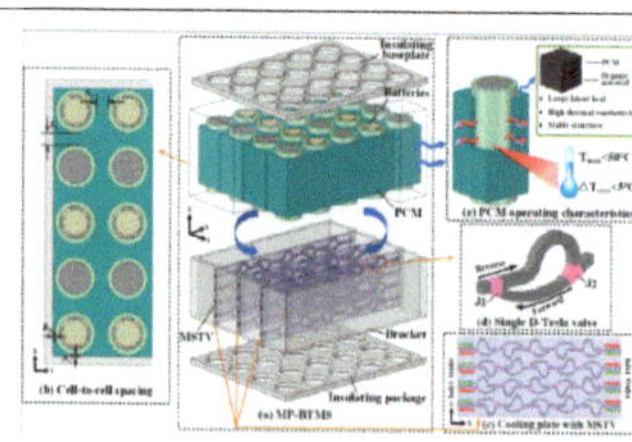	[64]	PCM-liquid
12	Novel BTMS with PCM, Air Cooling, and Fin Structures	CFD	Optimal BTMS parameters are d = 8.125 mm and l = 65 mm, with Tmax = 318.01 K, ΔT = 0.0135 K, and W = 5.13 kg. Increasing fin height from 5 mm to 65 mm enhances PCM utilization by 20.4% and reduces Tmax by 9.3 K. As d increases from 5 mm to 20 mm, PCM utilization decreases by 90.9% and BTMS weight increases by 146.7%. Fin addition improves thermal efficiency and compactness, significantly lowering battery temperature.	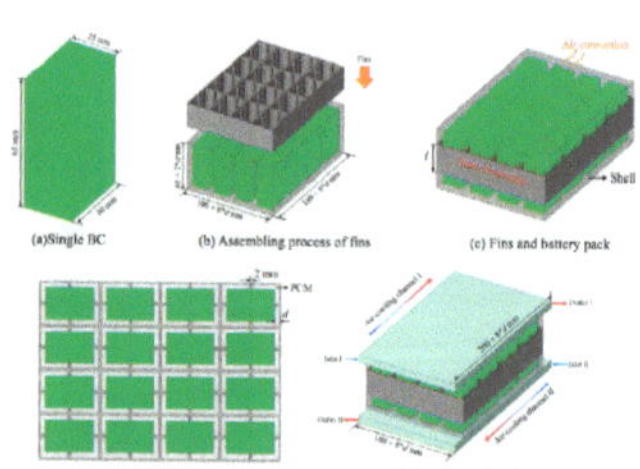	[65]	PCM-air
13	BTMS with PCM and Optimized Aluminium Fins	CFD	The identified optimal parameters are d = 8.125 mm and l = 65 mm, resulting in Tmax = 318.01 K, ΔT = 0.0135 K, and W = 5.13 kg. By increasing the fin height, the utilization of PCM is improved by 20.4% and the maximum temperature (Tmax) is reduced by 9.3 K. On the other hand, increasing the PCM thickness results in a significant decrease in utilization by 90.9% and a substantial increase in weight by 146.7%.	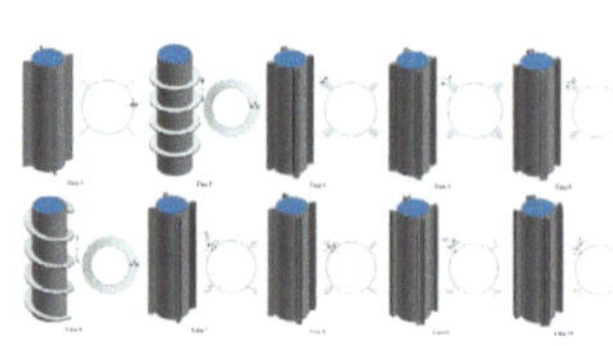	[66]	PCM-air

Table 3. *Cont.*

No.	Modification/ Novelty	Type of Study	Remarks	Geometry	Ref.	Modification Method
14	Innovative BTMS with PCM, Metal Foam, and Fin Shapes	CFD	The fourth case of the BTMS, which combines PCM, metal foam, and fins, exhibited the best thermal performance, keeping the battery surface temperature at the lowest level. This design achieved a maximum reduction of 3 K in battery surface temperature compared to pure PCM systems. Different fin shapes (rectangular, triangular, trapezoidal, I-shape, and wavy) were analyzed. The results showed that, while the triangular fins were most effective before the PCM melting began, the I-shape fins provided the lowest battery surface temperature after the PCM started melting.	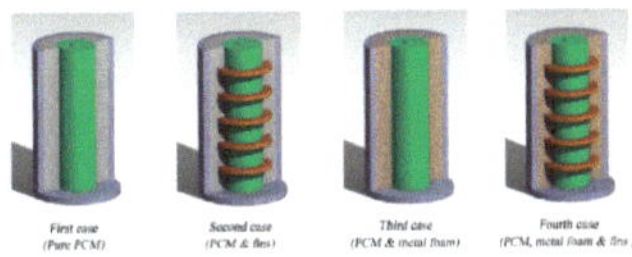	[67]	PCM-air
15	Hybrid BTMS with U-Shaped Micro Heat Pipe Array and Composite PCM	Exp.	The hybrid Battery Thermal Management System (BTMS), which combines a U-shaped micro heat pipe array (U-MHPA), composite phase change material (cPCM), and liquid cooling, significantly improves cooling performance. It effectively controls the maximum temperature and temperature difference within the battery module, even under extreme conditions. The system maintains the maximum temperature below 50 °C and the temperature difference below 5 °C.	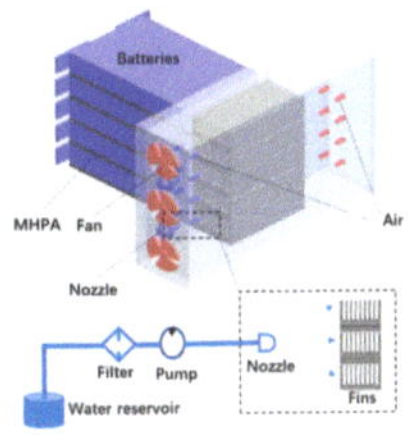	[68]	PCM-liquid
16	Comparison of Hydrogel and PCM-Based BTMS	CFD	Hydrogel-based BTMS demonstrated superior cooling performance compared to PCM-based systems. It showed a significant decrease of 5.27 °C in maximum temperature under specific conditions.	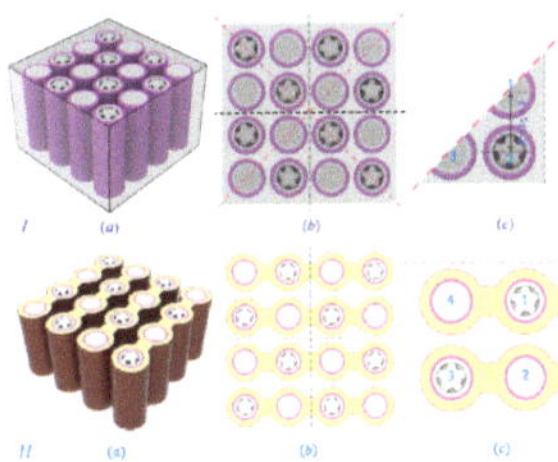	[69]	PCM-air
17	Electrochemical–Thermal Modeling of Li-ion Batteries with Fin-Intensified PCM BTMS	CFD	Without any thermal management system (TMS), the peak temperature during a 4C discharge reached 336 K. The use of base PCM alone reduced the peak temperature increase by almost 71%. The finned designs further improved this, with Design D1 reducing the temperature increase by 72% and Design D2 by 75.58%.	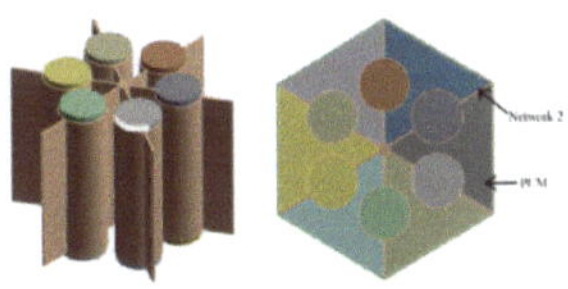	[70]	PCM-air
18	High-Performance Anisotropic CPCM Composite for Thermal Management	Exp.	The CPCM showed excellent thermal management performance. Under a 2C high discharge rate, the maximum temperature of the battery module with CPCM decreased by 21.9 °C (29.9%). The temperature difference decreased by 7.8 °C (55.3%), compared to modules without CPCM.	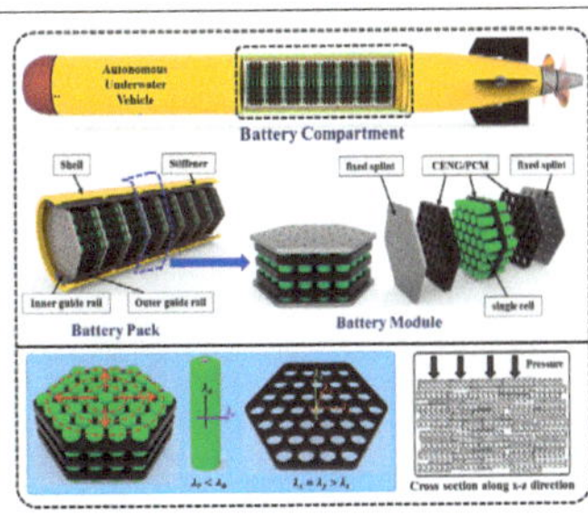	[71]	PCM-liquid

Table 3. *Cont.*

No.	Modification/ Novelty	Type of Study	Remarks	Geometry	Ref.	Modification Method
19	Aluminium Nitride Enhanced PCM for Superior Passive Cooling	Exp.	Managing battery temperatures within the range of 25 °C to 45 °C is crucial for optimizing the performance of the thermal regulator. When the temperature is below 30 °C, the batteries can function without the need for active cooling methods, thanks to the use of PCM or cPCM. When the ambient temperature reaches 35 °C, it is important to use a thermal regulator with cooling water to ensure that the battery temperature remains below 38.13 °C. When the ambient temperature exceeds 40 °C, the thermal regulator equipped with cPCM can lower the battery temperature to 35.02 °C.	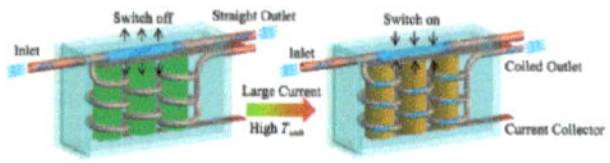	[55]	PCM-liquid
20	Snowflake Fin Design for Enhanced PCM Heat Transfer	CFD	With the addition of snowflake fins, the heat-transfer efficiency of PCM is greatly improved, resulting in a noticeable reduction in battery temperatures. With a 3C discharge rate, the Batteries–PCM–Fins design effectively keeps the battery module temperature below 45 °C, regardless of whether the ambient temperatures are 25 °C or 40 °C. The temperature difference remains less than 3 °C, ensuring optimal performance. These snowflake fins are designed to enhance thermal management efficiency, particularly during the initial cycles. They achieve this by minimizing the maximum temperature and temperature difference within the battery module.	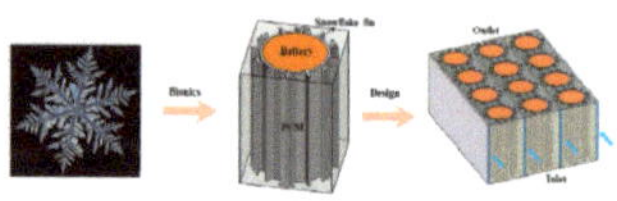	[72]	PCM-liquid

2.4. Thermoelectric Cooler BTMS

Traditional cooling technologies such as air cooling and liquid cooling have reached their maximum cooling potential. TECs remain a favoured cooling technology for various applications due to their affordability and environmentally friendly features [73]. There are two main categories of thermoelectric devices: thermoelectric generators (TEGs) and thermoelectric coolers (TECs) [74]. Thermoelectric generators (TEGs) use the Seebeck effect to convert heat energy into electrical energy. The Seebeck effect is a fascinating phenomenon that occurs when there is a temperature difference between two electrical conductors or semiconductors, resulting in the generation of a voltage differential. It is truly remarkable how temperature can have such a profound impact on electrical properties. TECs utilize electricity to disperse heat from a medium through the Peltier effect. This effect occurs when an electric current passes across the interface of two materials, causing the absorption or dissipation of heat [75]. TECs offer several advantages, such as a lightweight design, compact size, minimal noise, straightforward operation, and a long lifespan. TECs are utilized in a wide range of industries, such as microelectronics, communications, laser diodes, superconductor systems, the aerospace industry, healthcare, and the food sector, among numerous others [76].

TEC-based BTMS was first proposed by Li et al. [77] for controlling a battery pack temperature. In this study, an innovative battery pack design was presented, incorporating an acrylic container and copper holders, combined with a thermoelectric cooling system integrated with liquid and air circulations, which demonstrated a significant thermal management improvement, achieving a reduction of approximately 20 °C under a 40 V input, compared to conventional liquid cooling, and maintaining battery temperatures below critical thresholds during extreme discharge conditions. Hameed et al. [78] introduced a novel TEC–TEG BTMS. In this study, a new hybrid battery thermal management system (BTMS) was developed, which combined thermoelectric cooling (TEC) and thermoelectric generation (TEG) with forced air. The system effectively decreased the maximum surface temperature of a single $LiFePO_4$ battery cell by around 7 °C. Table 4 shows the newest findings in thermoelectric cooling-based BTMSs.

Table 4. An outline of the work that has been done on TEC-cooled BTMS.

No.	Modification/ Novelty	Type of Study	Remarks	Geometry	Ref.
1	Enhanced BTMS Cooling: Using TEG to Power TEC for Improved Heat Management	CFD	To increase the heat dissipation of TEC–TEG, high-quality heat sinks are used to make the temperature of BTMS as close to ambient temperature as possible. The findings revealed that the maximum battery surface temperature decreased from 38 °C to 33.1584 °C with the introduced BTMS.	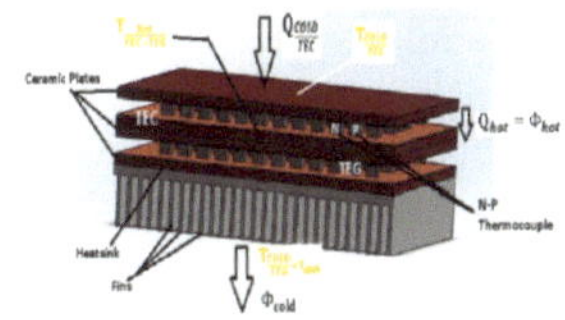	[78]
2	Novel BTMS: Integrating TECs and PCMs with Cooling Plate for Efficient Lithium-Ion Battery Temperature Control	CFD	Optimal fin length and thickness are 7 mm and 3 mm, respectively. At a 3 A TEC input current, the maximum temperature, temperature difference, and PCM liquid fraction were 315.10 K, 2.39 K, and 0.002 in Case 1. At a discharge rate of 5C, the maximum temperature, temperature difference, and PCM liquid fraction were 318.24 K, 3.60 K, and 0.181.	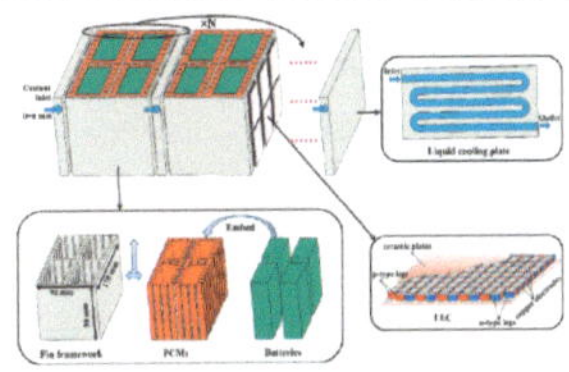	[79]
3	EV BTMS: Efficient Heat Transport from Battery Pack to TEC Cold Side and Release via Heatsink	Exp.	Forced air, and thermoelectric cooling (TEC) and -generation (TEG), are used to reduce the maximum surface temperature of a $LiFePO_4$ battery cell by approximately 7 °C.	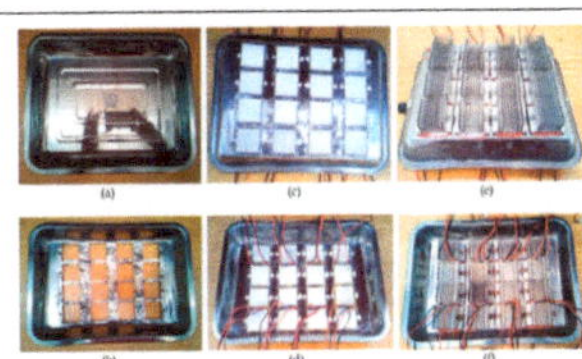	[80]
4	Active–Passive Hybrid BTMS: Combining TEC and PCM with CFD Simulation for Optimal Thermal Performance	CFD	The findings reveal that, even under a high discharge rate of 3C, the system can maintain the maximum temperature of batteries below 45 °C, while ensuring that the maximum temperature difference during the discharge process remains within 3 °C.	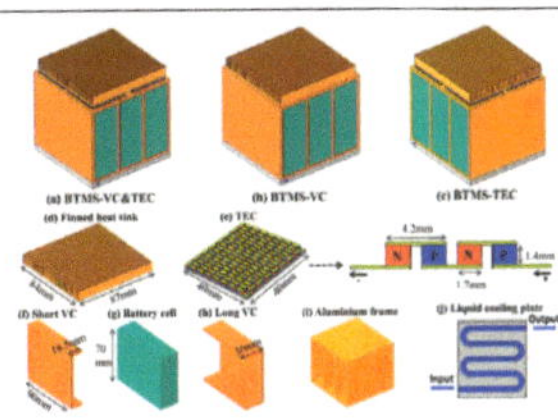	[81]
5	Hybrid TEC–PCM BTMS with Circular and Axial Fin Arrangements: Evaluating Temperature in Middle and Top Sections of Battery	Exp.	The use of aluminium circular fins, PCM, and a thermoelectric cooling system significantly reduced battery temperature to 65 °C. Changing the fin configuration from circular to axial led to an average temperature of 48 °C. This study found that axial fins regulate LIB temperature better than ring fins. The battery body's top temperature is always higher than the middle, regardless of mode or time interval.	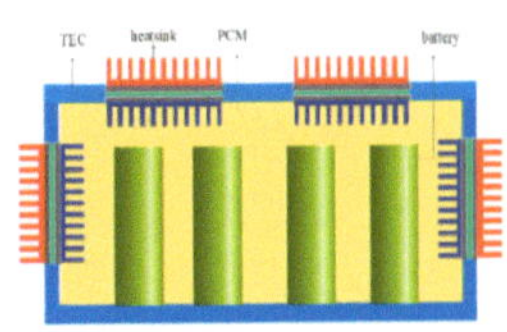	[82]

Table 4. *Cont.*

No.	Modification/ Novelty	Type of Study	Remarks	Geometry	Ref.
6	BTMS for Battery Module Heating: Dual TECs (6 V Bottom, 9 V Top) and 10 V Fan with Cold Plate Integration	Exp.	The total power consumption of the system is 81.2 W, with the bottom TEC consuming 31.21 W, the top TEC using 47 W, and the electronic fan requiring 2.994 W. This demonstrates the overall effectiveness and energy efficiency of the system. The battery capacity increases by 9.1% as the ambient temperature rises from −5 °C to 5 °C. At a bottom TEC voltage of 6 V, it takes 1631 s to heat the battery from −5 °C to 5 °C without top TEC voltage, with a temperature difference normally below 5.5 °C.	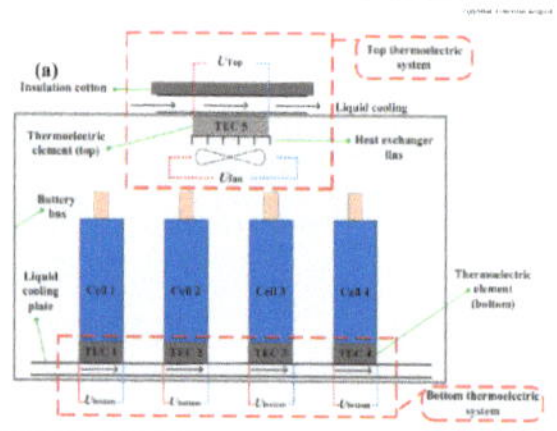	[83]
7	BTMS with PC-Wrapped Cells in Aluminum Case: TEC and Fins for Enhanced Heat Transfer, Tested at High Discharge Rates	CFD	Increasing TEC current lowers battery temperature and decreases uniformity and cooling efficiency; a 2A current keeps the battery below 40 °C. Delaying TEC current at 80% PCM melting rate improves temperature homogeneity, while a 2A current maintains a temperature gradient under 5 °C and extends effective temperature-control time. While the transient PCM + TEC model maintains temperature control and uniformity during 4C discharge, pulsed TEC current improves cooling power and thermal performance.	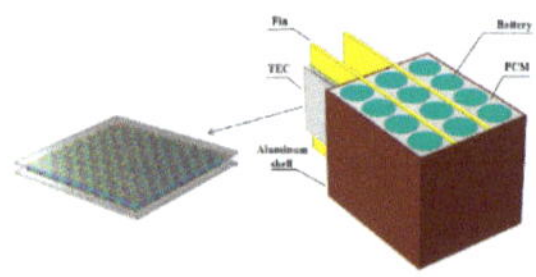	[84]
8	Proposed Battery Pack: Four Heat Sinks, 12 TECs, Honeycomb Framework, and Water Channel-Integration	CFD	Thermoelectric coolers (TECs) improve the thermal management of batteries by providing the best cooling performance when a current of 5A is applied. However, the efficiency of TECs decreases when the current exceeds this value. Air and water cooling enhance the efficiency of TEC, with recommended values for optimal performance being 50 W/m^{2}K for air convection and 0.11 m/s for water flow. Optimizing thermal performance and energy efficiency requires precise control strategies to balance TEC input, air cooling, and water cooling-parameters.	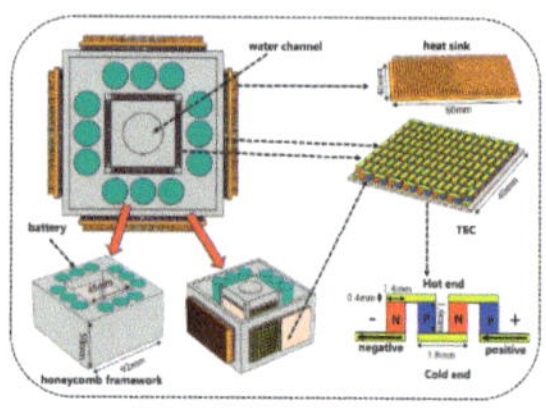	[85]
9	Dual Active BTMS: Incorporating Thermal Insulation Plate, TECs, and Liquid Cold Plates for Enhanced Cooling	CFD Exp.	TEC helps batteries maintain a reasonable temperature range in harsh environments and at a 3C discharge rate. Dual active cooling reduces energy consumption more efficiently than pure TEC cooling. Pure liquid cooling uses less energy, but TEC ensures acceptable battery performance under challenging conditions.	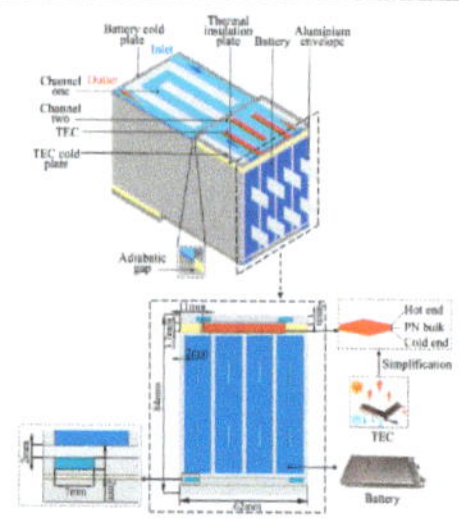	[86]
10	Hybrid Active–Passive BTMS: PCM, TEC, Liquid Cooling, and Fins for Enhanced Heat Transfer	CFD	Increasing fin thickness from 2 mm to 8 mm extends temperature control by 12%, enhancing thermoelectric cooler (TEC) cooling power and COP. Fins improve cooling capacity and temperature uniformity, with 4 mm fins being particularly effective. TEC input currents from 1A to 6A boost temperature control by 87.42%.	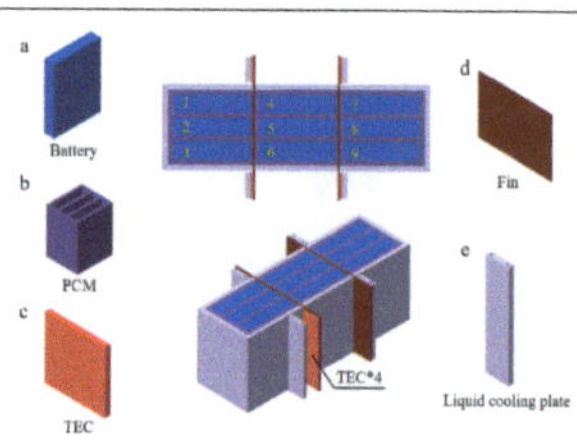	[87]

3. Discussion and Conclusions

The importance of effective battery thermal management systems (BTMS) for Li-ion batteries cannot be overstated, especially given their critical role in electric vehicles (EVs) and renewable energy-storage systems. In this section, after presenting the main findings for each method, a techno-economic comparative analysis of the four primary cooling methods—air-cooling, liquid-cooling, phase-change material (PCM)-cooling, and thermoelectric cooling—is presented. Furthermore, the main advantages and disadvantages of each method are stated.

3.1. Summary of Key Findings

Air cooling: Air-cooling methods, categorized into free and forced convection, offer simplicity, low cost, and ease of maintenance. However, their efficiency decreases significantly at high ambient temperatures. Researchers have proposed various modifications, such as air-channel modification, adding fin structures, and changing cell arrangements to enhance the performance of air-cooled BTMS.

- Air channel modifications are the most impactful, with designs like the X-type, honeycomb structures, and multiple inlet/outlet air cooling significantly improving temperature management.
- Cell arrangement modifications also contribute to better thermal performance, with innovative designs such as the modified Z-shaped system and the staggered-arranged system.
- Adding fin structures greatly enhances cooling efficiency by increasing the heat-dissipation surface area using radial fins for air-cooling and direct cooling with baffles.

These findings emphasize that modifying air channels, cell arrangements, and adding fin structures can significantly improve the performance of air-cooled BTMS. These improvements help maintain battery temperatures within optimal ranges, thereby enhancing the overall performance and safety of Li-ion batteries in electric vehicles.

Liquid-cooling: Liquid-cooling methods, including liquid indirect cooling (LIDC-BTMS) and liquid direct cooling (LDC-BTMS), are highly effective for demanding applications like EVs. Liquid cooling offers superior heat-transfer capabilities compared to air-cooling, making it suitable for high-power density scenarios.

- Cold plate modifications show significant improvements in cooling efficiency, with designs like the spider web liquid channel, mini channel plates with circular and zigzag channels, and the diamond flow type channel maintaining optimal temperatures under high-load conditions.
- Cooling channel modifications enhance thermal management by optimizing fluid flow, wavy microchannels, increasing surface area for heat dissipation, and using advanced materials such as nanofluids and dielectric fluids.
- Hybrid systems combining PCM and liquid-cooling methods provide superior temperature control, ensuring consistent battery performance, even under extreme conditions.

These findings highlight that liquid-cooling methods, particularly with advanced cold plate designs and optimized cooling channels, offer superior thermal management for Li-ion batteries. Hybrid systems that integrate PCM with liquid cooling further enhance cooling performance, making them highly suitable for demanding applications such as electric vehicles.

PCM Cooling: PCM-based cooling provides passive thermal management by absorbing heat during phase transitions. This method is effective for maintaining consistent battery temperatures but requires supplementary active cooling systems for prolonged high-power operations.

- PCM configurations such as square, circular, and rectangular designs optimize thermal management for specific temperature ranges, with materials like copper shells providing enhanced temperature control.
- Hybrid PCM systems that integrate PCM with air- or liquid-cooling methods offer improved temperature regulation and energy efficiency. These systems are particularly effective at managing thermal loads during high discharge rates.
- Innovative PCM designs featuring advanced fin structures and optimized configurations significantly enhance heat-transfer efficiency, reduce temperature differences, and extend battery operating time.

These findings highlight the effectiveness of PCM-based cooling methods in providing passive thermal management for Li-ion batteries. By incorporating advanced designs and hybrid systems, PCM cooling can maintain optimal battery temperatures, improving performance and safety in various applications, including electric vehicles.

TEC cooling: Thermoelectric coolers (TECs) utilize the Peltier effect for efficient temperature control. TECs offer advantages such as compact size, low noise, and long lifespan, making them suitable for various applications, including BTMS.

- Hybrid TEC–TEG systems demonstrate substantial improvements in cooling efficiency by leveraging the benefits of both TEC and TEG. These systems are effective at significantly reducing battery temperatures.
- Active and passive hybrid systems that integrate TEC with PCM or fins show enhanced temperature regulation and energy efficiency. These systems are particularly effective at managing thermal loads during high discharge rates.
- Innovative TEC designs offer advanced solutions for thermal management, such as dual active cooling systems and TEC with cold plates. These designs provide efficient cooling, and maintain battery temperatures within optimal ranges, even under extreme conditions.

These findings highlight the potential of thermoelectric cooling methods for providing efficient and reliable thermal management for Li-ion batteries. By integrating TEC with other cooling technologies and optimizing system designs, thermoelectric cooling can significantly enhance battery performance and safety, making it a viable option for applications in electric vehicles and other high-demand scenarios.

3.2. Techno-Economic Comparative Analysis of Cooling Methods

To compare the four cooling methods (air-cooling, liquid-cooling, PCM-cooling, and thermoelectric cooling), several key terms are defined that will help to evaluate their performance, efficiency, and suitability for different applications:

- Cooling Efficiency (CE): measures how effectively a cooling method maintains the battery temperature within the optimal range.
- Temperature Uniformity (TU): assesses the ability of the cooling method to maintain a uniform temperature distribution across the battery pack.
- Maximum Temperature Reduction (MTR): the extent to which the cooling method can lower the maximum temperature of the battery.
- Energy Consumption (EC): the amount of energy required by the cooling method to maintain optimal battery temperatures.
- System Complexity (SC): the level of complexity involved in implementing and maintaining the cooling method.
- Response Time (RT): the speed at which the cooling method can adapt to changes in thermal load.
- Cost-Effectiveness (C-E): the overall cost of implementing and operating the cooling method, relative to its performance benefits.
- Scalability (S): the ease with which the cooling method can be scaled up or down to accommodate different battery sizes and configurations.
- Safety and Reliability (SR): the degree to which the cooling method enhances battery safety and operational reliability.

Table 5 summarises the comparison of these methods from various aspects.

Using the data presented in Table 5, the technical performance and economic feasibility of air-cooling, liquid-cooling, PCM-cooling, and thermoelectric cooling methods is evaluated.

Table 5. Summary table of comparative study.

	Air Cooling	Liquid Cooling	PCM Cooling	TEC Cooling
CE	Moderate	High	Moderate/High	High
TU	Good	Excellent	Good	Excellent
MTR	High	High	Moderate	High
EC	Low	Moderate	Low	High
SC	Low	High	Moderate	High
RT	Moderate	Fast	Slow	Fast
CE	High	Moderate	High	Moderate
S	High	Moderate	High	Moderate
SR	Moderate	High	High	High

Air-cooling systems are highly cost-effective due to their low initial and maintenance costs. They are suitable for applications where moderate cooling performance is acceptable, and system simplicity is desired. While liquid-cooling systems have higher costs associated with their complexity and energy consumption, their high cooling efficiency and reliability make them suitable for high-demand applications like electric vehicles. PCM-based cooling methods are cost-effective, with low operational costs and moderate system complexity. They are ideal for applications requiring consistent thermal management with minimal energy consumption. Thermoelectric cooling methods, while offering high performance and precise control, come with higher costs and energy consumption. They are suitable for high-end applications where precise temperature management is critical.

Using this techno-economic analysis, a summary is presented in Table 6 showing the advantages and disadvantages of each method as well.

Table 6. Techno-Economic Comparison of Cooling Methods.

Cooling Method	Advantages	Disadvantages
Air-cooling	Cost-effective, uncomplicated design and maintenance	Less efficient at high thermal loads
Liquid-cooling	Excellent performance, high cooling efficiency	Higher cost, complex design and maintenance
PCM-cooling	Satisfactory performance, low energy consumption and cost	Suitable for passive cooling, slower response time
TEC-cooling	Precise temperature control, quick response	More expensive, high-energy consumption

From a techno-economic perspective, each cooling method has distinct advantages and disadvantages. Air-cooling is cost-effective and simple but less efficient at high thermal loads. Liquid-cooling provides excellent performance but at higher costs and complexity. PCM-cooling offers a balance of good performance with low energy consumption and cost, suitable for passive cooling needs. Thermoelectric cooling provides precise control but is more expensive and energy-intensive.

Selecting the appropriate cooling method depends on specific application requirements, including performance, cost constraints, system complexity, and operational conditions. Hybrid systems that combine multiple cooling methods can leverage the strengths of each approach to achieve optimal performance and cost-effectiveness.

4. Future Directions and Recommendations

The future of Battery Thermal Management Systems (BTMS) looks promising, thanks to new trends and technological developments. The focus of research could be on innovative materials such as nanomaterials and Phase Change Materials (PCMs), hybrid cooling systems that combine several approaches, and intelligent–adaptive BTMS with real-time control. Innovations in microchannel designs, nanofluids, and thermoelectric and magnetocaloric cooling technologies aim to improve efficiency and reduce size. High-fidelity computational models and artificial intelligence integration are planned to improve BTMS operation. Sustainability and cost-effectiveness remain important factors, as do technical obstacles and regulatory demands. Multidisciplinary co-operation and long-term testing will be critical for the successful implementation of next-generation BTMS, which will provide dependable, efficient, and environmentally friendly thermal management for lithium-ion batteries.

Author Contributions: Conceptualization, A.R.; investigation, A.R.; writing—original draft preparation, A.R.; writing—review and editing, M.A. and M.D.; visualization, A.R.; supervision, M.A. and M.D.; project administration, M.A. and M.D. All authors have read and agreed to the published version of the manuscript.

Funding: This research received no external funding.

Data Availability Statement: No new data is generated for this review study.

Conflicts of Interest: The authors declare no conflict of interest.

References

1. Hassan, Q.; Viktor, P.; Al-Musawi, T.J.; Ali, B.M.; Algburi, S.; Alzoubi, H.M.; Al-Jiboory, A.K.; Sameen, A.Z.; Salman, H.M.; Jaszczur, M. The renewable energy role in the global energy Transformations. *Renew. Energy Focus* **2024**, *48*, 100545. [CrossRef]
2. Rallabandi, S.; Selvaraj, R.V.I. *Advancements in Battery Cooling Techniques for Enhanced Performance and Safety in Electric Vehicles: A Comprehensive Review*; John Wiley and Sons Inc.: Hoboken, NJ, USA, 2024. [CrossRef]
3. Li, K.; Sun, C.; Zhang, M.; Wang, S.; Wei, B.; Cheng, Y.; Ju, X.; Xu, C. A Study of the Thermal Management and Discharge Strategies of Lithium-Ion Batteries in a Wide Temperature Range. *Energies* **2024**, *17*, 2319. [CrossRef]
4. Li, J.-X.; Li, P.-Z.; Wang, M.; Chen, C.; Gao, Y.; Yan, L.-Y.; Yang, S.-C.; Zhao, C.; Chen, M.-M.; Zhang, J.-S.; et al. Study on the temperature rise characteristics of aging lithium-ion batteries under different cooling methods. *Appl. Therm. Eng.* **2024**, *240*, 122235. [CrossRef]
5. Luo, L.; Liu, Y.; Liao, Z.; Zhong, J. Optimal structure design and heat transfer characteristic analysis of X-type air-cooled battery thermal management system. *J. Energy Storage* **2023**, *67*, 107681. [CrossRef]
6. Yang, W.; Zhou, F.; Chen, X.; Li, K.; Shen, J. Thermal performance of honeycomb-type cylindrical lithium-ion battery pack with air distribution plate and bionic heat sinks. *Appl. Therm. Eng.* **2023**, *218*, 119299. [CrossRef]
7. Duan, L.; Zhou, H.; Xu, W.; Li, L.; Liu, X.; Du, Z.; Jiang, H. Design method of multiple inlet/outlet air cooling frame of pouch lithium-ion battery based on thermal-fluid coupling topology optimization. *Int. J. Heat Mass Transf.* **2023**, *215*, 124496. [CrossRef]
8. Jilte, R.D.; Kumar, R.; Ahmadi, M.H.; Chen, L. *Battery Thermal Management System Employing Phase Change Material with Cell-To-Cell Air Cooling*; Elsevier Ltd.: Amsterdam, The Netherlands, 2019. [CrossRef]
9. Shen, X.; Cai, T.; He, C.; Yang, Y.; Chen, M. Thermal analysis of modified Z-shaped air-cooled battery thermal management system for electric vehicles. *J. Energy Storage* **2023**, *58*, 106356. [CrossRef]
10. Kashyap, P.; Panda, B.; Gao, L.; Garg, A. Design optimization of staggered-arranged battery thermal management system using an integrated approach of physics-based simulations and evolutionary algorithms. *J. Energy Storage* **2024**, *79*, 110229. [CrossRef]
11. Ahmad, S.; Liu, Y.; Khan, S.A.; Hao, M.; Huang, X. Hybrid battery thermal management by coupling fin intensified phase change material with air cooling. *J. Energy Storage* **2023**, *64*, 107167. [CrossRef]
12. Chaudhari, J.; Singh, G.K.; Rathod, M.K.; Ali, H.M. Experimental and computational analysis on lithium-ion battery thermal management system utilizing air cooling with radial fins. *J. Therm. Anal. Calorim.* **2024**, *149*, 203–218. [CrossRef]
13. Luo, Y.; Qiu, X.; Wang, S.; Jia, Z. Optimizing a direct flow cooling battery thermal management with bod baffles for electric vehicles: An experimental and simulation study. *J. Energy Storage* **2023**, *74*, 109410. [CrossRef]
14. Tuğan, V.; Yardımcı, U. Numerical study for battery thermal management system improvement with air channel in electric vehicles. *J. Energy Storage* **2023**, *72*, 108515. [CrossRef]
15. Yang, C.; Xi, H.; Wang, M. Structure optimization of air cooling battery thermal management system based on lithium-ion battery. *J. Energy Storage* **2023**, *59*, 106538. [CrossRef]
16. Suo, Y.; Tang, C.; Yang, H. Optimization design of the forced air-cooled battery thermal management system with a stepped divergence plenum. *J. Energy Storage* **2023**, *73*, 108904. [CrossRef]
17. Sutheesh, P.; Atul, A.; Rohinikumar, B. Numerical and experimental investigations of thermal performance of lithium-ion battery with hybrid cooling system under dry-out condition. *J. Energy Storage* **2024**, *84*, 110889. [CrossRef]
18. Lin, Y.; Chen, Y.-W.; Yang, J.-T. Optimized thermal management of a battery energy-storage system (BESS) inspired by air-cooling inefficiency factor of data centers. *Int. J. Heat Mass Transf.* **2023**, *200*, 123388. [CrossRef]
19. Zhang, S.-B.; He, X.; Long, N.-C.; Shen, Y.-J.; Gao, Q. Improving the air-cooling performance for lithium-ion battery packs by changing the air flow pattern. *Appl. Therm. Eng.* **2023**, *221*, 119825. [CrossRef]
20. Zhao, G.; Wang, X.; Negnevitsky, M.; Li, C.; Zhang, H.; Cheng, Y. A High-Performance Vortex Adjustment Design for an Air-Cooling Battery Thermal Management System in Electric Vehicles. *Batteries* **2023**, *9*, 208. [CrossRef]

21. Fan, H.; Wang, L.; Chen, W.; Liu, B.; Wang, P. A J-Type Air-Cooled Battery Thermal Management System Design and Optimization Based on the Electro-Thermal Coupled Model. *Energies* **2023**, *16*, 5962. [CrossRef]
22. Hasan, H.A.; Togun, H.; Abed, A.M.; Biswas, N.; Mohammed, H.I. Thermal performance assessment for an array of cylindrical Lithium-Ion battery cells using an Air-Cooling system. *Appl. Energy* **2023**, *346*, 121354. [CrossRef]
23. Yang, T.-F.; Yan, W.-M.; Lin, P.-Y.; Lin, C.-Y.; Yang, C.-C.; Sajjad, U. Thermal management of 21700 Li-ion battery packs: Experimental and numerical investigations. *Appl. Therm. Eng.* **2024**, *236*, 121518. [CrossRef]
24. Kummitha, O.R. Thermal cooling of li-ion cylindrical cells battery module with baffles arrangement for airflow cooling numerical analysis. *J. Energy Storage* **2023**, *59*, 106474. [CrossRef]
25. Gao, Q.; Lei, Z.; Huang, Y.; Zhang, C.; Chen, Y. Performance investigation of a liquid immersion cooling system with fish-shaped bionic structure for Lithium-ion battery pack. *Int. J. Heat Mass Transf.* **2024**, *222*, 125156. [CrossRef]
26. Zhou, H.; Niu, J.; Guo, X.; Xu, L.; Song, Z.; Yin, X. Thermal performance of a hybrid thermal management system based on the half helical coil coupled with air jet cooling for cylindrical Lithium-ion battery. *Appl. Therm. Eng.* **2023**, *225*, 120231. [CrossRef]
27. Yao, F.; Guan, X.; Chen, Q.; Lin, L. Research on thermal management system of lithium-ion battery with a new type of spider web liquid cooling channel and phase change materials. *J. Energy Storage* **2024**, *81*, 110447. [CrossRef]
28. Li, W.; Wang, Y.; Yang, W.; Zhang, K. Design and optimization of an integrated liquid cooling thermal management system with a diamond-type channel. *Therm. Sci. Eng. Prog.* **2024**, *47*, 102325. [CrossRef]
29. Yates, M.; Akrami, M.; Javadi, A.A. Analysing the performance of liquid cooling designs in cylindrical lithium-ion batteries. *J. Energy Storage* **2021**, *33*, 100913. [CrossRef]
30. Chen, X.; Shen, J.; Xu, X.; Wang, X.; Su, Y.; Qian, J.; Zhou, F. Performance of thermal management system for cylindrical battery containing bionic spiral fin wrapped with phase change material and embedded in liquid cooling plate. *Renew. Energy* **2024**, *223*, 120087. [CrossRef]
31. Zhao, J.; Du, W.; Xiang, H.; Gu, L. Heat transfer characteristics of liquid cooling system for lithium-ion battery pack. *Proc. Inst. Mech. Eng. Part. D J. Automob. Eng.* **2024**, 09544070231220750. [CrossRef]
32. Amalesh, T.; Narasimhan, N.L.; Reddy, G.R. Numerical and experimental studies on novel minichannel cold plates for lithium-ion battery thermal management. *J. Energy Storage* **2023**, *73*, 109167. [CrossRef]
33. Feng, Z.; Shen, X.; Li, P.; Zhao, J.; Zhang, H.; Xu, Y.; Yuan, J. Performance optimization and scheme evaluation of liquid cooling battery thermal management systems based on the entropy weight method. *J. Energy Storage* **2024**, *80*, 110329. [CrossRef]
34. Zhong, Q.; Chandra, P.K.; Li, W.; Gao, L.; Garg, A.; Lv, S.; Tai, K. A comprehensive numerical study based on topology optimization for cooling plates thermal design of battery packs. *Appl. Therm. Eng.* **2024**, *236*, 121918. [CrossRef]
35. Ye, J.; Aldaher, A.Y.M.; Tan, G. Thermal performance analysis of 18,650 battery thermal management system integrated with liquid-cooling and air-cooling. *J. Energy Storage* **2023**, *72*, 108766. [CrossRef]
36. Chavan, S.; Liu, J.; Venkateswarlu, B.; Joo, S.W.; Kim, S.C. Numerical simulation of lithium-ion battery thermal management systems: A comparison of fluid flow channels and cooling fluids. *J. Energy Storage* **2023**, *73*, 108940. [CrossRef]
37. Zhou, H.; Guo, X.; Xu, L.; Cui, Y.; Guo, S.; Song, Z. Thermal performance of a hybrid thermal management system that couples PCM/copper foam composite with air-jet and liquid cooling. *J. Energy Storage* **2023**, *74*, 109408. [CrossRef]
38. Kumar, K.; Sarkar, J.; Mondal, S.S. Multi-scale-multi-domain simulation of novel microchannel-integrated cylindrical Li-ion battery thermal management: Nanoparticle shape effect. *J. Energy Storage* **2024**, *84*, 110824. [CrossRef]
39. Zeng, W.; Ma, C.; Hu, S.; Li, S.; Zhang, Y. The performance investigation and optimization of reciprocating flow applied for liquid-cooling-based battery thermal management system. *Energy Convers. Manag.* **2023**, *292*, 117378. [CrossRef]
40. Tousi, M.; Najafi, M. Innovative hybrid nano/dielectric fluid cooling system for the new cylindrical shaped Li-ion batteries. *Int. J. Therm. Sci.* **2024**, *195*, 108634. [CrossRef]
41. Xin, S.; Wang, C.; Xi, H. Thermal management scheme and optimization of cylindrical lithium-ion battery pack based on air cooling and liquid cooling. *Appl. Therm. Eng.* **2023**, *224*, 120100. [CrossRef]
42. Zhao, L.; Li, W.; Wang, G.; Cheng, W.; Chen, M. A novel thermal management system for lithium-ion battery modules combining direct liquid-cooling with forced air-cooling. *Appl. Therm. Eng.* **2023**, *232*, 120992. [CrossRef]
43. Wang, J.; Mei, W.; Mao, B.; Wang, Q. Investigation on the temperature control performance and optimization strategy of a battery thermal management system combining phase change and liquid cooling. *Appl. Therm. Eng.* **2023**, *232*, 121080. [CrossRef]
44. Xie, N.; Zhang, Y.; Liu, X.; Luo, R.; Liu, Y.; Ma, C. Thermal performance and structural optimization of a hybrid thermal management system based on MHPA/PCM/liquid cooling for lithium-ion battery. *Appl. Therm. Eng.* **2023**, *235*, 121341. [CrossRef]
45. Dai, W.; Lai, H. Comparative study of flow-channel layout schemes in liquid cooling plates of a prismatic battery module. *Appl. Therm. Eng.* **2024**, *236*, 121501. [CrossRef]
46. Zhang, F.; Huang, Z.; Li, S.; Sun, S.; Zhao, H. Design and thermal performance analysis of a new micro-fin liquid cooling plate based on liquid cooling channel finning and bionic limulus-like fins. *Appl. Therm. Eng.* **2024**, *237*, 121597. [CrossRef]
47. Shen, J.; Chen, X.; Xu, X.; Kong, J.; Song, Z.; Wang, X.; Zhou, F. Thermal performance of a hybrid cooling plate integrated with microchannels and PCM. *Appl. Therm. Eng.* **2024**, *236*, 121917. [CrossRef]
48. Yan, G.; Alizadeh, A.; Rahmani, A.; Zarringhalam, M.; Shamsborhan, M.; Nasajpour-Esfahani, N.; Akrami, M. Natural convection of rectangular cavity enhanced by obstacle and fin to simulate phase change material melting process using Lattice Boltzmann method. *Alex. Eng. J.* **2023**, *81*, 319–336. [CrossRef]

49. Rahmani, A.; Dibaj, M.; Akrami, M. Computational investigation of magnetohydrodynamic flow and melting process of phase change material in a battery pack using the lattice Boltzmann method. *J. Energy Storage* **2024**, *78*, 110046. [CrossRef]
50. Chen, X.; Yang, W.; Shen, J.; Xu, X.; Zhou, F. Thermal performance of hybrid battery thermal management system with air cooling and phase change material embedding biomimetic variable section fins. *Appl. Therm. Eng.* **2023**, *231*, 120985. [CrossRef]
51. Rahmani, A.; Dibaj, M.; Akrami, M. Enhancing Heat Storage Cooling Systems via the Implementation of Honeycomb-Inspired Design: Investigating Efficiency and Performance. *Energies* **2024**, *17*, 351. [CrossRef]
52. Suo, Y.; Tang, C.; Jia, Q.; Zhao, W. Influence of PCM configuration and optimization of PCM proportion on the thermal management of a prismatic battery with a combined PCM and air cooling structure. *J. Energy Storage* **2024**, *80*, 110340. [CrossRef]
53. Tang, H.; Wang, S.; Li, H. *Flexibility Categorization, Sources, Capabilities and Technologies for Energy-Flexible and Grid-Responsive Buildings: State-Of-The-Art and Future Perspective*; Elsevier Ltd.: Amsterdam, The Netherlands, 2021. [CrossRef]
54. Wang, Y.; Gao, T.; Zhou, L.; Gong, J.; Li, J. A parametric study of a hybrid battery thermal management system that couples PCM with wavy microchannel cold plate. *Appl. Therm. Eng.* **2023**, *219*, 119625. [CrossRef]
55. Li, K.; Yao, X.; Li, Z.; Gao, T.; Zhang, W.; Liao, Z.; Ju, X.; Xu, C. Thermal management of Li-ion batteries with passive thermal regulators based on composite PCM materials. *J. Energy Storage* **2024**, *89*, 111661. [CrossRef]
56. Zhao, Y.; Zhang, X.; Yang, B.; Cai, S. *A Review of Battery Thermal Management Systems Using Liquid Cooling and PCM*; Elsevier Ltd.: Amsterdam, The Netherlands, 2024. [CrossRef]
57. Wang, H.; Guo, Y.; Ren, Y.; Yeboah, S.; Wang, J.; Long, F.; Zhang, Z.; Jiang, R. Investigation of the thermal management potential of phase change material for lithium-ion battery. *Appl. Therm. Eng.* **2024**, *236*, 121590. [CrossRef]
58. Shahid, S.; Agelin-Chaab, M. Investigation of thermal properties of phase change materials for novel hybrid thermal management strategies for cylindrical Li-ion cells. *Appl. Therm. Eng.* **2024**, *242*, 122471. [CrossRef]
59. Shahid, S.; Agelin-Chaab, M. Investigation of Heat Transfer Enhancement Techniques on a Scalable Novel Hybrid Thermal Management Strategy for Lithium-Ion Battery Packs. *Batteries* **2024**, *10*, 32. [CrossRef]
60. Kadam, G.; Kongi, P. Battery thermal management system based on PCM with addition of nanoparticles. *Mater. Today Proc.* **2023**, *72*, 1543–1549. [CrossRef]
61. Li, Y.; Chen, Z.; Feng, Y.; Liu, M.; Kang, C.; Yang, K.; Yuan, J.; Qiu, C.; Shi, H.; Jiang, Y. A novel petal-type battery thermal management system with dual phase change materials. *Int. J. Heat Mass Transf.* **2023**, *207*, 123989. [CrossRef]
62. Xu, B.; Xia, F.; Wang, Y.-L.; Xie, X.; Gan, W.-T. A battery thermal management scheme suited for cold regions based on PCM and aerogel: Demonstration of performance and availability. *Appl. Therm. Eng.* **2023**, *227*, 120378. [CrossRef]
63. Zhang, F.; Lu, F.; Liang, B.; Zhu, Y.; Gou, H.; Xiao, K.; He, Y. Thermal performance analysis of a new type of branch-fin enhanced battery thermal management PCM module. *Renew. Energy* **2023**, *206*, 1049–1063. [CrossRef]
64. Fan, Y.; Wang, Z.; Xiong, X.; Zhu, J.; Gao, Q.; Wang, H.; Wu, H. Novel concept design of low energy hybrid battery thermal management system using PCM and multistage Tesla valve liquid cooling. *Appl. Therm. Eng.* **2023**, *220*, 119680. [CrossRef]
65. Shi, H.; Liu, M.; Li, Y.; Wang, S.; Qiu, C.; Cheng, M.; Yuan, J.; Yang, K.; Kang, C. Multi-objective optimization of integrated lithium-ion battery thermal management system. *Appl. Therm. Eng.* **2023**, *223*, 119991. [CrossRef]
66. Liu, H.; Jin, C.; Li, H.; Ji, Y. A numerical study of PCM battery thermal management performance enhancement with fin structures. *Energy Rep.* **2023**, *9*, 1793–1802. [CrossRef]
67. Khaboshan, H.N.; Jaliliantabar, F.; Abdullah, A.A.; Panchal, S. Improving the cooling performance of cylindrical lithium-ion battery using three passive methods in a battery thermal management system. *Appl. Therm. Eng.* **2023**, *227*, 120320. [CrossRef]
68. Luo, T.; Zhang, Y.; Chen, X.; Jia, T.; Yu, H.; Mao, B.; Ma, C. A hybrid battery thermal management system composed of MHPA/PCM/Liquid with a highly efficient cooling strategy. *Appl. Therm. Eng.* **2024**, *251*, 123617. [CrossRef]
69. Mehryan, S.; Jannesari, H. Analyzing battery thermal management systems: A comparative study of hydrogel and various PCMs in different configurations and convection conditions. *J. Energy Storage* **2024**, *77*, 109998. [CrossRef]
70. Dey, H.; Pati, S.; Randive, P.R.; Baranyi, L. Effect of finned networks on PCM based battery thermal management system for cylindrical Li-ion batteries. *Case Stud. Therm. Eng.* **2024**, *59*, 104572. [CrossRef]
71. Li, B.; Mao, Z.; Song, B.; Wang, X.; Tian, W.; Sun, Q.; Wang, Y.-F.; Jin, Z. Experimental investigation on efficient thermal management of autonomous underwater vehicle battery packs using anisotropic expanded graphite/paraffin composite materials. *Appl. Therm. Eng.* **2024**, *242*, 122477. [CrossRef]
72. Luo, M.; Zhang, Y.; Wang, Z.; Niu, Y.; Lu, B.; Zhu, J.; Zhang, J.; Wang, K. Thermal performance enhancement with snowflake fins and liquid cooling in PCM-based battery thermal management system at high ambient temperature and high discharge rate. *J. Energy Storage* **2024**, *90*, 111754. [CrossRef]
73. Fouladvand, J. Energy Security of Thermal Energy Communities. Ph.D. Thesis, Delft University of Technology, Delft, The Netherlands, 2022.
74. Guclu, T.; Cuce, E. *Thermoelectric Coolers (TECs): From Theory to Practice*; Springer: New York, NY, USA, 2019. [CrossRef]
75. Cheng, K.; Qin, J.; Jiang, Y.; Zhang, S.; Bao, W. Performance comparison of single- and multi-stage onboard thermoelectric generators and stage number optimization at a large temperature difference. *Appl. Therm. Eng.* **2018**, *141*, 456–466. [CrossRef]
76. Lin, S.; Yu, J. Optimization of a trapezoid-type two-stage Peltier couples for thermoelectric cooling applications. *Int. J. Refrig.* **2016**, *65*, 103–110. [CrossRef]
77. Li, X.; Zhong, Z.; Luo, J.; Wang, Z.; Yuan, W.; Zhang, G.; Yang, C.; Yang, C. Experimental Investigation on a Thermoelectric Cooler for Thermal Management of a Lithium-Ion Battery Module. *Int. J. Photoenergy* **2019**, *2019*, 3725364. [CrossRef]

78. Hameed, M.M.; Bin Mansor, M.; Azau, M.A.M.; Alshara, A.K. Computational design and analysis of LiFePO4 battery thermal management system (BTMS) using thermoelectric cooling/thermoelectric generator (TEC–TEG) in electric vehicles (EVs). *J. Energy Storage* **2023**, *72*, 108394. [CrossRef]
79. Luo, D.; Wu, Z.; Yan, Y.; Cao, J.; Yang, X.; Zhao, Y.; Cao, B. Performance investigation and design optimization of a battery thermal management system with thermoelectric coolers and phase change materials. *J. Clean. Prod.* **2024**, *434*, 139834. [CrossRef]
80. Bayendang, N.P.; Balyan, V.; Kahn, M.T. The question of thermoelectric devices (TEDs) in/efficiency—A practical examination considering thermoelectric coolers (TECs). *Results Eng.* **2024**, *21*, 101827. [CrossRef]
81. Luo, D.; Wu, H.; Cao, J.; Yan, Y.; Yang, X.; Cao, B. Numerical investigation of a battery thermal management system integrated with vapor chamber and thermoelectric refrigeration. *J. Clean. Prod.* **2024**, *434*, 140089. [CrossRef]
82. Alghamdi, H.; Rosdi, M.F.M.; Mukhtar, A.; Yasir, A.S.H.M.; Alviz-Meza, A. Controlling thermal runaway by simultaneous use of thermoelectric module and phase change material in the lithium-ion batteries of electric vehicles. *Case Stud. Therm. Eng.* **2023**, *52*, 103697. [CrossRef]
83. Pan, Y.; Tang, A.; Liu, Z.; Shan, C. Experimental analysis of power battery preheating system based on thermoelectric elements. *Appl. Therm. Eng.* **2023**, *230*, 120860. [CrossRef]
84. Liu, X.; Yao, L.-C.; Su, C.-Q.; Xiong, X.; Wang, Y.-P. A hybrid battery thermal management system coupling with PCM and optimized thermoelectric cooling for high-rate discharge condition. *Case Stud. Therm. Eng.* **2023**, *49*, 103269. [CrossRef]
85. Luo, D.; Zhao, Y.; Cao, J.; Chen, W.-H.; Zhao, Y.; Cao, B. Performance analysis of a novel thermoelectric-based battery thermal management system. *Renew. Energy* **2024**, *224*, 120193. [CrossRef]
86. An, Z.; Liu, H.; Gao, W.; Gao, Z. Cooling and preheating performance of dual-active lithium-ion battery thermal management system under harsh conditions. *Appl. Therm. Eng.* **2024**, *242*, 122421. [CrossRef]
87. Liu, X.; Zhang, C.-F.; Zhou, J.-G.; Xiong, X.; Wang, Y.-P. Thermal performance of battery thermal management system using fins to enhance the combination of thermoelectric Cooler and phase change Material. *Appl. Energy* **2022**, *322*, 119503. [CrossRef]

batteries

MDPI

Review

Recent Progress and Prospects in Liquid Cooling Thermal Management System for Lithium-Ion Batteries

Jiahao Liu [1,*], Hao Chen [1], Silu Huang [1], Yu Jiao [1] and Mingyi Chen [2]

[1] College of Ocean Science and Engineering, Shanghai Maritime University, Shanghai 201306, China; yujiao@shmtu.edu.cn (Y.J.)
[2] School of Environment and Safety Engineering, University of Jiangsu, Zhenjiang 212013, China; chenmy@ujs.edu.cn
* Correspondence: jiahao@shmtu.edu.cn

Abstract: The performance of lithium-ion batteries is closely related to temperature, and much attention has been paid to their thermal safety. With the increasing application of the lithium-ion battery, higher requirements are put forward for battery thermal management systems. Compared with other cooling methods, liquid cooling is an efficient cooling method, which can control the maximum temperature and maximum temperature difference of the battery within an acceptable range. This article reviews the latest research in liquid cooling battery thermal management systems from the perspective of indirect and direct liquid cooling. Firstly, different coolants are compared. The indirect liquid cooling part analyzes the advantages and disadvantages of different liquid channels and system structures. Direct cooling summarizes the different systems' differences in cooling effectiveness and energy consumption. Then, the combination of liquid cooling, air cooling, phase change materials, and heat pipes is examined. Later, the connection between the cooling and heating functions in the liquid thermal management system is considered. In addition, from a safety perspective, it is found that liquid cooling can effectively manage thermal runaway. Finally, some problems are put forward, and a summary and outlook are given.

Keywords: battery thermal management system; liquid cooling; indirect liquid cooling; direct liquid cooling; composite cooling

Citation: Liu, J.; Chen, H.; Huang, S.; Jiao, Y.; Chen, M. Recent Progress and Prospects in Liquid Cooling Thermal Management System for Lithium-Ion Batteries. *Batteries* **2023**, *9*, 400. https://doi.org/10.3390/batteries9080400

Academic Editor: Thomas Wetzel

Received: 13 June 2023
Revised: 26 July 2023
Accepted: 30 July 2023
Published: 1 August 2023

1. Introduction

With the energy crisis and environmental pollution becoming increasingly prominent, vigorously developing clean energy, promoting environmental improvement, and advancing green and low-carbon construction has become an important task. Traditional fuel vehicles mainly use non-renewable fossil energy as a power source, which not only consumes more fossil energy, but also produces exhaust gas, which contributes to the greenhouse effect. In this context, electric vehicles have received a lot of attention because of their advantages, such as low pollution and high efficiency [1]. The key task in the development of electric vehicles is to find a suitable energy storage system that allows battery vehicles to have a long driving range and fast acceleration [2]. Lithium-ion batteries (LIBs) have been widely used in energy storage systems of electric vehicles due to their high energy density, high power density, low pollution, no memory effect, low self-discharge rate, and long cycle life [3–6]. Studies have shown that the performance of LIBs is closely related to the operating temperature [7,8]. Generally, the optimum operating temperature range for Li-ion batteries is 15–35 °C [9], and the maximum temperature difference between batteries should be controlled within 5 °C [5,10]. Therefore, a reasonable and effective battery thermal management system (BTMS) is necessary to enable the battery module to work safely and exhibit good charge and discharge performance [11].

The currently popular BTMSs can be divided into air cooling, liquid cooling, phase change material (PCM), heat pipe, and composite cooling. With its simple structure and low cost, air cooling has been widely used in early BTMSs. However, it is challenging to meet the demand for battery heat dissipation under the circumstance of rapid charging due to the low specific heat capacity and heat transfer coefficient of air, and limited cooling capacity [12,13]. PCM has become a research hotspot in battery thermal management due to its large latent heat of phase change and no need to consume additional energy, but its low thermal conductivity and easy leakage problems hinder its application in electric vehicle BTMSs [14]. Studies have shown that the thermal conductivity of PCM can be improved by adding foam metal, expanded graphite, carbon fiber, and other materials to PCM [8]. The heat pipe has the advantages of high thermal conductivity, excellent stability, and low maintenance cost, and has broad application prospects, but it is not currently widely used in battery thermal management due to its small contact area and large system volume [12]. Composite cooling is a combination of two or more cooling methods based on a single cooling method, used to meet higher heat dissipation requirements, as well as to improve the temperature distribution of the battery module, showing outstanding cooling effects and attracting the attention of researchers.

Compared with other cooling methods, liquid cooling has been used commercially in BTMSs for electric vehicles for its high thermal conductivity, excellent cooling effect, ability to meet high heat dissipation requirements, and more uniform battery temperature distribution. For example, the Tesla Model S electric vehicle uses indirect liquid cooling, and the coolant is a mixture of water and ethylene glycol [15]. The Chevrolet Volt and BMW i3 and i8 also use liquid cooling systems for battery thermal management to avoid excessive battery temperature [16]. In addition, 3M has developed a battery direct liquid cooling system for electric vehicles, which immerses the battery module directly into the coolant, showing an excellent cooling effect [5].

In recent years, many scholars have studied and reviewed the BTMS, which mainly focuses on summarizing the progress and achievements of the whole BTMS, but rarely makes a separate analysis and integration of liquid cooling. The liquid cooling system has unique advantages, and commercial applications of liquid cooling are increasing. Therefore, it is necessary to review and summarize the research on the liquid cooling of LIBs, and put forward reasonable suggestions for liquid cooling problems, to provide a reference for the development of the BTMS. In this paper, the heat generation mechanism of LIBs is analyzed, and the influence of temperature on battery performance is summarized. Secondly, the research results on liquid cooling by scholars in recent years are reviewed, starting with both indirect liquid cooling and direct liquid cooling. Subsequently, the battery preheating technology in BTMS is studied. Then, the effect of liquid cooling on the thermal runaway of the battery is discussed. Finally, some problems in the liquid cooling system are summarized, and the future research direction of LIB liquid cooling is prospected.

2. Heat Generation Mechanism and Temperature Effects

2.1. Heat Generation Mechanism

The LIB consists of four parts: the anode (negative electrode), the cathode (positive electrode), the electrolyte, and the separator [12,17]. During the charging and discharging process of the battery, as shown in Figure 1, LIBs are de-embedded back and forth between the positive and negative electrodes through the electrolyte and the separator, hence the name "rocking-chair battery" [18]. During charging, lithium ions receive energy from the outside, move to the anode, and embed into the anode, forming a Li-rich ion state at the anode; during discharge, lithium ions carry energy from the negative electrode to the positive electrode, while electrons move back and forth through the external circuit to generate current during charge and discharge [19].

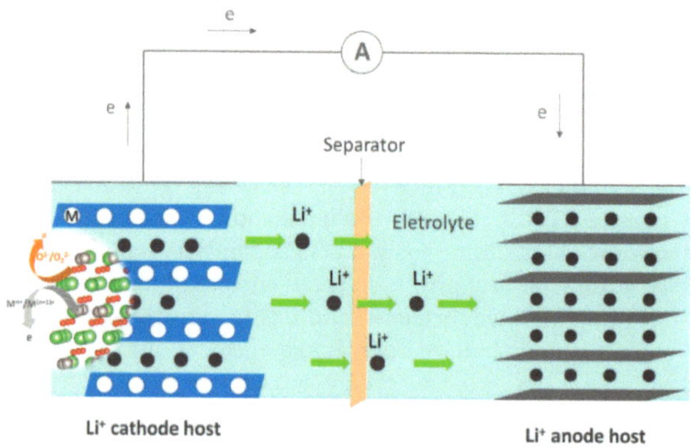

Figure 1. Diagram of the working principle of a lithium-ion battery (LIB) [20].

The following reactions occur during the charging and discharging process [21]:

$$\text{Positive electrode}: \ LiMO_2 \rightleftarrows Li_{1-x}MO_2 + xLi^+ + xe^- \tag{1}$$

$$\text{Negative electrode}: \ C + xLi^+ + xe^- + LiMO_2 \rightleftarrows Li_xC \tag{2}$$

Overall:

$$LiMO_2 + C \rightleftarrows Li_xC + Li_{1-x}MO_2 \tag{3}$$

$LiMO_2$ is the metal oxide used for the positive electrode, such as $LiCoO_2$, $LiNiO_2$, $LiMn_2O_4$, $LiFePO_4$, etc. C is the carbon material used for the negative electrode.

During the normal charging and discharging process of the battery, a large number of chemical reactions take place inside, and these complex chemical reactions are usually accompanied by the generation of heat. With the continuous generation and accumulation of heat, if the heat cannot be dissipated in time, the temperature of the battery will rise rapidly, which will cause a series of abnormal side reactions, leading to a deterioration of the battery's performance and even triggering thermal runaway. Therefore, it is necessary to study the mechanism of LIB heat generation. Generally, the heat generation of LIB in work can be divided into two parts: reversible heat and irreversible heat. Reversible heat is the heat released when lithium ions are deintercalated between the positive and negative electrodes. It comes from the reversible entropy change caused by the electrochemical reaction inside the battery, also known as reaction heat or entropy heat [22]. Irreversible heat includes polarization heat and ohmic heat [22,23]. Polarization heat is caused by lithium ions overcoming impedance during internal movement, while ohmic heat or joule heat is related to the resistance inside the battery.

In order to better analyze the thermal behavior of LIBs and predict the performance of batteries, many scholars have studied the thermal model of LIBs [24–26]. One of the most commonly used battery heat production equations, proposed by Bernardi [27], is used to predict the heat production rate of a single battery. The expression is:

$$Q = I(E - U) - IT\frac{dE}{dT} \tag{4}$$

where Q, I, E, U, and T are expressed as heat producing power, operating current, open-circuit voltage, operating voltage, and operating temperature, respectively. The first term $I\ (E - U)$ on the right represents irreversible heat. The second term $-IT\frac{dE}{dT}$ represents reversible heat, which is negative when charging and positive when discharging, while the entropy heat coefficient ($\frac{dE}{dT}$) is related to density, state of charge (SOC), and battery temperature [8]. The heat generation model has been widely used because of its convenient use and reasonable accuracy, but it does not take into account the existence of phase change heat and mixing heat. For irreversible heat, $(E - U)$ can be replaced by IR; R denotes the internal resistance of the battery, and the formula can be simplified as:

$$Q = I^2R - IT\frac{dE}{dT} \quad (5)$$

Compared with irreversible heat, reversible heat can be negligible in practical hybrid electric vehicle and pure electric vehicle applications [28]. Therefore, the formula can also be expressed as [29]:

$$Q = I^2R - IT\frac{\Delta S}{F} \quad (6)$$

where ΔS represents the entropy change, which is negative when charging and positive when discharging. F is the Faraday constant.

The heat generation model of LIB can be divided into electro-thermal models and electrochemical thermal models [30]. Ping et al. [31] established an electro-thermal coupling model to study the phenomenon from the normal cycle to thermal runaway. It was found that as the battery discharge rate increased, the thermal runaway time of the battery was advanced. Zhang et al. [23] established a one-dimensional electrochemical thermal model to analyze the heat generation of the battery during charging and discharging, and found that ohmic heat, electrochemical reaction heat, and joule heat are the main heat sources. Liebig et al. [32] proposed a model that can simulate the electrochemical and thermal behavior of the battery with high precision. It is found that the thermal behavior of the battery is mainly reversible heat at low current rates, while at high current rates, irreversible heat becomes the main. They then coupled an electrochemical model with a three-dimensional thermal model to simulate the thermal behavior of LIBs in a real environment [33]. Panchal et al. [34] proposed an electrochemical thermal model to study the transient temperature distribution of 18650 LIBs. It is discovered that the model accurately forecasts the data when compared with the outcomes of the experiments. Accordingly, it is possible to understand the thermal and electrochemical behavior of LIBs using a thermal model, which is useful for understanding the internal structure of the battery and can accurately predict the working temperature of the battery.

2.2. Temperature Effects

The optimal operating temperature range of LIBs is relatively narrow, and too low or too high temperature will affect the performance of the battery. As shown in Figure 2, the LIB shows the best performance only when it works within 15–35 °C. At lower temperatures, the ability of lithium ions to diffuse is limited, and it increases the internal impedance, and the discharge capacity of the battery decreases [7]. In contrast, the active chemistry in the battery can be damaged at high temperatures, which can result in irreparable damage to the capacity and life of the battery and possibly cause thermal runaway [35]. Thermal runaway is one of the main safety issues facing LIBs. It refers to the uncontrollable rise in battery temperature caused by a chain reaction of heat release from a single battery cell. When the battery temperature continues to rise and the heat cannot be dissipated in time, if no measures are taken to control it, thermal runaway will occur when the temperature threshold is exceeded [36]. In the process of thermal runaway, a large amount of heat and harmful gases are often generated, and even lead to battery fire and explosion [2].

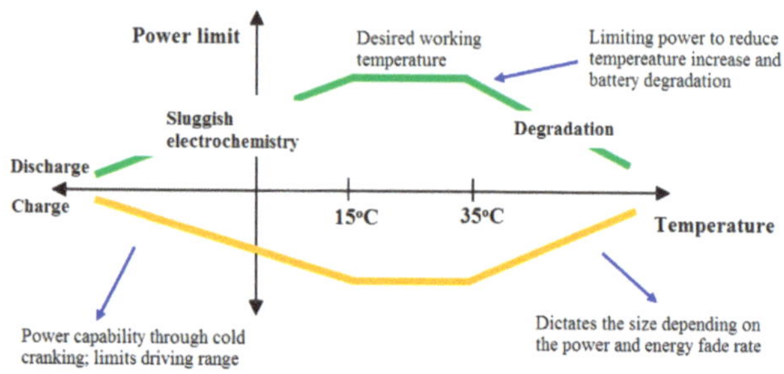

Figure 2. Operating range of LIBs [37].

According to a study by Jaguemont et al. [35] on the impact of low temperatures on LIB performance, battery capacity will drop off rapidly as the operating temperature drops. This is because LIBs will lose capacity as they mature at low temperatures. Zhang et al. [38] also investigated the aging mechanism of batteries at low temperatures and came to similar conclusions. According to the experimental results, the battery impedance increases at low temperatures, and its capacity is significantly reduced. Another point worth noting is that LIBs operating at low temperatures may undergo lithium-ion plating and dendrites, which can lead to severe degradation of the battery's performance [39,40]. At high temperatures, with the temperature continuing to rise, LIBs undergo a number of reactions that affect their performance. As shown in Figure 3, the solid electrolyte interface film (SEI) will first decompose at 90 °C or even lower temperatures. During the first charge–discharge cycle of the LIB, a passivation layer called SEI is created on the surface of the negative electrode. It can successfully stop the organic solvent in the electrolyte from damaging the electrode material, which improves the performance of the battery. Secondly, Li reacts with the electrolyte. Then, after 135 °C, the separator melts, which leads to a short circuit between the two electrodes [2]. When the temperature rises to 200 °C, it will trigger a series of complex reactions. As lithium ions and electrolytes are consumed in large quantities, the capacity and performance of the battery decline dramatically. Zhao et al. [41] emphasized that for every 1 °C increase in temperature between 30 °C and 40 °C, the battery life will be reduced by two months. Ramadass et al. [42] compared the charge–discharge cycles at room temperature, 45 °C, 50 °C, and 55 °C. The results showed that the battery capacity loss was 30% after 800 cycles at room temperature, and the capacity loss was as high as 70% after 490 cycles at 55 °C. As the temperature and number of cycles increased, this reduced capacity and a higher impedance resulted in a drop in battery power.

In fact, LIBs are frequently employed as battery packs to reach the necessary voltage and capacity. The battery pack might accumulate heat more easily, which can lead to localized overheating. As a result, another important aspect impacting battery performance is the temperature consistency between cells. For battery packs, the temperature difference between cells should be controlled to within 5 °C to prevent rapid deterioration occurring in a single cell and reducing the life of the battery pack [43]. Yang et al. [44] quantified the relationship between the temperature difference and the capacity loss rate of the parallel battery pack and studied the temperature difference in the battery capacity. They also discovered that when the temperature difference between the batteries grew, the rate of capacity loss grew. Yan et al. [10] pointed out that the small temperature difference between the batteries is conducive to promoting the balance of the battery during the cycle and maintaining the charging and discharging behavior of the entire battery pack. However, as the temperature difference increases, it can accelerate battery degradation, leading to safety issues, such as overheating and thermal runaway. Moreover, the performance of a battery pack made up of batteries with different capacities connected in series is constrained by the capacity of the smallest battery [45]. The consistency across batteries must be regarded to prevent overcharging individual batteries inside the battery pack [46].

In summary, it is essential to keep the temperature of the battery and the temperature difference between batteries within the ideal range for better performance and longer service life. The BTMS can not only dissipate the heat generated by the battery in time to avoid the high temperature of the battery, but also reduce the temperature gradient inside the battery and maintain a good temperature uniformity.

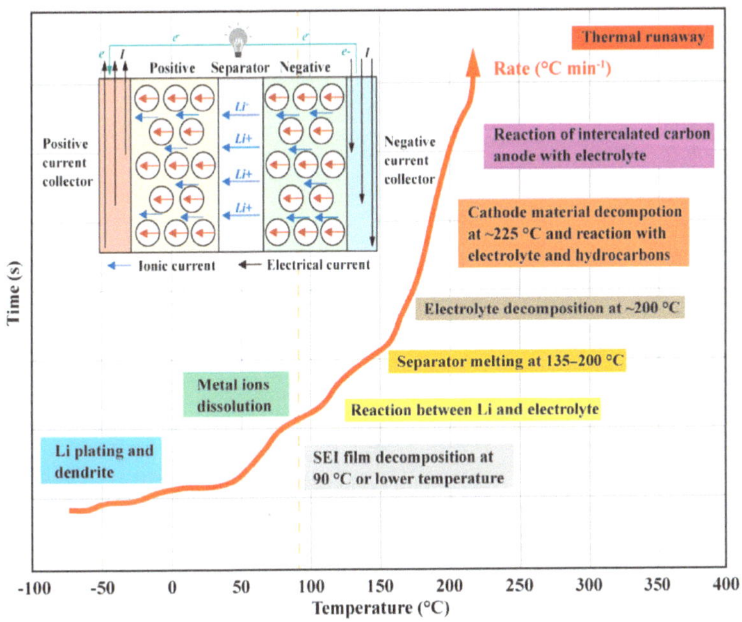

Figure 3. The reaction of LIBs at different temperatures [47].

3. Liquid-Cooled BTMS

Liquid cooling can be divided into indirect cooling and direct cooling (also known as immersion cooling), depending on whether the coolant is in contact with the battery, as shown in Figure 4. Indirect liquid cooling usually involves placing cooling plates [48], discrete tubes [8], or jackets [49] on the surface of the cell. This cooling technique moves the heat produced by the battery to the outside with the flowing coolant, avoiding direct contact between the coolant and the battery [50]. In direct liquid cooling, the coolant is in direct contact with the battery, which requires the use of non-conductive medium fluid as coolant [7]. Direct liquid cooling greatly improves the contact area between the battery and the coolant, thereby obtaining an extremely high heat transfer rate. Direct liquid cooling can be divided into single phase and two phase, according to whether the coolant has phase change. Compared with indirect liquid cooling, direct liquid cooling shows a better cooling effect and can improve the uniformity of temperature distribution. As the battery and coolant are in direct contact, this reduces the need for complex flow path designs and reduces the risk of accidental leakage of coolant, which could cause short circuits in the battery [51]. Although direct liquid cooling is considered to be a better choice for the BTMS in the future, it has not been commonly used in electric vehicle [52]. In this section, the main research progress of indirect liquid cooling and direct liquid cooling is introduced. Secondly, the research of different coolants in each part is summarized and analyzed. Finally, attention is paid to the composite cooling system based on liquid cooling.

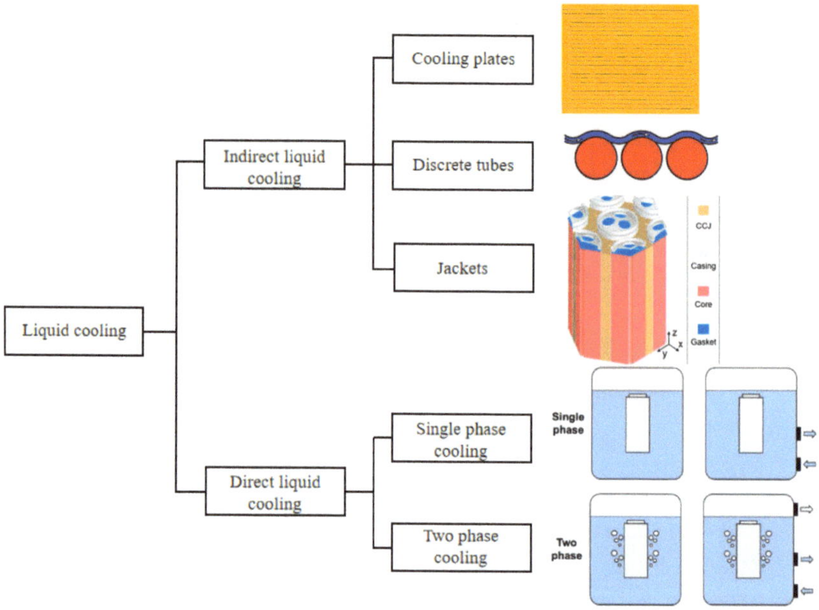

Figure 4. Classification of liquid cooling and corresponding examples [8,48,49,53].

3.1. Indirect Liquid Cooling

In indirect liquid cooling, different shapes of batteries are suitable for different cooling devices. Cylindrical batteries usually use discrete tubes or jackets in order to have a larger contact area between the surface and the coolant. For prismatic batteries or pouch batteries, flat cold plates are usually the best choice. The air gap between the cooling plate or tube and the battery will contribute to heat insulation and lower heat transfer efficiency as a result of the coolant and battery not making direct contact with each other. A high-precision cold metal plate, as well as a high-thermal-conductivity grease or epoxy bonding agent, are required to eliminate air gaps, thereby reducing thermal contact resistance [5,9,53–55]. In order to improve the heat dissipation of the battery, researchers have conducted many studies on indirect liquid cooling. The selection of coolants, the design of flow channels, and the optimization of system structures have received a lot of attention, and these studies are covered in this section.

3.1.1. Coolant

The viscosity, density, thermal conductivity, specific heat capacity, and flow rate of the coolant are significant variables determining the cooling capacity and heat transfer efficiency in liquid cooling. In order not to increase the pumping power and improve the cooling effect, a coolant with high thermal conductivity and low viscosity is needed. As the most accessible liquid in life, water has been widely used in indirect liquid cooling systems. It has been discovered that using water as a coolant may assure battery temperature consistency and enhance BTMS cooling performance [56,57]. Karimi and Dehghan [58] used air, water, and silicone oil as coolants to study the performance of the BTMS. According to the simulation results, using water as a coolant instead of air or silicone oil can not only use less power, but also keep the temperature within reasonable bounds. In addition, water has higher thermal conductivity and specific heat capacity than air, and lower viscosity than oil, which ensures the heat transfer capacity and helps reduce additional energy consumption. In order to reduce the freezing point of water, so that the BTMS can be used at a lower temperature (below zero), the water/glycol mixture is usually used as the coolant. Adding glycol to water can substantially extend the breadth of application of

BTMSs and lower the danger of battery damage caused by increased volume following water condensation. To eliminate short circuit issues brought on by contact with the battery following liquid leakage, the BTMS based on water or water/glycol liquid should strictly avoid direct contact between the coolant and the battery [59].

Recently, nanofluids have attracted the attention and research of scholars due to their excellent thermal conductivity in BTMSs [60–62]. The thermal conductivity of the liquid is improved by adding metal particles to the water to make it a nanofluid. The concentration and size of nanoparticles are two important factors affecting the thermal conductivity of nanofluids. Zakaria et al. [63] investigated the effect of Al_2O_3 nanoparticles with different volume concentrations (0.1, 0.3, and 0.5%) on thermal conductivity in a water/glycol mixture. The results showed that the thermal conductivity increased with the concentration of nanoparticles in the water/glycol mixture. In the water/glycol mixture with a volume ratio of 1:1, the volume concentrations of nanoparticles were 0.1%, 0.3%, and 0.5%, respectively, and the thermal conductivity increased by 2%, 4.2%, and 7.5%, respectively. Maheshwary et al. [64] investigated the effects of TiO_2 nanoparticle concentration, size, and shape on the thermal conductivity of nanofluids. Thermal conductivity rises with concentration, according to experimental data. Thermal conductivity can also be improved by altering the shape and particle size, although the concentration influences this property the most. In addition, Teng et al. [65] studied the effect of Al_2O_3 particle size on nanofluids and found that the enhancement effect of nanofluids on thermal conductivity became larger at higher temperatures, and for small-sized nanoparticles, the effect of temperature rise on the thermal conductivity of nanofluids was more obvious. In other studies [66,67], it was also found that smaller nanoparticles are beneficial to enhance the thermal conductivity of nanofluids.

Tousi et al. [68] studied the thermal management of 18650/21700 LIBs based on AgO nanofluids. It was revealed that the maximum temperature and temperature difference of the battery pack were greatly lowered after adding nanoparticles in water, while the pressure drop in the channel was increased. The findings demonstrate that, even at a high discharge rate of 7C, the maximum temperature and temperature difference of the battery can be kept within 33.5 °C and 0.67 °C at an appropriate flow rate and concentration of the nanofluid. Sarchami et al. [69] designed an indirect liquid cooling system combining a cooling channel and a copper sleeve, using Al_2O_3 nanofluids with volume fractions of 1% and 2% as coolants. Experiments showed that the maximum temperature and maximum temperature difference of the battery can decrease significantly with the increase of nanoparticle concentration. At a 5C charge–discharge rate, the maximum temperature and maximum temperature difference of the battery pack can be reduced below 305.13 K and 2.01 K, respectively. Guo et al. [70] similarly highlighted that the cooling performance of the BTMS can be achieved with more pressure loss. Additionally, they investigated the effect of nanoparticle shape on cell performance and discovered that brick-shaped nanoparticles could reduce cell capacity loss, while spherical-shaped nanoparticles preferred lower pressure loss. Therefore, it can be seen that the addition of nanoparticles is beneficial to improve the thermal conductivity of the coolant and the temperature uniformity in the battery pack, but nanofluids may have problems with particle deposition and blockage. Meanwhile, as nanoparticles are added and the fluid flow rate increases, this can lead to larger pressure drops and energy consumption in the system, which needs to be taken into account when choosing a nanofluid as a coolant.

Liquid metal is used in BTMSs as another coolant with high thermal conductivity. Yang et al. [71] proposed using liquid metal instead of water as a coolant for BTMS. It was found that the liquid metal cooling system has better cooling performance, which can control the battery pack temperature to a lower and more uniform range, while consuming less pumping power. In addition, liquid metals can be applied to extreme conditions, such as rapid charge–discharge processes and high-temperature environments. Liu et al. [72] also found that liquid metal has better cooling capacity and energy consumption than water. A variable flow velocity strategy (VFVS) BTMS was used in Liu's study in order to modify the liquid flow rate in response to various circumstances. The results demonstrated that under the three conditions of constant ambient temperature with variable C-rate (CTVC), constant C-rate with variable ambient temperature (CCVT), and variable ambient temperature with variable C-rate (VCVT), the pump power is reduced by 47%, 25.6%, and 47%, respectively, in comparison to the conventional constant flow velocity strategy (CFVS). This shows that the liquid metal used in indirect liquid cooling can display good cooling performance and effectively reduce energy consumption. Therefore, liquid metal can be considered to replace water as a coolant in a high-temperature environment and a high discharge rate, but experiments are lacking to verify the effectiveness of liquid metal for use in BTMSs.

In addition to studying coolants with different thermal conductivities to improve the cooling performance of the BTMS, some scholars have also studied the temperature of the coolant inlet. Panchal et al. [21] designed a microchannel cold plate to cool the prismatic battery and studied the water cooling effect at the three operating temperatures of 5 °C, 15 °C, and 25 °C. It was found that the coolant temperature would affect the cooling performance of the cold plate. Malik et al. [73] studied the performance of the battery pack at different coolant temperatures (10 °C, 20 °C, 30 °C, and 40 °C) based on the cold plate. Experimental results showed that the performance of the battery pack was improved with the increase of the coolant temperature. They found that when the coolant temperature is 30 °C, the maximum and average temperature of the battery pack can be controlled within 25 °C to 40 °C, whether at a low discharge rate or a high discharge rate. In Shang's study [74], it was discovered that the maximum temperature of the battery module decreased with the decrease of the coolant inlet temperature, but the change of temperature uniformity was not noticeable. Therefore, the temperature of the coolant is also an important factor affecting the performance of the BTMS. Appropriately reducing the inlet temperature of the coolant is beneficial to improve the temperature of the battery.

3.1.2. Flow Channel Optimization Design

In indirect liquid cooling, the coolant needs to pass through specific channels. However, different channel designs can have different effects on the thermal performance and energy consumption of the BTMS, so researchers have carried out many studies on channels. For the geometry of the cold plate channels, typical designs are shown in Figure 5. Huo et al. [75] designed a multichannel parallel cold plate (Figure 5a). The effects of channel number, flow direction, mass flow rate, and ambient temperature on battery temperature were studied by numerical simulation. Jarrett et al. [59] used the CFD method to numerically simulate the serpentine channel cold plate (Figure 5b); determined the optimal design of the objective function of cooling hydraulic drop, temperature uniformity, and average temperature; and studied the influence of boundary condition changes on the performance of the cold plate. For the cold plate channel, its shape is mostly linear along the flow direction, and the resistance will increase when the coolant flows in the pipeline. In this case, Huang et al. [76] used streamlined channels for indirect liquid cooling, as shown in Figure 5c. They found that this channel can effectively reduce the flow resistance of the liquid and improve the temperature distribution of the battery. On the other hand, Deng et al. [77] used a U-shaped serpentine channel cold plate (Figure 5d) to reduce the resistance of liquid flow in the channel.

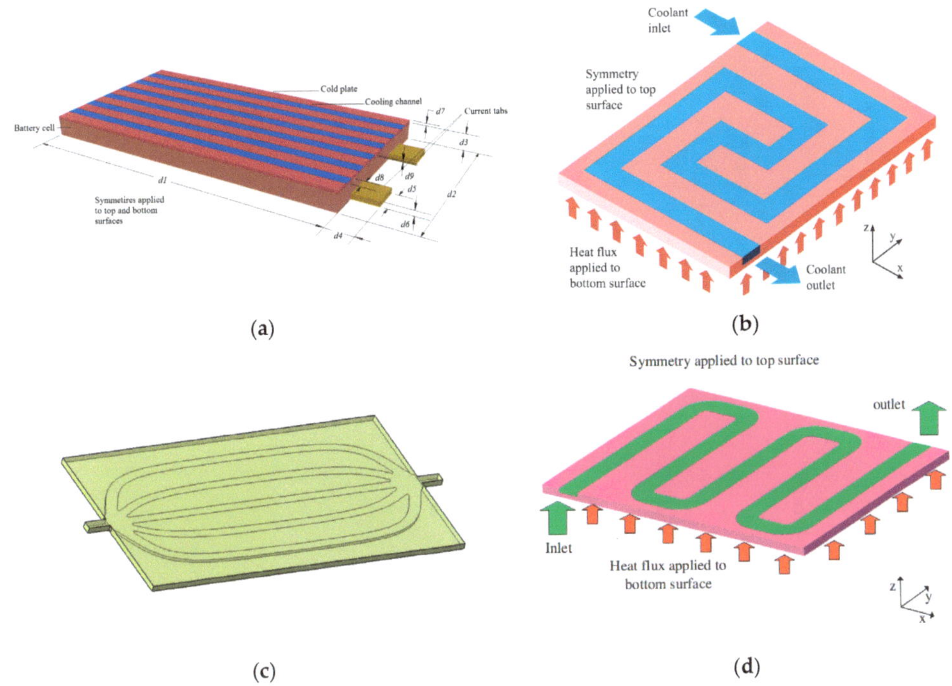

Figure 5. Typical design of the channel: (**a**) parallel channel [75]; (**b**) serpentine channel [59]; (**c**) streamlined channel [76]; (**d**) U-shaped serpentine channel [77].

Based on the typical channel structure, parallel channels and serpentine channels are gradually replaced by channels with a better cooling effect. Monika et al. [78] compared six cold plate designs with the same channel volume, as shown in Figure 6a, which are straight channel, serpentine channel, U-shaped channel, pumpkin-shaped channel, spiral channel, and hexagonal structure channel. Through numerical analysis, it was discovered that, despite having a high pressure drop, the snake channel and hexagonal channel have good cooling properties and can significantly improve the temperature distribution of the battery pack. In contrast, the pumpkin-shaped channel is useful for lowering the pressure drop and pumping power. In another study, comparing the performance of the cold plate of the snake channel with that of the parallel channel, it was also found that the pressure drop of the snake channel was greater than that of the parallel channel [79]. Moreover, Chen et al. [80] studied the cooling performance of the BTMS based on a parallel mini-channel cold plate (PMCP), and designed three different flow channels of PMCP (type I, Z, and U). It was discovered that the cooling performance of PMCP was greatly enhanced by altering the edge width and channel width of the cold plate. Three symmetrical PMCPS are also built in Chen's study, as seen in Figure 6b. The results showed that the symmetrical cold plate can control the maximum temperature and temperature difference of the battery module within a low range, and is conducive to reducing the energy consumption of the system.

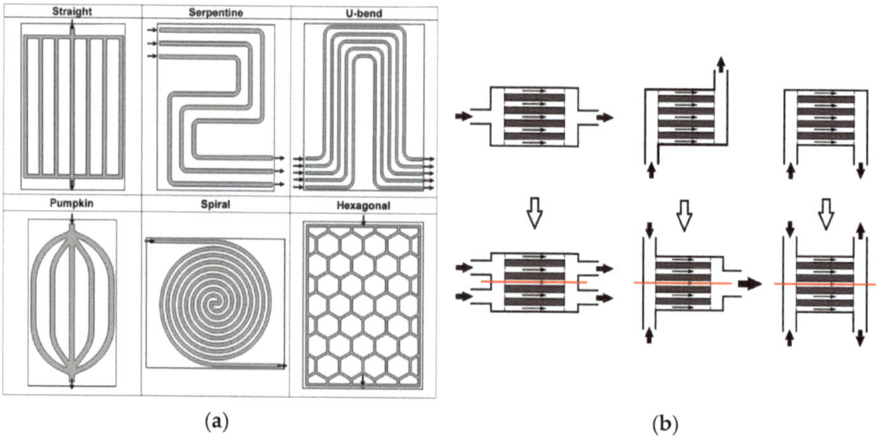

Figure 6. (**a**) Design of six cold plate channels [78]; (**b**) Three symmetric PMCPs [80].

With regard to the geometry of the flow channel, topology optimization (TO) can also be used to effectively improve the heat transfer efficiency of the cold plate and reduce the flow loss [81]. TO enables greater flexibility in optimization on the basis of pre-designed channels, thus improving the cooling performance of the system [82]. Chen et al. [83] designed a cooling plate with a new channel structure by TO, as shown in Figure 7. Numerical analysis showed that compared with the conventional rectangular channel cold plate (RCP) and serpentine channel cold plate (SCP), the topology optimization cold plate (TCP) has a higher heat transfer coefficient. At the same inlet pressure, the maximum temperature, temperature difference, and temperature standard deviation of TCP-RCP are 0.27%, 19.5%, and 24.66% lower, respectively, than those of the rectangular channel cold plate (RCP). Sun et al. [84] and Wang et al. [85] also used the TO method to study the cold plate, and found that TO can effectively improve the cooling efficiency of the BTMS. However, the application and research of TO in BTMSs are still relatively few. Most of the studies still analyze the cooling effect through numerical simulation. To explore the feasibility of the cooling plate by TO, TO should be experimentally verified in future research.

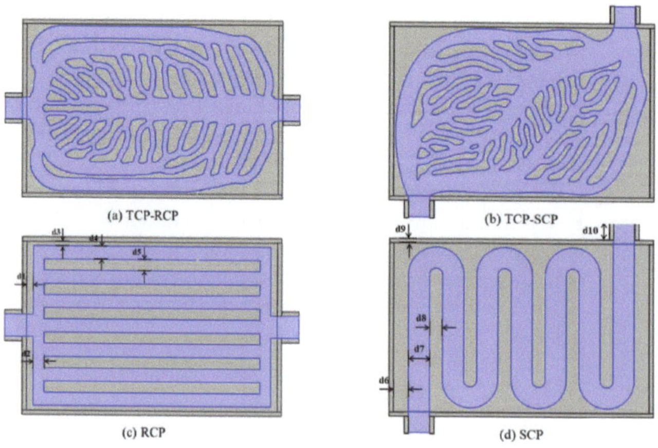

Figure 7. Cold plate diagram [83].

In the study of channel width, Jarrett and Kim [59] studied the influence of channel structure on its performance based on the serpentine channel cold plate and proposed several channel designs (Figure 8). Through a numerical optimization algorithm, it is concluded that the minimum average temperature and pressure drop can be optimized by increasing the channel width to the maximum extent. To optimize the design of uniform temperature, the flow channel with a narrow inlet and wide outlet should be adopted. In addition, Qian et al. [86] used numerical analysis to study the performance of parallel channel cold plates and found that when the channel width increased from 3 mm to 6 mm, the pressure drop of the system decreased by 55%. On the other hand, the distance between channels is also important to improve the performance of the cold plate. Choi et al. [87] used the CFD method to assess the design of the serpentine channel cold plate and found that narrowing the channel spacing near the outlet of the cold plate is beneficial to improve the cooling performance. Based on the above research, appropriately increasing the width of the channel and reducing the spacing between the channels may be an optimal choice to improve the performance of the cold plate.

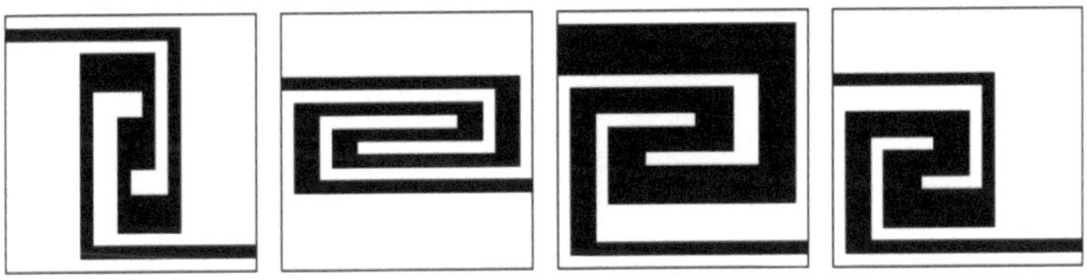

Figure 8. Optimization design of flow channel: black area represents the liquid channel [59].

The application of microchannels to indirect liquid cooling can increase the heat transfer area and thermal conductivity, reducing the average and differential temperatures of the battery and enhancing the efficiency of the thermal management system [86,88]. Lan et al. [89] designed a novel BTMS. Aluminum microchannel tubes were placed on the three surfaces of the prismatic battery, and water flowed in the channel as a coolant. According to the results, the BTMS with microchannels can demonstrate a greater cooling effect and lower the maximum temperature and maximum temperature difference of the battery without causing pressure drop or requiring excessive pumping power. For the study of the internal structure of the microchannel, Lee et al. [90] proposed a heat transfer enhancement scheme, using inclined fins to adjust the flow in the microchannel. This design promotes the redevelopment of the boundary layer and the generation of secondary flow and improves the heat transfer performance. Subsequently, Jin et al. [91] proposed a microchannel ultra-thin liquid cold plate (LCP) with inclined fins. As shown in Figure 9, the heat transfer performance of the traditional straight channel will weaken along the axial direction, and the thickness of the boundary layer will increase, resulting in a higher maximum temperature and larger temperature difference. The design of these inclined fins can initialize the flow boundary layer. The research shows that the cold plate with inclined fins has a better cooling effect than the traditional straight channel cold plate, which can significantly improve the performance of the BTMS.

Obviously, the cooling performance of indirect liquid cooling can be improved by adjusting the structure and configuration of the channel, including the channel geometry, channel width, channel spacing, number of inlets and outlets, microchannels, etc. At present, most of the research focuses on numerical simulation and theoretical analysis. In future research, manufacturing technology and cost need to be considered in order to obtain satisfactory results in practical applications.

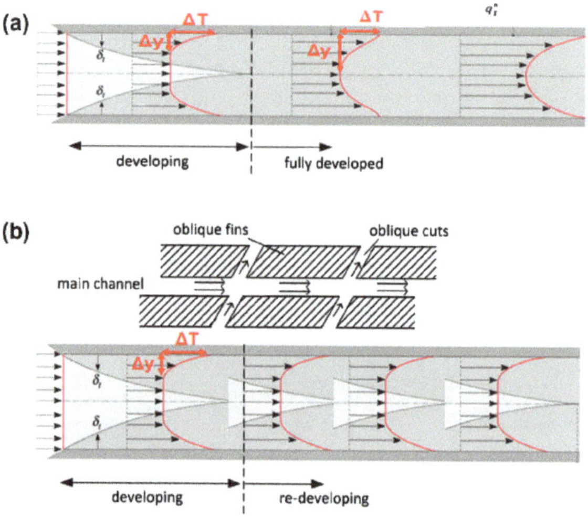

Figure 9. Boundary layer development: (**a**) traditional straight channel; (**b**) inclined fin microchannel [91].

3.1.3. Structure Optimization of Liquid Cooling System

The location of the cold plate, the contact area between the cooling structure and the battery, the number of cooling channels, and the coolant flow rate have an important influence on the cooling performance of the system. According to the position of the cold plate, it can be divided into bottom cooling and side cooling. Darcovich et al. [92] examined the cooling effectiveness of the two cold plate locations by putting the cold plate on the side and bottom of the prismatic battery. Because side cooling has more contact areas with the battery than bottom cooling, it has a lower maximal temperature and greater temperature uniformity. Tang et al. [93] designed a multichannel wavy tube for cylindrical battery packs. As shown in Figure 10, the two sides of the wave tube have radians that can be in close contact with the side of the battery. The maximum temperature and maximum temperature difference of the battery pack were studied by changing the wavy contact angle (α) and mass flow. The wave tube can reduce the uneven temperature in the battery pack and has a good heat dissipation capability at low flow rates, according to simulation and experimental results. Similarly, Zhao et al. [94] also designed a side cooling wave channel for the battery pack. Through simulation, it was discovered that even at a charge–discharge rate of 5C, the wave channel was still practical for the thermal control of the battery.

In addition to studying the location of the cooling device, changing the number of cold plates can also improve the temperature distribution of the battery pack. Qian et al. [86] designed two cooling systems based on different numbers of cold plates. The simulation results showed that the three cold plates helped to improve the temperature distribution. When the coolant flow rate is 1×10^{-3} kg/s, the maximum temperature and temperature difference of the battery pack are reduced by 13.3% and 43.3%, respectively. In another study [95], the author designed six different arrangement schemes of cold plates to study the heat dissipation effect of the battery pack. The maximum temperature of the battery pack decreased with the increase of the number of cold plates. When the number of cold plates increased to five, the heat dissipation performance of the system was the best. The author emphasized that when the number of cold plates is the same, placing the cold plates evenly near the intermediate battery is beneficial to reduce the maximum temperature. Therefore, within the allowable range of cost and power, side cold plates should be selected as much as possible, and the number of cold plates should be increased.

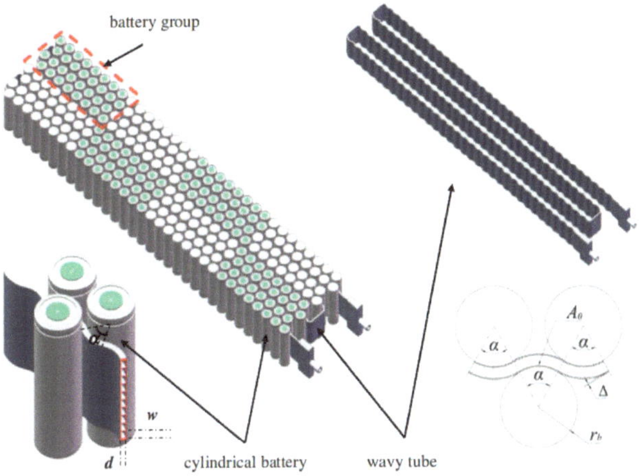

Figure 10. Battery module structure [93].

A key problem in indirect liquid cooling is that there is an air gap between the battery surface and the cooling structure, resulting in a decrease in cooling efficiency. Therefore, it is crucial to design a suitable contact area to increase the effectiveness of heat transfer. Zhou et al. [96] investigated a liquid cooling system based on the half-helical duct, as shown in Figure 11a. They pointed out that this structure is more efficient due to the small volume of liquid and no stagnation zone. Rao et al. [97] proposed a variable contact surface cylindrical BTMS, as shown in Figure 11b. According to the simulation results, the BTMS with a changeable contact surface helps to increase the battery's temperature uniformity. Siruvuri et al. [98] designed an s-shaped cooling channel and analyzed the influence of the contact area between the channel and the battery on BTMS. They pointed out that increasing the contact area can improve the cooling effect of the battery. In addition, the contact area with the battery can be controlled by changing the width of the cold plate [74] and the angle between the wave tube and the battery [93,94], thereby improving the cooling performance of the BTMS.

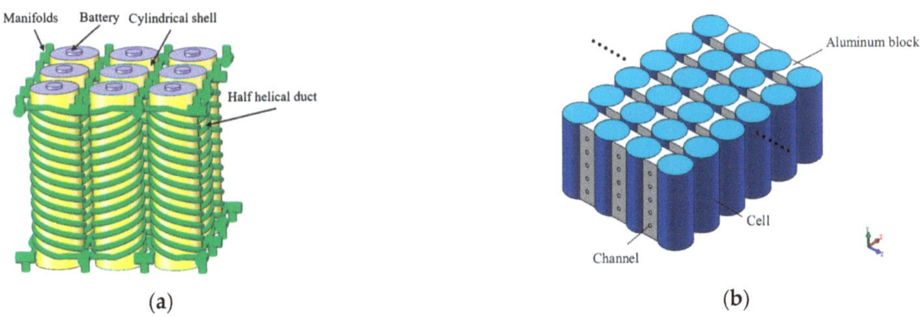

Figure 11. The contact surface design of the liquid cooling system: (**a**) half-helical duct [96]; (**b**) variable contact surface [97].

The number of cooling channels and the direction of flow are also two key factors in improving cooling performance. Zhao et al. [99] designed a new liquid cooling system for cylindrical LIBs. A liquid-cooled cylinder (LCC) with microchannels was positioned around the battery, as seen in Figure 12a. Numerical simulations have found that the maximum cell temperature can be controlled to within 40 °C when there are no less than 4 channels and a mass flow rate of 1×10^{-3} kg/s. For the flow direction, the reverse flow of some channels has little effect on the maximum temperature and temperature difference of the battery. Qian et al. [86] established a three-dimensional liquid cooling model based on the microchannel cold plate. The results showed that the best cooling effect was obtained when the number of channels was five. They designed four cases for different flow directions (Figure 12b). It is found that the maximum temperature of cases 1 and 3 is almost the same, which is 8% lower than that of cases 2 and 4. Deng [77] similarly found that cooling efficiency continued to improve as the number of channels increased, but after the number of channels exceeded five, the improvement became insignificant after increasing the number of channels further. Deng et al. [95] investigated the effect of liquid flow direction in the cold plate channel on cell temperature and found that changing the flow direction of the liquid facilitated the uniformity of the cell temperature.

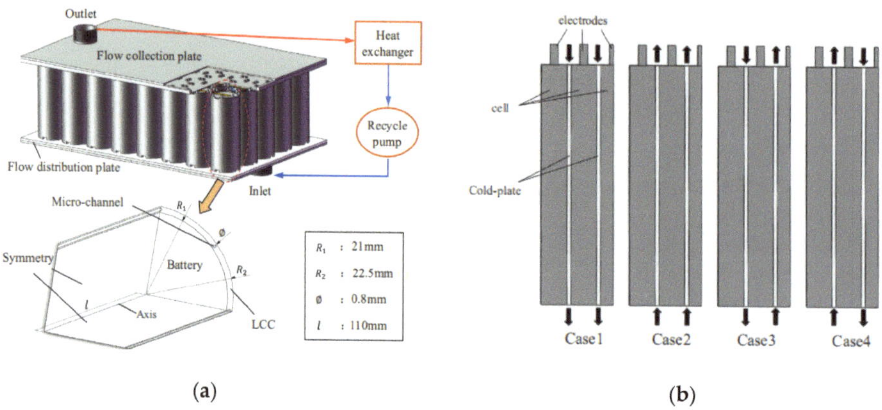

Figure 12. (**a**) Battery pack schematic [99]; (**b**) Different flow directions of liquids [86].

Therefore, the number of cold plate channels is not as high as possible, and an optimal number of channels exists. As the number of channels increases, it may lead to an uneven distribution of flow rates within the channels, making the temperature difference between cells increase. The change of flow direction is beneficial to reduce the maximum temperature of the battery, but its improvement on the cooling performance is affected by the coolant flow rate [75].

A lot of research has been conducted on the flow rate of coolant in the channel. By increasing the flow rate of the liquid, it is beneficial to dissipate the heat generated by the battery quickly, but it is the same as choosing the number of channels, and we cannot blindly increase the flow rate to improve the heat distribution of the system. It has been found that when the flow rate is increased to a certain value, a higher flow rate has no significant effect on the cooling effect, but instead increases the additional power of the pump. This phenomenon is called the limited cooling capacity when designing BTMS related to the liquid cooling structure of a specific cold plate [100]. Therefore, many studies have considered the balance between cooling effectiveness and energy consumption to find the optimum flow rate [9,13,75,101]. Pan et al. [102] studied the effects of the number of channels, thickness of the cold plate panel, coolant mass flow rate, and thickness of the cooling tube on the cooling performance of the battery system. Based on an orthogonal test created using these four variables, it was discovered that the thickness of the cooling plate or cooling tube had a minor impact on the battery's ability to dissipate heat, whereas the number of cooling channels and coolant mass flow rate had a significant impact. Increasing the mass flow rate was an efficient way to improve battery heat dissipation. Jia et al. [103] similarly found that the number of pipes and coolant flow rate were the main factors affecting cell temperature uniformity, while pipe width and pipe height were secondary factors.

Mohammadian et al. [104] proposed an internal cooling system, in which the electrolyte acts as a coolant through the microchannel between the positive and negative electrodes of the battery, as shown in Figure 13a. To evaluate the internal and external cooling techniques used by the BTMS, a two-dimensional and three-dimensional transient thermal analysis of the prismatic battery pack was carried out. The results show that internal cooling systems are superior to external cooling systems, with internal cooling reducing the internal cell temperature more than external cooling for the same pumping power, and significantly reducing the standard deviation of the internal cell temperature field. Although internal cooling helps to increase cooling efficiency and decrease the temperature gradient, it will also add weight to the system and be more expensive to manufacture due to its high manufacturing difficulty. In order to improve the safety of indirect liquid cooling, Basu et al. [16] designed an aluminum wave heat conduction element based on the liquid cooling system to replace the contact between the cooling channel and the battery (Figure 13b). The aluminum heat dissipation element transfers the heat generated by the battery to the cooling channels on both sides, and the coolant takes the heat away through the channels on both sides. In the event of coolant leakage, the heat conduction element can be used as a baffle between the coolant and the battery, thereby improving the safety of the BTMS.

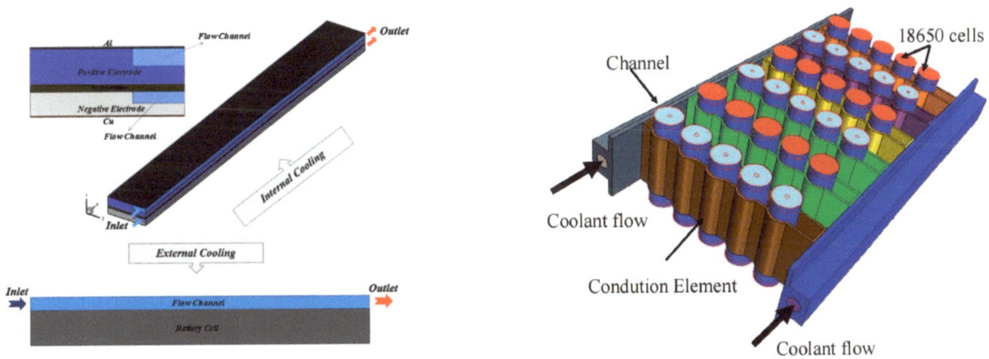

(a) (b)

Figure 13. (**a**) Internal cooling structure diagram [104]; (**b**) System architecture diagram [16].

Recently, many innovative structures have been proposed in the design of battery packs for electric vehicles (Figure 14) to improve the energy density and volume utilization of the battery system. Conventional battery packs (Figure 14a) are composed of cells, modules, and battery packs, and this method not only results in a waste of space and weight, but also leads to the group efficiency being only about 50%, which limits the driving range [105]. The cell-to-pack (CTP) battery system (Figure 14b,c) skips the module and goes directly from cell to pack, which improves both energy density and volume utilization by 15–20% [106]. However, while the battery system is still a separate structure for CTP, the cell-to-chassis (CTC) battery system (Figure 14d) integrates the battery directly into the vehicle chassis, which not only improves the energy density, but also enables a further increase in space utilization [107]. Compared with conventional battery packs, both CTP and CTC technologies simplify the system structure to increase the energy density of the battery, but from the safety point of view, the batteries are placed closely together, which is easy to cause heat accumulation, thus causing thermal safety problems [108].

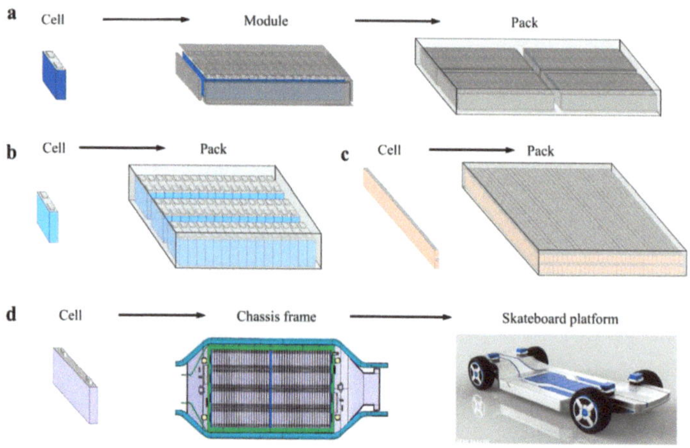

Figure 14. Schematic configuration of various battery systems: (**a**) conventional battery pack; (**b**) CATL cell-to-pack; (**c**) BYD blade battery cell-to-pack; (**d**) cell-to-chassis [109].

The heat dissipation of batteries in CTP and CTC systems mainly relies on cold plates, which brings new challenges to the BTMS. For example, Shen et al. [105] developed a 3D model of the CTP battery system and investigated the heat dissipation performance of the cold plate at the bottom of the battery pack. The simulation results showed that under extreme conditions (4C charging), the maximum temperature of the battery reached 79 °C, implying that the cold plate was not able to meet the heat dissipation needs of the system, and the battery might experience thermal runaway. In order to solve the heat dissipation problem in the CTP battery system, Sun et al. [110] optimized the structure of indirect liquid cooling under fast charging to study the effects of channel height, channel width, coolant flow, and coolant temperature on the battery temperature. The simulation results found that the coolant flow rate and temperature are the main factors affecting the battery temperature, while the channel height and width have little influence, and the temperature of the battery pack can be controlled at 30–35 °C under an effective configuration. In addition, the authors proposed a coolant inlet direction exchange method to reduce the temperature difference of the battery system, and achieved good results. From the perspective of crashworthiness and heat dissipation, Wang et al. [111] added negative Poisson's ratio (NPR) tubular structures (including an anti-collision layer and heat dissipation layer) on both sides of the CTP battery pack. The numerical simulation results indicated that the CTP-NPR system has superior structural safety and heat dissipation compared with the CTP system. Besides studying the cooling system structure, the safety of BTMS can be improved by optimizing the arrangement structure of batteries. In this regard, Jin et al. [109] investigated the influence of the battery arrangement structure on thermal runaway propagation in CTC systems. The results indicated that the safety of the battery system can be improved by adjusting the battery structure. After changing the battery arrangement from the conventional in-line configuration into the brick configuration, the thermal runaway propagation can be effectively prevented due to the lower peak heat flow in the brick configuration and the fact that the heat can be transferred to more adjacent batteries.

In summary, the performance of battery thermal management can be improved by adjusting the structure of indirect liquid cooling, but as the energy density of the battery continues to increase, this will create higher heat dissipation requirements for BTMS.

3.2. Direct Liquid Cooling

In the direct liquid cooling system, the coolant and the battery are in direct contact, which makes the heat transfer process more effective and simplifies the structure of the system and reduces the contact thermal resistance. The coolant in direct liquid cooling systems should be well-insulated, non-flammable, and environmentally friendly. Commonly used fluids include silicone oil, transformer oil, hydrofluoroether ether, etc. Compared with indirect liquid cooling, it can save space and costs and reduce overall weight, but from the perspective of energy consumption, direct liquid cooling systems require more energy since the coolant has a high viscosity [2]. Several investigators [91,112] have noted that direct liquid cooling systems for BTMSs may not be appropriate for application because of their greater viscosity, which uses more energy than indirect liquid cooling. However, oil-immersed direct contact is a mature transformer cooling technology, which can provide some references for the design of the BTMS [15]. Karimi et al. [58] pointed out that although silicone oil needs high pumping power, it can provide effective cooling performance with an appropriate flow configuration. In addition, Patil et al. [5] showed that dielectric fluid immersion cooling technology is safe, because the voltage and current of LIBs are not affected by immersion cooling and can provide similar electrical properties. Liu et al. [52] suggest that immersion cooling may be a better option for future battery thermal management. In summary, the battery thermal management based on direct liquid cooling has great research significance. The research on direct cooling is introduced below.

3.2.1. Coolant

A typical coolant used for direct cooling is oil. Oil has higher thermal conductivity and electrical insulation than air and water. The temperature of the battery can be reasonably controlled by completely submerging it in oil. Single-phase oils that are frequently used in direct liquid cooling are silicone oil and mineral oil. They have some benefits over conventional coolants, including high dielectric strength, great thermal conductivity, good fire resistance, low corrosion resistance, and difficulty in burning [113]. Chen et al. [9] compared air cooling, direct liquid cooling, and indirect liquid cooling using mineral oil as the coolant in direct contact with the cell and found mineral oil cooling to be more effective than air cooling. However, under the same temperature rise conditions, mineral oil requires a greater flow rate and consumes more energy than indirect cooling with water due to its high viscosity. Similar conclusions were reached by Karimi et al. [58] for a silicone oil study. In order to balance the cooling effect and energy consumption, Liu et al. [101] designed a BTMS based on mineral oil immersion to investigate the improvement of heat dissipation performance with different flow rates. At a 2C discharge rate, it was found that the battery immersed in mineral oil exhibited a better temperature distribution, even when the cooling medium was not flowing, with maximum and maximum temperature differences of 37.35 °C and 2.64 °C for the battery, respectively. As seen in Figure 15, increasing the flow rate of mineral oil reduced the cell temperature and kept the maximum temperature within 35 °C, but the improvement in cooling performance diminished at flow rates above 15 mL/min. Therefore, when using insulating oil as a coolant, as with indirect liquid cooling, there may be a cooling limit. Once the flow rate has increased to a certain value, any further increase in the flow rate will have a more limited improvement in the cooling effect, which requires a balance of energy consumption and cooling effect to be considered in order to select the optimum coolant flow rate.

The BTMS based on boiling heat transfer has attracted attention because of its ability to effectively control the temperature of the battery pack and maintain a uniform battery temperature. Commonly used in the boiling cooling of batteries is the Novec 7000 from 3M in St. Paul, MN, USA, which is non-flammable and has a boiling point of 34 °C, just within the optimum operating temperature range for LIBs [51]. When the temperature is below the boiling point, boiling liquid cooling is a single-phase cooling method that has a much greater cooling capacity than air and can improve the heat distribution of the cell even if boiling does not occur. Once the temperature exceeds the boiling point, the liquid will absorb a large amount of latent heat during evaporation into the gaseous state, keeping the temperature around the boiling point (Figure 16). However, once the liquid has evaporated into a gaseous state, an additional condensation system is required to collect the boiling liquid for recycling, which adds to the complexity of the BTMS. Wu et al. [114] proposed a boiling cooling BTMS, in which Novec 7000 is used as a coolant in direct contact with a 20 Ah $LiFePO_4$ LIB. The surface temperature of the battery can still be kept within 36 °C when the coolant is not flowing, and the temperature uniformity of the battery is also significantly enhanced. The authors proposed an intermittent flow mode, in which the maximum temperature and maximum temperature difference of the cell are controlled to below 36 °C and 2 °C, respectively. Wang et al. [115] and Gils et al. [116] studied Novec 7000 as a coolant in direct contact with the battery, and also obtained a satisfactory cooling effect.

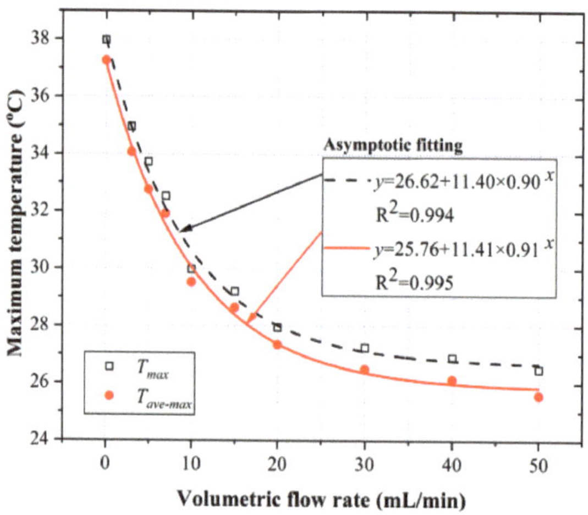

Figure 15. The maxima of the battery surface temperature (T_{max}) and the average battery surface temperature ($T_{ave\text{-}max}$) at different TO volumetric flow rates [101].

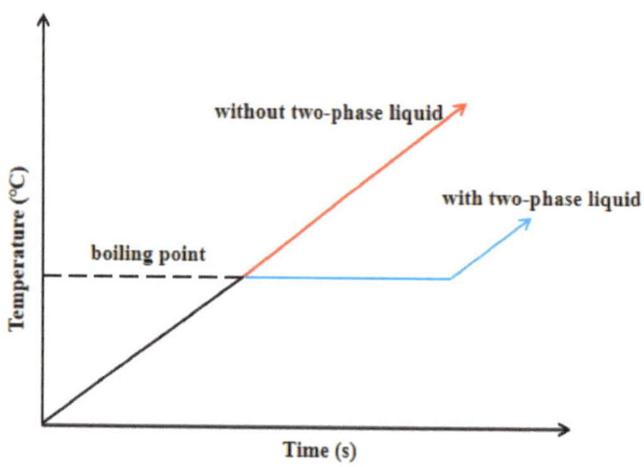

Figure 16. Temperature characteristics based on boiling liquids.

Esters, as another coolant that can be used in direct liquid cooling systems, can be divided into natural esters and synthetic esters. Natural esters are produced by vegetable oils containing a glycerol backbone, while synthetic esters are produced by the reaction between polyol and carboxylic acids [53]. Compared with synthetic esters, natural esters have the disadvantage of being prone to oxidation and requiring the addition of antioxidants. Esters are less harmful to the environment than mineral oils and silicone fluids, lower the risk of pollution even in the case of a leak, have high thermal conductivity, and are non-corrosive. Esters have a greater flash point and ignition point than mineral oils and a higher safety profile than oils, but they also have a higher viscosity, and more studies and evaluations are currently looking into using esters as transformer oils [117,118]. Madavan et al. [119] studied the performance of mixed esters, mixing several natural esters with mineral oils in proportion, and found that the breakdown voltage of the natural ester oils was superior to that of the mineral oils, but the viscosity and acidity of both the natural esters and the mixed oils were higher than that of the mineral oils. The authors noted that the higher-molecular acids are not harmful to the insulation system, but the high viscosity increases energy consumption. Another advantage of esters compared with mineral oil is their renewability. Due to the limited reserves of mineral oil, the feasibility of esters for thermal management systems has received a lot of attention and research. Moreover, the researchers also conducted a study to verify the feasibility of reusing the treated waste vegetable oil. In this regard, Chairul et al. [120] combined several treatments and were able to reduce the acidity of used edible vegetable oils to less than 0.06 mg KOH/g, and the density, viscosity, and electrical properties of the treated oils met the requirements of BS EN 62770.

In the area of BTMSs, M&I has introduced an ester-based immersion cooling medium fluid called MIVOLT with a high ignition point and easy biodegradability, including two products (DF7 for low viscosity and DFK for high viscosity) specifically for use in cooling electric vehicle batteries [53]. Qubeissi et al. [121] proposed a biodiesel-based immersion cooling system using four biodiesels (palm, karanja, jatropha, and mahua oils) in comparison to two conventional coolants (air and 3M Novec). The simulation results showed that using biodiesel for the BTMS was feasible because it could keep the battery pack temperature within a reasonable range. As seen in Figure 17, the cooling effect of palm oil and Novec is essentially the same. Moreover, because palm oil has a lower density than Novec, it facilitates a reduction in total weight. In conclusion, esters have been employed in transformers, but research on BTMSs is still lacking. Esters would be a good candidate coolant for BTMS if a low-viscosity and less oxidizable ester could be discovered as a cooling medium for immersion cooling.

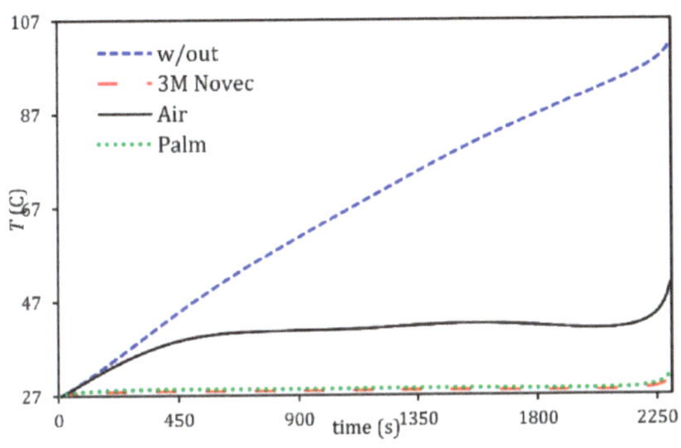

Figure 17. The cooling effect of different cooling methods at a 3C discharge rate [121].

3.2.2. Research on Cooling Performance

In recent years, research on immersion cooling has focused on the cooling effect of single batteries and battery packs. Pulugundla et al. [122] used experimental and numerical analysis to compare the cooling effects of immersion cooling and indirect liquid cooling on single cylindrical cells. They discovered that the cell temperature and temperature gradient of the immersion cooling system was much lower than those of the indirect cooling system at a 3C discharge rate. The authors also pointed out that immersion cooling can offer superior cooling performance for applications involving cylindrical LIBs. According to Bhattacharjee et al. [123], liquid cooling outperformed air cooling. The maximum temperature of the battery pack was decreased by 30.62% by air cooling and by 38.40% by indirect liquid cooling. The immersion cooling system exhibited remarkable cooling capacity, as it can reduce the battery pack's maximum temperature of 49.76 °C by

44.87% at a 2C discharge rate. Cold plate cooling and immersion cooling were compared by Dubey et al. [54], who came to the conclusion that the high thermal conductivity of immersion cooling at high discharge rates was the reason why the battery pack's maximum temperature and average temperature were significantly lower than those of the cold plate cooling system. Wang et al. [124] proposed immersion cooling with a compact structure and superior cooling performance (Figure 18). Experimental research was conducted on the effects of the immersion amount and flow rate on battery temperature using transformer oil as a coolant. It was found that the best cooling performance was achieved when the battery was fully submerged. Complete immersion displayed good cooling, even when the liquid was not flowing, with maximum temperatures falling from 58.3 °C to 44.9 °C and maximum temperature differences decreasing from 4.97 °C to 1.43 °C when compared with natural convection cooling. In addition, Sundin et al. [125] and Jithin et al. [126] both conducted studies based on single-phase immersion cooling fluids and similarly concluded that immersion cooling was effective in improving the temperature distribution of the batteries.

Figure 18. Schematic structure of liquid-immersed cooling [124].

Based on immersion cooling, Trimbake et al. [127] proposed a jet impingement immersion cooling of a 4-cell battery pack, using mineral oil as the coolant and designing both submerged jet and direct-jet impingement. Experimental results showed that the proposed jet impingement cooling was able to control the temperature difference between cells, as well as within a single cell, to within 1 °C under 2C charging and 3C discharging, respectively, showing excellent temperature uniformity. Liu et al. [128] examined the heat dissipation capabilities of mineral oil with silicone oil. Figure 19 shows that both insulating oils are able to control the maximum temperature and maximum temperature difference of the cell to within 35 °C and 5 °C, respectively, with the mineral oil providing better cooling than the silicone oil. Moreover, the coolant flow path in immersion cooling has garnered interest. Patil et al. [5] pointed out that the placement of a baffle in direct liquid cooling can alter the flow path of coolant. Compared with no baffle, the maximum temperature and maximum temperature difference of a single battery are decreased by 2.2% and 1.0%, respectively. Han et al. [129] stated that designing an optimal coolant flow path is necessary to improve the cooling effect on the battery thermal management performance and apply it to higher heat dissipation requirements.

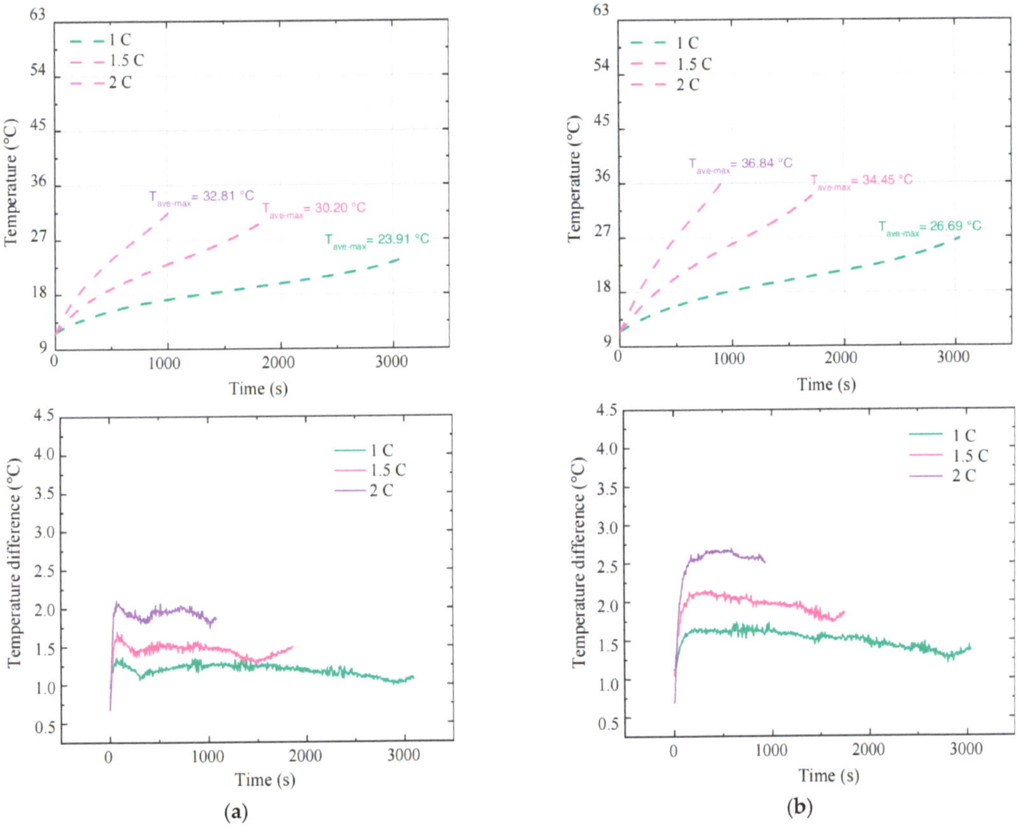

Figure 19. Maximum cell surface temperature and temperature difference curves at different discharge multipliers: (**a**) the battery immersed in the mineral oil; (**b**) the battery immersed in the silicone oil [128].

In the study of a two-phase immersion cooling system, Al Zareer et al. [130] used the phase change fluid R134a to partially immerse the battery pack in the liquid. The phase change fluid absorbed the heat generated by the battery and evaporated, and the vapor was returned to the cooling system through the condensation channel. Although this cooling method is quite simple, the BTMS based on phase change fluid shows superior cooling performance. It was found that at a 5C charge–discharge rate and a battery coverage of only 40%, the maximum temperature of the battery pack could be kept within 30 °C. To use liquid evaporation to cool the battery pack, the authors also suggested partially immersing the battery in pressurized, saturated liquid ammonia and propane. They discovered that the proposed two-phase immersion cooling system was capable of controlling the battery's temperature within the ideal temperature range [131,132]. To analyze the cooling effects of indirect cooling, single-phase immersion cooling, and two-phase immersion cooling on the battery pack, Li et al. [133] simulated a system consisting of 18 $LiFePO_4$ pouch cells. The results showed that neither indirect cooling nor single-phase immersion cooling could control the temperature of the battery pack to an acceptable range at a discharge rate of 10C, while in the two-phase immersion cooling system, not only did the maximum temperature go below 35 °C, but it also ensured better temperature uniformity. Recently, Li et al. [134] compared the cooling performance of hydrofluoroolefin (SF33) immersion cooling and forced air cooling, and studied four different depths of discharge (DOD) at different discharge rates. The experimental results showed that the maximum temperature rise of a single cell under forced air cooling was 14.06 °C at a 4C discharge rate, while the maximum temperature rise in immersion cooling was only 4.97 °C. At the same time, they found that even at a 7C discharge rate (Figure 20), the two-phase immersion cooling system was able to control the cell temperature to within 34.5 °C. This is due to the higher discharge rate, which facilitates stronger boiling heat transfer and obtains a stronger heat dissipation capability.

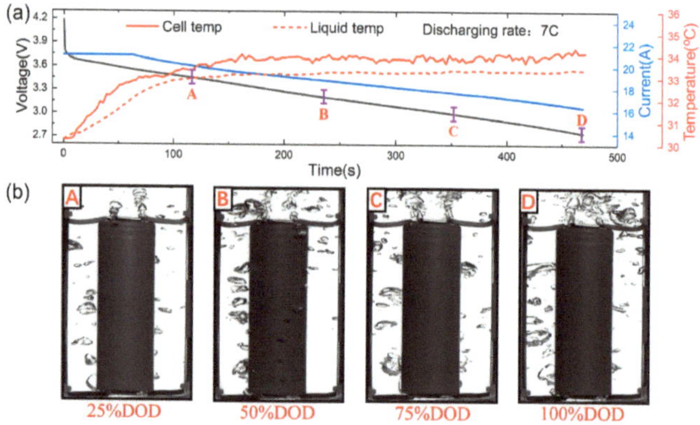

Figure 20. At a 7C discharge rate: (**a**) temperature of a single cell; (**b**) boiling of coolant [134].

Compared with air cooling, immersion cooling will increase the overall weight, and there is a risk of leaking coolant. Compared with indirect liquid cooling, the viscosity of the coolant is generally higher than that of water, and the optimal flow rate needs to be considered to reduce the additional power of the pump. Therefore, when evaluating the cooling performance of a cooling system, the overall energy consumption should be taken into account, in addition to the cooling effect. Karimi and Li [135] pointed out that the heat transfer rate of silicone oil is much higher than that of air at the same flow rate. At low flow rates, due to the higher viscosity of silicone oil, a higher pumping power is required, which is only 1.5 to 3 times higher than the thermal conductivity of air. Deng et al. [51] discussed the heat transfer efficiency of a high-viscosity submerged cooling oil compared with air at the same pressure drop and found that improvements could only be made to a limited extent. In another study [9] comparing flow rates, pressure drops, and energy consumption for different cooling methods at the same temperature rise, it was found that mineral oil-based immersion cooling was far more efficient than air-cooled systems in balancing pressure drop and heat transfer efficiency, even though a lower flow rate was chosen. Matsuoka et al. [136] studied five immersion cooling media: fluoride FC3283, fluoride FC43, 50cSt silicone oil, 20cSt silicone oil, and soybean oil. They found that the cooling performance in natural convection was influenced by the viscosity of the coolant, with lower-viscosity coolants being able to reduce temperatures significantly. Therefore, 20cSt silicone oil showed superior cooling performance in natural convection than 50cSt silicone oil. Dubey et al. [54] used hydrofluoroether for immersion cooling, which resulted in better pressure drop variation for immersion cooling than indirect cooling due to the lower viscosity and density of hydrofluoroether compared with cold plates using water/glycol. As a result, it is essential to try to choose a coolant with low viscosity in submerged cooling, not only to reduce pressure drop and energy consumption, but also to obtain better cooling performance. In addition to pressure drop and viscosity, Wu et al. [137] used silicone oil as a coolant in direct contact with the cell and found that a silicone oil-based submerged cooling system was able to ensure thermal safety issues for the cell and could reduce the volume occupied by the cell system and also improve the cooling efficiency. Chen et al. [9] and Patil et al. [5] also found that the use of direct liquid cooling could make the cooling system lighter and more compact.

According to the aforementioned research, both single-phase and two-phase immersion cooling systems have demonstrated satisfactory results. Immersion cooling has excellent cooling performance in BTMSs. In immersion cooling systems, more contact area is gained due to the direct contact between the cell and the cooling liquid, which facilitates the compactness of the cell thermal management and the design requirements for high heat transfer efficiency. The flow path of the coolant is worthy of investigation, which is mainly focused on simulations and lacks relevant experimental studies. At the same time, systems based on two-phase immersion cooling fluids have shown superior cooling performance and can be used for cooling under some extreme conditions.

3.3. Composite Cooling

With the rapid advancement of electric cars comes an increase in the demand for battery thermal management. Table 1 lists the benefits and drawbacks of each cooling technique. With high charge–discharge rates and high range requirements, it is difficult to achieve satisfactory results with a single cooling method, and combining two or more types of combined cooling has attracted the attention of researchers. In order to meet the need for high heat dissipation in battery systems under extreme conditions, a composite cooling system based on liquid cooling is discussed in this section.

Table 1. Comparison of different single cooling methods [8,12,138].

Cooling Type	Dominance	Problems
Air cooling	Simple structure, low cost, proven commercial application	Limited heat dissipation, low thermal conductivity, unable to meet high heat dissipation requirements
Liquid cooling	Good temperature uniformity, wide range of use, high thermal conductivity, high cooling efficiency, good compactness	Complex structure, leakage problems, high cost
PCM	High latent heat of phase change, no energy consumption, low cost, non-corrosive	Large volume changes, leakage problems after melting
Heat pipe	Compact, low weight, long life, high thermal conductivity, good stability	Small capacity, low cooling efficiency, small contact surface

Air cooling is simple in structure and can be achieved through the cell slits without the need for additional air ducts. The combination with liquid cooling increases the efficiency of thermal management and does not add additional complexity to the cell module structure. To investigate the impact of various assembly structures, air circulation, and liquid cooling conditions on the cell temperature, Wang et al. [139] suggested a composite cooling with a combination of forced air cooling and liquid cooling plates in an enclosed space (Figure 21). According to the research, a structure with a fan mounted beneath the cooling plate was the best option in terms of cooling effectiveness and cost. At a total heat generation of 576 W, the maximum temperature and temperature difference were decreased by 3.45 K and 3.88 K, respectively, and the temperature uniformity and temperature control efficiency were increased by 2.42 and 2.61 times, respectively, compared with the vacuum package's single cold plate cooling method. They also stated that this composite cooling system has a simple structure with good thermal performance, which can improve the performance and reliability of the battery module in the enclosed space. Yang et al. [140] combined air cooling and microchannel liquid cooling to investigate the thermal performance of a composite cooling system and found that the system facilitated improved battery performance and temperature uniformity. Li et al. [141] similarly found that a combination of air cooling and indirect liquid cooling could improve the performance of the BTMS. Simulations found that under extreme conditions (high temperature and high discharge rate), the maximum cell temperature could be controlled to within 45 °C, but the maximum temperature difference exceeded 5 °C. Traditional active air and liquid cooling works by using additional energy to circulate the cooling medium in order to produce a greater cooling effect, but this cooling effect may not always be as effective as desired.

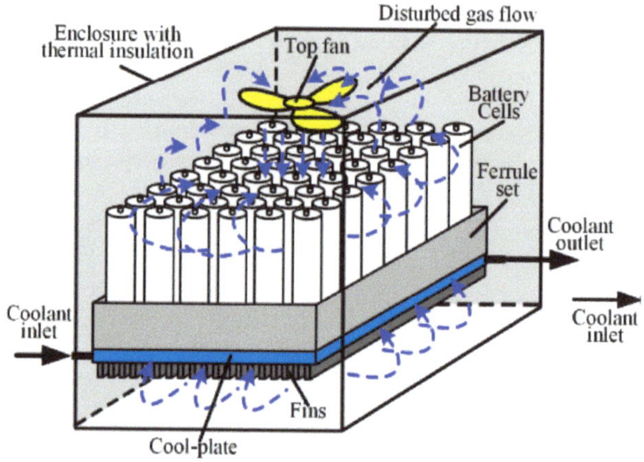

Figure 21. Air cooling structure based on liquid cooling [139].

The thermal management system coupled with liquid cooling and PCM can combine the advantages of the large convective heat transfer coefficient of liquid, large latent heat of PCM, and no energy consumption. It can not only reduce the energy consumption of the system, but also achieve a better cooling effect, and has a good development prospect. The PCM's moldable shape makes it ideal for embedding cooling channels or cooling plates for quick heat dissipation. Zheng et al. [142] developed a composite cooling system with PCM filling the cooling pipes and battery gaps, and, through simulation, found that it is difficult for PCM to utilize latent heat if the system is primarily liquid-cooled. They mentioned that PCM can be substituted with some high-thermal-conductivity materials because the system can still perform well at cooling, even in the absence of phase shift. In order to take advantage of the latent heat of PCM, Kong et al. [143] proposed a composite cooling system, in which PCM cooling is the mainstay, and liquid cooling is used to recover the latent heat of PCM. The efficacy and practicality of the composite system were confirmed through experiments. Zadeh et al. [144] investigated the BTMS of a 12-cell 18650 LIB pack and designed natural convection cooling, forced convection cooling, finned-natural convection cooling, PCM cooling, and combined cooling with PCM and liquid cooling. According to simulation findings, PCM in conjunction with liquid cooling is the only way to achieve the battery life requirements (≤45 °C). For a battery pack with 40 cylindrical cells, Cao et al. [145] suggested a delayed cooling device using PCM and a cooling plate combination. According to the simulation results, the delayed cooling system can improve the temperature difference between cells, while also reducing the power consumption of the system. The combined cooling can keep the temperature within 55 °C at a 4C discharge rate.

Heat pipes are a highly efficient heat transfer element based on gas–liquid phase change, with a high thermal conductivity that is tens or even hundreds of times higher than that of metals of the same size [146]. The heat pipe uses no extra power, similar to PCM. The evaporation end and the condensation end are the two extremities of the heat pipe, respectively. The evaporation end is in contact with the external heat source. As heat is absorbed by the coolant in the tube core, it will vaporize and travel to the condensation end, and then undergo exothermic liquefaction. Finally, through the capillary force, the cooled liquid travels to the evaporation end. In the BTMS, the condensing end of the heat pipe, which relies only on natural convection to cool it, is unable to meet the heat dissipation requirements of the battery [147]. The combination of heat pipes and liquid cooling to create a BTMS has garnered interest in order to achieve superior cooling. Mei et al. [148] studied the BTMS with heat pipe coupling liquid cooling. The simulation results showed that the liquid cooling thermal management based on the flat heat pipe can effectively improve the temperature uniformity of the battery. Yuan et al. [146] proposed a BTMS coupled with a heat pipe and a cold plate. The condensation end of the heat pipe is in contact with the cooling plate (Figure 22). The coolant flow rate, ambient and inlet coolant temperature, and discharge rate were studied. The experimental results showed that the structure can meet the heat dissipation requirements at a 2C discharge rate in the environment of 30 °C. The maximum temperature and temperature difference of the battery are maintained at 34.1 °C and 1 °C, respectively. Compared with single cooling, composite cooling can meet the heat dissipation requirements of batteries in continuous cycle operation. Liang et al. [149] designed a composite cooling system with the condensing end of the heat pipe embedded in a liquid cooling channel to study the thermal performance of intermittent and continuous cooling, and the results showed that the use of intermittent cooling not only met the thermal requirements of the battery, but also reduced the energy consumption of the system by reducing the operating time. In order to improve the temperature uniformity of the battery module, Wei et al. [150] proposed a reciprocating cooling strategy based on heat pipe-coupled liquid cooling. The simulation found that reciprocating cooling can significantly improve the temperature gradient in the battery module. In summary, the condensing end of the heat pipe can not only be coupled to the cold plate, but can also be submerged directly in the coolant. Although the combination of liquid cooling and heat pipes results in superior cooling, the coolant requires a special container, which increases the complexity of the construction and the weight of the system.

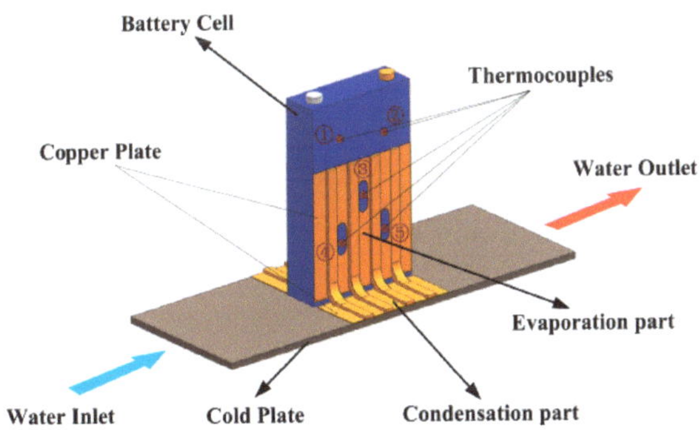

Figure 22. Structure of heat pipe and cold plate combination [146].

4. Battery Preheating Technology

The BTMS has three functions: preheating the battery at low temperatures, cooling the battery at high temperatures, and controlling the temperature difference between batteries, so that the battery system can maintain stable performance under different conditions. The current research on BTMS mainly focuses on cooling the battery and reducing the temperature difference between batteries, because the high temperature effect of the battery is not limited by the ambient temperature. The high temperature effect is mainly caused by the internal heat generation of LIBs, which usually occurs under fast charging and discharging, while the low temperature effect occurs at lower ambient temperatures [151]. It is worth noting that preheating the battery in cold environments is also an important function of the BTMS, and the heating and cooling will affect each other, so the design of the preheating function in the BTMS should not be neglected.

Battery preheating methods can be divided into internal heating and external heating. Internal heating includes self-heating and current excitation heating, while external heating includes techniques combined with cooling methods and thermoelectric element heating [152]. Although internal heating has higher heating efficiency and better temperature uniformity, the control mechanism of this method is more complicated and may lead to safety problems [153]. External heating is the more common heating method at low temperatures due to its simple construction. Combined with the above research on indirect liquid cooling and direct liquid cooling, this section focuses on the relevant literature based on liquid preheating.

The preheating system can be divided into non-contact heating and immersion heating, according to whether the battery is in contact with liquid or not [154]. Non-contact heating technology can be used to improve preheating performance by adjusting the structure of the system. For example, Tang et al. [155] designed a novel indirect liquid cooling thermal management system, which consists of a straight microchannel flat tube and heat-conducting blocks, and a three-dimensional transient heat transfer model was used to study the thermal performance of the battery system. It was found that the cooling performance of the system increased with the increase of contact surface angle and inlet liquid flow rate. For the preheating study of the battery system at subzero temperature, they found that a larger gradient angle increment was beneficial to improve the temperature uniformity. Chen et al. [156] proposed a double-direction liquid heating system for the CTC battery system to maintain the normal start-up and energy output of the battery at very low temperature (−40 °C). Numerical results showed that the system was able to heat the battery above 0 °C in a short time without consuming much energy. Wu et al. [157] designed a novel BTMS consisting of a vapor chamber (VC) and a microchannel cold plate (Figure 23a) to realize the cooling and heating of the battery system. Numerical simulations and experimental results showed that the proposed hybrid cooling system was able to control the temperature of the battery system within the optimal operating range. They also heated the battery system at subzero temperature and found that the hybrid system can heat the battery faster and maintain good temperature uniformity compared to the system with a microchannel cold plate.

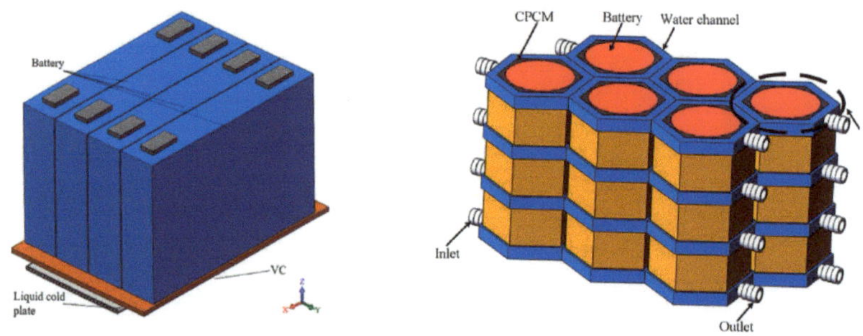

(a) (b)

Figure 23. The BTMS with heating and cooling functions: (**a**) hybrid microchannel system [157]; (**b**) honeycomb system [158].

In addition to using the fluid medium as a heat source to heat the battery at low temperatures, it can also be heated by the external thermoelectric element to investigate the effect of the BTMS on the temperature during the heating process. An et al. [158] designed a honeycomb BTMS (Figure 23b) with a combination of liquid cooling and PCM, which has the double function of heating and cooling, where the heating was achieved by the rubber heating belt. The results showed that this composite cooling system can control the maximum temperature and temperature difference of the battery module within 46.21 °C and 3.5 °C, respectively. And the temperature of the battery module can be increased from −15 °C to 10 °C in 6 min under the effect of heating belts with different power, and the maximum temperature difference can be controlled within 5 °C. In order to improve the temperature uniformity of the battery during the preheating process, Wang et al. [159] proposed an immersion preheating system, in which the battery pack was directly immersed in the coolant. The simulation results showed that the temperature rise rate of the system can be as high as 4.18 °C/min, and the temperature difference between batteries can be controlled within 4 °C, showing excellent preheating performance.

Compared with non-contact heating, the immersion preheating method can obtain higher heat transfer coefficient, so as to obtain better temperature rise rate and temperature uniformity [160]. However, considering the space constraints in vehicles, most liquid preheating methods are non-contact. For the liquid preheating system, there are not only the problems of high system tightness requirement and complex structure, but also the balance between energy consumption and heating performance of the BTMS. In the future research, we should further study various preheating systems, so as to realize the application of the battery system in the whole climate and improve the efficiency of BTMS.

5. Liquid Cooling Safety

The goal of BTMSs is to transfer the heat produced by the battery as quickly and efficiently as possible to prevent heat accumulation that could lead to thermal runaway. Nevertheless, some abuse of the battery (mechanical, electrical, and thermal abuse) can lead to thermal runaway, during which the battery may expand, eject, catch fire, or explode [53]. In recent years, there have been many studies on thermal runaway, including the effect of BTMSs on thermal runaway [161,162]. Compared with other cooling methods, liquid cooling has attracted wide attention because of its high thermal conductivity, which can effectively reduce the temperature and temperature difference of the battery and avoid the occurrence of thermal runaway.

To verify that liquid cooling can improve the safety of the system, Mohammed et al. [163] designed a dual-purpose cold plate for prismatic LIBs. They added pins inside the cold plate to increase the contact area and reduce the pressure drop. Through simulation, it is found that whether it is normal operation or thermal runaway, the cold plate can effectively control the temperature of the battery. By increasing the coolant flow rate, the temperature of the battery with thermal runaway can be reduced to 75 °C. Ke et al. [55] studied the effect of indirect liquid cooling on the thermal runaway propagation of the battery pack. The battery pack is composed of 10 batteries in parallel and placed in an indirect liquid cooling system based on a serpentine channel. A section is heated to cause thermal runaway, and the effect of indirect liquid cooling on thermal runaway propagation at different flow rates (0 L/h, 32 L/h, 64 L/h, and 96 L/h) is analyzed. The experimental results showed that the thermal management system could prevent the spread of thermal runaway, but the influence of a lower flow rate on thermal runaway was not obvious. Only when the flow rate is 96 L/h can the spread of thermal runaway be effectively suppressed.

In addition, the combination of PCM and liquid cooling can also be used to suppress the propagation of thermal runaway. In this aspect, Zhang et al. [164] proposed a BTMS with a combination of PCM and liquid cooling (Figure 24a). The results show that this hybrid cooling method can fully meet the heat dissipation requirements under extreme conditions and effectively prevent the spread of thermal runaway. However, the PCM with high thermal conductivity may accelerate thermal runaway propagation, which requires a balance between thermal performance during battery operation and thermal runaway propagation. Subsequently, the author proposed a BTMS composed of non-uniform thermal conductivity PCM and auxiliary liquid cooling (Figure 24b). The PCM with different thermal conductivity was filled around the battery. High-thermal-conductivity PCM (HPCM) was used for battery cooling, and low thermal conductivity PCM (LPCM) was used to slow down the propagation of thermal runaway. Finally, indirect liquid cooling transfers the heat absorbed by PCM to the external environment [165]. They pointed out that this hybrid cooling can meet the needs of system heat dissipation without reducing the energy density of the system, and control the thermal runaway in a specific direction and range to reduce the harm of thermal runaway. Indirect liquid cooling can play a role in managing the thermal runaway of the battery to a certain extent and can slow down the rapid spread of thermal runaway in the battery pack. However, conductive fluid is often used in indirect liquid cooling. Once thermal runaway occurs, it may lead to leakage caused by the rupture of a liquid pipeline, causing greater safety problems.

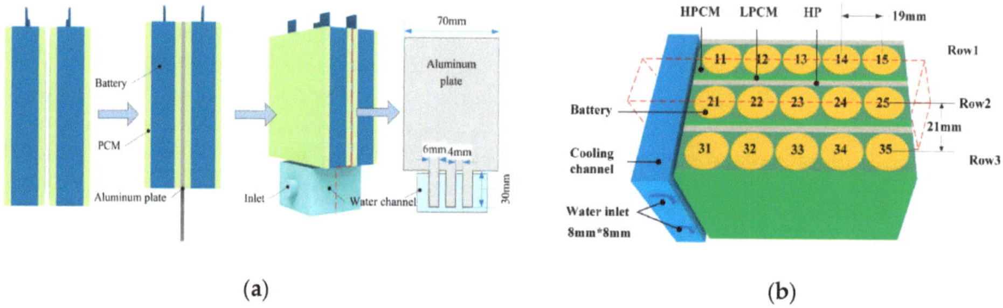

Figure 24. Structure schematic diagram of BTMSs: (**a**) PCM-Water system [164]; (**b**) PCM with non-uniform thermal conductivity combined with liquid cooling system [165].

Many of the cooling media used in immersion cooling can act as flame retardants in addition to being non-conductive, which can act as a barrier to increase system safety [54]. Patil et al. [5] proposed a BTMS based on dielectric fluid and tab cooling, which immersed the battery directly in the dielectric fluid to simulate the thermal runaway caused by the internal short circuit caused by the battery. The results show that this cooling method can effectively manage the thermal runaway of the battery pack. In addition to single-phase immersion cooling, immersion cooling based on two-phase liquids also shows excellent management of thermal runaway. Li et al. [166] designed a direct liquid cooling system based on fluorocarbons and studied the ability of five coolants to suppress the thermal runaway of the battery. There was no thermal runaway in the liquid cooling experiment, indicating that the battery immersed in fluorocarbons can effectively prevent or suppress thermal runaway. Additionally, they noted that since the battery and coolant are in constant contact, direct liquid cooling requires good material compatibility in applications. Studies have shown that immersion cooling has a prominent effect on the control of thermal runaway of the battery, especially based on two-phase cooling fluids, which can make full use of the latent heat of evaporation to control the temperature of the battery within a lower range, thus avoiding thermal runaway caused by excessive temperature. Although immersion cooling has these advantages, there are few studies on the effect of immersion cooling on thermal runaway, mainly focusing on some thermal runaway caused by mechanical abuse and thermal abuse, while electrical abuse has not been studied.

6. Future Developments of Liquid Cooling System

Compared with other cooling, liquid cooling is a promising cooling method. Through the summary and analysis of the above literature, the future development trends of liquid cooling are summarized, and the key problems to be solved are put forward.

As the most easily accessible medium, water has a wide range of applications in indirect liquid cooling. Although the addition of glycol can expand the applicable temperature range of the BTMS, it is difficult to meet the heat dissipation requirements of the battery under fast charging conditions. Nanoparticles and liquid metals can significantly improve thermal conductivity and become ideal candidate materials for BTMSs. Compared with water cooling systems, BTMSs based on nanofluid and liquid metal are able to keep the battery temperature in a much lower range and show excellent cooling performance under extreme conditions. However, the density of the coolant will increase with the addition of nanoparticles, resulting in a larger pressure drop when the liquid flows, which also puts forward higher requirements for the performance of the pump. A key issue for the nanofluid is the preparation of a stable fluid that avoids the occurrence of particle settling and clogging. In order to improve the cooling efficiency by using nanofluids in the future liquid cooling systems, this problem needs to be solved.

Compared with indirect liquid cooling, the single-phase oil immersion cooling system is really a more efficient way. Currently, a major problem in the single-phase oil immersion cooling system is the large energy consumption, which is due to the relatively high viscosity of the oil. At the same flow rate, it will consume more pumping power, resulting in increased energy consumption [43]. There is a limitation of cooling capacity in oil-immersed cooling systems, and energy consumption can be reduced by selecting the appropriate flow rate, but it is not easy to choose the best flow rate for different working conditions and systems. Two-phase immersion cooling is a very effective cooling method. It shows superior heat absorption capacity in the boiling process. Especially, the battery can reduce the temperature of the battery to an acceptable range under fast charging. However, it should be noted that the boiling liquid is expensive and easy to volatilize, which will increase the structural complexity of the entire system and limit its application in electric vehicles [126]. As the driving range and output of electric vehicles increase, a more efficient thermal management system is needed. If these problems can be solved, immersion cooling based on boiling liquid will be a preferred cooling method for high-energy-density batteries.

Electric vehicles should not only consider the cooling performance of the BTMS, but also control the overall cost. For the liquid cooling, the number of channels and the coolant flow rate play a major role in the cooling performance. A higher flow rate is beneficial to achieve a better temperature distribution of the battery, but this will increase the energy consumption and operating cost of the system. For the cold plate design, the complexity of the system will become larger as the number of channels increases, which will not only make it more difficult to manufacture the system, but also discourage maintenance. In the future thermal management system, the system coupled with liquid cooling and PCM seems to be a better way, because PCM does not need additional energy consumption, and the liquid cooling system can cool the PCM to avoid the PCM completely melting and failing. With the addition of PCM, the problem of uneven temperature distribution in liquid cooling systems can be improved, and the energy consumption of the system can also be reduced [167]. The system combining PCM and liquid cooling also offers the advantages of compactness, longer service life, and lower maintenance cost [168]. In addition, this composite thermal management system demonstrates good preheating performance at low temperatures. However, it is worth noting that the composite cooling system of immersion cooling and PCM has not been studied at present, which may be due to the high sealing requirement of the system and the volume change of PCM. In summary, the composite system of liquid cooling combined with PCM is a BTMS with low energy consumption and a good cooling effect.

The preheating function of the system should also be considered in the future liquid cooling research. In the study of battery preheating, although liquid preheating technology has been applied in electric vehicles, it is still a challenge to preheat batteries efficiently and safely. For the external heating technology, the heating method and heat transfer path should be considered, and the parameters can be continuously optimized through numerical simulation, so as to reduce the heat loss and improve the preheating efficiency of the system.

7. Summary and Prospects

According to the above research, liquid cooling is a very efficient thermal management technique that can control the temperature and temperature difference of the battery within an acceptable range. This paper reviews the literature on direct and indirect liquid cooling for BTMSs. The coolant commonly used in indirect liquid cooling is water or a water/glycol mixture, and has been commercialized in the field of electric vehicles. However, the cooling effect of water is limited, and the cooling performance and efficiency of BTMSs can be improved by adding nanoparticles. Liquid metal has the advantages of low viscosity and good thermal conductivity, which can replace water to meet the heat dissipation requirements under some extreme conditions. In addition, the cooling performance of BTMSs can be improved by changing the flow channel design and system structure, but the pressure drop, weight, and cost of the system need to be paid attention to.

The single-phase oil has been used more frequently for direct liquid cooling. However, when increasing the flow rate, care should be taken to balance the cooling effect and energy consumption. The two-phase immersion cooling liquid exhibits an excellent cooling effect and the boiling point of the liquid can be reduced by changing the pressure, thereby increasing the cooling efficiency of the system. Based on this review, some research weaknesses in direct liquid cooling have been identified. Esters have been applied in transformers, but there is no in-depth study in battery thermal management. Therefore, the feasibility of esters as coolants for direct liquid cooling can be explored, as well as esters that can replace transformer oil. On the other hand, the contact time between the battery and the coolant in direct liquid cooling is long, and material compatibility between the two needs to be considered. At the same time, few articles have studied the effect of immersion cooling on battery performance and thermal runaway.

The composite system of liquid cooling combined with other cooling methods can meet thermal management requirements under different conditions, especially in fast-charging or high-temperature environments. In the development of electric vehicles, the compactness and lightweightness of the battery system have always been concerned. However, liquid cooling brings additional integration challenges that need to be carefully considered when designing a BTMS. In summary, although liquid cooling still has these research gaps and challenges, it is still a promising BTMS.

Author Contributions: Conceptualization, J.L. and Y.J.; methodology, J.L.; software, H.C.; validation, H.C., S.H. and J.L.; formal analysis, H.C. and S.H.; investigation, J.L.; resources, J.L.; data curation, H.C.; writing—original draft preparation, H.C. and M.C.; writing—review and editing, J.L.; visualization, H.C.; supervision, J.L. and M.C.; project administration, Y.J.; funding acquisition, J.L. All authors have read and agreed to the published version of the manuscript.

Funding: This research was funded by the National Natural Science Foundation of China, grant number 51909152.

Data Availability Statement: Data will be available on request.

Conflicts of Interest: The authors declare no conflict of interest.

References

1. Zhao, C.; Zhang, B.; Zheng, Y.; Huang, S.; Yan, T.; Liu, X. Hybrid Battery Thermal Management System in Electrical Vehicles: A Review. *Energies* **2020**, *13*, 6257. [CrossRef]
2. Liu, H.; Wei, Z.; He, W.; Zhao, J. Thermal issues about Li-ion batteries and recent progress in battery thermal management systems: A review. *Energy Convers. Manag.* **2017**, *150*, 304–330. [CrossRef]
3. Ouyang, D.; Weng, J.; Chen, M.; Wang, J. Impact of high-temperature environment on the optimal cycle rate of lithium-ion battery. *J. Energy Storage* **2020**, *28*, 101242. [CrossRef]
4. Zhao, R.; Liu, J.; Gu, J.; Zhai, L.; Ma, F. Experimental study of a direct evaporative cooling approach for Li-ion battery thermal management. *Int. J. Energy Res.* **2020**, *44*, 6660–6673. [CrossRef]
5. Suresh Patil, M.; Seo, J.-H.; Lee, M.-Y. A novel dielectric fluid immersion cooling technology for Li-ion battery thermal management. *Energy Convers. Manag.* **2021**, *229*, 113715. [CrossRef]
6. Bravo Diaz, L.; He, X.; Hu, Z.; Restuccia, F.; Marinescu, M.; Barreras, J.V.; Patel, Y.; Offer, G.; Rein, G. Review—Meta-Review of Fire Safety of Lithium-Ion Batteries: Industry Challenges and Research Contributions. *J. Electrochem. Soc.* **2020**, *167*, 090559. [CrossRef]
7. An, Z.; Jia, L.; Ding, Y.; Dang, C.; Li, X. A review on lithium-ion power battery thermal management technologies and thermal safety. *J. Therm. Sci.* **2017**, *26*, 391–412. [CrossRef]
8. Xia, G.; Cao, L.; Bi, G. A review on battery thermal management in electric vehicle application. *J. Power Sources* **2017**, *367*, 90–105. [CrossRef]
9. Chen, D.; Jiang, J.; Kim, G.-H.; Yang, C.; Pesaran, A. Comparison of different cooling methods for lithium ion battery cells. *Appl. Therm. Eng.* **2016**, *94*, 846–854. [CrossRef]
10. Yan, J.; Li, K.; Chen, H.; Wang, Q.; Sun, J. Experimental study on the application of phase change material in the dynamic cycling of battery pack system. *Energy Convers. Manag.* **2016**, *128*, 12–19. [CrossRef]
11. Ouyang, D.; Chen, M.; Huang, Q.; Weng, J.; Wang, Z.; Wang, J. A Review on the Thermal Hazards of the Lithium-Ion Battery and the Corresponding Countermeasures. *Appl. Sci.* **2019**, *9*, 2483. [CrossRef]
12. Jaguemont, J.; Van Mierlo, J. A comprehensive review of future thermal management systems for battery-electrified vehicles. *J. Energy Storage* **2020**, *31*, 101551. [CrossRef]
13. Wang, H.; Tao, T.; Xu, J.; Mei, X.; Liu, X.; Gou, P. Cooling capacity of a novel modular liquid-cooled battery thermal management system for cylindrical lithium ion batteries. *Appl. Therm. Eng.* **2020**, *178*, 115591. [CrossRef]

14. Qin, P.; Liao, M.; Zhang, D.; Liu, Y.; Sun, J.; Wang, Q. Experimental and numerical study on a novel hybrid battery thermal management system integrated forced-air convection and phase change material. *Energy Convers. Manag.* **2019**, *195*, 1371–1381. [CrossRef]
15. Wu, W.; Wang, S.; Wu, W.; Chen, K.; Hong, S.; Lai, Y. A critical review of battery thermal performance and liquid based battery thermal management. *Energy Convers. Manag.* **2019**, *182*, 262–281. [CrossRef]
16. Basu, S.; Hariharan, K.S.; Kolake, S.M.; Song, T.; Sohn, D.K.; Yeo, T. Coupled electrochemical thermal modelling of a novel Li-ion battery pack thermal management system. *Appl. Energy* **2016**, *181*, 1–13. [CrossRef]
17. Chawla, N.; Bharti, N.; Singh, S. Recent Advances in Non-Flammable Electrolytes for Safer Lithium-Ion Batteries. *Batteries* **2019**, *5*, 19. [CrossRef]
18. Zhu, H.; Qin, X.; Sun, X.; Yan, W.; Yang, J.; Xie, Y. Rocking-chair configuration in ultrathin lithium vanadate-graphene hybrid nanosheets for electrical modulation. *Sci. Rep.* **2013**, *3*, 1246. [CrossRef]
19. Wang, Q.; Mao, B.; Stoliarov, S.I.; Sun, J. A review of lithium ion battery failure mechanisms and fire prevention strategies. *Prog. Energy Combust. Sci.* **2019**, *73*, 95–131. [CrossRef]
20. Li, B.; Xia, D. Anionic Redox in Rechargeable Lithium Batteries. *Adv. Mater.* **2017**, *29*, 1701054. [CrossRef]
21. Panchal, S.; Khasow, R.; Dincer, I.; Agelin-Chaab, M.; Fraser, R.; Fowler, M. Thermal design and simulation of mini-channel cold plate for water cooled large sized prismatic lithium-ion battery. *Appl. Therm. Eng.* **2017**, *122*, 80–90. [CrossRef]
22. Lai, Y.; Du, S.; Ai, L.; Ai, L.; Cheng, Y.; Tang, Y.; Jia, M. Insight into heat generation of lithium ion batteries based on the electrochemical-thermal model at high discharge rates. *Int. J. Hydrog. Energy* **2015**, *40*, 13039–13049. [CrossRef]
23. Zhang, X. Thermal analysis of a cylindrical lithium-ion battery. *Electrochim. Acta* **2011**, *56*, 1246–1255. [CrossRef]
24. Zhao, R.; Gu, J.; Liu, J. An investigation on the significance of reversible heat to the thermal behavior of lithium ion battery through simulations. *J. Power Sources* **2014**, *266*, 422–432. [CrossRef]
25. Christensen, J.; Cook, D.; Albertus, P. An Efficient Parallelizable 3D Thermoelectrochemical Model of a Li-Ion Cell. *J. Electrochem. Soc.* **2013**, *160*, A2258–A2267. [CrossRef]
26. Guo, M.; Sikha, G.; White, R.E. Single-Particle Model for a Lithium-Ion Cell: Thermal Behavior. *J. Electrochem. Soc.* **2011**, *158*, A122. [CrossRef]
27. Bernardi, D.; Pawlikowski, E.; Newman, J. A General Energy Balance for Battery Systems. *J. Electrochem. Soc.* **2019**, *132*, 5–12. [CrossRef]
28. Smith, K.; Wang, C.-Y. Power and thermal characterization of a lithium-ion battery pack for hybrid-electric vehicles. *J. Power Sources* **2006**, *160*, 662–673. [CrossRef]
29. Fathabadi, H. A novel design including cooling media for Lithium-ion batteries pack used in hybrid and electric vehicles. *J. Power Sources* **2014**, *245*, 495–500. [CrossRef]
30. Kalaf, O.; Solyali, D.; Asmael, M.; Zeeshan, Q.; Safaei, B.; Askir, A. Experimental and simulation study of liquid coolant battery thermal management system for electric vehicles: A review. *Int. J. Energy Res.* **2020**, *45*, 6495–6517. [CrossRef]
31. Ping, P.; Wang, Q.; Chung, Y.; Wen, J. Modelling electro-thermal response of lithium-ion batteries from normal to abuse conditions. *Appl. Energy* **2017**, *205*, 1327–1344. [CrossRef]
32. Liebig, G.; Gupta, G.; Kirstein, U.; Schuldt, F.; Agert, C. Parameterization and Validation of an Electrochemical Thermal Model of a Lithium-Ion Battery. *Batteries* **2019**, *5*, 62. [CrossRef]
33. Liebig, G.; Kirstein, U.; Geißendörfer, S.; Schuldt, F.; Agert, C. The Impact of Environmental Factors on the Thermal Characteristic of a Lithium–ion Battery. *Batteries* **2020**, *6*, 3. [CrossRef]
34. Panchal, S.; Mathew, M.; Fraser, R.; Fowler, M. Electrochemical thermal modeling and experimental measurements of 18650 cylindrical lithium-ion battery during discharge cycle for an EV. *Appl. Therm. Eng.* **2018**, *135*, 123–132. [CrossRef]
35. Jaguemont, J.; Boulon, L.; Dubé, Y. A comprehensive review of lithium-ion batteries used in hybrid and electric vehicles at cold temperatures. *Appl. Energy* **2016**, *164*, 99–114. [CrossRef]
36. Wang, Q.; Ping, P.; Zhao, X.; Chu, G.; Sun, J.; Chen, C. Thermal runaway caused fire and explosion of lithium ion battery. *J. Power Sources* **2012**, *208*, 210–224. [CrossRef]
37. Wang, Q.; Jiang, B.; Li, B.; Yan, Y. A critical review of thermal management models and solutions of lithium-ion batteries for the development of pure electric vehicles. *Renew. Sustain. Energy Rev.* **2016**, *64*, 106–128. [CrossRef]
38. Zhang, Y.; Ge, H.; Huang, J.; Li, Z.; Zhang, J. A comparative degradation study of commercial lithium-ion cells under low-temperature cycling. *RSC Adv.* **2017**, *7*, 23157–23163. [CrossRef]
39. Petzl, M.; Kasper, M.; Danzer, M.A. Lithium plating in a commercial lithium-ion battery—A low-temperature aging study. *J. Power Sources* **2015**, *275*, 799–807. [CrossRef]
40. Broussely, M.; Biensan, P.; Bonhomme, F.; Blanchard, P.; Herreyre, S.; Nechev, K.; Staniewicz, R.J. Main aging mechanisms in Li ion batteries. *J. Power Sources* **2005**, *146*, 90–96. [CrossRef]
41. Zhao, R.; Gu, J.; Liu, J. An experimental study of heat pipe thermal management system with wet cooling method for lithium ion batteries. *J. Power Sources* **2015**, *273*, 1089–1097. [CrossRef]
42. Ramadass PH BW RP, B.; Haran, B.; White, R.; Popov, B.N. Capacity fade of Sony 18650 cells cycled at elevated temperatures Part I. Cycling performance. *J. Power Sources* **2002**, *112*, 606–613. [CrossRef]
43. Tan, X.; Lyu, P.; Fan, Y.; Rao, J.; Ouyang, K. Numerical investigation of the direct liquid cooling of a fast-charging lithium-ion battery pack in hydrofluoroether. *Appl. Therm. Eng.* **2021**, *196*, 117279. [CrossRef]

44. Yang, N.; Zhang, X.; Shang, B.; Li, G. Unbalanced discharging and aging due to temperature differences among the cells in a lithium-ion battery pack with parallel combination. *J. Power Sources* **2016**, *306*, 733–741. [CrossRef]
45. Chiu, K.-C.; Lin, C.-H.; Yeh, S.-F.; Lin, Y.-H.; Huang, C.-S.; Chen, K.-C. Cycle life analysis of series connected lithium-ion batteries with temperature difference. *J. Power Sources* **2014**, *263*, 75–84. [CrossRef]
46. Bandhauer, T.M.; Garimella, S.; Fuller, T.F. A Critical Review of Thermal Issues in Lithium-Ion Batteries. *J. Electrochem. Soc.* **2011**, *158*, R1. [CrossRef]
47. Liu, L.; Zhang, X.; Lin, X. Recent Developments of Thermal Management Strategies for Lithium-Ion Batteries: A State-of-The-Art Review. *Energy Technol.* **2022**, *10*, 2101135. [CrossRef]
48. E, J.; Xu, S.; Deng, Y.; Zhu, H.; Zuo, W.; Wang, H.; Chen, J.; Peng, Q.; Zhang, Z. Investigation on thermal performance and pressure loss of the fluid cold-plate used in thermal management system of the battery pack. *Appl. Therm. Eng.* **2018**, *145*, 552–568. [CrossRef]
49. Sheng, L.; Zhang, H.; Su, L.; Zhang, Z.; Zhang, H.; Li, K.; Fang, Y.; Ye, W. Effect analysis on thermal profile management of a cylindrical lithium-ion battery utilizing a cellular liquid cooling jacket. *Energy* **2021**, *220*, 119725. [CrossRef]
50. Thakur, A.K.; Prabakaran, R.; Elkadeem, M.R.; Sharshir, S.W.; Arıcı, M.; Wang, C.; Zhao, W.; Hwang, J.-Y.; Saidur, R. A state of art review and future viewpoint on advance cooling techniques for Lithium–ion battery system of electric vehicles. *J. Energy Storage* **2020**, *32*, 101771. [CrossRef]
51. Deng, Y.; Feng, C.; E, J.; Zhu, H.; Chen, J.; Wen, M.; Yin, H. Effects of different coolants and cooling strategies on the cooling performance of the power lithium ion battery system: A review. *Appl. Therm. Eng.* **2018**, *142*, 10–29. [CrossRef]
52. Liu, J.; Li, H.; Li, W.; Shi, J.; Wang, H.; Chen, J. Thermal characteristics of power battery pack with liquid-based thermal management. *Appl. Therm. Eng.* **2020**, *164*, 114421. [CrossRef]
53. Roe, C.; Feng, X.; White, G.; Li, R.; Wang, H.; Rui, X.; Li, C.; Zhang, F.; Null, V.; Parkes, M.; et al. Immersion cooling for lithium-ion batteries—A review. *J. Power Sources* **2022**, *525*, 231094. [CrossRef]
54. Dubey, P.; Pulugundla, G.; Srouji, A.K. Direct Comparison of Immersion and Cold-Plate Based Cooling for Automotive Li-Ion Battery Modules. *Energies* **2021**, *14*, 1259. [CrossRef]
55. Ke, Q.; Li, X.; Guo, J.; Cao, W.; Wang, Y.; Jiang, F. The retarding effect of liquid-cooling thermal management on thermal runaway propagation in lithium-ion batteries. *J. Energy Storage* **2022**, *48*, 104063. [CrossRef]
56. Wang, C.; Zhang, G.; Li, X.; Huang, J.; Wang, Z.; Lv, Y.; Meng, L.; Situ, W.; Rao, M. Experimental examination of large capacity liFePO4 battery pack at high temperature and rapid discharge using novel liquid cooling strategy. *Int. J. Energy Res.* **2018**, *42*, 1172–1182. [CrossRef]
57. Panchal, S.; Dincer, I.; Agelin-Chaab, M.; Fraser, R.; Fowler, M. Thermal modeling and validation of temperature distributions in a prismatic lithium-ion battery at different discharge rates and varying boundary conditions. *Appl. Therm. Eng.* **2016**, *96*, 190–199. [CrossRef]
58. Karimi, G.; Dehghan, A.R. Thermal analysis of high-power lithium-ion battery packs using flow network approach. *Int. J. Energy Res.* **2014**, *38*, 1793–1811. [CrossRef]
59. Jarrett, A.; Kim, I.Y. Design optimization of electric vehicle battery cooling plates for thermal performance. *J. Power Sources* **2011**, *196*, 10359–10368. [CrossRef]
60. Tran, L.; Lopez, J.; Lopez, J.; Uriostegui, A.; Barrera, A.; Wiggins, N. Li-ion battery cooling system integrates in nano-fluid environment. *Appl Nanosci* **2017**, *7*, 25–29. [CrossRef]
61. Liu, H.; Chika, E.; Zhao, J. Investigation into the effectiveness of nanofluids on the mini-channel thermal management for high power lithium ion battery. *Appl. Therm. Eng.* **2018**, *142*, 511–523. [CrossRef]
62. Wiriyasart, S.; Hommalee, C.; Sirikasemsuk, S.; Prurapark, R.; Naphon, P. Thermal management system with nanofluids for electric vehicle battery cooling modules. *Case Stud. Therm. Eng.* **2020**, *18*, 100583. [CrossRef]
63. Zakaria, I.; Azmi, W.H.; Mohamed, W.A.N.W.; Mamat, R.; Najafi, G. Experimental Investigation of Thermal Conductivity and Electrical Conductivity of Al2O3 Nanofluid in Water—Ethylene Glycol Mixture for Proton Exchange Membrane Fuel Cell Application. *Int. Commun. Heat Mass Transf.* **2015**, *61*, 61–68. [CrossRef]
64. Maheshwary, P.B.; Handa, C.C.; Nemade, K.R. A comprehensive study of effect of concentration, particle size and particle shape on thermal conductivity of titania/water based nanofluid. *Appl. Therm. Eng.* **2017**, *119*, 79–88. [CrossRef]
65. Teng, T.-P.; Hung, Y.-H.; Teng, T.-C.; Mo, H.-E.; Hsu, H.-G. The effect of alumina/water nanofluid particle size on thermal conductivity. *Appl. Therm. Eng.* **2010**, *30*, 2213–2218. [CrossRef]
66. Das, P.K. A review based on the effect and mechanism of thermal conductivity of normal nanofluids and hybrid nanofluids. *J. Mol. Liq.* **2017**, *240*, 420–446. [CrossRef]
67. Wu, J.; Zhao, J.; Lei, J.; Liu, B. Effectiveness of nanofluid on improving the performance of microchannel heat sink. *Appl. Therm. Eng.* **2016**, *101*, 402–412. [CrossRef]
68. Tousi, M.; Sarchami, A.; Kiani, M.; Najafi, M.; Houshfar, E. Numerical study of novel liquid-cooled thermal management system for cylindrical Li-ion battery packs under high discharge rate based on AgO nanofluid and copper sheath. *J. Energy Storage* **2021**, *41*, 102910. [CrossRef]
69. Sarchami, A.; Najafi, M.; Imam, A.; Houshfar, E. Experimental study of thermal management system for cylindrical Li-ion battery pack based on nanofluid cooling and copper sheath. *Int. J. Therm. Sci.* **2022**, *171*, 107244. [CrossRef]

70. Guo, Z.; Wang, Y.; Zhao, S.; Zhao, T.; Ni, M. Investigation of battery thermal management system with considering effect of battery aging and nanofluids. *Int. J. Heat Mass Transf.* **2023**, *202*, 123685. [CrossRef]
71. Yang, X.-H.; Tan, S.-C.; Liu, J. Thermal management of Li-ion battery with liquid metal. *Energy Convers. Manag.* **2016**, *117*, 577–585. [CrossRef]
72. Liu, Z.; Wang, H.; Yang, C.; Zhao, J. Simulation study of lithium-ion battery thermal management system based on a variable flow velocity method with liquid metal. *Appl. Therm. Eng.* **2020**, *179*, 115578. [CrossRef]
73. Malik, M.; Dincer, I.; Rosen, M.A.; Mathew, M.; Fowler, M. Thermal and electrical performance evaluations of series connected Li-ion batteries in a pack with liquid cooling. *Appl. Therm. Eng.* **2018**, *129*, 472–481. [CrossRef]
74. Shang, Z.; Qi, H.; Liu, X.; Ouyang, C.; Wang, Y. Structural optimization of lithium-ion battery for improving thermal performance based on a liquid cooling system. *Int. J. Heat Mass Transf.* **2019**, *130*, 33–41. [CrossRef]
75. Huo, Y.; Rao, Z.; Liu, X.; Zhao, J. Investigation of power battery thermal management by using mini-channel cold plate. *Energy Convers. Manag.* **2015**, *89*, 387–395. [CrossRef]
76. Huang, Y.; Mei, P.; Lu, Y.; Huang, R.; Yu, X.; Chen, Z.; Roskilly, A.P. A novel approach for Lithium-ion battery thermal management with streamline shape mini channel cooling plates. *Appl. Therm. Eng.* **2019**, *157*, 113623. [CrossRef]
77. Deng, T.; Zhang, G.; Ran, Y. Study on thermal management of rectangular Li-ion battery with serpentine-channel cold plate. *Int. J. Heat Mass Transf.* **2018**, *125*, 143–152. [CrossRef]
78. Monika, K.; Datta, S.P. Comparative assessment among several channel designs with constant volume for cooling of pouch-type battery module. *Energy Convers. Manag.* **2022**, *251*, 114936. [CrossRef]
79. Chen, F.C.; Gao, Z.; Loutfy, R.O.; Hecht, M. Analysis of Optimal Heat Transfer in a PEM Fuel Cell Cooling Plate. *Fuel Cells* **2003**, *3*, 181–188. [CrossRef]
80. Chen, K.; Chen, Y.; Song, M.; Wang, S. Multi-parameter structure design of parallel mini-channel cold plate for battery thermal management. *Int. J. Energy Res.* **2020**, *44*, 4321–4334. [CrossRef]
81. Li, H.; Ding, X.; Jing, D.; Xiong, M.; Meng, F. Experimental and numerical investigation of liquid-cooled heat sinks designed by topology optimization. *Int. J. Therm. Sci.* **2019**, *146*, 106065. [CrossRef]
82. Guo, C.; Liu, H.-l.; Guo, Q.; Shao, X.-d.; Zhu, M.-l. Investigations on a novel cold plate achieved by topology optimization for lithium-ion batteries. *Energy* **2022**, *261*, 125097. [CrossRef]
83. Chen, F.; Wang, J.; Yang, X. Topology optimization design and numerical analysis on cold plates for lithium-ion battery thermal management. *Int. J. Heat Mass Transf.* **2022**, *183*, 122087. [CrossRef]
84. Sun, Y.; Bai, R.; Ma, J. Design and thermal analysis of a new topological cooling plate for prismatic lithium battery thermal management. *Appl. Therm. Eng.* **2023**, *219*, 119547. [CrossRef]
85. Wang, Y.; Duong, X.Q.; Zadeh, P.G.; Chung, J.D. Liquid-cooled cold plate for a Li-ion battery thermal management system designed by topology optimization. *J. Mech. Sci. Technol.* **2023**, *37*, 2079–2086. [CrossRef]
86. Qian, Z.; Li, Y.; Rao, Z. Thermal performance of lithium-ion battery thermal management system by using mini-channel cooling. *Energy Convers. Manag.* **2016**, *126*, 622–631. [CrossRef]
87. Choi, J.; Kim, Y.-H.; Lee, Y.; Lee, K.-J.; Kim, Y. Numerical analysis on the performance of cooling plates in a PEFC. *J. Mech. Sci. Technol.* **2008**, *22*, 1417–1425. [CrossRef]
88. Du, X.; Qian, Z.; Chen, Z.; Rao, Z. Experimental investigation on mini-channel cooling-based thermal management for Li-ion battery module under different cooling schemes. *Int. J. Energy Res.* **2018**, *42*, 2781–2788. [CrossRef]
89. Lan, C.; Xu, J.; Qiao, Y.; Ma, Y. Thermal management for high power lithium-ion battery by minichannel aluminum tubes. *Appl. Therm. Eng.* **2016**, *101*, 284–292. [CrossRef]
90. Lee, Y.J.; Lee, P.S.; Chou, S.K. Enhanced Thermal Transport in Microchannel Using Oblique Fins. *J. Heat Transf.* **2012**, *134*, 101901. [CrossRef]
91. Jin, L.W.; Lee, P.S.; Kong, X.X.; Fan, Y.; Chou, S.K. Ultra-thin minichannel LCP for EV battery thermal management. *Appl. Energy* **2014**, *113*, 1786–1794. [CrossRef]
92. Darcovich, K.; MacNeil, D.D.; Recoskie, S.; Cadic, Q.; Ilinca, F. Comparison of cooling plate configurations for automotive battery pack thermal management. *Appl. Therm. Eng.* **2019**, *155*, 185–195. [CrossRef]
93. Tang, Z.; Min, X.; Song, A.; Cheng, J. Thermal Management of a Cylindrical Lithium-Ion Battery Module Using a Multichannel Wavy Tube. *J. Energy Eng.* **2019**, *145*, 04018072. [CrossRef]
94. Zhao, C.; Cao, W.; Dong, T.; Jiang, F. Thermal behavior study of discharging/charging cylindrical lithium-ion battery module cooled by channeled liquid flow. *Int. J. Heat Mass Transf.* **2018**, *120*, 751–762. [CrossRef]
95. Deng, T.; Zhang, G.; Ran, Y.; Liu, P. Thermal performance of lithium ion battery pack by using cold plate. *Appl. Therm. Eng.* **2019**, *160*, 114088. [CrossRef]
96. Zhou, H.; Zhou, F.; Zhang, Q.; Wang, Q.; Song, Z. Thermal management of cylindrical lithium-ion battery based on a liquid cooling method with half-helical duct. *Appl. Therm. Eng.* **2019**, *162*, 114257. [CrossRef]
97. Rao, Z.; Qian, Z.; Kuang, Y.; Li, Y. Thermal performance of liquid cooling based thermal management system for cylindrical lithium-ion battery module with variable contact surface. *Appl. Therm. Eng.* **2017**, *123*, 1514–1522. [CrossRef]
98. Siruvuri, S.D.V.S.S.V.; Budarapu, P.R. Studies on thermal management of Lithium-ion battery pack using water as the cooling fluid. *J. Energy Storage* **2020**, *29*, 101377. [CrossRef]

99. Zhao, J.; Rao, Z.; Li, Y. Thermal performance of mini-channel liquid cooled cylinder based battery thermal management for cylindrical lithium-ion power battery. *Energy Convers. Manag.* **2015**, *103*, 157–165. [CrossRef]
100. Huang, Y.; Wang, S.; Lu, Y.; Huang, R.; Yu, X. Study on a liquid cooled battery thermal management system pertaining to the transient regime. *Appl. Therm. Eng.* **2020**, *180*, 115793. [CrossRef]
101. Liu, J.; Fan, Y.; Xie, Q. Feasibility study of a novel oil-immersed battery cooling system: Experiments and theoretical analysis. *Appl. Therm. Eng.* **2022**, *208*, 118251. [CrossRef]
102. Pan, C.; Tang, Q.; He, Z.; Wang, L.; Chen, L. Structure Optimization of Battery Module With a Parallel Multi-Channel Liquid Cooling Plate Based on Orthogonal Test. *J. Electrochem. Energy Convers. Storage* **2020**, *17*, 021104. [CrossRef]
103. E, J.; Han, D.; Qiu, A.; Zhu, H.; Deng, Y.; Chen, J.; Zhao, X.; Zuo, W.; Wang, H.; Chen, J.; et al. Orthogonal experimental design of liquid-cooling structure on the cooling effect of a liquid-cooled battery thermal management system. *Appl. Therm. Eng.* **2018**, *132*, 508–520. [CrossRef]
104. Mohammadian, S.K.; He, Y.-L.; Zhang, Y. Internal cooling of a lithium-ion battery using electrolyte as coolant through microchannels embedded inside the electrodes. *J. Power Sources* **2015**, *293*, 458–466. [CrossRef]
105. Shen, K.; Sun, J.; Zheng, Y.; Xu, C.; Wang, H.; Wang, S.; Chen, S.; Feng, X. A comprehensive analysis and experimental investigation for the thermal management of cell-to-pack battery system. *Appl. Therm. Eng.* **2022**, *211*, 118422. [CrossRef]
106. Wang, H.; Wang, S.; Feng, X.; Zhang, X.; Dai, K.; Sheng, J.; Zhao, Z.; Du, Z.; Zhang, Z.; Shen, K.; et al. An experimental study on the thermal characteristics of the Cell-To-Pack system. *Energy* **2021**, *227*, 120338. [CrossRef]
107. Fichtner, M. Recent Research and Progress in Batteries for Electric Vehicles. *Batter. Supercaps* **2021**, *5*, e202100224. [CrossRef]
108. Hu, G.; Huang, P.; Bai, Z.; Wang, Q.; Qi, K. Comprehensively analysis the failure evolution and safety evaluation of automotive lithium ion battery. *eTransportation* **2021**, *10*, 100140. [CrossRef]
109. Jin, C.; Sun, Y.; Yao, J.; Feng, X.; Lai, X.; Shen, K.; Wang, H.; Rui, X.; Xu, C.; Zheng, Y.; et al. No thermal runaway propagation optimization design of battery arrangement for cell-to-chassis technology. *eTransportation* **2022**, *14*, 100199. [CrossRef]
110. Sun, J.; Chen, S.; Shen, K.; Zheng, Y. Liquid cooling system optimization for a cell-to-pack battery module under fast charging. *Int. J. Energy Res.* **2022**, *46*, 12241–12253. [CrossRef]
111. Wang, W.; Dai, S.; Zhao, W.; Wang, C.; Ma, T. Design optimization of a novel negative Poisson's ratio non-module battery pack system considering crashworthiness and heat dissipation. *Compos. Struct.* **2021**, *275*, 114458. [CrossRef]
112. Chen, K.; Li, X. Accurate determination of battery discharge characteristics—A comparison between two battery temperature control methods. *J. Power Sources* **2014**, *247*, 961–966. [CrossRef]
113. Amalesh, T.; Lakshmi Narasimhan, N. Cooling of a lithium ion battery using phase change material with air/dielectric fluid media: A numerical study. *Proc. Inst. Mech. Eng. Part A J. Power Energy* **2019**, *234*, 722–738. [CrossRef]
114. Wu, N.; Ye, X.; Yao, J.; Zhang, X.; Zhou, X.; Yu, B. Efficient thermal management of the large-format pouch lithium-ion cell via the boiling-cooling system operated with intermittent flow. *Int. J. Heat Mass Transf.* **2021**, *170*, 121018. [CrossRef]
115. Wang, Y.-F.; Wu, J.-T. Thermal performance predictions for an HFE-7000 direct flow boiling cooled battery thermal management system for electric vehicles. *Energy Convers. Manag.* **2020**, *207*, 112569. [CrossRef]
116. van Gils, R.W.; Danilov, D.; Notten, P.H.L.; Speetjens, M.F.M.; Nijmeijer, H. Battery thermal management by boiling heat-transfer. *Energy Convers. Manag.* **2014**, *79*, 9–17. [CrossRef]
117. Rafiq, M.; Lv, Y.Z.; Zhou, Y.; Ma, K.B.; Wang, W.; Li, C.R.; Wang, Q. Use of vegetable oils as transformer oils—A review. *Renew. Sustain. Energy Rev.* **2015**, *52*, 308–324. [CrossRef]
118. Ab Ghani, S.; Muhamad, N.A.; Noorden, Z.A.; Zainuddin, H.; Abu Bakar, N.; Talib, M.A. Methods for improving the workability of natural ester insulating oils in power transformer applications: A review. *Electr. Power Syst. Res.* **2018**, *163*, 655–667. [CrossRef]
119. Madavan, R.; Saroja, S.; Karthick, A.; Murugesan, S.; Mohanavel, V.; Velmurugan, P.; Al Obaid, S.; Alfarraj, S.; Sivakumar, S. Performance analysis of mixed vegetable oil as an alternative for transformer insulation oil. In *Biomass Conversion and Biorefinery*; Springer: New York, NY, USA, 2022. [CrossRef]
120. Chairul, I.S.; Bakar, N.A.; Othman, M.N.; Ghani, S.A.; Khiar, M.S.A.; Talib, M.A. Potential of Used Cooking Oil as Dielectric Liquid for Oil-Immersed Power Transformers. *IEEE Trans. Dielectr. Electr. Insul.* **2021**, *28*, 1400–1407. [CrossRef]
121. Al Qubeissi, M.; Mahmoud, A.; Al-Damook, M.; Almshahy, A.; Khatir, Z.; Soyhan, H.S.; Raja Ahsan Shah, R.M. Comparative Analysis of Battery Thermal Management System Using Biodiesel Fuels. *Energies* **2023**, *16*, 565. [CrossRef]
122. Pulugundla, G.; Dubey, P.; Wu, Z.; Wang, Q.; Srouji, A.K. Thermal Management of Lithium Ion Cells at High Discharge Rate Using Submerged-Cell Cooling. In Proceedings of the 2020 IEEE Transportation Electrification Conference & Expo (ITEC), Chicago, IL, USA, 23–26 June 2020; IEEE: Chicago, IL, USA, 2020; pp. 1–5.
123. Bhattacharjee, A.; Mohanty, R.K.; Ghosh, A. Design of an Optimized Thermal Management System for Li-Ion Batteries under Different Discharging Conditions. *Energies* **2020**, *13*, 5695. [CrossRef]
124. Wang, H.; Tao, T.; Xu, J.; Shi, H.; Mei, X.; Gou, P. Thermal performance of a liquid-immersed battery thermal management system for lithium-ion pouch batteries. *J. Energy Storage* **2022**, *46*, 103835. [CrossRef]
125. Sundin, D.W.; Sponholtz, S. Thermal Management of Li-Ion Batteries With Single-Phase Liquid Immersion Cooling. *IEEE Open J. Veh. Technol.* **2020**, *1*, 82–92. [CrossRef]
126. Jithin, K.V.; Rajesh, P.K. Numerical analysis of single-phase liquid immersion cooling for lithium-ion battery thermal management using different dielectric fluids. *Int. J. Heat Mass Transf.* **2022**, *188*, 122608. [CrossRef]

127. Trimbake, A.; Singh, C.P.; Krishnan, S. Mineral Oil Immersion Cooling of Lithium-Ion Batteries: An Experimental Investigation. *J. Electrochem. Energy Convers. Storage* **2022**, *19*, 021007. [CrossRef]
128. Liu, J.; Fan, Y.; Yang, M.; Wang, J.; Xie, Q. Experimental investigation on the cooling effectiveness of an oil-immersed battery cooling system. *J. Therm. Anal. Calorim.* **2022**, *147*, 14841–14857. [CrossRef]
129. Han, J.-W.; Garud, K.S.; Hwang, S.-G.; Lee, M.-Y. Experimental Study on Dielectric Fluid Immersion Cooling for Thermal Management of Lithium-Ion Battery. *Symmetry* **2022**, *14*, 2126. [CrossRef]
130. Al-Zareer, M.; Dincer, I.; Rosen, M.A. Heat and mass transfer modeling and assessment of a new battery cooling system. *Int. J. Heat Mass Transf.* **2018**, *126*, 765–778. [CrossRef]
131. Al-Zareer, M.; Dincer, I.; Rosen, M.A. Novel thermal management system using boiling cooling for high-powered lithium-ion battery packs for hybrid electric vehicles. *J. Power Sources* **2017**, *363*, 291–303. [CrossRef]
132. Al-Zareer, M.; Dincer, I.; Rosen, M.A. Electrochemical modeling and performance evaluation of a new ammonia-based battery thermal management system for electric and hybrid electric vehicles. *Electrochim. Acta* **2017**, *247*, 171–182. [CrossRef]
133. Li, Y.; Zhou, Z.; Su, L.; Bai, M.; Gao, L.; Li, Y.; Liu, X.; Li, Y.; Song, Y. Numerical Simulations for Indirect and Direct Cooling of 54 V LiFePO4 Battery Pack. *Energies* **2022**, *15*, 4581. [CrossRef]
134. Li, Y.; Zhou, Z.; Hu, L.; Bai, M.; Gao, L.; Li, Y.; Liu, X.; Li, Y.; Song, Y. Experimental studies of liquid immersion cooling for 18650 lithium-ion battery under different discharging conditions. *Case Stud. Therm. Eng.* **2022**, *34*, 102034. [CrossRef]
135. Karimi, G.; Li, X. Thermal management of lithium-ion batteries for electric vehicles. *Int. J. Energy Res.* **2013**, *37*, 13–24. [CrossRef]
136. Matsuoka, M.; Matsuda, K.; Kubo, H. Liquid Immersion Cooling Technology with Natural Convection in Data Center. In Proceedings of the 2017 IEEE 6th International Conference on Cloud Networking (CloudNet), Prague, Czech Republic, 25–27 September 2017; IEEE: Prague, Czech Republic, 2017; pp. 1–7.
137. Wu, S.; Lao, L.; Wu, L.; Liu, L.; Lin, C.; Zhang, Q. Effect analysis on integration efficiency and safety performance of a battery thermal management system based on direct contact liquid cooling. *Appl. Therm. Eng.* **2022**, *201*, 117788. [CrossRef]
138. Kim, J.; Oh, J.; Lee, H. Review on battery thermal management system for electric vehicles. *Appl. Therm. Eng.* **2019**, *149*, 192–212. [CrossRef]
139. Wang, S.; Li, Y.; Li, Y.-Z.; Mao, Y.; Zhang, Y.; Guo, W.; Zhong, M. A forced gas cooling circle packaging with liquid cooling plate for the thermal management of Li-ion batteries under space environment. *Appl. Therm. Eng.* **2017**, *123*, 929–939. [CrossRef]
140. Yang, W.; Zhou, F.; Zhou, H.; Wang, Q.; Kong, J. Thermal performance of cylindrical lithium-ion battery thermal management system integrated with mini-channel liquid cooling and air cooling. *Appl. Therm. Eng.* **2020**, *175*, 115331. [CrossRef]
141. Li, M.; Liu, F.; Han, B.; Guo, J.; Xu, Y. Research on temperature control performance of battery thermal management system composited with multi-channel parallel liquid cooling and air cooling. *Ionics* **2021**, *27*, 2685–2695. [CrossRef]
142. Zheng, Y.; Shi, Y.; Huang, Y. Optimisation with adiabatic interlayers for liquid-dominated cooling system on fast charging battery packs. *Appl. Therm. Eng.* **2019**, *147*, 636–646. [CrossRef]
143. Kong, D.; Peng, R.; Ping, P.; Du, J.; Chen, G.; Wen, J. A novel battery thermal management system coupling with PCM and optimized controllable liquid cooling for different ambient temperatures. *Energy Convers. Manag.* **2020**, *204*, 112280. [CrossRef]
144. Zadeh, P.G.; Wang, Y.; Chung, J.D. Thermal management modeling for cylindrical lithium-ion battery packs considering safety and lifespan. *J. Mech. Sci. Technol.* **2022**, *36*, 3727–3733. [CrossRef]
145. Cao, J.; Ling, Z.; Fang, X.; Zhang, Z. Delayed liquid cooling strategy with phase change material to achieve high temperature uniformity of Li-ion battery under high-rate discharge. *J. Power Sources* **2020**, *450*, 227673. [CrossRef]
146. Yuan, X.; Tang, A.; Shan, C.; Liu, Z.; Li, J. Experimental investigation on thermal performance of a battery liquid cooling structure coupled with heat pipe. *J. Energy Storage* **2020**, *32*, 101984. [CrossRef]
147. Tran, T.-H.; Harmand, S.; Sahut, B. Experimental investigation on heat pipe cooling for Hybrid Electric Vehicle and Electric Vehicle lithium-ion battery. *J. Power Sources* **2014**, *265*, 262–272. [CrossRef]
148. Mei, N.; Xu, X.; Li, R. Heat Dissipation Analysis on the Liquid Cooling System Coupled with a Flat Heat Pipe of a Lithium-Ion Battery. *ACS Omega* **2020**, *5*, 17431–17441. [CrossRef]
149. Liang, J.; Gan, Y.; Li, Y. Investigation on the thermal performance of a battery thermal management system using heat pipe under different ambient temperatures. *Energy Convers. Manag.* **2018**, *155*, 1–9. [CrossRef]
150. Wei, T.; Xiaoming, X.; Hua, D.; Yaohua, G.; Jicheng, L.; Hongchao, W. Sensitivity Analysis of the Battery Thermal Management System with a Reciprocating Cooling Strategy Combined with a Flat Heat Pipe. *ACS Omega* **2020**, *5*, 8258–8267. [CrossRef]
151. Ma, S.; Jiang, M.; Tao, P.; Song, C.; Wu, J.; Wang, J.; Deng, T.; Shang, W. Temperature effect and thermal impact in lithium-ion batteries: A review. *Prog. Nat. Sci. Mater. Int.* **2018**, *28*, 653–666. [CrossRef]
152. Wang, Y.; Zhang, X.; Chen, Z. Low temperature preheating techniques for Lithium-ion batteries: Recent advances and future challenges. *Appl. Energy* **2022**, *313*, 118832. [CrossRef]
153. Wu, S.; Xiong, R.; Li, H.; Nian, V.; Ma, S. The state of the art on preheating lithium-ion batteries in cold weather. *J. Energy Storage* **2020**, *27*, 101059. [CrossRef]
154. Chen, M.; Li, J. Experimental study on heating performance of pure electric vehicle power battery under low temperature environment. *Int. J. Heat Mass Transf.* **2021**, *172*, 121191. [CrossRef]
155. Tang, Z.; Wang, S.; Liu, Z.; Cheng, J. Numerical analysis of temperature uniformity of a liquid cooling battery module composed of heat-conducting blocks with gradient contact surface angles. *Appl. Therm. Eng.* **2020**, *178*, 115509. [CrossRef]

156. Chen, S.; Zhang, G.; Wu, C.; Huang, W.; Xu, C.; Jin, C.; Wu, Y.; Jiang, Z.; Dai, H.; Feng, X.; et al. Multi-objective optimization design for a double-direction liquid heating system-based Cell-to-Chassis battery module. *Int. J. Heat Mass Transf.* **2022**, *183* 122184. [CrossRef]
157. Cheng, J.; Shuai, S.; Tang, Z. Thermal performance of a lithium-ion battery thermal management system with vapor chamber and minichannel cold plate. *Appl. Therm. Eng.* **2023**, *222*, 119694. [CrossRef]
158. An, Z.; Zhang, C.; Luo, Y.; Zhang, J. Cooling and preheating behavior of compact power Lithium-ion battery thermal management system. *Appl. Therm. Eng.* **2023**, *226*, 120238. [CrossRef]
159. Wang, Y.; Rao, Z.; Liu, S.; Li, X.; Li, H.; Xiong, R. Evaluating the performance of liquid immersing preheating system for Lithium-ion battery pack. *Appl. Therm. Eng.* **2021**, *190*, 116811. [CrossRef]
160. Guo, J.; Jiang, F. A novel electric vehicle thermal management system based on cooling and heating of batteries by refrigerant *Energy Convers. Manag.* **2021**, *237*, 114145. [CrossRef]
161. Wilke, S.; Schweitzer, B.; Khateeb, S.; Al-Hallaj, S. Preventing thermal runaway propagation in lithium ion battery packs using a phase change composite material: An experimental study. *J. Power Sources* **2017**, *340*, 51–59. [CrossRef]
162. Liu, J.; Fan, Y.; Xie, Q. Temperature mitigation effect of phase change material on overcharging lithium-ion batteries: An experimental study. *J. Therm. Anal. Calorim.* **2021**, *147*, 5153–5163. [CrossRef]
163. Mohammed, A.H.; Esmaeeli, R.; Aliniagerdroudbari, H.; Alhadri, M.; Hashemi, S.R.; Nadkarni, G.; Farhad, S. Dual-purpose cooling plate for thermal management of prismatic lithium-ion batteries during normal operation and thermal runaway. *Appl. Therm. Eng.* **2019**, *16*, 1141060. [CrossRef]
164. Zhang, W.; Liang, Z.; Yin, X.; Ling, G. Avoiding thermal runaway propagation of lithium-ion battery modules by using hybrid phase change material and liquid cooling. *Appl. Therm. Eng.* **2021**, *184*, 116380. [CrossRef]
165. Zhang, W.; Huang, L.; Zhang, Z.; Li, X.; Ma, R.; Ren, Y.; Wu, W. Non-uniform phase change material strategy for directional mitigation of battery thermal runaway propagation. *Renew. Energy* **2022**, *200*, 1338–1351. [CrossRef]
166. Li, X.; Zhou, Z.; Zhang, M.; Zhang, F.; Zhou, X. A liquid cooling technology based on fluorocarbons for lithium-ion battery thermal safety. *J. Loss Prev. Process Ind.* **2022**, *78*, 104818. [CrossRef]
167. Li, Q.; Cho, J.-R.; Zhai, J. Optimization of Thermal Management System with Water and Phase Change Material Cooling for Li-Ion Battery Pack. *Energies* **2021**, *14*, 5312. [CrossRef]
168. Leng, Z.; Yuan, Y.; Cao, X.; Zeng, C.; Zhong, W.; Gao, B. Heat pipe/phase change material thermal management of Li-ion power battery packs: A numerical study on coupled heat transfer performance. *Energy* **2022**, *240*, 122754. [CrossRef]

MDPI AG
Grosspeteranlage 5
4052 Basel
Switzerland
Tel.: +41 61 683 77 34

Batteries Editorial Office
E-mail: batteries@mdpi.com
www.mdpi.com/journal/batteries